ADVANCES IN POSTHARVEST TECHNOLOGIES OF VEGETABLE CROPS

Postharvest Biology and Technology

ADVANCES IN POSTHARVEST TECHNOLOGIES OF VEGETABLE CROPS

Edited by
Bijendra Singh
Sudhir Singh
Tanmay K. Koley

Apple Academic Press Inc.	Apple Academic Press Inc.
3333 Mistwell Crescent	9 Spinnaker Way
Oakville, ON L6L 0A2	Waretown, NJ 08758
Canada	USA

© 2018 by Apple Academic Press, Inc.

First issued in paperback 2021

Exclusive worldwide distribution by CRC Press, a member of Taylor & Francis Group

No claim to original U.S. Government works

ISBN-13: 978-1-77463-057-0 (pbk)
ISBN-13: 978-1-77188-619-2 (hbk)

Library and Archives Canada Cataloguing in Publication

Advances in postharvest technologies of vegetable crops / edited by Bijendra Singh, Sudhir Singh, Tanmay K. Koley.

(Postharvest biology and technology book series)
Includes bibliographical references and index.
Issued in print and electronic formats.
ISBN 978-1-77188-619-2 (hardcover).--ISBN 978-1-315-16102-0 (PDF)

1. Vegetables--Postharvest technology. I. Singh, Bijendra, Dr., editor II. Series: Postharvest biology and technology book series

| SB360.A38 2018 | 631.5'6 | C2018-900497-5 | C2018-900498-3 |

Library of Congress Cataloging-in-Publication Data

Names: Singh, Bijendra, (Ph. D.), editor.
Title: Advances in postharvest technologies of vegetable crops / editors: Bijendra Singh, Sudhir Singh, Tanmay K. Koley.
Description: Waretown, NJ : Apple Academic Press, 2018. | Series: Postharvest biology and technology | Includes bibliographical references and index.
Identifiers: LCCN 2018002175 (print) | LCCN 2018011402 (ebook) | ISBN 9781315161020 (ebook) | ISBN 9781771886192 (hardcover : alk. paper)
Subjects: LCSH: Vegetables--Postharvest technology.
Classification: LCC SB324.85 (ebook) | LCC SB324.85 A38 2018 (print) | DDC 635/.3--dc23
LC record available at https://lccn.loc.gov/2018002175

Apple Academic Press also publishes its books in a variety of electronic formats. Some content that appears in print may not be available in electronic format. For information about Apple Academic Press products, visit our website at **www.appleacademicpress.com** and the CRC Press website at **www.crcpress.com**

CONTENTS

LIST OF CONTRIBUTORS

A. B. Rai
Division of Crop Protection, ICAR-Indian Institute of Vegetable Research, Varanasi 221305, UP, India

A. Nath
Project Directorate for Farming Systems Research, Modipuram, Meerut, Uttar Pradesh, India,E-mail: amitnath2005@gmail.com

Abdel Gawad Saad
Agriculture Structure and Environmental Control Division, Central Institute of Postharvest Engineering and Technology (CIPHET), Ludhiana 141004, Punjab, India

Achuit K. Singh
Division of Crop Improvement, ICAR—Indian Institute of Vegetable Research, Varanasi 221305, UP, India, E-mail: achuits@gmail.com

Ajay Tripathi
ICAR—Indian Institute of Vegetable Research, Kelabela, UP, India

Anuradha Srivastava
ICAR Research Complex for Eastern Region (ICAR-RCER), Patna, India

Arti Maurya
ICAR—Indian Institute of Vegetable Research, Kelabela, UP, India

Arup Chattopadhyay
AICRP on Vegetable Crops, Bidhan Chandra Krishi Viswavidyalaya, Kalyani Centre, Mohanpur 741252, Nadia, West Bengal, India

Ashutosh Rai
Division of Crop Improvement, ICAR—Indian Institute of Vegetable Research, Varanasi 221305, UP, India

Avinash Chandra Rai
Division of Crop Improvement, ICAR—Indian Institute of Vegetable Research, Varanasi 221305, UP, India, E-mail: singhvns@gmail.com

B. Jirli
Department of Extension Education, BHU, Varanasi 221005, UP, India

B. Singh
ICAR—Indian Institute of Vegetable Research, Kelabela, UP, India

D. V. Sudhakar Rao
Division of Post-Harvest Technology, Indian Institute of Horticultural Research, Hessaraghatta Lake Post, Bengaluru, India

G. K. Srivastava
Division of Basic Sciences, ICAR-Indian Institute of Pulses Research, Kanpur 208024, UP, India

I. N. Doreyappa Gowda
ICAR-IIHR-Central Horticultural Experiment Station, Kodagu 571248, India, Tel: 91 8212473879;
fax: +91 821 2473468

Ivi Chakraborty
Department of Postharvest Technology of Horticultural Crops, Faculty of Horticulture, Bidhan
Chandra Krishi Viswavidyalaya, Mohanpur 741252, Nadia, West Bengal, India

Jagdish Singh
Division of Basic Sciences, ICAR-Indian Institute of Pulses Research, Kanpur 208024, UP, India
E-mail:jagdish1959@gmail.com

K. Chitravathi
Defence Food Research Laboratory, Siddarthanagar, Mysore, India

K. Narsaiah
AS & EC Division, Central Institute of Post-Harvest Engineering and Technology (CIPHET),
Ludhiana141004, Punjab, India

Kalyan Barman
Department of Horticulture, Institute of Agricultural Sciences, Banaras Hindu University, Varanasi
221005, Uttar Pradesh, India

Kaushik Banerjee
National Referral Laboratory, ICAR—National Research Centre for Grapes, Pune 412307,
Maharashtra, India

L. E. Unni
Defence Food Research Laboratory, Siddarthanagar, Mysore, India

M. H. Kodandaram
Division of Crop Protection, ICAR-Indian Institute of Vegetable Research, Varanasi 221305, UP,
India, E-mail: kodandaram75@gmail.com

M. Manjunath
ICAR—Indian Institute of Vegetable Research, Jakhini, Shahanshapur, Varanasi 221305, UP, India

Major Singh
Division of Crop Improvement, ICAR—Indian Institute of Vegetable Research, Varanasi 221305, UP,
India

O. P. Chauhan
Defence Food Research Laboratory, Siddarthanagar, Mysore, India, E-mail: opchauhan@gmail.com

P. S. Badal
Department of Agricultural Economics, BHU, Varanasi 221005, UP, India

Pranita Jaiswal
Agriculture Structure and Environmental Control Division, Central Institute of Postharvest
Engineering and Technology (CIPHET), Ludhiana 141004, Punjab, India

Pritam Kalia
Division of Vegetable Science, Indian Agricultural Research Institute, New Delhi 110012, India

R. K. Pal
National Research Centre on Pomegranate, Kegaon, Solapur 413255, Maharashtra, India, E-mail:
rkrishnapal@gmail.com

Rajani Kanaujia
Division of Basic Sciences, ICAR–Indian Institute of Pulses Research, Kanpur 208024, UP, India

Rakesh Singh
Department of Agricultural Economics, Institute of Agricultural Sciences, BHU, Varanasi 221005, UP, India, E-mail: rsingh66bhu@gmail.com

Ram Asrey
Division of Food Science and Postharvest Technology, Indian Agricultural Research Institute, New Delhi 110012, India, E-mail: ramasrey@iari.res.in

Ranjitha K.
Division of Post-Harvest Technology, Indian Institute of Horticultural Research, Hessaraghatta Lake Post, Bengaluru, India

Rekha Rawat
AS & EC Division, Central Institute of Post-Harvest Engineering and Technology (CIPHET), Ludhiana141004, Punjab, India

Ruchi Mishra
Indian Institute of Vegetable Research, Post Bag No. 1, Jakhini (Shahanshapur), Varanasi 221305, UP, India

S. N. Jha
Agriculture Structure and Environmental Control Division, Central Institute of Postharvest Engineering and Technology (CIPHET), Ludhiana 141004, Punjab, India, Email: snjha_ciphet@yahoo.co.in

Sheetal Bhadwal
AS & EC Division, Central Institute of Post-Harvest Engineering and Technology (CIPHET), Ludhiana141004, Punjab, India

Shrawan Singh
Division of Vegetable Science, Indian Agricultural Research Institute, New Delhi 110012, India

Shruti Sethi
Division of Food Science and Postharvest Technology, ICAR—Indian Agricultural Research Institute, New Delhi 110012, India

Shweta Pal
Division of Basic Sciences, ICAR-Indian Institute of Pulses Research, Kanpur 208024, UP, India

Sudhir Singh
Indian Institute of Vegetable Research, Post Bag No. 1, Jakhini (Shahanshapur), Varanasi 221305, UP, India

Sunita Singh
Division of Food Science and Postharvest Technology, ICAR—Indian Agricultural Research Institute, New Delhi 110012, India

Suresh Kumar P.
ICAR—National Research Centre for Banana, Tiruchirapalli, Tamil Nadu, India

Tanmay Kumar Koley
ICAR—Indian Institute of Vegetable Research, Kelabela, UP, India

V. R. Sagar
Division of Postharvest Technology, Indian Agricultural Research Institute, New Delhi, India

Y. Bijen Kumar
Division of Crop Protection, ICAR-Indian Institute of Vegetable Research, Varanasi 221305, UP, India

LIST OF ABBREVIATION

2D	two-dimensional
ACC	1-aminocyclopropane-1-carboxylic acid
ACC-S	1-aminocyclopropane-1-carboxylic acid synthase
AICRP-VC	All India Coordinated Research Programme on Vegetable Crops
ANCs	anti-nutritional compounds
ATP	adenosine triphosphate
AVA	applied vibro-acoustics
CCPR	Codex Committee on Pesticide Residues
CD	cyclodextrins
cDNA	complementary DNA
CFU	colony forming unit
CT	computed tomography
DFD	delayed fruit deterioration
DH	doubled haploids
DM	dry matter
DPPH	2,2-diphenyl-1-picrylhydrazyl
DSC	differential scanning calorimetry
EC	evaporative cooling
ERFs	ethylene response factors
EST	expression sequence tag
ETL	economic threshold level
EU	European Union
FDA	Food and Drug Administration
FSSAI	Food Safety and Standards Authority of India
FW	fresh weight
GAPs	Good Agricultural Practices
GCA	general combining ability
GHI	Global Hunger Index
GMP	good manufacturing practices
HPCD	high-pressure carbon dioxide

HPP	high pressure processing
HSP	heat-shock proteins
HSTFs	heat-shock transcription factors
HUS	haemolytic uremic syndrome
IGRs	insect growth regulators
IPM	integrated pest management
ISSR	inter-simple sequence repeats
LAB	lactic acid bacteria
MAP	modified atmospheric packaging
MCP	1-methylcyclopropene
MDGs	Millennium Development Goals
MP	minimally processed
MRI	magnetic resonance imaging
MRL	maximum residue limit
NILs	near isogenic lines
NIR	near infrared
NNMB	National Nutrition Monitoring Bureau
NO	nitric oxide
NOAEL	No-Observed-Adverse-Effect Level
NOCC	N,O-carboxy methyl chitosan
NSKE	neem seed kernel extract
OMF	oscillating magnetic fields
ORAC	oxygen radical absorbing capacity
PABA	pteridine, p-aminobenzoate
PCR	principal components regression
PG	polygalacturonase
PHI	pre-harvest interval
PLS	partial least squares
QTL	quantitative trait loci
RBCs	red blood cells
RFID	radio frequency identification
RILs	recombinant inbreed lines
ROI	return on investment
ROS	reactive oxygen species
SAM	S-adenosyl-l-methionine
SCA	specific combining ability
SCAR	sequence-characterized regions

SDGs	Sustainable Development Goals
SMEs	small and marginal enterprises
SPs	synthetic pyrethroids
SqMV	squash mosaic virus
TSS	total soluble solids
TTI	time–temperature integrator
TTP	thrombocytopenic purpura
UN-FAO	United Nation-Food and Agriculture Organization
UV	ultraviolet
VAD	vitamin A deficiency
WHO	World Health Organization
WMV-2	watermelon mosaic virus 2

INTRODUCTION

Vegetables play a significant role in nutrition and food security. The demand for fresh vegetables is increasing due to the presence of novel bioactive compounds and because they aid in controlling many lifestyle diseases. Subsequently, vegetable production is increasing with high productivity due to hybrid seeds and improved production technologies, which has resulted in availability of vegetables throughout the year. However, the increased production in peak seasons has led to market surplus and creates more pressure on postharvest management and processing technologies. The magnitude of postharvest losses of vegetables in India is 4.58–12.44%. Effective pre- and postharvest management practices can retain the quality attributes of vegetables for longer periods of time. This book reviews these pre- and postharvest management practices and the recent advances in processing to deliver the novel quality vegetable processed products to consumers.

Chapter 1 deals with natural pigments in vegetables, which are of considerable importance in prevention of chronic diseases such as neurological disorders, cardiovascular diseases, cancer, diabetes and so on. Chapter 2 discusses the causes of postharvest losses that directly linked with abiotic, biotic factors, harvesting and postharvest practices. Chapter 3 highlights the importance of maturity indices in vegetables. Harvesting at the optimum stage of maturity is of significant importance in determining the vegetable quality and shelf life. Chapter 4 focuses on non-destructive quality evaluation of vegetables with the aim of developing effective quality evaluation systems for maintaining acceptable quality standards to consumers. These quality evaluation standards would be very helpful in acceptability of vegetables globally. Chapter 5 looks into breeding aspects for improving nutritional qualities and shelf life of vegetables for fulfilling the vegetable requirement for growing population along with ensuring good health to consumers. Chapter 6 provides the importance of biotechnological applications in controlling senescence, effective preharvest management practices in maintaining nutritional quality and novel biotechnological approaches for commercialization of genetically modified vegetables. Chapter 7 points to the importance of various edible coating substances suitable for reducing the respiratory activities, prolonging the senescence stage and thus retaining the quality attributes for a longer time. Chapter 8 describes the strategies for low-temperature storage of vegetables with special reference to pre-storage treatments, farm-storage techniques, modified and controlled atmospheric storage and its commercial application for increasing the shelf life. Chapter 9 deals with active and smart packaging systems for prevention of microbial spoilage with special reference to spoilage indicators and pathogen sensors. Chapter 10 highlights the presence of anti-nutritional compounds in different solanaceous, cucurbits, cruciferous, leguminous and leafy vegetables. Chapter 11 deals with different steps in minimal processing operations and their effect on pathogenic spoilage organisms. Chapter 12 describes the novel non-thermal processes to develop the shelf stable processed products free from chemical additives and as an alternative to drying, canning and freezing processes. Chapter 13 deals with the importance of encapsulation of bioactive compounds, which acts as an effective means in improving the delivery of various bioactive compounds, of vegetables on industrial applications. Chapter 14 shows the importance of fermentation technology in vegetables as fermentation processes in vegetables enhance the protein quality in vegetables due to increased bioavailability of various vitamins and minerals. Chapter 15 shows various advancements in low cost drying technology to develop the dried vegetables with good consumer preference and market acceptability along with retention of nutrients. Chapter 16 deals with different sources of food-borne diseases, outbreaks, detection of easy methods of food-borne pathogens and various microbial prevention methods in vegetables so that safe, fresh vegetables can be supplied to consumers. Chapter 17 provides the effective usage of pesticides in vegetables to avoid pre- and postharvest pesticide contamination during vegetable production. Chapter 18 describes the entrepreneurship opportunities through public–private partnership modes and establishment of Mega Food Parks integrated with cold chain and food safety management systems.

PREFACE

Globally vegetable production is increasing due to growing demand in both urban and rural populations as a result of health benefits and controlling many lifestyle diseases. Further, poor postharvest management and postharvest processing facilities in terms of inadequate cold chain management, transport, handling and processing has led to unacceptable levels of wastages both quantitatively as well as qualitatively. The magnitude of postharvest losses is 15–20% in the case of fruits and vegetables. Value addition is the easiest way to reduce the postharvest losses on one hand and improving the economic status of poor farmers on the other hand, and it also plays a significant role in nutrition and food security throughout the year. At present, many low-cost processing technologies are available that can be effectively utilized in cost-effective value-added processed products. However, innovative processing technologies can be helpful in retaining the inherent nutritional and functional qualities with safe processed vegetables to consumers. Innovative processing can also cater to the demand of convenient forms of vegetables in the form of easy-to-cook and ready-to-eat vegetables in urban cities.

With the increasing demand for comprehensive information on postharvest technologies on vegetables, this book is an attempt to summarize the recent progress in this field. Thus, *Advances in Postharvest Technologies of Vegetable Crops* brings together many research scientists contributing to this field. The aim of the book is to increase the knowledge in the area of postharvest technologies in vegetables to research scientists, technologists, researchers, extension workers and students. It would be also helpful for teachers to increase the practical experience in relevant areas of postharvest management and processing of vegetables.

This book is devoted to various traditional low-cost and innovative vegetable processing technologies. The book covers most of the important areas of pre- and postharvest management practices, vegetable processing, nutritional quality and safety issues related to vegetable processing. Besides this, pre- and postharvest pesticide contamination on quality, microbial safety and quality assurance and biotechnological applications in postharvest management and entrepreneurship opportunities in processed vegetables in emerging economics are unique and important components in the book.

The editors acknowledge the expert researchers for significantly contributing to the book. The editors would also like to thank the former Director Dr. P. S. Naik for initiating the work.

ABOUT THE EDITORS

Dr. Bijendra Singh has developed 23 cultivars including 11 cultivars in okra (seven varieties and four hybrids), seven varieties in garden pea, two varieties each in cauliflower and radish, and one variety in ash gourd. Of these varieties/hybrids, 18 were identified through All India Coordinated Research Projects on vegetable crops at a national level, and five were identified at the state level by UP State Varieties Release Committee. In addition to high yields, the varieties/hybrids emphasize important desirable traits like YVMV resistance in the okra hybrid Shitla Uphar, Shitla Jyoti, Kashi Vaibhav, Kashi Mahima, and the tall and short intermodal variety Kashi Pragati (VRO-6), bushy habit in okra variety Kashi Vibhuti (VRO-5) and dark green and small fruited variety Kashi Kranti (VRO-22). Kashi Nandani, a short duration and heat tolerant pea variety, is widely adopted by growers under rice–wheat cropping systems. Kashi Kranti in okra and Kashi Surabhi in ash gourd showed better quality attributes for export and processing industries. Dr. B. Singh has also registered seven genetic stocks of bushy and short intermodal length in okra, paper leaf curl virus resistant in chilli, downy mildew resistant in snape melon, and long and thin fruited line in okra, triple-podded vegetable pea and multiple disease resistant in pea with NBPGR for unique character.

Apart from developing varieties and establishing disease-resistance traits, Dr. B. Singh has developed 27 vegetable production technologies for the mid-altitude of the North Eastern Hill Region. Dr. Singh has also contributed to the transfer of technologies to farmers and other stakeholders through capacity building, FLD, adoption of villages, development of seed villages and promoting kitchen gardens among the rural people.

Dr. Singh has published 135 research papers in national and international reputed journals, seven books, 33 technical bulletins, 14 manuals, and several book chapters and popular articles.

The impact of varieties/hybrids and vegetable production technologies developed by Dr. Singh is visible at farmers' fields by their wider adoption. Dr B. Singh is actively associated in developing technologies on processed vegetable products. Technologies on bitter gourd chips, instant *moringa* soup, instant *moringa* drink and instant bottle gourd *kheer* mix have wider acceptance among the masses.

Dr. Singh is a recipient of an Outstanding Multidisciplinary Team Research Award (2003–2004). He is also the recipient of the Dr. Kirti Singh Gold Medal Award of Horticulture Society of India of the year 2012 for his outstanding contribution and leadership in the field of vegetable science. He also received the Dr. Biswajit Choudhury Memorial Award 2014 for outstanding vegetable scientist by the Indian Society of Vegetable Science and the Dr. Punjab Singh Vishisht Krishi Vaigyanik Puraskar Award (2014) for his outstanding contribution in the field of Agricultural Research Management by UPPAS. Dr. Singh is also on active member on the assessment committee for selection of members of scientific committee/panels under the Food Safety Standards Authority of India.

Dr. Sudhir Singh has standardized low-cost drying technologies of bitter gourd, ivy gourd, cauliflower, okra, broccoli, green chilli powder and tomato powder. He focussed on low-cost steeping preservation technologies of cauliflower, pointed gourd and red carrot slices with extended shelf life of 5–6 months at room temperature and convenience vegetables such as easy-to-cook spinach, bathua, amaranth and fenugreek leaves have also been tested at pilot scale. Further, greater consumer preferences have been shown on bitter gourd chips for retention of bitter gourd taste and crispness texture. Vegetable-based convenience dessert such as bottle gourd *kheer* mix, carrot *kheer* mix and carrot dessert have also been standardized for large scale trials.

He has published 50 research papers in national and international high impact factor journals, four technical bulletins, six book chapters, and two practical manuals. He has applied for five patents. One patent has been granted. Dr. Sudhir Singh has guided 44 MSc students and 1 PhD student. He has been awarded the Prof. G. S. Bains award (2015) for his contributions in food science and technology, and Dr. J. S. Pruthi Award in 2010 for his contribution in fruits and vegetable technology from the Associations of Food Scientists and Technologists, CFTRI, Mysore, India.

Dr. Tanmay Kumar Koley recently joined in The Indian Council of Agricultural Research system where he has worked on antioxidants in many horticultural crops. He identified antioxidant-rich genotypes from zizyphus, carrot, radish, capsicum, brinjal, fenugreek, bitter gourd, pumpkin and others. Recently, he started work on bioactive pigments, their purification and use as ingredient for the development of functional foods. Dr. Koley has published six research articles in internationally reputed journal. In addition to this he also published five book chapters and several articles. He has been awarded the Jawaharlal Nehru Memorial Award 2011 for his outstanding in PhD thesis. He is also a recipient of a Gold Medal in BSc for best student of the batch and PhD for outstanding research work in PhD program.

ABOUT THE BOOK SERIES: POSTHARVEST BIOLOGY AND TECHNOLOGY

As we know, preserving the quality of fresh produce has long been a challenging task. In the past, several approaches were in use for the postharvest management of fresh produce, but due to continuous advancement in technology, the increased health consciousness of consumers, and environmental concerns, these approaches have been modified and enhanced to address these issues and concerns.

The Postharvest Biology and Technology series presents edited books that address many important aspects related to postharvest technology of fresh produce. The series presents existing and novel management systems that are in use today or that have great potential to maintain the postharvest quality of fresh produce in terms of microbiological safety, nutrition, and sensory quality.

The books are aimed at professionals, postharvest scientists, academicians researching postharvest problems, and graduate-level students. This series is a comprehensive venture that provides up-to-date scientific and technical information focusing on postharvest management for fresh produce.

Books in the series address the following themes:

- Nutritional composition and antioxidant properties of fresh produce
- Postharvest physiology and biochemistry
- Biotic and abiotic factors affecting maturity and quality
- Preharvest treatments affecting postharvest quality
- Maturity and harvesting issues
- Nondestructive quality assessment
- Physiological and biochemical changes during ripening
- Postharvest treatments and their effects on shelf life and quality
- Postharvest operations such as sorting, grading, ripening, de-greening, curing, etc.
- Storage and shelf-life studies

- Packaging, transportation, and marketing
- Vase life improvement of flowers and foliage
- Postharvest management of spice, medicinal, and plantation crops
- Fruit and vegetable processing waste/byproducts: management and utilization
- Postharvest diseases and physiological disorders
- Minimal processing of fruits and vegetables
- Quarantine and phytosanitory treatments for fresh produce
- Conventional and modern breeding approaches to improve the postharvest quality
- Biotechnological approaches to improve postharvest quality of horticultural crops

We are seeking editors to edit volumes in different postharvest areas for the series. Interested editors may also propose other relevant subjects within their field of expertise, which may not be mentioned in the list above.We can only publish a limited number of volumes each year, so if you are interested, please email your proposal wasim@appleacademic-press.com at your earliest convenience.

We look forward to hearing from you soon.

Editor-in-Chief:
Mohammed Wasim Siddiqui, PhD
Scientist-cum-Assistant Professor | Bihar Agricultural University
Department of Food Science and Technology | Sabour | Bhagalpur | Bihar | INDIA
AAP Sr. Acquisitions Editor, Horticultural Science
Founding/Managing Editor, *Journal of Postharvest Technology*
Email: wasim@appleacademicpress.com
wasim_serene@yahoo.com

BOOKS IN THE POSTHARVEST BIOLOGY AND TECHNOLOGY SERIES

Postharvest Biology and Technology of Horticultural Crops: Principles and Practices for Quality Maintenance
Editor: Mohammed Wasim Siddiqui, PhD

Postharvest Management of Horticultural Crops: Practices for Quality Preservation
Editor: Mohammed Wasim Siddiqui, PhD, Asgar Ali, PhD

Insect Pests of Stored Grain: Biology, Behavior, and Management Strategies
Editor: Ranjeet Kumar, PhD

Innovative Packaging of Fruits and Vegetables: Strategies for Safety and Quality Maintenance
Editors: Mohammed Wasim Siddiqui, PhD,
Mohammad Shafiur Rahman, PhD, and Ali Abas Wani, PhD

Advances in Postharvest Technologies of Vegetable Crops
Editors: Bijendra Singh, PhD, Sudhir Singh, PhD, and
Tanmay K. Koley, PhD

Plant Food By-Products: Industrial Relevance for Food Additives and Nutraceuticals
Editors: J. Fernando Ayala-Zavala, PhD, Gustavo González-Aguilar, PhD, and Mohammed Wasim Siddiqui, PhD

Emerging Postharvest Treatment of Fruits and Vegetables
Editors: Kalyan Barman, PhD, Swati Sharma, PhD, and
Mohammed Wasim Siddiqui, PhD

CHAPTER 1

BIOACTIVE PIGMENTS IN VEGETABLES

TANMAY KUMAR KOLEY[1], KAUSHIK BANERJEE[2], ARTI MAURYA[1], AJAY TRIPATHI[1], and B. SINGH[1]

[1]ICAR—Indian Institute of Vegetable Research, Kelabela, UP, India

[2]ICAR—National Research Centre for Grapes, Pune, India

CONTENTS

1.1   INTRODUCTION

Colour is considered one of the most important quality parameters associated with food preference by the human and other animals, insects, birds and so forth. Presence of natural pigments at different proportions is responsible for providing a wide range of colour to food with different tones and shades. Significantly, most of the natural pigments possess functional properties which play vital roles in preventing chronic diseases such as neurological illness, cardiovascular illness, cancer, diabetes and so forth. Thus, all these naturally occurring pigments are considered bioactive pigments. Bioactive pigments majorly found in plant kingdom are carotenoids, betalains and flavonoids, including anthocyanins. Significantly, all these pigments are present in vegetables in various concentrations. As vegetables are commonly consumed across the world, on the basis of experiences and awareness, the anti-nutritional and toxic compounds in vegetables have been largely eliminated through natural and human selection pressure. Thus, vegetables are considered rich sources of bioactive pigments without purification.

1.2   CAROTENOIDS

Carotenoids represent a group of bioactive secondary plant metabolite which makes plant parts, such as leaves, stem, flower and fruits, appear in different colours. They are usually C_{40} tetraterpenoids, which are built from eight C_5 isoprenoid units, and form lipid-soluble yellow, orange and red pigments in plants. They are biosynthesized from geranyl–geranyl phosphate pathway. In plants, they are the most important part of light-harvesting system in chloroplasts. They quench the excess energy of excited chlorophyll or singlet oxygen and thus play a crucial role in protecting plants against photo-oxidative damage (Gross, 1991). They are one of major groups of natural pigments in flowers, which helps attract the pollinators.

Animals as well as humans cannot synthesize carotenoids at their own and thus have to depend on dietary sources from plants. More than 600 carotenoids are found in nature, with 40 dietary carotenoids regularly being consumed by human (Bendich, 1993), and among these, six carotenoids are often quantified in human serum, namely lutein, zeaxanthin, β-cryptoxanthin, lycopene, and α- and β-carotene.

Carotenoids are classified into two broad groups: carotenes and xanthophylls. Carotenes are composed of carbon and hydrogen atoms, and in this group, β-carotene, α-carotene and lycopene are the predominant members in nature. Unlike carotenes, xanthophylls carry at least one oxygen atom in addition to carbon and hydrogen atoms. In nature, zeaxanthin, lutein, α- and β-cryptoxanthin are the important members of the xanthophylls group.

1.2.1   CAROTENE

In vegetables, β-carotene is most predominant carotene pigments. In addition to this, lycopene is observed in few vegetables. α-Carotene exists in few vegetables in meagre amount.

1.2.1.1 β-CAROTENE

β-Carotene is the most abundant carotenoid in nature. Chemically, it is a terpenoid (isoprenoid) hydrocarbon carrying eight isoprene units. It is distinguished with beta rings at both ends of a molecule. Similar to other carotenoids, it also acts as lipid soluble, so its absorption can be enhanced by taking it with fats. Its molecular formula and molecular weight are $C_{40}H_{56}$ and 536.87264 g/mol, respectively (Fig. 1.1).

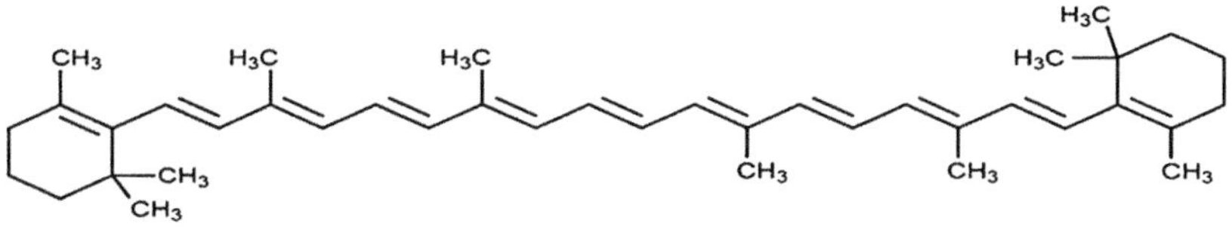

FIGURE 1.1 Molecular structure of β-carotene.

1.2.1.1.1 Sources of β-Carotene from Vegetables

β-Carotene is abundantly available in yellow/orange vegetables such as carrots, ripe pumpkins and winter squash. In addition to this, dark green leafy vegetables also prove to be good sources of β-carotene (Table 1.1).

1.2.1.2 LYCOPENE

Lycopene is another important carotenoid found in a limited number of vegetables. Chemically, it is also a terpenoid hydrocarbon without any rings at the end of its structure. Due to absence of an adequate number of beta rings, it does not possess provitamin A activity. On the other hand, due to presence of a large number of conjugated double bonds in its structure (i.e. 11) among carotenoids, it exhibits a strong antioxidant property (Palozza and Krinsky, 1992; Burton, 1989; Krinsky, 1989). The molecular formula and molecular weight of lycopene are $C_{40}H_{56}$ and 536.87, respectively (Fig. 1.2).

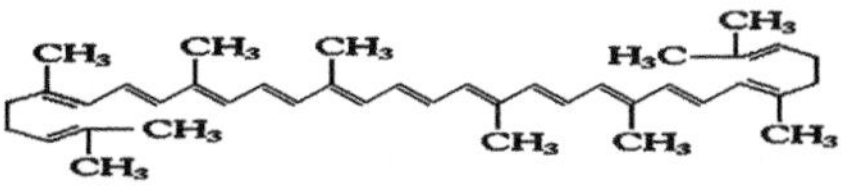

FIGURE 1.2 Molecular structure of lycopene.

TABLE 1.1   Carotenoids Contents (mg/100 g) in Some Leafy Vegetables.

S. No	Crop name	Carotenoids		References
	Leaf	α-Carotene	β-Carotene	
1	*Allium cepa*	ND	16.90	Raju et al. (2007)
2	*Amaranthus gangeticus*	ND	18.67	
3	*Allmania nodiflora*	ND	9.63	
4	*Alternanthera pungens*	ND	34.66	
5	*Alternanthera sessils*	ND	27.07	
6	*Amaranthus tristis*	ND	16.76	
7	*Amaranthus viridis*	6.75	58.95	
8	*Basella alba*	18.23	43.82	
9	*Boerhavia diffusa*	ND	27.67	
10	*Brassica oleraceae var. botrytis*	ND	3.85	
11	*Beta vulgaris*	1.54	12.50	
12	*Chenopodium album*	ND	114.61	
13	*Cucurbita maxima*	ND	10.27	
14	*Coriandrum. Sativum*	ND	67.50	
15	*Daucus carota* (Leaf)	21.53	12.09	
16	*Gynandropsis pentaphylla*	ND	37.04	
17	*Hudrocotyl asiatica*	ND	9.02	
18	*Lactuca sativa*	ND	42.72	
19	*Mentha spicata*	ND	7.48	
20	*Piper betle*	18.42	13.35	
21	*Phyllanthus niruri*	ND	60.88	
22	*Portulaca oleracea*	ND	27.05	
23	*Rumex acetosella*	ND	70.83	
24	*Solanum nigrum*	ND	50.11	
25	*Talinum cuniefolium*	18.77	42.44	
26	*Trianthema portulacastrum*	ND	37.76	
27	*Tribulus terrestris*	6.1	30.81	
28	*Cichorium intybus*, cv. 'Anivip'	ND	7.31 ± 1.56	Žnidarčič et al. (2011)
29	*Cichorium intybus* cv. Monivip	ND	3.94 ± 0.65	
30	*Taraxacum officinale*	ND	6.34 ± 0.94	
31	*Eruca sativa*	0.28 ± 0.04	7.96 ± 1.43	
32	Wild rocket—*Diplotaxis tenuifolia*	0.17 ± 0.06	7.01 ± 1.04	

1.2.1.2.1   Sources of Lycopene

Lycopene exists in a number of vegetables such as tomato, processed tomato products, red carrot, watermelon, red pepper (Holden et al., 1999) and so forth (Table 1.2). In addition to this, placenta of ripe cucurbitaceous vegetables such as bitter gourd, snake gourd, gac fruit and so forth are good sources of lycopene. The concentration of lycopene in the ripe aril of gac fruit (*Momordica cochinchinensis*) is about 380 µg/g of aril (Aoki et al., 2002), which is about 10fold higher than that in known lycopene-rich fruits and vegetables.

TABLE 1.2   Lycopene (mg/kg) Concentration in Vegetables.

Vegetables	Scientific name	*Lycopene* content	Reference
Watermelon	*Citrullus lanatus*	45.32	USDA, 2005
Ripe tomatoes	*Lycopersicon esculentum*	25.73	USDA, 2005
Red carrot	*Daucus carota*	61.00	Surles et. al., 2004
Red peppers	*Capsicum annuum*	3.08	USDA, 2005
Asparagus (cooked)	*Asparagus officinalis*	0.30	USDA, 2005
Process food			
Tomato paste		287.64	USDA, 2005a
Tomato juice, canned		90.37	USDA, 2005a
Tomato soup		54.60	USDA, 2005a
Canned tomato sauce		151.52	USDA, 2005a

1.2.2   XANTHOPHYLL

Xanthophylls are yellow pigments that are widely found in nature. Lutein and zeaxanthin are predominant xanthophyll, commonly found in vegetables. Besides these, other important xanthophyll found in vegetables are neoxanthin, violaxanthin, flavoxanthin, β- and α-cryptoxanthin. The predominant isomeric xanthophylls form is *trans-* for all foods. However, processed foods contain more *cis-* xanthophylls isomers than fresh food.

1.2.2.1 CHEMISTRY

The molecular structure of xanthophylls is more or less similar to that of carotenes with a little difference. Xanthophylls contain oxygen atoms, whereas carotenes are purely hydrocarbons without any oxygen atoms in its structure. The oxygen atoms are present in xanthophylls as hydroxyl groups or as pairs of hydrogen atoms that are substituted by oxygen atoms acting as a bridge (epoxide). This chemical feature makes xanthophyll more polar than the purely hydrocarbon carotenes, and thus it can be easily chromatographically separated from the carotenes. The molecular formula and molecular weight of xanthophylls are $C_{40}H_{56}O_2$ and 568.871 g/mol, respectively.

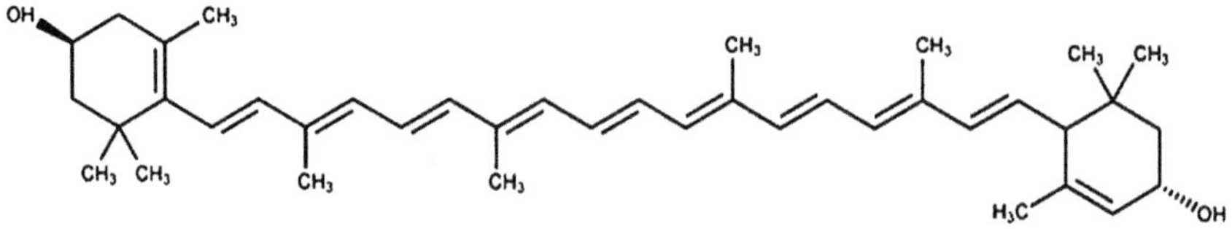

FIGURE 1.3   Molecular structure of xanthophyll.

1.2.2.2 DISTRIBUTION

Xanthophylls are abundantly available in yellow-pigmented vegetables such as pumpkin, carrot and yellow capsicum. Besides these, green leafy vegetables are also richer sources of xanthophylls (Table 1.3).

1.2.3 BIOLOGICAL FUNCTION

Humans and other animals are not capable of synthesizing carotenoids, so they have to depend on dietary sources. Fruits and vegetables are the primary sources of carotenoids in diet, and their consumption provides huge health benefits.

1.2.3.1 PROVITAMIN A ACTIVITY

The most important physiological function of carotenoids in human nutrition is provitamin A activity. For this activity, the occurrence of an

TABLE 1.3 Xanthophyll Content (mg/100 g) in Leafy Vegetables.

S No	Crop name	Xanthophyll				References
		Lutein	Zeaxanthin	Violaxanthin	Neoxanthin	
1	*Allium cepa*	30.21	0.21	1.83	6.61	Raju et al. (2007)
2	*Amaranthus gangeticus*	32.02	0.34	26.50	1.46	
3	*Allmania nodiflora*	23.10	0.18	8.90	7.74	
4	*Alternanthera pungens*	71.86	0.67	42.03	ND	
5	*Alternanthera sessils*	32.47	0.26	23.93	13.89	
6	*Amaranthus tristis*	30.30	0.23	19.15	1.15	
7	*Amaranthus viridis*	90.43	1.04	84.06	12.63	
8	*Basella alba*	113.82	1.76	6.27	7.74	
9	*Boerhavia diffusa*	26.83	0.19	3.93	1.41	
10	*Brassica oleraceae var. botrytis*	4.50	ND	1.45	0.52	
11	*Beta vulgaris*	26.86	0.14	3.97	6.39	
12	*Chenopodium album*	187.59	5.00	142.59	ND	
13	*Cucurbita maxima*	27.18	0.25	15.58	ND	
14	*Coriandrum. Sativum*	9.92	ND	83.43	5.47	
15	*Daucuscarota* (Leaf)	40.17	0.59	7.00	2.09	
16	*Gynandropsis pentaphylla*	42.65	1.28	ND	49.31	

TABLE 1.3 *(Continued)*

S No	Crop name	Lutein	Zeaxanthin	Xanthophyll Violaxanthin	Neoxanthin	References
17	*Hudrocotyl asiatica*	15.93	ND	0.65	0.89	
18	*Lactuca sativa*	87.12	ND	29.09	5.75	
19	*Mentha spicata*	17.74	0.26	5.62	2.11	
20	*Piper betle*	36.43	0.47	0.89	0.82	
21	*Phyllanthus niruri*	77.55	1.63	3.67	33.60	
22	*Portulaca oleracea*	50.84	0.94	11.47	0.73	
23	*Rumex acetosella*	144.30	ND	1.45	7.70	
24	*Solanum nigrum*	84.86	ND	22.17	2.79	
25	*Talinumcuniefolium*	89.79	1.22	10.51	7.95	
26	*Trianthema portulacastrum*	41.51	0.44	5.00	2.50	
27	*Tribulus terrestris*	56.39	0.04	8.84	1.34	
28	*Cichoriumintybus,* cv. 'Anivip'	5.91	0.09	1.67	–	Žnidarčič et al. (2011)
29	cv. Monivip	3.87	0.07	0.69	–	
30	*Taraxacum* 1 officinale,	5.25	0.08	0.65	–	
31	*eruca sativa*	7.44	0.06	1.56	–	
32	*Diplotaxis tenuifolia*	5.82	0.05	0.71	–	

unsubstituted β ring with a C_{11} polyene chain is the minimum requirements. Structurally, vitamin A (retinol) is one half of the molecule of β-carotene. Thus, β-carotene is a potent provitamin A with 100% activity. Other carotenoids, namely γ-carotene, α-carotene, β-cryptoxanthin, α-cryptoxanthin and β-carotene-5,6-epoxide has one unsubstituted ring; thus, they have about half the bioactivity of β-carotene. The carotenoids which are devoid of β rings (acyclic carotenoids, such as lycopene) and in which the β rings have hydroxy, epoxy and carbonyl substituents (xanthophylls) do not have any provitamin A activity.

Vitamin A is essential for the promotion of growth, normal reproductive performance, embryonal development and visual function. The contribution of provitamin A carotenoids to daily intake of vitamin A depends on dietary food habits and available food sources. In developing countries, the vegetable carotenoids contribute to greater than 80% of the available vitamin A. However, bioequivalence of the vegetable carotenoids is much less than what is anticipated from the studies using dietary supplementation (Stahl and Sies, 2005). The efficacy of cleavage, substrate specificity for various provitamin A compounds, as well as genetic variations and factors with an impact on the expression of carotenoid-metabolizing enzymes are individual variants determining vitamin A supply via carotenoids (Stahl and Sies, 2005).

1.2.3.2   BONE HEALTH

Osteoporosis is a major public health problem in elderly people. Aging bone induces osteoporosis. Recent epidemiological studies reveal that nutritional factors have a potential impact on osteoporosis. Nutritional compounds present in food including vegetables may help to prevent osteoporosis.

Among the various carotenoids, β-cryptoxanthin has unique anabolic effect on bone metabolism. Culture with β-cryptoxanthin caused a significant increase in calcium content and alkaline phosphatase activity in the femoral diaphyseal and metaphyseal tissue under in vitro condition (Yamaguchi and Uchiyama, 2004). Alkaline phosphatase helps in mineralization in the bone tissue. It has also been observed that β-cryptoxanthin has inhibitory effect on bone resorption in bone tissue culture. Recent epidemiological studies reveal that β-cryptoxanthin modulates gene expression of various proteins that involves in bone formation in osteoblast and bone resorption in osteoclast (Yamaguchi, 2008). Intake of β-cryptoxanthin has proved to have effect in bone loss in animal model (Uchiyama et al., 2004) and restorative effect on bone metabolism in menopausal women (Yamaguchi et al., 2006).

Significantly, β-cryptoxanthin with zinc at lower concentrations has been found to have synergistic effect on osteoblastic bone formation and additive suppressive effect on osteoclastic cell formation (Uchiyama et al., 2005). It necessitates identifying new vegetables having higher β-cryptoxanthin and zinc.

1.2.3.3 EYE HEALTH

Age-related macular degeneration (ARMD) and cataract are the major concerns in the elderly population. Scientific investigation reveals that intake of carotenoids-rich food reduces occurrences of ARMD and cataract. Among the carotenoids, the study of lutein and zeaxanthin finds genuine beneficial nutrients for preventing ARMD and cataract (Wisniewska and Subczynski, 2006). It has been observed that long-term supplementation with lutein improves the visual performance in the patient with ARMD and cataract. Properties of xanthophylls for protecting eyes are considered to be associated with their antioxidant activities. Lutein and zeaxanthin, the major xanthophylls, are selectively accumulated in the macular region of the eye. They play a crucial role to protect against light-initiated oxidative damage through quenching singlet oxygen, scavenging free radicals, inhibiting peroxidation of membrane phospholipids and reducing lipofuscin formation.

The other eye-protecting property of xanthophylls is associated with their blue light–filtering properties. Excessive exposure to blue light leads to retinal damage and increases the rate of photoreceptor apoptosis (Roca et al., 2004). Research indicates that xanthophylls in macular region prevent light-induced photoreceptor damage and thus retard age-related visual loss (Haegerstrom-Portnoy, 1988). Actually lutein and zeaxanthin have a broad absorption band, with a peak approximately at 450 nm. Therefore, they can absorb and attenuate the blue light before it reaches the photoreceptors (Greenstein et al., 2007).

1.2.4 INDUSTRIAL USE

The Food and Drug Administration has approved a petition from Lycored seeking the use of higher levels of tomato lycopene to restore colour to processed meats. Lycopene-based Lycored pigment gives the alternative option to the manufacturers of sausages, deli meats and jerky, who use synthetic FD&C Red #40 and bug-derived carmine to colour the meat. Xanthophylls are used for the improvement of egg yolk colour by feeding the chicken with xanthophylls-rich vegetables/foods.

1.3 BETALAINS

Betalains are another class of bioactive secondary metabolite, commonly found in plants in the order of caryophyllales in plant kingdom. They are water-soluble red, reddish violet, yellow pigment, present in cell vacuole. They are biosynthesized from tyrosine. Their role in plant physiology is still obscure. However, they played a crucial role as attractants to pollinators and seed dispersers.

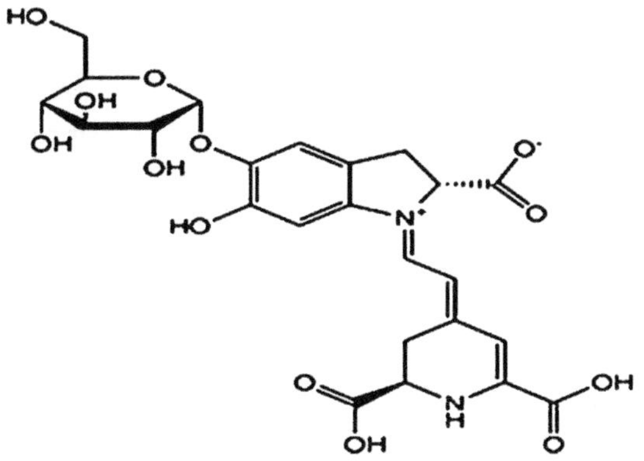

FIGURE 1.4 Chemical structure of betanin.

1.3.1 CHEMISTRY

Chemically betalains are aromatic indole derivatives. Although, similar to anthocyanins, they are water soluble in nature, structurally they are different. In addition to carbon, hydrogen, oxygen atom, betalains also carry nitrogen atom in its structure, whereas anthocyanin do not contain nitrogen. The molecular formula of betalians is $C_{24}H_{26}N_2O_{13}$, and the molecular weight of betalian is 550.46884 g/mol.

Betalains are categorized into two broad groups—betacyanins and betaxanthins. Betacyanins are reddish to violet in colour and glycoside of glyconebetanidin or isobetanidin. Some of the betacyanin are acylated with hydroxycinnamic acid, namely ferulic acid, *p*-coumaric acid and aliphatic acid, namely 3-hydro-3methyl glutryl acid. Some of the common betcyanins in vegetables are amaranthin, betanin and gomphrenin. Betaxanthins come yellow to orange consisting of betalamic acid with amines or amino acids. Some common betaxanthins available in vegetables are vulgaxanthin, miraxanthin, portulaxanthin andindicaxanthin.

1.3.1.1   CLASSIFICATION OF BETALAINS

Betalains	
Betacyanin Betaxanthin	
Amaranthin group	**Amino acid-derived conjugates**
Amaranthin	Portulacaxanthin II
Isoamaranthin	Portulacaxanthin III
	Tryptophan-betaxanthin
	Miraxanthin V
	Vulgaxanthin I
Betanin group	**Amine-derived conjugates**
Betanin	3-Methoxytyramine-betaxanthin
Isobetanin	
Gomphrenin group	
Gomphrenin	
Isogomphrenin	

1.3.2   DISTRIBUTION

Among higher plants, the occurrence of betalains is restricted to the Caryophyllales (Mabry et al., 1963). Many crops belonging to this order are commonly used as vegetables such as spinach (*Spinacia oleracea*), beetroot (*Beta vulgaris*), swish chard (*Beta vulgaris* var*cicla*), palak (*Beta vulgaris* var*bengalensis*), amaranthus (*Amaranthus cruentus, A. blitum, A. dubius and A. tricolor*), bathua (*Chenopodium album*), quinoa (*Chenopodium quinoa*), basella (*Basselaalba*) and so on. Although various vegetables contain betalains, due to relative abundance most common commercially exploited vegetables are found as a source of betalains. Nilsson (1970) was the first researcher who quantified the betalains content of various red beetroots cultivar and reported that the betacyanin and betaxanthin contents range from 0.04 to 0.21% and 0.02–0.14%, respectively. Gasztonyi et al. (2001) studied betalains content in five beetroot cultivar and identified four betacyanins, namely betanin, isobetanin, betanidin and isobetanidin, and two betaxanthins, namely vulgaxanthin I and vulgaxanthin II. Girod and Zryd (1991) identified another two types of betaxanthins, namely dopaxanthin and miraxanthin V in orange callus cultures of beetroot cultivar, bikoresmonogerm. Kugler et al., (2004) identified 25 different betaxanthins in Swiss chard. Cai et al. (2001) investigated the distribution of betalains in 37 species in the family of amaranthaceae and identified a total of 16 kinds of betacyanins and three kinds of betaxanthins. Among betacyanins, eight amaranthin type, six gomphrenin type and two betanin type of pigments were identified. Six betacyanins were identified as simple non-acylated glycosides and 10 acylated glycosides which were glucoacylated with ferulic, *p*-coumaric or 3-hydroxy-3-methylglutaric acids. Three betaxanthins were identified as immonium conjugates of betalamic acid with dopamine, 3-methoxytyramine and (*S*)-tryptophan.

1.3.3   BIOLOGICAL FUNCTION

1.3.3.1   ANTIOXIDANT ACTIVITY

There has been growing interest among the consumers on dietary antioxidants as nutraceuticals. Escribano et al. (1998) were the first researchers to report the antiradical activity of betalains from beetroots. Later, Kanner et al. (2001) reported betalains as a new class of dietary antioxidants. Cai et al. (2003) were the first researchers who comprehensively studied antioxidant activity of 19 different betalains from the plants in the family Amaranthaceae using the 1,1-diphenyl-2-picrylhydrazyl method and observed that all the betalains exhibited strong antioxidant activity with EC_{50} values ranging from 3.4 to 8.4 mM. They reported that gomphrenin-type betacyanins (mean Z 3.7 mM) and betaxanthins (mean Z 4.2 mM) showed high antioxidant activity. To find the relationship between the chemical structure and activity of the betalains, Cai et al. (2003) observed that the free radical-scavenging activity of betalains usually increases with the increase in number of hydroxyl/imino groups. It has also been observed that radical-scavenging activity depends on the position of hydroxyl groups and glycosylation of aglycones in the betalain molecules. They conclude that presence of the hydroxyl group in the C-5 position on aglycones in the betalain molecules significantly improves antioxidant activity, whereas the increase in the number of glycosylation to aglycones reduces the antioxidant activity (Cai et al., 2003).

1.3.3.2   ANTIMICROBIAL PROPERTIES

Betalains-rich *Amaranthus spinosus* showed significant antimalarial activity in mice. The antimalarial function was linked to the metal chelation properties of betacyanin. As for nucleic acids synthesis, the plasmodium enzyme ribonucleotide reductase needs magnesium and iron ions as cofactors, researchers postulated that orthodiphenol and several carboxy groups in betanin chelate the indispensable inner cations Ca^{2+}, Fe^{2+} and Mg^2 and thus suppressed the growth of plasmodium (Hilou et al., 2006). Antimalarial properties of amaranthus were also attributed to the quaternary nitrogen group present in betalains (Hilou et al., 2006). Ancelin and Vial (1986) reported that quaternary nitrogen group inhibits plasmodium growth by blocking the intracellular transport of choline. Choline is necessary for the biosynthesis of the phosphatidylcholines, which are essential molecules for the survival of plasmodium.

1.3.4   COLOUR PROPERTIES

Betalains provide wide hue of colour patterns due to its chemical structure which influences its absorption maxima. Betacyanin provides red-to-violet colour. In general, two absorption maxima have been observed for betacyanin—one absorption maxima is observed within the UV ranges of 270–280 nm, and the other is observed within visible ranges of 535–540 nm. In addition to this, the third absorption maxima have been observed within UV ranges of 300–330 nm when betacyanins are acylated with hydroxycinnamic acid. However, acylation with aliphatic acid did not alter the absorption spectra of betcyanin (Stintzing and Carle, 2004). A hypsochromic shift of about 6 nm has been observed when glycosylation occurred with betanidin moiety, whereas additional sugar moiety to the first one merely affects the colour (Cai et al., 1998; Stintzing and Carle, 2004). In the case of betaxanthins amino acid or amine, side chains influenced the colour properties, and small structural changes lead to hypsochromicor bathochromic shifts (Strack et al., 2003). Betaxanthin containing amine group as side chain displays relatively lower absorption maxima compared with amino acid counterparts (Stintzing et al., 2002).

1.3.5 INDUSTRIAL USES

Betalains have been successfully explored as natural pigments. Commercially, betalains are known as beetroot red as they extracted mainly from beetroot. Its E number is E162. In the European Union and the United States of America, it is successfully commercialized (Castellar et al., 2006) and are available in the form of juice concentrates or freeze-dried or spray-dried powders (Cerezal et al., 1994). As they are unstable when subject to light, heat or oxygen, they are generally recommended for colouring of frozen food or food with relatively short shelf-life and sold in dry state. They are successfully used for colouring ice cream, sugar confectionary, fruit desert. They are also used in dry tomato soup mix for improvement of colour. Synthetic colour used for colouring of meat and sausage product can be successfully replaced with betalains. Hopi red dye that is produced from amarathus flower is used as natural colourant by Hopi tribes in America.

1.4 ANTHOCYANINS

Anthocyanins are another most important group of bioactive pigments in vegetables. The name 'anthocyanins' derived from the Greek word 'Anthos' means flower, and 'kianos' means blue. Similar to betalains, they are selectively accumulated in vacuoles of cells and contribute orange to red, purple and blue colouration to the flower, fruits, leaves, stem and roots of plants. Pigmentation properties of anthocyanin play a paramount contribution in plant evolution as these are the prime pigments in flower which play a crucial role as attractant to pollinators. In addition to this, in photosynthetic tissues, they act as a sunscreen by protecting the tissues from photoinhibition. They absorb blue-green and ultraviolet light from sun and thus protect the cells from high-light induced damage. In cell, they played an important role as powerful antioxidants. However, contribution of anthocyanin towards the scavenging activity of free radicals produced as a by-product of metabolic processes in plants is still obscure. As they are present in vacuole and thus spatially separated from reactive oxygen, species generated as a results of metabolic reaction. Recently some studies however revealed that hydrogen peroxide produced during metabolic reaction in other organelles were nutralysed by anthocyanins. Currently, they are considered one of the most important nutraceuticals because of their strong antioxidant activity and other beneficial physiological properties and thus gaining importance as one of the potential functional food ingredient.

1.4.1 CHEMISTRY

Chemically anthocyanin is flavillium 2-phenyl chloromenylium. They are the byproducts of phenyl propanoid pathway having C6–C3–C6-skeleton. It composed of carbon, oxygen and hydrogen atom. The molecular formula of the basic structure of anthocyanin is $C_{15}H_{11}O^+$ and molecular weight is 207.24724 g/mol. Anthocyanin is composed of aglyconeanthocyanidin and sugar moiety as glycone. Anthocyanidinis composed of an aromatic ring A which bonded to oxygen containing heterocyclic ring C, which is further bonded to another aromatic ring B by a carbon–carbon bond (Konczak and Zhang, 2004). Glycone either pentose or hexose are attached to the C-3 or C-6 position of aglycone depending on the type of anthocyanin. In nature, 18 different aglycones have been observed, among which six are the most commonly found in higher plants that are, namely pelargonidin, cyanidin, delphinidin, peonidin, petunidin and malvidin. Anthocyanidins, namely pelrgonidin, cyanidin and delphinidin are non-methylated, whereas petunidin, peonidin and malvidin are methylated either at C-3′ or C-5′ position of C ring.

Among these, cyanidin, delphinidin and pelrgonidin are the most widespread in nature and are available in 80% of pigmented leaves, 69% of fruits and 50% of flowers (Kong et al., 2003).

1.4.1.1 STABILITY OF ANTHOCYANIN

The isolated anthocyanins are highly unstable and susceptible to degradation during processing treatments, formulation and storage conditions (Giusti and Wrolstad, 2003), and they have limited scope as natural colourants in foods. Various factors such as pH of media, surrounding storage temperature, their own chemical structure, exposure to light, oxygen and solvents, the presence of enzymes, flavonoids, proteins and metallic ions affect the stability of anthocyanin. Acylation with hydroxycinamic acid (ferulic acid, caffeic acid or p-coumaric acid) or hydroxybenzoic acid to glycone moiety improved the stability of anthocyanin. In addition to this, intra- and inter-molecular co-pigmentation, self-association, metal complexion with Al^{3+} or Fe^{2+} and presence of inorganic salts improved the stability of anthocyanin (Malien Aubert et al., 2001).

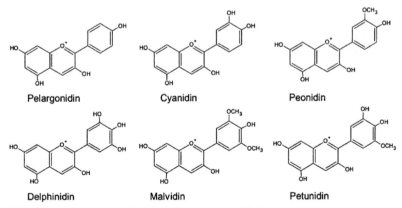

FIGURE 1.5   Most commonly occurring anthocyanidin in nature.

1.4.2 DISTRIBUTION

In nature, anthocyanins are widely distributed in leaves, stems, flowers, fruits and roots. About 80% of coloured leaves, 69% in fruits and 50% in flowers are known to contain three major glycoside derivatives of the non-methylated anthocyanidins, namely cyanidin, delphinidin and pelargonidin (Dey and Harborne, 1993). In vegetables, wide ranges of anthocyanins have been observed which provide colouration from deep purple to blue, pink and red. Li et al. (2012) reported that cyanidin, petunidin, pelargonidin and delphinidin are the main aglyconein highly red or purple pigmented vegetables. Some of the common vegetables rich in anthocyanins are red potato, purple potato, red cabbage, purple broccoli, purple capsicum, purple brinjal, purple radish, pink radish, black carrot and forth. In some of the vegetables such as purple French bean, purple colouration suddenly occurs as a result of mutation which is stabilized by breeding methods. In some of vegetables, such as tomato, gene responsible for synthesis of anthocyanin is cloned from flower.

1.4.3 BIOLOGICAL FUNCTION

1.4.3.1 ANTIOXIDANT ACTIVITY

Under in vitro condition, anthocyanins displayed strong antioxidant potential which was even higher than ascorbic acid and tocopherol (Bagchi et al., 1998). The conjugated structure makes them suitable as good antioxidant because it allows electron delocalization which results in very stable radical products during the reaction (van Acker et al., 1996). The type of functional group, their degree and position and glycosylation pattern modify the antioxidant potential of anthocyanin. Although they are good antioxidant under in vitro condition, they showed variable results which may be because of their complex mechanism of action. In vitro antioxidant potential of vegetable rich in anthocyanins is widely studied, and in many cases a positive linear correlation has been established between the anthocyanins content and antioxidant activity (Koley et al., 2014).

1.4.3.2 ANTI-DIABETIC PROPERTY

Epidemiological studies indicate that the consumption of a diet including anthocyanin-rich vegetable decreases the incidence of type-2 diabetes. Nizamutdinova et al. (2009) studied the anti-diabetic effect of anthocyanin from soybean in steptozotocin-induced diabetic rat and reported that anthocyanins activate the insulin receptors phosphorylation, enhance and thus reduce glucose concentration. Another way to control diabetes is to prevent postprandial hyperglycaemia. It has been observed that anthocyanin delays digestion of carbohydrates by inhibiting α-glucosidase and thus prevents post-prandial hyperglycaemia. Anthocyanins from vegetables were found to be effective as potent α-glucosidase inhibitors (Matsui et al., 2002).

1.4.4 COLOUR PROPERTIES

Anthocyanins provide wide hue of colour patterns due to the presence of flavylium ion (Wrolstad et al., 2005) in its chemical structure. The flavylium ion, the principle part of the molecules anthocyanin that absorbs light, is known as chromophore. The hydroxyl group, sugar moiety and acyl group (hydroxycinnamic and hydroxybenzoic acid) that are responsible to fine tune the colour are known as auxochrome. Acylation pattern influenced the colour and spectral character of the anthocyanin extract. Presence of cinnamic acid is always resulted in a bathochromic as well as hyperchromic shift in antocyaanin extract. As per example, orange hue of pelargonidin aglycones shifts into an orange-red colour hue after acylation with hydroxycinnamic acid (Giusti et al., 1999). Likewise sugar moiety quantitatively influences the spectral and colour properties of anthocyanin. Molar absorptivity of the flavylium cation increases with the increase in sugar substitution. Increase in sugar substitution also results in hypsochromic shift with increase in hue (Giusti and Wrolstand, 2003). The pH of the surrounding media influences the colour character of the anthocyanin. At pH 1, protonated form of anthocyanin flvilium ion is pre-dominated which provides deep red and purple, depending on antocyanidin. When pH rises between 2 and 4 due to deprotonation, quinoidal bases are formed which give blue colouration. When pH further rises to 5, a colourless carbinol pseudo-base is formed in which chromophore is no longer capable of absorbing light. At pH 6, yellow-coloured chalcones are formed. When pH values of surrounding media go off higher than 7, the anthocyanins are degraded (Castañeda-Ovando et al., 2009).

(i) Chalcone (Yellow)

(ii) Carbinol pseudo-base (colourless)

(iv) Quinoidal Base (coloured)

(iii) Flavylium cation (coloured)

1.5 INDUSTRIAL USES

Similar to carotenoids and betalains, anthocyanins are successfully explored commercially as a natural colourant. Its E no. is E163. In the European Union and the United States of America, it is permitted as commercial food additives. In recent time Food Safety and Standards Authority of India, the supreme authority in food safety in India approved anthocyanin as food additives. However, they barred its use in infant food. Although various vegetables are good sources of anthocyanin pigments, their instability and high production cost make majority of vegetable unsuitable for colourant sources. However, vegetables such as black carrot, red cabbage, purple potato, sweet potato and red radish are found to be suitable as source of colourant due to relative abundance and higher percentage of acylated anthocyanin. Among them, red cabbage and black carrot are globally commercially explored in the form of juice concentrate. They have been used for colouring of ice cream, sugar confectionary, jelly, soft drink and so forth. Anthocyanins from red radish were found to be suitable alternative to Allura red used to colour maraschino cherries.

REFERENCES

Ancelin, M. L.; Vial, H. J. Quaternary Ammonium Compounds Efficiently Inhibit *Plasmodium falciparum* Growth in vitro by Impairment of Choline Transport. *Antimicrob. Agents Chemother.* **1986,** *29,* 814–820.

Aoki, H.; Kieu, N. T. M.; Kuze, N.; Tomisaka, K. Carotenoid Pigments in Gac Fruit (*Momordica cochinchinensis* Spreng). *Biosci. Biotechnol. Biochem.* **2002,** *66*(11), 2479–2482.

Bagchi D., Garg, A.; Krohn, R. L.; Bagchi, M.; Bagchi, B. J.; Balmoori, J. Protective Effects of Grape Seed Proanthocyanidins and Selected Antioxidants Against TPA-Induced Hepatic and Brain Lipid Peroxidation and DNA Fragmentation, and Peritoneal Macrophage Activation in Mice. *Gen. Pharmacol.* **1998,** *30*(5), 771–776.

Bendich, A. Biological Functions of Dietary Carotenoids. *Ann. N. Y. Acad. Sci.* **1993,** *691*(1), 61–67.

Burton, G. W. Antioxidant Action of Carotenoids. *J. Nutr.* **1989,** *119*(1), 109–111.

Cai, Y.; Sun, M.; Wu, H.; Huang, R.; Corke, H. Characterization and Quantification of Betacyanin Pigments from Diverse Amaranthus Species. *J. Agric. Food Chem.* **1998,** *46,* 2063–2070.

Cai, Y. Z.; Sun, M.; Corke, H. Identification and Distribution of Simple and Acylated betacyanin Pigments in the Amaranthaceae. *J. Agric Food Chem.* **2001,** *49,* 1971–1978.

Cai, Y.; Sun, M.; Corke, H. Antioxidant Activity of Betalains from Plants of the Amaranthaceae. *J. Agric. Food Chem.* **2003,** *51*(8), 2288–2294.

Castellar, M. R.; Obon, J. M.; Fernandez-Lopez, J. A. The Isolation and Properties of a Concentrated Red-Purple Betacyanin Food Colourant from Opuntia stricta fruits. *J. Sci. Food Agric.* **2006,** *86,* 122–128.

Castañeda-Ovando, A.; Pacheco-Hernández, M. L.; Páez-Hernández, M. E.; Rodríguez, J. A.; Galán-Vidal, C. A. Chemical Studies of Anthocyanins: A Review. *Food Chem.* **2009,** *113*(4), 859–871.

Cerezal, M. P.; Pino, A. J.; Salabarria, Y. Red Beet (*Beta vulgaris* L.) Colourant Stability in the form of a Concentrated Liquor. *Tecnología Alimentaria,* **1994,** *29,* 7–16.

Dey, P. M.; Harborne, J. B. *Plant Phenolics Methods in Plant Niochemistry (2nd printing);* Academic Press Limited: London, 1993; pp 326–341.

Escribano, J.; Pedreno, M. A.; Garcia-Carmona, F.; Munoz, R. Characterization of the Antiradical Activity of Betalains from *Beta Vulgaris* L. Roots. *Phytochem. Anal.* **1998,** *9,* 124–127.

Gasztonyi, M. N.; Daood, H.; Hajos, M. T.; Biacs, P. Comparison of Red Beet (Beta vulgaris var. conditiva) Varieties on the Basis of Their Pigment Components. *J. Sci. Food Agric.* **2001,** *81,* 932–933.

Girod, P. A.; Zryd, J. P. Secondary Metabolism in Cultured Red Beet (Beta vulgaris L.) Cells: Differential Regulation of Betaxanthin and Betacyanin Biosynthesis. *Plant Cell Tissue Organ Cult.* **1991,** *25*(1), 1–12.

Giusti, M. M.; Wrolstad, R. E. Acylated Anthocyanins from Edible Sources and Their Applications in Food Systems. *Biochem. Eng. J.* **2003,** *14,* 217–225.

Giusti, M. M.; Rodriguez-Saona, L. E.; Wrolstad, R. E. Molar Absorptivity and Color Characteristics of Acylated and Non-acylatedpelargonidin-Based Anthocyanins. *J. Agric. Food Chem.* **1999,** *47,* 4631–4637.

Greenstein, V. C.; Chiosi, F.; Baker, P.; Seiple, W.; Holopigian, K.; Braunstein, R. E.; Sparrow, J. R.; Scotopic Sensitivity and Color Vision with a Blue-Light-Absorbing Intra-ocular Lens. *J. Cataract Refractive Surg.* **2007,** *33*(4), 667–672.

Gross, J. *Pigments in Vegetables: Chlorophylls and Carotenoids;* Van Nostrand Reinhold: New York, 1991.

Haegerstrom-Portnoy, G. Short-Wavelength-Sensitive-Cone Sensitivity Loss with Aging: A Protective Role for Macular Pigment? *J. Opt. Soc. Am. A* **1988,** *5*(12), 2140–2144.

Hilou, A; Nacoulma, O. G.; Guiguemde, T. R. In vivo Antimalarial Activities of Extracts from Amaranthusspinosus L. and Boerhaaviaerecta L. in Mice. *J. Ethnopharmacol.* **2006,** *103*, 236–240.

Holden, A. L.; Eldridge, A. L.; Beecher, G. R. Carotenoids Content of U.S. Food: An Update of the Database. *J. Food Compos. Anal.* **1999,** *12,* 169–196.

Kanner, K.; Harel, S.; Granit, R. Betalains-A New Class of Dietary Cationizedantioxidants. *J. Agric. Food Chem.* **2001,** *49,* 5178–5185.

Koley, T. K.; Singh, S.; Khemariya, P.; Sarkar, A.; Kaur, C.; Chaurasia, S. N. S.; Naik, P. S. Evaluation of Bioactive Properties of Indian Carrot (Daucuscarota L.): A Chemometric Approach. *Food Res. Int.* 2014, *60*(2014), 76–85.

Konczak, I.; Zhang, W. Anthocyanins-More than Nature's Colours. *J. Biomed. Biotechnol.* **2004,** *5,* 239–240.

Kong, J. M.; Chia, L. S.; Goh, N. K.; Chia, T. F.; Brouillard, R. Analysis and Biological Activities of Anthocyanins. *Phytochemistry* **2003,** *64,* 923–933.

Krinsky, N. I. Antioxidant Functions of Carotenoids. *Free Radical Biol. Med.* **1989,** *7*(6), 617–635.

Kugler, F.; Stintzing, F. C.; Carle, R. Identification of Betalains from Differently Coloured Swiss Chard (Beta Vulgaris ssp. cicla [L.] Alef.cv. "Bright Lights") by High-Perfor-mance Liquid Chromatography-Electrospray Ionisation Mass Spectrometry. *J. Agric. Food Chem.* **2004,** *52,* 2975–2981.

Li H.; Deng, Z.; Zhu, H.; Hu, C.; Liu, R.; Young, J. C.; Tsao, R. Highly Pigmented Vege-tables: Anthocyanin Compositions and Their Role in Antioxidant Activities. *Food Res Int.* **2012,** *46*(1), 250–259.

Mabry, T. J.; Ann, T.; Turner, B. L. The Betacyanins and Their Distribution. *Phytochemistry* **1963**, *2*(1), 61–64.

Malien Aubert, C.; Dangles, O.; Amiot, M. J. Color Stability of Commercial Anthocyanin-Based Extracts in Relation to the Phenolics Composition. Protective Effects by Intramolecular and Intermolecular Copigmentation. *J. Agric. Food Chem.* **2001**, *49*, 170–176.

Matsui, T.; Ebichi, S.; Kobayashi, M.; Fukui, K.; Sugita, K.; Terahara, N.; Matsumoto, K. Anti-Hyperglycemic Effect of Diacylated Anthocyanin Derived from Ipomoea Batatas Cultivar Ayamurasaki can be Achieved Through the α-glucosidase Inhibitory Action *J. Agric. Food Chem.* **2002**, *50*, 7244–7248.

Nizamutdinova, I. T.; Jin, Y. C.; Chung, J.; Shin, S. C.; Lee, S. J.; Seo, H. G.; Lee, J. H.; Chang, K. C.; Kim, H. J. The Anti-Diabetic Effect of Anthocyanins in Streptozotocin-Induced Diabetic Rats Through Glucose Transporter 4 Regulation and Prevention of Insulin Resistance and Pancreatic Apoptosis. *Mol. Nutri. Food Res.* **2009**, *53*, 1419–1429.

Nilsson, T. Studies into the Pigments in Beetroot (Beta vulgaris L. ssp. vulgaris var. rubra L.). *Lantbrukshogskolansannaler* **1970**, *36*, 179–219.

Palozza, P.; Krinsky, N. I. Antioxidant Effects of Carotenoids in vivo and in vitro: An Overview. *Methods Enzymol.* **1992**, *213*, 403–420.

Raju, M.; Varakumar, S.; Lakshminarayana, R.; Krishnakantha, T. P.; Baskaran, V. Carotenoid Composition and Vitamin A Activity of Medicinally Important Green Leafy Vegetables. *Food Chem.* **2007**, *101*, 1598–605.

Roca, A.; Shin, K. J.; Liu, X.; Simon, M. I.; Chen, J. Comparative Analysis of Transcriptional Profiles Between Two Apoptotic Pathways of Light-Induced Retinal Degeneration. *Neuroscience* **2004**, *129*(3), 779–790.

Stahl, W.; Sies, H. Bioactivity and Protective Effects of Natural Carotenoids. *BBA Mol. Basis Dis.* **2005**, *1740*(2), 101–107.

Stintzing, F. C.; Carle, R. Functional Properties of Anthocyanins and Betalains in Plants, Food, and in Human Nutrition. *Trends Food Sci. Technol.* **2004**, *15*, 19–38.

Stintzing, F. C., Schieber, A.; Carle, R. Identification of Betalains from Yellow Beet (Beta vulgaris L.) and Cactus Pear [Opuntiaficus-indica (L.)Mill.] by High-Performance Liquid Chromatography–Electrospray Ionization Mass Spectroscopy. *J. Agric. Food Chem.* **2002**, *50*, 2302–2307.

Strack, D.; Vogt, T.; Schliemann, W. Recent Advances in Betalain Research. *Phytochemistry* **2003**, *62*, 247–269.

Surles, R. L.; Weng, N.; Simon, P. W.; Tanumihardjo, S. A. Carotenoid Profiles and Consumer Sensory Evaluation of Specialty Carrots (*Daucuscarota*, L.) of Various Colors. *J. Agric. Food Chem.* **2004**, *52*(11), 3417–3421.

United States Department of Agriculture, Agric, Research Service. USDA Nutrients database for Standard References Release 18. 2005a, http://www.nal.usda.gov/fnic/foodcomp.

United States Department of Agriculture USDA National Nutrient Database for Standard Reference—Release 18. 2005, http://www.ars.usda.gov/ba/bhnrc/ndl (accessed Jan 11, 2006).

Uchiyama, S.; Ishiyama, K.; Hashimoto, K.; Yamaguchi, M. Synergistic Effect of 6-cryptoxanthin and Zinc Sulfate on the Bone Component in Rat Femoral Tissues in vitro: The Unique Anabolic Effect with Zinc. *Biol. Pharma. Bull.* **2005**, *28*, 2142–2145.

Uchiyama, S.; Sumida, T.; Yrmaguchi, M. Oral Administration of 6-cryptoxanthin Induces Anabolic Effects on Bone Components in the Femoral Tissues of Rats in vivo. *Biol. Pharm. Bull.* **2004,** *27,* 232–235.

van Acker, S. A. B. E.; van den Berg, D.-J.; Tromp, M. N. J. L.; Griffioen, D. H.; van Bennekom, W. P.; van der Vijgh, W. J. F.; Bast, A. Structural Aspects of Antioxidant Activity of Flavonoids. *Free Radical Biol. Med.* **1996,** *20,* 331–342

Wisniewska, A.; Subczynski, W. K. Distribution of Macular Xanthophylls Between Domains in a Model of Photoreceptor Outer Segment Membranes. *Free Radical Biol. Med.* **2006,** *41*(8), 1257–1265.

Wrolstad, R. E.; Durstand, R. W.; Lee, J. Tracking Color and Pigment Changes in Antho-cyanin Products. *Trends Food Sci. Technol.* **2005,** *16*(9), 423–428.

Yamaguchi, M. β-cryptoxanthin and Bone Metabolism: The Preventive Role in Osteopo-rosis. *J. Health Sci.* **2008,** *54*(4), 356–369.

Yamaguchi, M.; Uchiyama, S. Beta Cryptoxanthin Stimulates Bone Formation and Inhibits Bone Resorption in Tissue Culture in vitro. *Mol Cell Biochem.* **2004,** *258,* 137–144.

Yamaguchi, M.; Igarashi, A.; Uchiyama, S.; Sugawara, K.; Sumida, T.; Morita, S.; Ogawa, H.; Nishitani, M.; Kajimoto, Y Effect of β-Cryptoxanthin on Circulating Bone Metabolic Markers: Intake of Juice (*Citrus Unshiu*) Supplemented with β-Cryptoxanthin has an Effect in Menopausal Women. *J. Health Sci.* **2006,** *52,* 758–768.

Žnidarčič, D.; Ban, D.; Šircelj, H. Carotenoid and Chlorophyll Composition of Commonly Consumed Leafy Vegetables in Mediterranean Countries. *Food chem.* **2011,** *129*(3), 1164–1168.

CHAPTER 2

PRE- AND POST-HARVEST LOSSES IN VEGETABLES

IVI CHAKRABORTY[1] and ARUP CHATTOPADHYAY[2*]

[1]*Department of Postharvest Technology of Horticultural Crops, Faculty of Horticulture, Bidhan Chandra Krishi Viswavidyalaya, Mohanpur 741252, Nadia, West Bengal, India*

[2]*AICRP on Vegetable Crops, Bidhan Chandra Krishi Viswavidyalaya, Kalyani Centre, Mohanpur 741252, Nadia, West Bengal, India*

CONTENTS

2.1   INTRODUCTION

Immense growth and development have been witnessed in horticultural sector since last decade, particularly in Asia and the Pacific regions. Substantial changes in socio-economic status lead to an increased demand for protective food, including vegetables throughout the year that could be met with efficient utilization of natural resources. At the same time, fine-tuning of technologies could provide an opportunity for intensifying the production of different horticultural crops. However, inadequate storage facilities, lack of processing and defective marketing systems contributes to a high proportion of waste which varies between 10 and 40% in developing countries. Suitable post-harvest handling protocols could not successfully be implemented in the region owing to preponderance of small and marginal farmers, who lack resources and are unable to timely market their produce, resulting considerable wastage. Thus, spoilage of produce is very common and being accelerated by the sultry climate of the region. The World Food Conference held in Rome during 1974 recommended reducing post-harvest food loss (PHL) as a tangible output to increase food availability. Since then, several committees were formed with expert groups to prevent loss of perishable crops, particularly fruits and vegetables.

Vegetables play an important role for overcoming micronutrient deficiencies and provide immense opportunity to small and marginal farmers with higher return per unit of land than any other staple crops. Over the past quarter century, the production of vegetables in the world has increased many fold and the value of global trade in vegetables recently exceeds that of cereals. Although India is one of the leading producers of vegetable crops, many Indians are still unable to obtain daily dietary allowances, and thus, the Human Development Index is quite low. Huge quantities of vegetables produced in many countries go to waste due to improper post-harvest handling operations and the lack of processing and storage facilities. This leads to a considerable gap between gross food production and net availability. It would be worthwhile to note that the production of food crop is of significance only when they reach the consumer in good condition and at a reasonable price. The concept of placing exclusive emphasis on increased production of vegetables is self-defeating. It is important to see how much quantity of produce goes through different marketing channels and finally reaches to the end users in acceptable condition. Efforts should be made to integrate production with post-harvest management, as post-harvest loss reduction and utilization have considerable bearing on food availability. As a matter of fact, food loss reduction is normally less costly than equivalent increase in food production. The success of production lies in the proper distribution of produce and its subsequent utilization by the consumer with least waste in the process. Immense opportunities exist in both domestic and international markets for fresh and processed vegetables. There is an urgent need for developing sound foundation in integrated post-harvest management systems for vegetables, with proper infrastructural facilities and logistical support.

2.2   STATUS: NATIONAL AND GLOBAL

Approximately half of the population in the Third World countries does not have adequate food supplies in spite of remarkable progress made in increasing food production at the global level. Food losses occurring during the post-harvest handling, transportation, storage and marketing stages are the major reasons for such pathetic scenario. These losses are higher in countries where the requirement of food and malnutrition both is in their highest degree. Post-harvest losses are very high in most of the horticultural crops, including vegetables as they are highly perishable due to the presence of excess moisture, and if care is not taken in their harvesting, handling and transport, they soon decay and become unfit for consumption. However, the projected production of fruits and vegetables would only cater to domestic demand, leaving no scope for growth on export front because the huge wastage would continue to rise simultaneously in absence of on-farm processing facilities. India incurs post-harvest fruits and vegetable losses worth over two lakh crores each year largely owing to the absence of modern cold storage facilities and lack of proper food processing units. West Bengal is the leading producer of horticultural crops, particularly vegetables and fruits across India. Unfortunately, the state incurs the highest post-harvest losses worth over Rs. 13,600 crores annually followed by Gujarat (Rs. 11,400 crores), Bihar (Rs. 10,700 crores), Uttar Pradesh (Rs. 10,300 crores) and Maharashtra (Rs. 10,100 crores) according to recent estimates of the Associated Chambers of Commerce and Industry of India. The trend in Jharkhand too is almost similar, where farmers have to sell their vegetables at throwaway prices during peak season. The total storage capacity in India is over 300 lakh tonnes, and there is an additional requirement of cold storage of about 370 lakh tonnes for fruit and vegetable storage alone. The existing cold storage capacity in India is confined mostly to wholesale markets and meant for storing potato, whereas the majority of fruit and vegetables are sold at local or regional markets which do not have cold storage facility. Storage and handling facilities need to be strengthened in the fruit and vegetable markets, thereby providing infrastructure facilities to promote increased productivity and to reduce post-harvest losses. A lot of positive steps have been initiated by the Government of India through launching of the National Horticulture Mission. Different stakeholders like growers, entrepreneur, technologists, private sector, government should come forward together to initiate a consolidated approach on storage, marketing, transportation, technological support and processing facilities for horticultural crops, which play an important role in Indian economy.

In India, there is a vast scope for growing diversified vegetable throughout the year because of favourable soil and climatic conditions. Vegetables are one of the most important commodities for balanced diet due to their high nutritional values and anti-oxidant properties. They are the cheapest source of protective foods supplied in fresh or processed or preserved form throughout the year. Vegetables are available in surplus only in certain seasons at different regions. In peak growing season, vegetables are spoilt in various stages to the extent of 20–25% due to improper post-harvest handling practices, marketing and storage problems. As vegetables are highly perishable, proper post-harvest handling management and processing is required. A variety of fresh vegetables can be made available in plenty due to diversified and suitable agro-climatic situations in India. Hence, there is no dearth for raw material for processing. At present product profile being developed in India is limited to few fruits and vegetables for example mango, pineapple, grapes, tomato, pea, potato, cucumber and others, but there is a greater scope for the development of value-added products from cauliflower, carrot, bitter gourd, onion, garlic, watermelon, muskmelon and others in the country.

In recent years, India has given major thrust on post-harvest wastage and has been looking for way-out on how to reduce losses on the farms and in the market places. Different workers in India have investigated the extent of losses of vegetables at different stages of production system through sampling or interviews.

Losses may vary greatly among commodities and production areas and seasons. In the United States, the losses of fresh fruits and vegetables are estimated to range from 2 to 23%, depending on the commodity, with an overall average of about 12% losses between production and consumption sites (Cappellini and Ceponis, 1984; Harvey, 1978). Kantor et al. (1997) estimated a total loss of 25% of vegetables in the United States. Estimates of post-harvest losses in developing countries vary greatly from 1 to 50% or even higher (National Academy of Sciences, 1978).

2.3   CAUSES FOR POST-HARVEST LOSSES

2.3.1   PRE-HARVEST

2.3.1.1   VARIETAL/GENOTYPE

The potential quality of vegetables is dependent on selection of varieties/ cultivars. Cultivars are characterized by different quality parameters making them more desirable to the producers as well as consumers. A producer must take a vital decision for the appropriate choice of high-yielding cultivar with desired qualities and longer shelf life (Hanna, 2009). Failure to select an appropriate cultivar may lead to lower yield, low-quality fruits or less market acceptability. Fruits of different cultivars differ in size, shape, colour, texture and flavour as well as storage potential. Cultivars also influence some post-harvest fresh as well as processing qualities of tomatoes stored under different conditions (Getinet et al., 2008; Chakraborty et al., 2007; Chattopadhyay et al., 2013). Selection of cultivar is therefore critical to the post-harvest storage life and eating qualities of tomatoes. Sometimes natural mutations cause genetic aberrations of the product, resulting defective appearance in sweet potato vis-a-vis peculiar forms in tomato, pepper, squash and potato. Pre-harvest losses in vegetables can be either quantitative or qualitative. Even though emphasis in crop research now-a-days is increasing shifting from quantity to quality of produce (Oko-Ibom and Asiegbu, 2007), there is still little improvement in the overall quality (appearance, nutritional quality etc.) of commercially produced vegetable varieties/hybrids, hence resulting in high amount of qualitative losses.

2.3.1.2   ABIOTIC FACTORS

2.3.1.2.1   Temperature

The productivity and quality of many crops depend on temperature. Elevated temperature prevailing during the growing season in the tropics is often caused increased heat stress. High temperature stress affects the biochemical reactions responsible for normal cell function in plants. It primarily disrupts the photosynthetic activities of crops (Weis and Berry, 1988). Poor fruit setting, smaller size and inferior quality tomato fruits are of common occurrence under heat stress condition (Stevens and Rudich, 1978). Temperature stress before flower opening causes irregularities in the epidermis and endothecium, lack of opening of the stromium and poor pollen formation (Sato et al., 2002). Dropping of flower bud, abnormal flower development, poor pollen production, dehiscence, and viability, ovule abortion and poor viability, reduced carbohydrate availability and other reproductive abnormalities are the possible reasons for the failure to set fruit at high temperature in tomato (Hazra et al., 2007). Erickson and Markhart (2002) observed that high post-pollination temperatures inhibit fruit set in pepper. In addition, photosynthetic activities are disrupted at high temperature resulting in significant loss of potential productivity. Lettuce is basically a cool season crop, if exposed to high temperatures or drought, bitter flavours develop and the leaves become less tender (Peirce, 1987). French bean grown at relatively low day temperatures is more tolerant to ozone than grown at higher temperatures under controlled conditions. This is just opposite in studies performed with cucumber seedlings (Agrawal et al., 1993). The accumulation of sucrose in cabbage is entirely temperature dependent and likely as a result of cessation of leaf production in autumn season (Nilsson, 1988). More accumulation of ascorbic acid in plant tissues occurs at a temperature below 20°C, but exceptions do exist. The B-vitamins accumulate in higher concentrations at temperature 10–15 °C in leafy greens, but in tomatoes, the accumulation was maximum at 27–30°C (Shewfelt, 1990). One of the most significant effects of temperature on the growth and development of produce quality is the initiation of flowers in higher plants (Beverly et al., 1993). Early initiation of flower stalk (bolting) before crop maturity is undesirable character, particularly in cole crops and root vegetables. These crops will be subjected to bolt with prolonged exposure to cool weather as days lengthen and nights shorten (Peirce, 1987). Fruit length of cucumber is affected by very high and low temperatures, whereas fruit curvature is increased with increasing temperature (Kanahama, 1989). Seasonal changes even in the production of chayote (Sechium edule (Jacq.) Swartz.) result in 20% rejection of fruit in packhouse (Saenz and Valverde, 1986).

2.3.1.2.2   Light

Environmental factors are often unmanageable in field condition, but have shown to play a significant role on the nutritional quality of vegetables.

Significant effects of light intensity have been envisaged on yield and quality of many crops over the course of time. Sometimes, the appearance of the product alters due to insufficient or excess light. Degradation of the plant pigment in the affected area is common when exposed to excess solar energy, and if the duration of exposure or intensity is sufficiently high, cellular death and collapse of the tissue follows. The effect of high light intensity stress is predominantly thermal in nature, though light bleaching of chlorophyll can occur. Excess light results in sun scald that is a significant problem in tomato, and bell pepper. In general, the ascorbic acid content in plant tissues is comparatively less in low light intensity (Kader, 1987). However, this concentration increases with increase in light intensity in leafy greens (Shewfelt, 1990). Light is also essential for the formation of ß-carotene in tomato (Raymundo et al., 1976). More accumulation of sugar and dry matter contents has also been found in tomato grown in full sunlight than those grown in shade (Winsor, 1979). Low light causes an increased incidence of puffy fruit and blotchy ripening in tomato (Rylski et al., 1994) and poor colour development in cucumber (Janse, 1984).

2.3.1.2.3   Wind

Wind injury can cause cold injury or winter injury, especially when the humidity is low at the time of the wind/cold period. Some of the ill effects of wind injury include sandblasting, plant parts rubbing together causing surface scarring, reduction in seed yield by shattering when the crop is about to harvest, more evaporation from the leaf surface on a hot day which aggravates heat stress factor. But wind can also help in pollination process, drying plants off and hardening of plants. On the basis of the severity and frequency wind damage may be of two categories: damage caused by severe storms like typhoon/hurricane and that caused by much more frequent winds of intermediate strength. High-intensity winds result in leaf damage and defoliation, particularly in leafy vegetables which have a disastrous effect on product appearance and marketability. Use of windbreaks has long been advocated in vegetable production areas that are subject to excessive wind (Holmes and Koekemoer, 1994).

2.3.1.2.4 Salinity

Vegetable production is impaired by gradual increase in soil salinity, particularly in irrigated croplands which provide 40% of the world's food (FAO, 2001). Excessive soil salinity causes significant loss in productivity of most vegetables which are particularly sensitive throughout their growing period. Among the vegetables, onion is the most sensitive to saline soils, whereas cucumbers, eggplants, peppers, and tomatoes are moderately sensitive. High evapo-transpiration during hot and dry environments results in excessive water loss, thus leaving salt around the plant roots which interferes the uptake of water from the soil. Salinity imparts an initial water deficit which results from the relatively high solute concentrations in the soil, causes ion-specific stresses, resulting from altered $K^+/$ Na^+ ratios, and leads to a build-up in Na^+ and Cl^- concentrations that are detrimental to plants (Yamaguchi and Blumwald, 2005). Plants suffering from salt stress are reflected by many ways such as loss of turgor, reduction in growth, wilting, leaf curling and epinasty, leaf abscission, decreased photosynthesis, respiratory changes, loss of cellular integrity, tissue necrosis and ultimate death of the plant (Jones, 1986; Cheeseman, 1988). Salinity problem is common occurrence in coastal regions which are characterized by low-quality and high-saline irrigation water due to contamination of the groundwater and intrusion of saline water due to natural or man-made events. Salinity becomes high in dry season and low during rainy season when freshwater flushing is prevalent. Moreover, coastal areas are threatened by natural disaster like tsunamis which can cause agricultural lands unproductive. Although the seawater rapidly recedes, the groundwater contamination and subsequent osmotic stress causes crop losses and affects soil fertility. In the inland areas, traditional water wells are commonly used for irrigation water in many countries. The bedrock deposit contains salts and the water from these wells are becoming more saline, thus affecting irrigated vegetable production in these areas.

2.3.1.2.5 Hail Damage

Hail damage is sporadic in nature, but sometimes it causes huge damage to the crop. Size of the hail stone, crop growth stage and the duration of exposure are critical factors affecting the degree of damage (Duran et al., 1994). The direct effect of hail damage not only results on the physical appearance of the product like increased deformed fruit (Hong et al., 1989) and anatomical alterations (Visai and Marro, 1986; Fogliani et al., 1985), but also increases the incidence of disease like bacterial spot of bell pepper (Kousik et al., 1994). Where the incidence of hail is frequent, methods (the use of hail nets over the crops) to minimize damage may be cost effective (Reid and Innes, 1996).

2.3.1.2.6   Physiological/Nutritional Disorders

Physiological disorders refer to the breakdown of tissue, resulting particularly from temperature fluctuation or macro/micro nutrient deficiency during any growth stages of vegetables (Table 2.1). Adequate supply of nutrients, pollution-free soil environment and optimum temperature and moisture favour normal growth of the plants. Any deviation from these conditions results in expression of disorders of various magnitudes. The deficiency or excess of any of the nutrient element, heavy metals, soluble toxic salts in the irrigation water, toxic gaseous pollution in the air, unsuitable prevalent temperature, moisture and soil pH have direct effect on plant growth.

TABLE 2.1   Physiological Disorders of Vegetables.

Name of the crop	Disorder	Symptoms	Causal agent/factor
Tomato, chilli, bell pepper, watermelon	Blossom-end rot	Water-soaked spots on the blossom end of the fruit followed by enlarged black spots.	Calcium deficiency in the developing fruit, extreme fluctuations in moisture, root pruning and excessive nitrogen fertilization in ammonia form
Tomato, chilli, bell pepper	Sunscald	Fruits at maturity when exposed to the sun are prone to scald. The tissue has blistered water-soaked appearance. Rapid desiccation leads to sunken area which usually has white or grey colour in green fruit or yellowish in red fruits.	Exposure of fruits to intense sunlight, particularly during the month of May and June in summer crop

TABLE 2.1   *(Continued)*

Name of the crop	Disorder	Symptoms	Causal agent/ factor
Tomato	Catface	Malformation and scarring of fruits, particularly at blossom ends. Affected fruits are puckered with swollen protuberances and can have cavities extending deep into the flesh. Generally, any disturbance to flowers can lead to abnormally shaped fruits.	Extreme heat, drought, low temperature and contact with hormone-type herbicide sprays may cause flower injury.
Tomato	Puffiness	Fruit appears somewhat bloated and angular. When cut, cavities may be present that lack the normal 'gel' and the fruit as a whole is not as dense.	Puffiness results from incomplete pollination, fertilization, or seed development often as a result of cool temperatures that negatively impact fertilization. High nitrogen and low potassium can also lead to puffiness.
Tomato	Cracking	Cracks results from extremely rapid fruit growth brought on by periods of abundant rain and high temperatures, especially when these conditions take place following periods of stress. Cracks of varying depth radiate from the stem end of the fruit, blemishing the fruit and providing an entrance for decay-causing organisms. Radial cracking is more likely to develop in full ripe fruit than in mature green. Fruits exposed to sun develop more concentric cracking than those, which are covered with foliage.	Deficiency of boron or moisture imbalance

TABLE 2.1 *(Continued)*

Name of the crop	Disorder	Symptoms	Causal agent/ factor
Tomato	Greywall or blotchy ripening	Greenish yellow and whitish patches appear on ripened fruit, particular on the stem end portion sometimes white or brown tissues are present in blotched area.	Imbalance of nitrogen and potasic nutrient in soil, water deficiency in the excessive transpiration.
Tomato	Internal white tissue	When ripe fruits are cut, white hard areas especially in the vascular region are present in the outer walls. Under severe conditions, fruit may also show white tissue in cross-wall and centre of fruit.	High temperatures during the ripening period in the field
Tomato	Irregular ripening	Green fruit show no symptoms but as fruit ripens, colour fails to develop uniformly. Colour often develops along locule walls with intermediate areas remaining green or yellow, producing a star-burst appearance. With sufficient time, nearly normal external colour develops on most fruit but internal areas remain hard with little or no colour development.	Feeding of nymphs of the Silver leaf whitefly (*Bemisia argentifolii*) on the tomato foliage
Tomato	Rain check	Rain check can be described as tiny cracks that develop on the shoulder of the fruit. These cracks can vary from just a few to almost complete coverage of the shoulder. The cracks feel rough to the touch and affected areas can take on a leathery appearance and not develop proper colour as fruit ripens. Green fruits are most susceptible, followed by breakers and are not affected at all. Damage occurs most often on exposed fruit after a rain.	Heavy rains after a long dry period

TABLE 2.1 *(Continued)*

Name of the crop	Disorder	Symptoms	Causal agent/ factor
Tomato	Pox and fleck	In most cases when a fruit is affected, both disorders are found together but are considered separate problems. Pox is described as small cuticular disruptions found at random on the fruit surface. The number can vary from a few to many. Fleck, also known as Gold Fleck, develops as small irregular-shaped green spots at random on the surface of immature fruit which become golden in colour as fruit ripens. Number of spots can vary from few to many.	Genetical defects of the varieties
Tomato	Zebra stripe	Zebra stripe can be characterized as a series of dark green spots arranged in a line from the stem end to the bloom end. At times, it seems the spots coalesce together and form elongated markings. Many times the dark green areas will disappear when fruit ripens.	Genetic defect that only develops under certain environmental conditions
Tomato	Zippering	Zippering is described as a fruit having thin scars that extend partially or fully from the stem scar area to the blossom end. The longitudinal scar has small transverse scars along it. At times, there may be open holes in the locules in addition to the zipper scar. Usually an anther that is attached to the newly forming fruit causes the zipper scar. Some people feel that a zipper is formed when the 'blooms' stick to the fruit and do not shed properly, but this may not be a cause.	Genetic defect in some of the varieties

TABLE 2.1   *(Continued)*

Name of the crop	Disorder	Symptoms	Causal agent/ factor
Chilli	Fruit splitting or skin cracking	Cracking observe around the shoulder of fruits.	Fluctuations in temperature and humidity
Chilli and bell pepper	Flower and fruit drop (unfruitfulness)	Dropping of flowers and fruits	High temperature and low humidity, low light intensity, short day and high temperature
Potato	Black heart	Internal breakdown of tissues in tubers and become black	Lack of oxygen is the probable cause. It is also probable that at high storage temperature (35–40°C) accumulation of CO_2 at the centre of the tuber might contribute to damage and the subsequent decrease in respiration and development of black heart.
Potato	Hollow heart	Irregular cavity develops in the centre of tubers. In tissue surrounding the cavity, there is no decay or discolouration.	Appears often in varieties which bulk rapidly and produce large sized tubers
Potato	Greening	When tubers are exposed to sun, they develop green colour due to formation of chlorophyll.	Proper earthing up

TABLE 2.1   *(Continued)*

Name of the crop	Disorder	Symptoms	Causal agent/ factor
Cauliflower	Browning	In early stage, the water-soaked areas appear on the stem and curd surface. As the plant grows, the stem becomes hollow with water-soaked tissue covering the internal walls of the cavity. In advanced stage of deficiency, brown or pink coloured areas are seen on curd surface, and therefore, it is also called brown rot or red rot or browning of the curd. Sometimes the stem may become hollow even without brown areas on the curd. The affected curds are bitter in taste.	Boron deficiency
Cauliflower	Whiptail	In young plants, chlorosis of leaf margins and the whole leaves may turn white. The leaf blades do not develop properly, only the midribs develop. The growing point of the plant is also deformed which prevents the curd development.	Molybdenum deficiency
Cauliflower and broccoli	Buttoning	The development of small pre-mature curds or buttons while the plants are young. The button heads are exposed and the plants showing this condition have usually small poorly developed leaves.	Poor nitrogen supply, planting of over-aged seedlings, unfavourable climatic conditions (for 10 days or more at 4–10°C) and improper time of planting.
Cauliflower	Riceyness	When the surface of the curd is loose and has velvety appearance due to elongation of pedicel and formation of small white flower buds at the curding stage, such curds are called ricey.	High temperature during curd development, heavy application of nitrogenous fertilizer and high humidity.

TABLE 2.1 *(Continued)*

Name of the crop	Disorder	Symptoms	Causal agent/ factor
Cabbage and Brussels sprouts	Blindness	During the early stage of plant growth, damage to growing point by insects, low temperature or frost causes blindness. Plants grow without terminal bud and fail to form any curd. The leaves of blind plants become thicker and leathery owing to accumulation of carbohydrates.	Damage to the terminal growing point due to low temperature, cutworm damage or rough handling of transplants.
Cauliflower	Chlorosis	It shows an inter-veinal, yellow mottling of lower older leaves.	Magnesium deficiency when grown on high acidic soil
Cauliflower and broccoli	Hollow stem	Develop hollow stem and curd	Heavy application of nitrogenous fertilizer and boron deficiency
Cauliflower and cabbage	Frost injury	Leaves of young seedling turns yellowish-white on both surface, petioles become flaccid and white. Fully grown curds are more sensitive to frost than smaller one. In cabbage, the younger leaves mainly sensitive, at the centre of the head turn brown, whereas outwardly the head appears healthy.	Appearance of frost during curd development
Cauliflower	Fuzziness	Fuzziness appears as flower pedicels of velvety curds elongate	Sowing at abnormal time
Cauliflower	Leafy curds	Development of small green leaves (bracts) inside the segments of the curd makes them leafy. Prevalence of high temperature especially after curd initiation or fluctuation in temperature at curding stage aggravates leafy curds.	High temperature or fluctuation of temperature and delayed harvesting.

TABLE 2.1 *(Continued)*

Name of the crop	Disorder	Symptoms	Causal agent/ factor
Cabbage, Brussels sprouts and cauliflower	Tip burn	Necrosis on leaf-margins, edges or tips. Brown spots which develop have a tendency to break down during transport and in storage, thereby providing ingress to opportunistic decay causing organisms.	Ca deficiency, rapid growth due to excessive nitrogen, high temperature and water stress.
Cabbage	Bursting/split-ting of heads	Rapid growth causes splitting of heads.	Over-watering and non-uniform growth
Cabbage	Black speck or pepper spot	Varying degrees of severity on the outer leaves of the head but can often be seen well into the centre of the head (Becker and Bjorkman, 1991; Studstill et al., 2007). Symptoms can appear in the field or after harvest. Black speck becomes visible after a period of cold storage. Midribs look as if ground black pepper was sprinkled on them, thus the name 'pepper spot' (Studstill et al., 2007).	Not known
Cabbage	Black petiole or black midrib	As heads approach maturity, the dorsal side of the internal leaf petioles or midribs turns black at or near the point where the midrib attaches to the core (Becker and Bjorkman, 1991).	P–K imbalance and relatively higher N levels in the soil
Cabbage	Vein streaking necrosis	Necrotic lesions similar in size to those of pepper spot sometimes occur on the surface of the leaf in the vicinity of leaf midribs	Not known

TABLE 2.1   *(Continued)*

Name of the crop	Disorder	Symptoms	Causal agent/ factor
Cabbage	Necrotic spot	This disorder is characterized by lesions similar to those of pepper spot. Unlike pepper spot, groups of cells die simultaneously, and they are frequently associated with vascular tissue, rather than with stomata Unlike vein streaking, necrotic spots affect parenchyma cells, whereas the epidermal cells appear unaffected. Symptoms appear as sunken oval spots on leaves or on the midrib. The disorder usually develops during storage, whereas symptoms are not shown at harvest.	Controlled atmosphere storage
Carrot	Cavity spot	It is characterized by elliptical lesions present on the surface of the roots.	Deficiency of calcium and excess of potassium as K reduces the uptake of Ca
Carrot	Splitting	Appearance of longitudinal cracks on the root	Genetic factors and other factors (higher dose of ammoniac form of nitrogen, irregular irrigation and fertigation at the time of root development, wider spacing)
Radish and carrot	Forking	There is secondary elongating growth in the roots that gives a look of fork like structure to the root.	Excessive moisture during root development, heavy soil due to compactness of soil, use of undecomposed organic matter

TABLE 2.1 *(Continued)*

Name of the crop	Disorder	Symptoms	Causal agent/ factor
Radish	Pithiness	Pithiness is characterized by the death of xylem and collapse of paranchymatous tissues in roots. It may lead to production of hollow roots. Pithiness is the sign of senescence and its degree varies from varieties to varieties.	Excess N, P and K; high temperature prevailing before harvesting and delay in harvesting.
Radish	Akashin	The growth of the root is checked.	Deficiency of boron and fluctuation in day and night temperature
Beet	Zoning	Alternate dark and light coloured rings are formed on the beet root.	High temperature above 30°C, wide-range fluctuation in day and night temperature and irregular supply of moisture during root growth and development.
Beet	Internal black spot (brown heart or break-down of beet)	Plants remain dwarf and stunted. The leaves remain smaller than normal and may assume a variegated appearance. The roots do not grow to full size and under severe deficiency conditions the roots become distorted and very small and have rough, unhealthy greyish appearance instead of clean and smooth. The surface is often wrinkled and cracked.	Deficiency of boron in neutral or alkaline soils

TABLE 2.1 *(Continued)*

Name of the crop	Disorder	Symptoms	Causal agent/ factor
Cucurbits	Unfruitfulness	Common problems of both monoecious and dioecious cucurbits. Pistillate flowers in monoecious and female plants in dioecious crops are shed due to lack of pollination and fertilization. In some cases, ovary of the unfertilized flower may flow a bit due to paratheno-carpic stimulation which also abscises after a few days.	Lack of pollinator visit in the field due to indiscriminate use of pesticides; less number of male plants in the population of dioecious crops.
Cucumber	Pillow	It is a fruit disorder of processing cucumber. An abnormal white Styrofoam like porous textured tissue is formed in the mesocarp of the fleshy harvested fruits. Vascular tissue with some pillow areas may collapse and become necrotic.	Low calcium level in the tissue
Muskmelon and watermelon	Delay in fruit ripening	Delay in repining is sometime associated with less sweetness and cracking of fruits	High moisture level and temperature fluctuation at ripening stage
Summer squash	Leaf silvering	The leaves become silver coloured and contain less chlorophyll; photosynthesis is hampered in the silvered leaves.	Moisture scarcity
French bean	Hypocotyl necrosis	Necrosis of hypocotyl is associated with low Ca content in seed after germination.	Deficiency of calcium
French bean	Cotyledon cracking	Cotyledon cracks transversely	Sowing dry seed in wet soil, white seeded cultivars are more prone to this disorder. Seed moisture content below 12% also enhances cracking of cotyledons

TABLE 2.1   *(Continued)*

Name of the crop	Disorder	Symptoms	Causal agent/factor
French bean	Blossom drop and ovule abortion	Bean blossoms drop off without producing a pod	High temperature above 29 °C
French bean	Delayed flowering and pod setting	Flower initiation and development are greatly delayed under suboptimal temperatures, especially below 10°C, where fertilization may not occur, producing small and misshapen pods.	Low temperature
Lettuce	Tip burn	It is the marginal collapse and necrosis, at or near the leaf margins, of rapidly expanding inner lettuce leaves. The disorder usually occurs near harvest, when it can result in complete crop loss. Early symptoms include vein discolouration and/or the development of dark brown to black spots near or at the leaf margins. As the disorder progresses, the necrotic areas coalesce, forming a lesion up to several centimetres long.	Deficiency of calcium
Lettuce	Russet spotting	Tissue browning or lesions appear as oxidative membrane injury	Deleterious effects of ethylene
Celery, lettuce, endive, spinach, artichoke	Black heart	Discolouration of the tender young leaves at the centre of the plant. These affected leaves turn black and die.	Deficiency of calcium

2.3.1.3   BIOTIC FACTORS

2.3.1.3.1   Insect Infestation

The vegetable crops are attacked by number of insect pests in every season, and these cause high damage to the production. Insects and pests generally attack vegetables because of their liking and to complete their life cycle. They damage and use every part of the plant and ultimate cause high economic loss to the farmers. Insects injure plants by chewing leaves, stems and roots, sucking juices, egg laying and transmitting diseases. Most vegetable crops are subjected to pest damage, particularly seeds, roots, leaves, stems and fruits are all susceptible to damage effecting plant vigour resulting severe crop losses. Like other insects, the cucurbit fruit fly *Bactrocera cucurbitae*, is one of the most important pests of vegetables, and depending on the environmental conditions and susceptibility of the crop species, the extent of losses can vary between 30 and 100% (Dhillon et al., 2005; Sapkota et al., 2010; Sarwar et al., 2013). Along with other insects, the vegetable crops are heavily attacked by aphids. Among 14 different winter vegetables, there are 14 species of aphids found to attack on all these species of plants, out of these only *Pentalonia nigronervosa* is found as monophagus species, whereas other are polyphagus. Among them *Aphis craccivora, Aphis corrianderi, Aphis gossypii, Acrythosiphon pisi, Bravicoryne brassicae, Lipaphis erysimi, Myzus persicae, P. nigronervosa*, have been observed as a regular pests of 12 plant species, whereas *A. gossypii, Aphis fabae, M. persicae* are also serious pests of four plant species though sporadic in nature. *Macrosiphoniella sanborni, Hydaphis corriandri, P. nigronervosa, L. erysimi, A. pisi, Macrosiphum rosaeformes* are regular pests of six plant species. *A. craccivora, Uroleucon compositae, A. fabae, Aphis spiraecola, A. gossypii, Rhopalosiphum nymphae* have been observed as sporadic pests of seven plant species (Rafi et al., 2010). These pests are a great concern for the farming community, as they could threaten the agricultural and horticultural industries by increasing the price of production and cost to the consumers. Insect pest problems during the growing season can also affect pre-harvest quality, both obvious and not so obvious ways. Visible blemishes on the vegetable surface caused by insect feeding can have a negative effect on the appearance of vegetables, thus decreasing their appeal to consumers. Feeding injury on vegetables by insects can lead to surface injury and punctures, creating entry points for decay organisms and increasing the probability of post-harvest diseases. In addition, the presence of insect pests on vegetables entering storage leads to the possibility of these insects proliferating in storage and becoming an issue. Apart from causing direct damage, many of them also act as vectors for several viral diseases. The crop yield losses to the tune of 11–100% have been reported (Shivalingaswamy et al., 2002; Dhillon et al., 2005; Satpathy et al., 2005; Raju et al., 2007; Singh et al.,

2007; Ghosal et al., 2012) in different vegetable crops grown in India as depicted in Table 2.2.

TABLE 2.2   Yield Losses due to Major Insect Pests of Vegetable Crops in India.

Name of the crop	Name of the pest	Yield loss (%)
Tomato	Fruit borer (*Helicoverpa armigera*)	24–73
Cabbage	Diamondback moth (*Plutella xylostella*)	17–99
	Cabbage caterpillar (*Pieris brassicae*)	69
	Cabbage leaf webber (*Crocidolomia binotalis*)	28–51
	Cabbage borer (*Hellula undalis*)	30–58
Brinjal	Fruit and shoot borer (*Leucinodes orbonalis*)	11–93
Chillies	Thrips (*Scirtothrips dorsalis*)	12–90
	Mites (*Polyphagotarsonemus latus*)	34
Okra	Fruit borer (*H. armigera*)	22
	Leafhopper (*Amrasca biguttula biguttula*)	54–66
	Whitefly (*Bemisia tabaci*)	54
	Shoot and fruit borer (*Earias vittella*)	23–54
Cucurbits	Fruit fly (*Bactrocera cucurbitae*)	
	Bitter gourd	60–80
	Cucumber	20–39
	Ivy gourd	63
	Musk melon	76–100
	Snake gourd	63
	Sponge gourd	50

2.3.1.3.2   Diseases

A wide variety of fungal and bacterial pathogens cause pre-harvest disease in vegetables. Some of these infect produce before harvest and then remain quiescent until conditions are more favourable for disease development after harvest. Other pathogens infect produce during and after harvest through surface injuries. Direct crop losses caused by diseases may be measured as the proportion of crop not sold. In addition to losses in yield and quality in the field and later during storage and transport, there are many, less tangible ways in which diseases exact an economic toll. For example, the fungus *Botrytis cinerea* may cause multiple but almost imperceptible ghost spot lesions on tomato fruit, which, depending on the rigor of official or consumer inspection, may result in little or no financial loss to the grower. However, the same fungus causing a single, girdling lesion on the stem of an indeterminate tomato cultivar will result in the total loss in yield from that plant, as often happens in the greenhouse. Bacterial spot on processing tomatoes makes the skin very difficult to peel by standard factory procedures, so the skins have to be removed by hand, which is very expensive. On the other hand, buyers of fresh-market tomatoes at roadside stands may scarcely notice a few lesions of bacterial spot. Similarly, when cabbage is fermented to produce sauerkraut, or cooked, the lesions caused by thrips are very pronounced and unacceptable, whereas thrips damage may be of little consequence if the cabbage is finely chopped and used fresh in coleslaw. Nematode damage to roots may be mechanical or chemical, thereby reducing root capacity to absorb and translocate water and nutrients, even when soil moisture is adequate. Some vegetable crops are tolerant of nematode damage, whereas others are highly sensitive. Seedlings and young transplants usually are especially susceptible. The distribution of nematodes in the soil may be in the field or in the greenhouse, is normally uneven. Plant-parasitic nematodes may reduce crop yield and quality but other biotic and abiotic stresses on plants make it difficult to predict the impact of nematode damage. Losses may increase significantly if nematodes interact with other pathogens, such as fungi and viruses.

Gemini viruses cause significant crop losses in crops like tomato, okra, chilli and others. Despite the amount of effort that has gone into Gemini virus control research, no sustained resistance has been found. Plant viruses also cause considerable damage to various cucurbits including bottle gourd. Nearly, 30 viruses are known to infect cucurbit crops under field conditions (Lovisolo, 1980). Viral diseases result in losses through reduction in growth and yield and are responsible for distortion and mottling of fruits, making the product unmarketable. Fruit set can be dramatically affected by some viruses. With the exception of *Squash mosaic virus* (SqMV), which is seed borne in melon and transmitted by beetles, the other major viruses are transmitted by several aphid species in a non-persistent manner. Some major Cucurbit viruses include SqMV, *Cucumber mosaic virus*, *Watermelon mosaic virus 2* (WMV-2), Papaya ringspot virus—W (formerly WMV-1) and *Zucchini yellow mosaic virus*. *Tobacco ringspot virus*, *Tomato ringspot virus*, *Clover yellow vein virus* and *Aster yellow mycoplasma* are considered to be minor viruses that infect cucurbits. Bottle gourd is affected mainly by cucumber green mottle mosaic—tobamovirus, Melon necrotic spot—carmovirus and Zucchini yellow fleck—potyvirus. Bottle gourd mosaic disease is widely prevalent in almost all the bottle gourd growing states of India, causing losses through reduction in growth and yield.

2.3.1.3.3   Noxious Weeds

Vegetable crops vary widely in their response to weed competition, ranging from non-competitive crop, such as onion, to moderately competitive crops, such as potato and transplanted cabbage. Direct-seeded onion does not produce marketable bulbs, if weeds are not controlled. The time from seeding to the two-leaf stage in onion averages 46 days; during that time, wild mustard can emerge, complete its vegetative growth and flower. The critical period of weed competition in vegetable crops is the minimum time that weeds must be suppressed to prevent yield losses. Cucumber yield is reduced if the crop is not kept weed-free for up to 4 weeks after seeding, or if it is weedy for more than three to 4 weeks. The critical period for rutabaga is 2–4 weeks after crop emergence when barnyard grass and lamb's quarters are present. Similarly, the critical period for carrot crops grown on organic soils is found to be between 3 and 6 weeks after crop emergence. Although the concept of a critical weed period has practical limitations because crops and weeds grow at different rates from year to year, early removal of weeds is clearly important in reducing losses from competition. Because of their density and proximity to crop plants, weeds also provide microclimates conducive to infection by fungi and bacteria. Crop losses caused by competition from weeds can be assessed quite readily, but weeds also contribute to overall crop losses by acting as alternative hosts for pathogens and insects. For example, wild cucumber (*Echinocystis lobata* (Michx.) Torr & Gray) harbours the fungus *Didymella bryoniae*, which causes gummy stem blight in melon and cucumber. The universal pathogens *B. cinerea, Sclerotinia sclerotiorum* and *Sarracenia minor* are found on many weed species *B. cinerea* in particular having hundreds of hosts. Weeds also may act as a reservoir for many vegetable viruses and mycoplasma-like organisms, and of their insect and nematode vectors (Hazra et al., 2011). The passage of workers and machinery through weed-infested crops can transmit viruses from weeds to crop plants; weed canopies provide the humid and cool microclimate in which fungi and bacteria infect their vegetable hosts; and finally, weeds provide shelter for pest insects and other types of animals, such as rabbits and rodents (Table 2.3).

TABLE 2.3 Examples of Weed Hosts for Different Insect Pests and Pathogens in Vegetable Crops.

Name of the weed	Insect pest	Crop host
Wild mustard	Cabbage root maggot	Cabbage, broccoli
Chick weed (*Stelaria media*)	Melon aphids	Most cucurbits
Wild morning glory (*Convolvulus arvensis*)	Aphids	Watermelon, pumpkin, some gourds (mostly ridge gourd)
Wild carrot (*Daucus carota*)	Carrot rust fly	Carrot, celery
Night shade (*Solanum nigrum*)	Onion thrips	Onion
Lambsquarter (*Chenopodium album*)	Common stalk borer	Mostly tomato
Pig weed (*Amaranthus viridis*)	Spinach mildew	Spinach, Swiss chard
Night shade (*S. nigrum*)	Verticillium wilt, tomato mosaic virus	Tomato
Night shade (*S. nigrum*)	Potato mosaic	Potato
Mustards (*Brassica* sp.)	Clubroot	Cabbage, cauliflower

2.3.1.4 MATURITY STAGE

Physiological or harvestable maturity can have a pronounced influence on the appearance of vegetables. Most of the vegetable crops never reach physiological maturity. So harvestable maturity is more appropriate term in that sense they are harvested early in their developmental cycle. Harvest maturity varies in accordance with the crop concerned. Thus, maturity always has a considerable influence on the quality of fresh produce as well as the storage potential and occurrence of many storage disorders (Siddiqui and Dhua, 2010; Singh, 2009). There are three main stages in the life span of vegetables: maturation, ripening and senescence. For climacteric vegetables (e.g. tomato), maturation is indicative of the fruit being ready for harvest. At this point, the edible part of the vegetable is fully developed in size, although it may not be ready for immediate consumption. Ripening follows or overlaps maturation, rendering the produce edible, as indicated by texture, taste, colour and flavour. Senescence is the last stage, characterized by natural degradation of the vegetable, as in loss of texture, flavour, and so on. Here, ripening treatment (use of ethrel in tomato) is an important operation in many climacteric vegetables before retailing. For non-climacteric vegetables (e.g. chilli, watermelon, okra, cauliflower, broccoli etc.), development of eating quality before harvesting is indicative of the vegetable being ready for harvest. Therefore, non-climacteric vegetables should be harvested after attaining proper development of eating quality while still attached to the mother plant. In this case, ripening treatment is not given before retailing. The maturity of harvested vegetable has an important role on shelf life, quality and market price. Hence, certain standards of maturity must be kept in mind while harvesting the vegetable. During development, the individual units of a product exhibit a progressive increase in size, and in many instances, the size at harvest is a critical factor (e.g. okra, gherkins). Exceeding a required size class can greatly reduce the value and marketability of these vegetables. In other instances, size is less critical and a relatively wide range of sizes are acceptable (e.g. brinjal, the giant taro, *Alocasia macrorrhiza* (L.) G. Don.). With some products, there are distinct alterations in colour during maturation. For example, tomato undergoes marked changes in colour, and these changes in colour are largely dependent upon attachment to the parent plant. The maturity stage of tomato fruit at harvest is an important determinant of many quality traits (Beckles, 2012). Tomato, being a climacteric fruit, can be harvested at different stages during maturity, like green, breaker, turning, half ripe or red ripe stage. Each stage at harvest has its own postharvest attributes that the fruit will exhibit. It has been found that the shelf life of tomato cultivars is the maximum when harvested at green mature stage (Moneruzzaman et al., 2009). Although shelf life has been the most important aspect in loss reduction biotechnology of vegetables, other aspects may be of interest rather than shelf life. Fruit nutritional values (synthesis of lycopene) and appearance may be affected when harvested green. For instance, sugar transport to fruits in a vine-ripened tomato appears to increase during the latter part of maturity (Carrari et al., 2006), and therefore, when fruits are harvested immature or in a green state sugar import to fruits will be cut-off making post-harvest degradation of starch, the main source of carbohydrates, which is both undesirable and inadequate (Balibrea et al., 2006). Meanwhile, harvesting later also promotes higher sugar accumulation in ripe fruits which are susceptible to mechanical injuries with a shorter shelf life (Toivonen, 2007). The pH of tomatoes is an important parameter in the tomato processing industry. Tomatoes are processed as high-acid foods, and therefore, the higher the acidity the better for processing. Cultivars with high pH (more than 4.4) therefore may not be suitable for processing because of taking more heating time for cooking and contamination with *Bacillus coagulens*. The acidity of tomatoes is highest at the pink stage of maturity with a rapid decrease as the fruit ripens. Moneruzzaman et al.(2009) and Cliff et al.(2009) suggested tomato fruits can be harvested at mature green to give producers enough time for long-distance transportation but for local marketing harvesting at the fully ripe stage is preferred to maximize nutritional value. Likewise, most of the cole crops showed maximum nutritive value at harvest time and not after storage (Weston and Barth, 1997). Small turnip leaves contain more riboflavin and thiamin but less carotene than large leaves (Salunkhe et al., 1991). In snap bean and other green vegetables, ascorbic acid tends to increase with maturation and decrease with advanced maturation. Sometimes, days taken from anthesis to harvestable maturity also influence the overall fruit quality of some vegetables (okra, pointed gourd, cowpea etc.). In pumpkin, older fruits remain attached to the vine is more susceptible to storage rot than those that are not attached (Hawthorne, 1990). A very limited number of vegetables (e.g. cantaloupe, African horned cucumber) undergo significant alterations in surface topography during maturation which are of importance in achieving the final desired appearance of the product. The methods for measuring maturity must be simple and preferably non-destructive as it may need to be assessed indifferent places such as field or packing shed and or in the market (Singh, 2009).

2.3.1.5   MEASURES TO REDUCE PRE-HARVEST LOSSES

Post-harvest quality management of vegetables starts from the field and continues until it reaches to the consumer. The post-harvest quality status of vegetables depends on some pre-harvest practices carried out during production. The quality of any vegetable after harvest cannot be improved by the use of any post-harvest treatment or handling practices but can only be maintained. Understanding and managing the various roles of pre-harvest factors such as cultivar selection, judicious fertilizer application and irrigation management, adoption of IPM, IDM and IWM technologies, proper control of abiotic stresses, judging of proper maturity stage, adoption of proper harvesting procedures, the best physical handling procedures and so on can play significant role to produce high-quality vegetables at harvest. Thus, the quality and storage life of vegetables after harvest depends not only on the post-harvest factors alone but also on some pre-harvest factors during production and, until both factors are managed properly, quality loss will still be a major challenge for vegetable producers and handlers.

Magnitude of pre-harvest losses in vegetables is still to be minimized by proper cultural operations, harvesting, storage and transportation facilities. Creating and/or maintaining production conditions that minimize undesirable product appearance are essential. It is evident that a diverse range of biotic and abiotic factors can alter the appearance of vegetables prior to harvest. Even under optimum production conditions, a portion of every crop is invariably downgraded due to appearance defects. Although it is not possible to eliminate all pre-harvest losses due to appearance defects, the extent of these losses can be reduced through a better understanding of the nature of the problem and by being cognizant of potential solutions. Although field grading during harvest is utilized to eliminate a significant portion of product with substandard appearance, minimizing the occurrence of inferior product can significantly increase net profit.

Through the establishment of cold storage and other amenities at the growers and retailers level, there is a greater scope for vegetable processing industry. Presently tomatoes, peas, potatoes and cucumber are being processed on a large scale. There are about 4000 small- and large-scale processing units in India, which process only about 2.5% of the total fruit and vegetable as against 40–85% in developed countries (e.g. Malaysia–83%, Philippines–78%, Brazil and USA–70%).

However, our farmers have yet not been able to cater this opportunity and still follow traditional sowing and picking patterns. These results in highly volatile vegetable supply market wherein the market is flooded with seasonal vegetables irrespective of demand presence on one hand and very high priced vegetables in off-season on the other.

2.3.2   HARVESTING

2.3.2.1   MECHANICAL DAMAGE

Physical handling can have a drastic effect on the post-harvest quality or life of harvested vegetables. Rough handling during harvesting and after harvesting can result in mechanical injuries which affect quality. Typical industrial production systems associated with tomatoes may include mechanical harvesting, packing into crates, sorting, grading, washing and transporting over long distances. At each of these stages, there may be significant occurrence of mechanical injury which may be bruising, scarring, scuffing, cutting or puncturing the fruits. In small-scale tomato production, mechanical injuries may result from the use of inappropriate harvesting containers and packaging materials. The effects of mechanical injuries on fruit are cumulative (Miller, 2003). Injuries which are equivalent to or greater than the bioyield point lead to a total breakdown of the structure of the affected cells which is accompanied by unwanted metabolic activities which may include increased ethylene production, accelerated respiration rates and ripening (Miller, 2003; Sargent, 1992), which results in either reduced shelf life or poor quality. It is therefore important to handle tomato fruit with care during the harvest and post-harvest activities to minimize mechanical injuries to avoid losses.

Apart from causing damage, bruising and cracking make the vegetable more prone to attack by organisms and significantly increase of water loss and gaseous exchange. Many a times, the mechanical received by the vegetable due to the pressure thrust during transportation, though not visible, leads to rupture of inner tissues and cells and such produce is degraded faster during the natural senescence process. Processing operations such as spillage, abrasion, excessive polishing, peeling or trimming add to the loss of the commodity. Puncturing of the containers and defective seals also cause mechanical spoilage. Mechanical damage is extremely common during produce handling and is defined as plastic deformation, superficial rupture and destruction of vegetable tissue due to external forces (Sanches et al., 2008). Physical impact is of the most common cause of mechanical damage. According to Vignealt et al. (2002), impacts are transitory movements caused by sudden acceleration or deceleration causing great dissipation of energy and consequent damage to the fruit. During packing, vegetables are more exposed to impact and vibration than to compression forces (Garcia et al., 1988). Some of the effects of mechanical damage are related to internal loss in quality with flavour and nutritional alteration. For vegetable, such as tomato, loss of acid content (Moretti et al., 1998) and vitamin C (Moretti, 1999) occur during mechanical stress.

2.3.2.2   EARLY HARVESTING/DELAYED HARVESTING

Maturity at harvest has a very important influence on subsequent storage life and eating quality. Horticultural maturity is that stage of development at which a plant or plant part is ready for use by consumers for a particular purpose. This use can occur at any stage of development depending on the commodity (Watada et al., 1984). Harvesting is done at cooler times of the day preferably early in the morning, taking advantage of the lower temperature which minimizes produce heat load and increases work efficiency of harvesters. Harvesting starts when morning dew has evaporated to avoid damage to the plants due to turgidity, otherwise or if harvesting has to be done very early in the morning (e.g. dawn), care must be observed to avoid damage to the plants that still have fruits for subsequent harvesting. The produce should be shifted to shade as early as possible. Harvesting during hot periods raises field heat of the produce, causing wilting and shrivelling. It is not advisable to harvest when it is about to rain or just after a rain because disease incidence could be higher (Bautista and Acedo, 1987). Rain water can accumulate on the stem end of the fruit which is the main entry point of decay-causing microorganisms and could create a favourable condition for microbial growth. However, if harvesting cannot be avoided during rainy days, the harvested produce must be washed and dried properly before packaging. Washing and drying become doubly necessary for fruits from plants not staked or trellised as soil particles, which may contain decay organisms, could adhere to the fruit, especially during rainy season. Care in harvesting is necessary as any bruises or injuries may later manifest as black or brown patches making them unattractive.

Physiological processes occur in vegetables that permanently change their taste, appearance and quality if they are not harvested at the proper stage of maturity. Texture, fibre and consistency are greatly affected by stage of maturity. The stage of maturity at harvest, post-harvest handling and the time interval between harvesting and serving affect the quality of all vegetables. Some vegetables are more highly perishable than others. Sweet corn and peas are very difficult to maintain in an acceptable fresh state for even a very short time, whereas other vegetables may have a much longer shelf life. Even after harvest, respiration and other life processes continue, and in most cases, a slowing of these processes will increase the shelf life of the vegetable. Lowering the internal temperature helps to slow these processes. This is one reason for harvesting vegetables early in the day before the heat from the sun has warmed them. After harvest, most vegetables should be kept cool and out of direct sunlight until processed or consumed. Fruit-vegetables include two groups: (i) immature fruit-vegetables, such as green bell pepper, green chilli pepper, cucumber, summer (soft-rind) squash, chayote, lima beans, snap beans, sweet pea, edible-pod pea, okra, eggplant and sweet corn; and (ii) mature fruit-vegetables, such as tomato, red peppers, muskmelons (cantaloupe, casaba, crenshaw, honeydew, Persian), watermelon, pumpkin and winter (hard-rind) squash. For immature fruit-vegetables, the optimum eating quality is reached before full maturity and delayed harvesting results in lower quality at harvest and faster deterioration rate after harvest. For mature fruit-vegetables, most of the fruits reach peak eating quality when fully ripened on the plant and, with the exception of tomato, all are incapable of continuing their ripening processes once removed from the plant. Fruits picked at less than mature stages are subject to greater shrivelling and mechanical damage and are of inferior flavour quality. Overripe fruits are likely to become soft and/or mealy in texture soon after harvest. The necessity of shipping mature fruit-vegetables long distances has often encouraged harvesting them at less than ideal maturity, resulting in suboptimal taste quality to the consumer (Kader, 1995). Table 2.4 gives suggestions to aid in determining the proper stage of maturity for harvesting many vegetables. Harvesting too soon may result in only a reduction in yield. However, harvesting too late can result in poor quality due to development of objectionable fibre and the conversion of sugars into starches.

TABLE 2.4 Harvesting Stages in Vegetables.

Vegetable crop	Harvesting stage		
	Too early	Optimum	Too late
Artichoke, globe	Flower buds small	When buds are 2–4″ in diameter	Buds large with scales or bracts loose
Asparagus	Insufficient length	6–8″ long; no fibre	Excess woody fibre in stem

TABLE 2.4 *(Continued)*

Vegetable crop	Harvesting stage		
	Too early	**Optimum**	**Too late**
Beans, lima	Insufficient bean size	Bright green pod; seed good size	Pods turned yellow
Beans, pole green	Insufficient size	Bean cavity full; seed ¼ grown	Seed large; pods fibrous
Beans, snap bush	Insufficient size	Pods turgid; seeds just visible	Pods fibrous; seed large
Beets	Insufficient size	Roots 2–3″ in diameter	Roots pithy; strong taste
Broccoli	Insufficient size	Bright green colour; bloom still tightly closed	Head loose; some blooms beginning to show
Brussels sprouts	Insufficient size; hard to harvest	Bright green; tight head	Head loose; colour change to green yellow
Cabbage	Insufficient leaf cover	Heads firm; leaf tight	Leaf loose; heads cracked open
Cantaloupes	Stem does not want to separate from fruit	Stem easily breaks away clean when pulled	Background colour of melon is yellow; rind soft
Carrots	Insufficient size	½–¾″ at shoulder	Strong taste; oversweet
Cauliflower	Curd not developed	Head compact; fairly smooth	Curds open; separate
Celery	Stem too small	Plant 12–15″ tall; stem medium thick	Seed stalk formed; bitterness
Collards	Insufficient leaf size	Bright green colour; small midrib	Midrib large; fibrous
Corn, sweet	Grain watery; small	Grain plump; liquid in milk stage	Grain starting to dent; liquid in dough stage
Cucumber	Insufficient size	Skin dark green; seeds soft	Skin beginning to yellow; seeds hard
Eggplant	Insufficient size	High glossy skin; side springs back when mashed	Seeds brown; side will not spring back when mashed
Lettuce, head	Head not fully formed	Fairly firm; good size	Heads very hard
Okra	Insufficient size	2–3″ long; still tender	Fibre development; pods tough

TABLE 2.4 *(Continued)*

Vegetable crop	Harvesting stage		
	Too early	Optimum	Too late
Onions, dry	Tops all green	Tops yellow; ¾ fallen over	All tops down; bulb rot started
Peas	Peas immature and too small to shell	Peas small to medium; sweet bright green	Pods yellow; peas large
Pepper, pimiento	Insufficient size	Bright red and firm	Fruit shrivelled
Pepper, red bell	Insufficient size , fruits has a chocolate colour	Bright red and firm	Fruit shrivelled
Potato, Irish	Insufficient size	When tops begin to die back	Damaged by freezing weather
Potato, sweet	Size small; immature	Most roots 2–3″ in diameter	Early plantings get too large and crack; damaged by low soil temperature below 50 °F
Rhubarb	Size small; immature	Stem 8–15″ long best	Fleshy stem becomes fibrous
Squash, summer	Insufficient size	Rind can be penetrated by thumbnail	Penetration by thumbnail difficult; seed large
Squash, winter	Rind soft	Rind difficult to penetrate by thumbnail	Damaged by frost
Tomatoes	May be harvested in three stages:		
	Mature green—tomato firm, mature, colour change from green to light green, no pink colour showing on blossom end. These tomatoes will store 1–2 weeks in refrigerator.		
	Pink—pink colour on blossom end about the size of a dime. These tomatoes, at room temperature, will ripen in about three days.		
	Ripe—tomato full red but still firm. Should be used immediately.		
Watermelon	Flesh green; stem green and difficult to separate, producing metallic sound while tapping with hands	Melon surface next to ground turns from light straw colour to a richer yellow, drying of tendrils	Top surface has dull look

2.3.3   POST-HARVEST

2.3.3.1   PACKHOUSE OPERATION (SORTING, GRADING, PACKING ETC.)

A packhouse is usually a physical structure where harvested produces are accumulated (consolidation or collection centre) and prepared for storage or transport and/or distribution to markets. Generally tender or perishable or small volumes of vegetables are prepared in the field itself for nearby markets. Products need to be transported to a packinghouse or packing shed for large operations, for distant or demanding markets or for special operations like washing, brushing, waxing, controlled ripening, refrigeration, storage or any specific type of treatment or packaging. These two systems (field vs. packinghouse preparation) are not mutually exclusive. In many cases, partial field preparation is completed later in the packing shed. Because it is a waste of time and money to handle unmarketable units, primary selection of fruits and vegetables is always carried out in the field where products with severe defects, injuries or diseases are removed. Field preparation of lettuce is an example where a team of three workers cut, prepare and pack. For distant markets, boxes prepared in the field are delivered to packinghouses for palletizing, precooling and sometimes cold storage before shipping (Hasan et al., 2015). Packing of crop produce is the main activity from which the name 'packhouse' is derived. But in fact, there are another series of activities before and after packing like cleaning, sorting/grading, pretreatments, packing, cooling, storage and dispatch to market and altogether they are called packhouse operations. Practically most of such collection centres in developing countries hardly follow any packhouse operations. They serve only as collection points, where farmers bring their produce for traders to collect. Practically, rest all other activities can be undertaken on the farm or at the level of retail, wholesale or supermarket chain. A packhouse ought to serve as a hub for coordination and governance of a farm–packhouse–market organization in which production and packhouse activities will be mostly in accordance to the market demand and usually being managed by three teams: management team (overall leadership; members of the cooperative/group); production team (production and harvest scheduling); and marketing team who will take care the packhouse and market operations, market linking. However, a packhouse may be simple or modern. An individual farmer can also develop a simple packhouse. Generally, a packhouse should be designed to enable different operations for handling vegetables. It should be located as close to farms as possible, provide service to the greatest number of farmers, facilitate drop-off and pick-up of produce and containers, be easily accessible to markets or transport terminals. The site and premises of the packhouse should have minimal risk of contamination, protection from sun and rain, dependable water and electricity supplies, adequate drainage, provision for comfort and safety of workers.

There is an absolute lack of the concept of packing house establishments in India. Vegetables are generally packed in the field without any pre-treatment. Some are even transported without any packaging. In developed countries on the other hand, vegetables are generally selected, cut, placed in bulk containers and transported to packing stations where they are trimmed, sorted, graded, packed in cartons or crates and cooled. They are temporarily placed in cool storage for subsequent loading or are loaded directly onto refrigerated vehicles, and transported to market. A number of important operations are also carried out at packing stations. These include fungicidal dipping, surface coating with wax, ripening and conditioning, vapour heat treatment and so on. Due to the lack of proper packaging systems in India, large volumes of the inedible portions of vegetables such as cauliflower, peas and others are transported to wholesale markets from the field. They are discarded to various degrees and large quantities of biomass which could be used as value-added products are wasted. Removal of these inedible vegetable portions prior to marketing would reduce both transportation costs that ultimately undergo decomposition, cause sanitation problems and produce gases which are detrimental to the environment. Farmer's cooperatives and other agencies should, therefore, be encouraged to establish packing stations at nodal points to augment the marketing of fresh horticultural produce (Rolle, 2005).

After harvest, vegetables need to be prepared for market. Missing link of packhouse operations are to be followed.

2.3.3.1.1 Dumping

The first step of handling is known as dumping. It should be done gently either using water or dry dumping. Wet dumping can be done by immersing the produce in water. It reduces mechanical injury, bruising, abrasions on the fruits, as water is gentler on produce. The dry dumping is done by soft brushes fitted on the sloped ramp or moving conveyor belts. It will help in removing dust and dirt on the fruits.

2.3.3.1.2   Sorting and Grading

Sorting is practiced for most of the vegetables to remove injured, decayed, misshapen fruits, damaged, diseased and insect-infested produce on the basis of visual observation. It will save energy and money because culls will not be handled, cooled, packed or transported. Removing decaying fruits are especially important, because these will limit the spread of infection to other healthy fruits during handling. However, in the advanced countries, different types of sorters are used. The commonly used sorting equipment is belt conveyor, push-bar conveyor and roller conveyor (Kitinoja and Kader, 2003). Such infrastructure facilities in developing countries are hardly available to reduce post-harvest losses of vegetables. Washing is a standard post-harvest handling operation for many vegetables to remove adherences, dirt and external pathogenic structures. Unfortunately, in developing countries, vegetables are hardly being washed before entering into the marketing channel, and this contributes to poor quality and considerable losses of the produce. Grading is one of the important post-harvest handling operations, especially on the basis of shape, size, defect and colour. In developing countries, grading is practiced in limited scale on the basis of grading standard (Extra Class, Class-I and Class-II) for most of the vegetables due to lack of awareness, experiences and infrastructural facilities. Some of the vegetables like Brinjal, chilli and cucumber are not properly graded by the growers. Regarding the grading of okra, the middlemen prefer to purchase well-shaped okra pods. They rely on the visual/external quality parameters of okra pods like tenderness (by tip pinching), colour, size and shape of the pods. However, tenderness is found to be the common means of judging the quality of the okra pods prior to taking decision on whether the produce to be purchased or not. Automatic grading of vegetables is also a common practice in the developed countries and automatic rotary cylinder sizer is mostly used for this purpose (Reyes, 1988).

2.3.3.1.3   Washing and Cleaning

Washing with chlorine solution (100–150 ppm) can also be used to control inoculums build-up during post-harvest operations (Pirovani et al., 2001).

For best results, the pH of wash solution should be between 6.5 and 7.5. Hydrogen peroxide is also used for cleaning and phytosanitation of fruits and vegetables (Sapers et al., 1999). Practically in most of the cases, wash water is recycled without much filtration, frequently undesirable health hazardous banned chemicals like (copper sulphate, malachite green, Congo red etc.) are added to the wash water in the field, transit or in the market yard to extend the shelf life or making vegetables attractive to the consumer.

2.3.3.1.4 Packaging

Increased use of corrugated cartons for local distribution of produce could be accomplished with improvement in the quality of boxes produced in India. The ventilated CFB box which contains ventilated partitions, found to be ideal for the packaging and transportation of vegetables owing to the comparably minimal level of bruising observed in these boxes. Cushioning materials used in the packaging of fruits and vegetables in wooden boxes include dry grass, paddy straw, leaves, sawdust, paper shreds and so on, all of which end up as garbage and add to environmental pollution in cities. Moulded trays or cardboard partitions used in CFB boxes are, however, easily recycled.

2.3.3.1.5 Palletization

Loading and unloading are very important steps in the post-harvest handling of fruits and vegetables but are often neglected. The individual handling of packaged produce in India leads to mishandling and to high post-harvest losses in India. With the introduction of CFB boxes, serious consideration should be given to the introduction of palletization and mechanical loading and unloading of produce, particularly with the use of fork-lift trucks, to minimize produce mishandling.

2.3.3.1.6 Storage

The lowest temperature that does not cause chilling injury is the ideal storage temperature for fresh vegetables. Mechanical refrigeration is, however, energy intensive and expensive, involves considerable initial capital investment and requires uninterrupted supplies of electricity which are not always readily available, and cannot be quickly and easily installed. Available cold storage in developing countries is used primarily for the storage of potatoes. Appropriate multi-purpose cool storage technologies are therefore required.

2.3.3.1.7　On-farm Storage

On-farm storage is required in remote and inaccessible areas of India to reduce losses in highly perishable fresh horticultural produce. The high cost and high energy requirements of refrigeration, and the difficulty of installing and running refrigerated facilities in remote areas of India, preclude the use of refrigerated storage in many parts of India. Low-cost, low-energy, environmentally friendly cool chambers made from locally available materials, and which utilize the principles of evaporative cooling, were therefore developed in response to this problem. These cool chambers are able to maintain temperatures at 10–15°C below ambient, as well as at a relative humidity of 90%, depending on the season. Vegetables may store in plastic crates within the chamber. Information on production area, volume and quality of produce, and target market and quality requirements should be properly recorded and known by all involved in production and marketing.

Reliable data on the magnitude of post-harvest losses of vegetables, especially in the tropics and subtropics are meager. Post-harvest losses of vegetables were 25–40% as reported by Miaruddin and Shahjahan (2008) in Bangladesh. In India, research is still going on how to reduce losses on the farms and in the market places. Data on post-harvest losses generally have been collected either via surveys/interviews or via sampling/direct measurements and are reported as physical and/or economic loss. Occasionally data are provided on qualitative losses (due to damage, disease, pests, appearance changes etc.).

Success in any horticultural enterprise depends on the ability to satisfy the needs, aspirations and indulgences of consumers. Producers must keep the consumer in mind at all times. Also, they must provide high-quality products that will make the consumer come back, repurchase and make propaganda about the better quality fruit, vegetables or flowers. Getting the produce to any market means involving different people and organizations who inevitably 'add value' as the produce passes through the chain.

However, all of these people have a different view of what quality means to them. Still, it is critically important that producers are fully aware of all the steps in the supply chain at which quality could be compromised. Producers have to ensure that they deliver only the highest quality vegetables into the supply chain as quality may only be maintained by post-harvest technologies, not improved. The implementation of modern supply chain management concepts (Bowersox et al., 2003) in the horticultural sector has the potential to increase efficiency of supply and distribution and marketing and brings about improved profitability to all chain participants (Hewett, 2003). Post-harvest losses of vegetables in supply chain are greater at the hands of the intermediaries, especially the middlemen and wholesalers. The higher post-harvest loss at the intermediary levels would possibly due to the lack of proper transportation and storage facilities as depicted in Table 2.5.

TABLE 2.5   Post-Harvest Losses of Selected Vegetables.

Vegetable crops	Post-harvest losses (%) at different levels of supply chain			
	Growers	Middlemen	Wholesalers	Retailers
Tomato	6.9	9.1	8.0	8.9
Cauliflower	4.2	9.2	10.3	10.7
Okra	9.4	9.8	4.9	8.3
Brinjal	6.9	7.4	8.4	6.6
Cucumber	7.2	4.5	10.7	4.7
Red amaranth	5.5	9.2	7.8	6.1

2.3.4   POST-HARVEST FOOD LOSS

It is defined as measurable qualitative and quantitative food loss along the supply chain, starting at the time of harvest till its consumption or other end uses (Hodges et al., 2011). This system comprises inter-connected activities from the time of harvest through crop processing, marketing and food preparation to the final decision by the consumer to eat or discard the food. PHLs can occur either due to food waste or due to inadvertent losses along the way. Thus, food waste is the loss of edible food due to human action or inaction such as throwing away wilted produce, not consuming available food before its expiry date, serving sizes beyond one's ability to consume, losses incur during social or religious occasions. Food loss, on the other hand, is the inadvertent loss in food quantity because of infra-structure and management limitations of a given food value chain. Food losses can either be the result of a direct quantitative loss or arise indirectly due to qualitative loss.

Food losses can be quantitative as measured by decreased weight or volume, or can be qualitative, such as reduced nutritive value and unwanted changes to taste, colour, texture or cosmetic features of food (Buzby and Hyman, 2012). However, quantitative food loss can be defined as unintended losses in weight of food along the supply chain rather than intentional weight loss through drying or other processing. The quantitative loss is caused by the reduction in weight due to factors such as spoilage, consumption by pest and also due to physical changes in temperature, moisture content and chemical changes (FAO, 1980). However, reduction in weight due to drying of starchy crops like bulb and tuber crops up to a limit is a necessary post-harvest process for them (FAO, 1980). Although this process involves considerable reduction in weight, there is no loss of food value, and therefore, should not be counted as loss.

The qualitative loss of vegetables may occur due to incidence of insect pest, mites, rodents and birds, or from handling, physical changes or chemical changes in the constituents and by contamination of mycotoxins, pesticide residues, insect fragments or excreta of rodents and birds and their dead bodies. When this qualitative deterioration makes food unfit for human consumption and is rejected, this contributes to food loss.

According to Bloom (2010), food waste occurs when an edible food item goes unutilized as a result of human action or inaction and is often the result of a decision made farm-to-fork by businesses, governments and farmers (Buzby and Hyman, 2012). The definitions of food waste and food losses are not consistent worldwide (Waarts et al., 2011). Wastes & Resources Action Programme final report in 2009 defines waste as 'any substance or object which the holder discards or intends or is required to discard' (Quested and Johnson, 2009).

2.3.5 MEASURING FOOD LOSSES

Unit loss measurements characterize losses of a commodity expressed by the percentage of units or a percentage loss in weight. Some problems with unit loss measurements include the following:

1. The point at which a commodity becomes inedible often depends upon the socio-economic level of the consumers and/or on local cultural preferences.
2. Reduction of quality, condition or appearance might involve serious monetary losses but would not be reflected in the food loss data as long as the produce was consumed.
3. Diversion of produce to a secondary or salvage market might represent a real loss in monetary terms but would not be considered a loss by this method because it would be consumed.
4. Moisture loss is an important factor in quality and consumer acceptability of fresh vegetables. Such loss of acceptability would be measured as a unit loss only if dehydration was so severe as to render the commodity unfit for human consumption.

There is simply no 'easy' way to measure PHLs. As vegetables are handled by many people, produce samples may be examined for loss at convenient points in the distribution chain. Many past measurements have targeted post-harvest losses occurring on the farm (at harvest), in the packinghouse, after storage, and at wholesale and retail markets. Likely included as culls are small sizes, immature and over-mature or overripe produce, and variously damaged or defective (deformed, hail or frost damaged etc.) units. Culls are a post-harvest loss unless there was an available alternate use or secondary market. For example, if culled fruits were processed to jams or candies, further measurements would be required to determine the extent to which losses in the processed products occurred. If long-term storage is involved in the value chain, post-harvest loss sampling may occur as packed produce is removed from cold or dry storage to be loaded into transit vehicles. Measurements of weight are commonly made before and after transportation, so weight loss can usually be determined in distribution centres or upon arrival at retail stores (Kitinoja and Kader, 2015).

Post-harvest losses for horticultural produce are still difficult to measure. In some cases, everything harvested by a farmer may end up being sold to consumers. In others, losses or waste may be considerable. Occasionally, losses may be 100%, for example when there is a price collapse, market glut and it would cost more to harvest and market the produce than to plough it back into the ground. Use of average loss figures is thus often misleading. There can be losses in quality, as measured both by the price obtained and nutritional value, as well as in quantity.

There are no reliable methods for evaluating post-harvest losses of fresh produce. Accurate records of losses at various stages of the marketing chain are also hardly maintained, particularly in tropical countries where considerable losses occur. Such losses are not only accounted as a waste of food, but they also represent a similar waste of human effort, farm inputs, livelihoods, investments and scarce resources such as soil fertility and water.

2.3.6   CAUSES OF POST-HARVEST LOSSES OF VEGETABLES

There are many causes of loss after harvest. It may be biological or due to poor post-harvest handling practices. The quality of the harvested vegetables depends on the condition of growth as well as the manipulation in the physical, physiological and biochemical changes they undergo after harvest. Major causes are

1. Acceleration of senescence (ageing)
2. Water loss
3. Mechanical injuries
4. Growth and development
5. Physiological disorders
6. Disease infections
7. Post-harvest insect infestation
8. Air flow at high speed

Acceleration of Senescence (Ageing)
Common symptoms of senescence in vegetables are excessive softening, tissue breakdown, loss of colour, loss of flavour, off-flavour, discolouration, fibrous, splitting and others. Fresh vegetables are living parts of plant and contain 65–95% water. As long as the plant or its part receives the requisite food element from its basic source, for example soil, it is easier to maintain the life processes and accompanying metabolic activities. After harvest, the life processes continue at the expense of reserve or stored substances existing in the plant parts. When food and water reserves are exhausted, losses occur in the produce terms of its quality and quantity. Different factors are responsible for increasing the rate of depletion of product's food and water reserves and consequently enhance the losses. The rate of deterioration of vegetables is also related to the respiration rate.

Generally mature plant parts have low respiration and actively growing plant parts have high respiration. Four categories of respiration rate have been found in vegetable crops, namely low, moderate, high and very high (Singh, 2009).

The respiration rate of vegetable is comparatively much higher in elevated temperature. Therefore, control of temperature is crucial for mini-mizing loss of quality through senescence by maintaining the accumulated substrates for prolong period. Rapid pre-cooling of the harvested produce is generally being done to remove field heat. However, to extend the shelf life, care should be taken so that the optimum temperature for particular crop sustenance should be maintained throughout the process of handling, storage and marketing (cool chain management).

Respiration is a continuous process in growing plant and harvested produce with the help of accumulated source of carbohydrate and being interrupted when food reserves of these are exhausted leading to ageing. Respiration requires a continuous supply of oxygen and any restriction of O_2 increases CO_2 concentration leading to alcoholic fermentation, which develops foul smell in the produce. More accumulation of carbon dioxide, resulting from poor ventilation during storage, transit and marketing, leads to complete spoilage of the produce. Most of the vegetables belong to the non-climacteric group (leafy vegetables, cabbage, cauliflower, okra, pointed gourd etc.), although very few are considered as climacteric (tomato). Therefore, the respiration rate must be reduced or suppressed as much as possible to extend the post-harvest life of the vegetables.

During storage ethylene injury is also a serious threat which could be reduced by holding the commodity at its lowest safe temperature and by keeping it under modified or controlled atmospheres. Some of the exam-ples of ethylene injury are russet spotting of lettuce, development of bitter flavour in carrots, leaf and flower abscission and yellowing of broccoli, cabbage, cauliflower.

Water Loss
Fresh produce continues to lose water after harvest causing shrinkage and loss in weight. The rate at which water is lost varies according to the product. The most significant factor is the ratio of the surface area of the fruit or vegetable to its volume. The greater the ratio, the more rapid will be the loss of water. The rate of loss is related to the difference between the water vapour pressure inside the produce and in the surrounding air.

Leafy vegetables lose water quickly because they have a thin skin with expanded surface area and many pores. Potatoes, on the other hand, have a thick skin with few pores and tomatoes are having waxy coating in the pericarp. But whatever the product, to extend shelf or storage life, the rate of water loss must be lowered. To escape the water loss, vegetables are to be kept under lower temperature and/or in a moist atmosphere (e.g. in plastic bags). Application of surface coating with wax also prevent water loss from vegetables provided they are having lower rate of respiration as because the coating may impede oxygen and carbon dioxide movement (Singh, 2009), resulting into anaerobic respiration.

Mechanical Injuries
Fresh fruits and vegetables are extremely susceptible to mechanical injury because of tender texture and high moisture content. Poor handling, unsuitable containers (e.g. bamboo basket), improper packaging (heavily loaded) and transportation may result in bruising, cutting, breaking, impact wounding and other forms of injury. Externally post-harvest losses are encouraged by any sort of physical injury causing internal bruising, abrasion, splitting and skin breaks, thus rapidly increasing the rate of respiration, ethylene production, water loss and entry of harmful pathogen.

When soft produce is bruised, generally leaving some external symptoms such as shape distortion, spots etc., contrary to the hard external surfaces where bruising may become visible not initially but on reaching to the consumer. Bruising may be caused by impact or pressure damage. Impact damage can occur dropping off of individual produce or package during harvesting and/or transit. Pressure damage can occur in the product stacked too high or packed in a container unable to support the required weight.

Abrasion (rubbing) of surface tissue leads to rupture of cells, tissue disintegration and may develop visible symptoms immediately or after a while. Common causes of abrasion injury are rubbing of produce against dirty or rough surfaces of containers and equipment and friction between loosely packed produce during transport. When single produce is dropped on to hard surface or loose produce bounce against each other during transport, heavy impacts to rigid or hard produce may cause cracking or splitting

Growth and Development
When harvested at a proper stage of maturity, the quality and storability of the vegetable is expected to be better. As the whole plant and its different

parts are consumed as vegetables like root, shoot, leaves, fruit, flower, pseudostem, modified stem and others, the edible or commercial maturity may or may not coincide with its physiological maturity. Most of the vegetables are preferred to consume when they are tender enough. Therefore, if harvested too early (underdeveloped) or too late (over mature), may be unacceptable.

Some vegetables continue their growth even after harvest at the expense of their reserve elements. Sprouting of potato, shooting of onion, carrot and others, elongation and changing shape of asparagus are examples of continued growth after harvest which deteriorates the quality of those vegetables. Formation of fibres can also occur in some vegetables, for example okra.

Physiological Disorders

Sometimes the active biological processes occurring in produce may disrupt or failed as consequence of some undesirable external or internal factors resulting in quality loss. Examples of these disorders are heat injury, chilling injury, ethylene damage, carbon dioxide damage and low oxygen (anaerobic) injury. The effect of high temperature is crop specific and generally heat damage occurs in produce above 30°C. Excess heat build-up can occur in stacks of produce with high respiration rate and poor or no ventilation or may be due to direct or indirect exposure to sunlight. Heat injury may cause brown patches; inhibit loss of green colour, excessive softness, off-flavours, yellowing of leaves, wilting and others. When a particular vegetable is exposed to low temperature below its tolerance limit, it started to suffer from chilling injury. Common symptoms of chilling injury are surface pitting, discoloured skin and water-soaked areas of flesh (e.g. brinjal). Symptoms of ethylene damage include surface pitting, increase disease incidence, yellowing and increased softening. Ethylene damage is typically caused by any sort of physical injury or the mixing of more ethylene producing and ethylene sensitive produce during storage and transport. Excessive carbon dioxide inside packaging or storage unit may develop water soaked patches; skin discolouration and product undergoes anaerobic respiration. It may happen when controlled and modified atmosphere storage and transport is incorrectly managed (Singh, 2009). Some vegetables such as lettuce and cabbage are sensitive to 2% CO_2, suffering from brown spot and brown vascular tissue. Vegetables, particularly fruit vegetables, suffer from injury when held at atmosphere below 2% oxygen.

Pest-Disease Infestation

Huge post-harvest losses are caused by the invasion of fungi, bacteria, insects and other organisms because after harvest the natural defence mechanism of a crop decreases and still possessing plenty of nutrients and moisture to support microbial growth. Post-harvest decay control is becoming a more difficult and challenging task, because large number of pesticides are banning rapidly because of consumer and environment concern. Spoilage organism may also be present in wash water, particularly when the water is not filtered or changed frequently. The susceptibility of the vegetables varies considerably and is affected by several factors such as genetically, mechanical injury (bruises, cuts, cracks etc. allow entry of organism), stress (excessively low or high temperature, high or low humidity or unsuitable atmospheric condition) and stage of development (increased susceptibility at advanced stage of senescence). Disease symptoms may range from invisible to small surface lesions to severe infections with external and internal breakdown of a substantial part of the produce.

Air Flow at High Speed

The microclimate of outer layer of vegetables where vapour pressure deficit is least protects the crop from several hazards like shrinkage, tissue softening and so on. Depletion of these humid layer invites rapid water loss, disease, insect damage and reduction in crops own resistance mechanism. This may occur due to high velocity air flow through this layer, mechanical injury like impact, bruising and others.

2.3.6.1   SOCIOECONOMIC FACTORS

Although the biological and environmental factors that contribute to post-harvest losses are well recognized, and relevant technologies also have been developed to overcome these losses, considerable gap prevails in reality regarding their application, and they have not yet been adopted due to one or more of the following socio-economic factors (Kader, 1983).

1. **Inadequate marketing systems:**
    a. Growers can produce good-quality vegetables, but in most of the cases, their accessibility to represent such commodities to the target market and consumer are limited and consequently

losses will be extensive. This problem exists in many locations, especially in developing countries. It is accentuated by lack of market information and communication between producers and receivers.

b. Marketing cooperatives should be encouraged among producers of major commodities in important production areas. Such organizations are especially needed in developing countries because of the relatively small farm size. Advantages of marketing cooperatives include: providing central accumulation points for the harvested commodity, feasibility to prepare for market properly and storage when needed, facilitating transportation to the markets and acting as a common selling unit for the members, coordinating the marketing program and distributing profits rationally.

c. Wholesale markets in most of the developing countries are in desperate need of improvement in terms of facilities and sanitation. These are overcrowded, unsanitary, unscientific, lack adequate facilities for loading, unloading, ripening, consumer packaging and temporary storage (Kader, 2005).

d. Alternative distribution systems, such as direct selling to the consumer (roadside stands, produce markets in cities, local farmers' market in the rural areas etc.), should be encouraged.

2. **Inadequate transportation facilities:** In most developing countries, roads are inadequate for proper transport of horticultural crops. Also, transport vehicles and other modes, especially those suited for fresh horticultural perishables, are in short supply. The majority of producers has small holdings and cannot afford their own transport vehicles. Therefore, production can be ensured closer to the major population centres to minimize transportation costs and hazards.

3. **Transport losses may be due to the following reason:**
   a. Unhygienic and unsuitable transport containers and mean
   b. Overloading of mixed fruits and vegetables (in some developing countries people and even animals ride on top of the vegetable load)
   c. Irresponsible driving
   d. Poor condition of roads, lack of feeder roads loading to highways or collection centres

    e. Heat and moisture accumulation due to poor ventilation within the transport vehicles and containers

    f. Virtual absence of refrigerated and insulated trucks

    g. Delays in product transport after harvesting or at collection centres

4. **Government regulations and legislations:** The degree of governmental controls, especially on wholesale and retail prices of fresh vegetables varies from one area to another and one market to the other in a same locality. In many cases, price controls are counter-productive. Although intended for consumer protection, looseness in monitoring and regulations encourages fraud and provides no incentive for producing high-quality produce or for post-harvest quality maintenance. This is one of the reasons for inadequacy in post-harvest handling practices among growers and marketers. On the other hand, regulations on covering proper handling procedures and public health aspects (food safety issues) during marketing, if enforced properly, would be blessing to the growers as well as consumers.

5. **Unavailability of needed tools and equipment:** Even if growers and handlers of fresh horticultural crops were convinced of the merits of using some special tools and/or equipment in harvesting and post-harvest handling (e.g. containers, equipment for cleaning, waxing, packing and pre-cooling facilities), unavailability of such tools locally or even in the domestic market at reasonable price, limit their use. Most of the tools are neither manufactured locally nor imported in sufficient quantity to meet demand. Various governmental regulations in some countries do not permit direct importation of required devices by producers. In many cases, such tools can be manufactured locally at much lower cost than those imported.

6. **Lack of awareness/information on post-harvest handling:** The human element in post-harvest handling of horticultural commodities is extremely important. Most of the growers and handlers involved directly in harvesting, handling and marketing process in developing countries have limited knowhow regarding quality maintenance of fresh produce (Kader, 2005). An effective extension program on these aspects is critically essential. The availability of needed information on the Internet (numerous websites including: http://www.fao.org/inpho; http://www.postharvest.com.au; http://

postharvest.ucdavis.edu; http://www.postharvest.ifas.ufl.edu; and www.postharvest.org) is an important step in the right direction, especially with the expanded access to the Internet worldwide.

7. **Poor maintenance:** In many cases, some good facilities that were established once by government or by private sector considering the urgency are not functioning properly because of lack of recurring fund, proper maintenance and unavailability of spare parts, skilled technical and managerial person to handle the facility. This problem is especially true of public-sector facilities. Any new project should include in its plan adequate funds for future maintenance to ensure its success and extended usefulness.

8. **Hygiene and sanitation:** Sanitation must be of great concern to produce handlers, not only to protect produce against post-harvest pathogen attack, but also to protect consumers from food borne illnesses. *Escherichia coli* 0157:H7, *Salmonella, Chryptosporidium, Hepatitis* and *Cyclospera* are among the disease-causing organisms that have been transferred via fresh fruits and vegetables. Use of a disinfectant in wash water can help to prevent both post-harvest diseases and food borne illnesses.

A more serious problem is the presence of human pathogens on produce. These may not be visible or detected because of changes in appearance, flavour, colour or other external characteristics. Specific pathogens are able to survive on produce sufficiently long to constitute a threat. Three types of organisms that can be transported on fruits and vegetables may constitute a risk to human health: virus, for example hepatitis A; bacteria, for example *Salmonella* spp., *Escherichia coli, Shigella* spp.; and parasites, for example *Giardia* spp. Mycotoxins and fungi do not usually constitute a problem, because fungi development is usually detected and eliminated well before the formation of mycotoxins (FAO, 2004).

2.3.6.2 UNDESIRABLE PHYSIOLOGICAL AND BIOCHEMICAL CHANGES

Both respiration and genetic control play a significant role in a series of events which lead to desirable and undesirable changes in the quality of vegetables after harvest. It has become evident that vegetables may synthesize several chemical compounds; however, many of these compounds detract from quality by virtue of undesirable changes following harvest and post-harvest periods and occurrence of stress related possible toxicity (Haard, 1983a). As for example, leguminous vegetables contain isoflavonoids, solanaceous vegetables synthesize diterpenes and steroidal alkaloids, and Umbelliferae tend to accumulate coumarins and furanoteruenoids. In potato, more than 24 diterpenes have been identified as stress metabolites. Both anabolic and catabolic processes are responsible for post-harvest biochemical changes as a consequence of 'biological clock', that is under some form of genetic control. The relationship between cellular respiration and gene expression, and between these primary events and secondary metabolism is still unclear. There exists a negative relationship between respiratory rate and storability of crops. Hence, vegetables having high respiration rate (peas, asparagus) deteriorate rapidly, whereas crop (turnip) having slow respiration rate could be stored for longer period.

The benefits of lowering storage temperature vary from commodity to commodity at least for two reasons. First, lower down of temperature in storage may have significant influence on both respiration and storability for certain crops. Second, some crops are sensitive to chilling temperatures, resulting anomalous metabolism that may lead to a wide array of physio-logical disorders (Haard, 1983b). Similarly, respiration rate and storability respond to modification of the storage atmosphere by reducing oxygen and increasing carbon dioxide partial pressure. Again, crops exhibit indi-viduality in their ability to cope with the stress of these changes, and for some commodities, physiological injury rather than extended storage life results from such treatments.

Studies with various crops have shown that manipulation of tissue calcium by either agronomic practice or post-harvest sprays or dips can retard an assortment of physiological injuries and extend storability. Gener-ally, ethylene gas can stimulate the respiration of post-harvest vegetables. The dependency of respiration on ethylene concentration and the type of effect vary with the commodity. Climacteric vegetables, which exhibit an autonomous burst in respiration coincident with ripening, respond to exog-enous ethylene concentrations above certain threshold levels by showing an earlier respiratory climacteric and ripening; that is, ethylene has an all-or-none effect on such crops. Non-climacteric vegetables exhibit an ethylene-concentration-dependent respiratory rise, that is the magnitude of the respiratory burst is greater as ethylene concentration is increased.

The respiration in senescing plant tissues is coupled to the conservation of energy in adenosine triphophate (ATP). However, the relationship of respiratory rate and metabolic changes in food is not entirely clear in terms of ATP formation and expenditure, indicating unknown fact about ATP-dependent processes in post-harvest systems.

The composition and nutritive quality of vegetables are affected by many pre- and post-harvest factors (Hulme, 1971). The varietal influence (Stevens, 1974), cultural practices such as mulching, rootstock and others, climatic condition, maturity at harvest, harvesting method and post-harvest handling conditions (time between harvest and consumption), extent of physical damage, processing methods (blanching, drying, freezing etc.) and cooking processes (roasting, boiling, frying) affect the nutritional quality of vegetables (Goddard and Mathews, 1979; Salunkhe et al., 1973). Some of these factors are responsible for post-harvest losses in vitamin, mineral, dietary fibres and others. Water soluble vitamins (vitamin C, thiamin, riboflavin, niacin, vitamin B_6, folacin, vitamin B_{12}, biotin and pantothenic acid) are much more susceptible to post-harvest losses than fat-soluble (vitamins A, D, E and K). Post-harvest losses in nutritional quality, particularly vitamin C content, can be substantial and are enhanced by physical damage, extended storage, higher temperature, low relative humidity and chilling injury of chilling-sensitive commodities. It is used as index stability to other nutrients. Post-harvest losses in other nutrients may occur due to degradation at higher temperature in the presence of oxygen (Kader, 1987).

Vegetables are mostly preferred by consumer with high textural integrity, especially when used as fresh cut or salad, and loss of water negatively affects their quality. Most of the vegetables contain a huge amount of water, which can often exceed 95% by fresh weight. Texture and the degree of softness are determined by the amount of water contained in the produce and the ability to retain that water during post-harvest storage (Hasan et al., 2015). Although handling and preparing of fresh produce after harvest and during storage, the nutritional value is affected in several ways, for example:

- Both dry-matter and moisture content is reduced during storage as the continuation of living processes within the produce uses up stored food reserves.
- Vitamin C content decreases with time after harvest.

- Consumption of fresh fruits and vegetables are particularly valuable provided they are grown and handled hygienically as cooking partially destroys vitamins C and B_1.
- Peeling may cause significant loss of food value because major amount of nutrients and dietary fibre remains on or just beneath the skin.
- Water used in cooking vegetables contains the dissolved minerals and vitamins and should not be thrown out but used in soups or in preparing other foods (Simson and Straus, 2010). Use of excessive water for blanching or boiling should be avoided.

2.4 MEASURE TO REDUCE POST-HARVEST LOSSES

After harvesting, the vegetables are to be cooled down, cleaned, sorted, sized and graded, packed, stored and/or transported for end use, may be either market as fresh produce or to the processing industry. Practically, all these steps are not followed carefully in most of the supply chain resulting in to huge post-harvest loss. To mitigate this loss, post-harvest horticulturist need to coordinate their efforts with production horticulturists, agricultural marketing economists, engineers, food technologists and others who may be involved in various aspects of the production and marketing system. In most of the cases, solutions to existing problems in the post-harvest handling system require utilization of available need based technologies and information at the appropriate scale rather than developing new technologies. Overcoming the socio-economic constraints is essential to achieve the goal of reducing PHLs.

The post-harvest treatments play an important role in extending the storage and marketable life of horticultural perishables. The most important post-harvest treatments include:

1. **Pre-cooling:** Pre-cooling of the produce soon after their harvest to remove the field heat is one of the important components of the cool chain as it is quite effective for the shelf life of the produce. Important methods of pre-cooling applicable for specific vegetables are mentioned below:
   a. *Room cooling:* In this method, produce is simply loaded into a room and cool air is allowed to circulate among the cartons,

sacks, bins or bulk load. It is low cost and slow method of cooling, for example beet, cabbage, garlic, ginger and so on.

b. *Forced air cooling:* It pulls or pushes air forcibly through the vents/holes in storage unit. This is the fastest method of pre-cooling commonly used for wide range of horticultural produce. Uniform and faster cooling of the produce can be achieved if the stacks of pallets are properly aligned. Cooling time depends on (i) the airflow, (ii) the temperature difference between the produce and the cold air and (iii) type of produce (respiratory behaviour and diameter etc.). For example, okra, brinjal, cucumber, summer squash, ginger, tomato and so on.

c. *Hydrocooling:* The use of cold water is an old and effective cooling method used for cooling easily a wide range of fruits and vegetables. For the packed commodities, it is less used because of difficulty in the movement of water through the containers and because of high cost involved in water tolerant containers. This method of cooling not only avoids water loss but may even add water to the commodity. Use of antimicrobials or sanitizers in cold water ensures both cooling and cleaning of produce. The hydrocooler normally used are of two types: shower and immersion type, for example broccoli, beet, cauliflower, brinjal, cucumber and so on.

d. *Vacuum cooling:* In this method, air is pumped out from the chamber in which the produce is loaded for pre-cooling helping water evaporation from the product at very low air pressure. Removal of air results in the reduction of pressure of the atmosphere around the produce, which further lowers, the boiling temperature of its water, for example peas, spinach, lettuce, carrot and so on. Vacuum cooling cause about 1% loss in weight of produces for each 6°C of cooling.

e. *Package icing:* Packaged or top icing can be used only with water tolerant, chilling insensitive products and with water tolerant packages such as waxed fibreboard, plastic or foam box (Dhatt and Mahajan, 2007). If icing is allowed to distribute throughout the container achieving better contact with the product, it keeps a high relative humidity with cooling effect around the product. As the ice comes in contact with the produce, it melts rapidly and the cooling rate slows considerably. Package ice may be finely crushed, flake or in slurry form, for example carrot, peas, spinach, leek, celery and so on.

    f. *Half-cooling time:* It is the interval in any unit time during which the initial temperature difference between product and coolant (air, water, ice etc.) is halved. The concept of half-cooling time is used for calculations involved in pre-cooling process. Heat is usually removed very rapidly during the first half cooing period, whereas the gradient between product and coolant is maximum, eventually it becomes nil as the temperature of the product and coolant converge. However, regardless of the volume, velocity or temperature of the cooling medium, the cooling rate will be unsatisfactory if the product to be cooled is not accessible to the coolant (Ryall and Lipton, 1972).

2. **Ethylene inhibitors/growth regulator/fungicide treatments:** 1-methyl cyclopropene, Amenoethoxyvinyl gycine, cyclohexi-mide, potassium permangenete ($KMnO_4$), benzothiadiazole and others are some of the chemicals which inhibit ethylene production and/or microbial activity during handling and storage of vegetables. The growth regulators such as GA_3 can effectively be used to extend/enhance the shelf life of vegetables.

3. **Calcium application:** The post-harvest application of $CaCl_2$ or $Ca(NO_3)_2$ plays an important role in enhancing the storage and marketable life of vegetables by maintaining their firmness and quality. Calcium application delays ageing or ripening, reduces post-harvest decay, controls the development of many physiological disorders and increases the calcium content, thus improving their nutritional value. The post-harvest application of $CaCl_2$ or $Ca(NO_3)_2$ at 2–4% for 5–10-min dip extend the storage life at lower temperature. Calcium infiltration reduces chilling injury and increase disease resistance in stored vegetables.

4. **Waxing:** Waxing of fruit vegetables is a common and safer post-harvest practice. Food grade waxes are used to replace some of the natural waxes removed during harvesting and sorting operations and may reduce water loss during handling and marketing. It also helps in sealing tiny injuries and scratches on surface of harvested vegetables. It improves cosmetic appearance and prolongs the storage life of fruits and vegetables. The wax coating must be allowed to dry thoroughly before packing (Dhatt and Mahajan, 2007).

5. **Irradiation:** Ionizing radiation can be applied to fresh fruits and vegetables to control both macro and microorganisms (insect pest) and inhibit or prevent cell reproduction and some chemical changes by exposing the crop to radiations from radioisotopes (normally in the form of gamma-rays 0.25–1 kGy), where 1 (Gray) Gy = 100 rads. It is successfully used as a growth inhibitor (about 10 krad) in retarding the sprouting of potato, sweet potatoes, onion and others (Ryall and Lipton, 1972).

6. **Packaging:** In India, fresh vegetables are usually transported in perforated plastic crates, gunny bags and bamboo baskets. Softer vegetables are given cushioning with dry grass, banana leaves or paper shreds, foam nets as a lining material. The consumer packages include plastic bags, shrink wraps with or without consumer trays made up of foam or plastic, packaging film with anti-fog features for better display of the produce are useful and innovative ideas of packaging.

Improved packaging also include modified atmospheric packaging (MAP) where the amount of oxygen in the package is decreased from 20 to 5% or less that suppresses the rate of respiration and growth of spoilage organisms with a portion of the removed oxygen replaced by carbon dioxide or by inert nitrogen. Fresh produce will continue to respire in the package, changing the gas composition over time. MAP may be used for fresh-cut salad greens, potatoes, asparagus, onions, broccoli, cucumbers, beans, peppers, shredded lettuce and cabbage and others, as these products remain metabolically active long after harvest (Raju et al., 2011).

Active packaging is also a new concept that is impregnated or coated with the active components designed to release various anti-microbial agents (e.g. chlorine dioxide, carbon dioxide or trap oxygen and ethylene) at a controlled rate over time. It may be a film or wrapper or specially made pouches or sachets that can be placed inside the package to release some active components.

General Strategies for post-harvest loss reduction of vegetables:

1. Good returns can only be expected from the appropriate pre- and post-harvest operations.
2. Pre-harvest parameters such as selection of good-quality germplasm and suitable varieties, proper planting material, crop

management and disease and pest control must be geared towards producing high-quality produce.

3. Poor harvesting practices can lead to irreparable damage to produce. It is therefore necessary to standardize maturity indices and harvesting techniques for specific vegetable to minimize damage at the time of harvest.

4. Farmers commonly pack the vegetables strategically placing the larger pieces at the top on the pack to attract the buyer on first sight. During packing of vegetables, care must be taken to minimize physical damage that results from impact bruises due to stacking and overfilling of bags, abrasion or vibration bruises due to close contact of vegetables against each other. Therefore, packages should be neither loose (to avoid vibration bruising during transport) nor overfilled, and should provide good aeration.

5. Palletization and containerization is an essential practice of local and international trade.

6. Certain essentials of cold storage such as calculation of refrigeration loads depending on the produce temperature, weight of the produce, the ambient temperature and storage performance of the produce need to be considered to optimize the use of cold storage.

7. Humidity-controlled cold storages with considerable storage capacities are more effective.

8. The rate of post-harvest losses of horticulture crops in a country can be obtained from the Food Balance Sheet by calculating the difference between total production and total consumption. An assessment of the rate of loss is needed for identification of crop specific problems.

9. Avoid below recommended temperatures in storage, because many vegetables are susceptible freezing or chilling injury.

10. Provide adequate ventilation in the containers and in storage room.

11. Storage facilities should be kept clean and protected from insect pest.

12. Containers should be strong enough to withstand stacking.

13. Monitoring of temperature in the storage room by placing thermometers at different locations.

14. Crop specific temperature, moisture content and gaseous composition must be maintained in the packaging and storage unit.

15. Avoid storing ethylene-sensitive commodities with ethylene producing and produce known for emitting strong odours (garlic, onions, turnips, cabbages and radish) with odour-absorbing commodities.
16. Regular inspection of stored produce for signs of injury, pest and disease attack and remove them immediately to prevent the spread of problems (Dhatt and Mahajan, 2007).

2.5 CONCLUSION

• Loss of vegetables are high in India as well as other developing countries because of the inherent problem of small farm holding and collection and transportation of small quantities of produce of highly variable in size and quality into a large enough quantity for efficient domestic marketing or for export. Therefore, quite often it is difficult to apply standardized grading and other handling procedures.

• The warm, humid climate that prevails in tropical and subtropical countries adds more stress and accelerates the spoilage of perishable crops

• To reduce the substantial pre- and post-harvest losses caused by biotic and abiotic factors, scientific knowhow on cultivation practices and post-harvest operations need to be imparted to the farmers and handlers. The boon of information and communication technology should be tapped to provide practical advice to mitigate their queries.

• The adoption of cold chain is an essential step for fresh produce up to wholesale and retail marketing, although practically it is quite difficult in Indian situation.

• High cost and the lack of abundant uninterrupted power supplies are the biggest constraints in developing cold chain systems in India. Development of cost and energy intensive cooling systems on the basis of evaporative cooling techniques may be considered as a viable alternative.

• Infusion of new technologies, better practices, coordination and development of crop and location-wise technical packages are essential for reducing the post-harvest losses of vegetables.

REFERENCES

Agrawal, M.; Krizek, D. T.; Agrawal, S. B.; Kramer, G. F.; Lee, E. H.; Mitecki, R. M.; Rowland, R. A. Influence of Inverse Day/Night Temperature on Ozone Sensitivity and Selected Morphological and Physiological Responses of Cucumber. *J. Am. Soc. Hortic. Sci.* **1993**, *118*, 649–654.

Balibrea, M. E.; Martinez-Andujar, C.; Cuartero, J.; Bolarin, M. C.; Perez-Alfocea, F. The High Fruit Soluble Sugar Content in Wild *Lycopersicon* Species and Their Hybrids with Cultivars Depends on Sucrose Import During Ripening Rather than on Sucrose Metabolism. *Func. Plant Biol.* **2006**, *33*(3), 279–288.

Bautista, O. K.; Acedo, A. L. Jr. *Postharvest Handling of Fruits and Vegetables*; National Bookstore: Manila, Philippines, 1987; p 27.

Beckles, D. M. Factors Affecting the Postharvest Soluble Solids and Sugar Content of Tomato (*Solanum lycopersicum* L.) fruit. *Postharvest Biol. Technol.* **2012**, *63*(1), 129–140.

Beverly, R. B.; Latimer, J. G.; Smittle, D. A. Preharvest physiological and cultural effects on postharvest quality, In *Postharvest handling: A system approach;* Shewfelt, R. L., Prussia, S. E., Eds.; Academic Press: New York, 1993; pp.73–98.

Beverly, R. B.; Latimer, J. G.; Smittle, D. A. Preharvest Physiological and Cultural Effects on Postharvest Quality. In *Postharvest Handling: A Systems Approach;* Shewfelt, R. L., Prussia, S. E., Eds.; Academic Press: New York, 1993; pp. 73–98.

Bloom, J. *American Wasteland;* Da Capo Press: Cambridge, MA, 2010.

Bowersox, D. J.; Closs, D. J.; Cooper, M. B. *Supply Chain Logistics Management;* New York: McGraw Hill Higher Education, 2003.

Buzby, J. C.; Hyman, H. Total and Per Capita Value of Food Loss in the United States. *Food Policy* **2012**, *37*, 561–570.

Cappellini, R. A.; Ceponis, M. J. Postharvest Losses in Fresh Fruits and Vegetables. In *Postharvest Pathology of Fruits and Vegetables: Postharvest Losses in Perishable Crops;* Moline, H. E., Ed.; 1984. *Univ. Calif. Bull.* **1914**, 24–30.

Carrari, F.; Baxter, C.; Usadel, B. Integrated Analysis of Metabolite and Transcript Levels Reveals Themetabolic Shifts that Underlie Tomato Fruit Development and Highlight Regulatory Aspects of Metabolic Network Behavior. *Plant Physiol.* **2006**, *142*(4), 1380–1396.

Chakraborty, I.; Chattopadhyay, V. A.; Hazra, P. Studies on Processing and Nutritional Qualities of Tomato as Influenced by Genotypes and Environment. *Veg. Sci.* **2007**, *34*(1), 26–31.

Chattopadhyay, V. A.; Chakraborty, I.; Siddique, W. Characterization of Determinate Tomato Hybrids: Search for Better Processing Qualities. *J. Food Process Technol.* **2013**, *4*, 222.

Cheeseman, J. M. Mechanisms of Salinity Tolerance in Plants. *Plant Physiol.* **1988**, *87*, 57–550.

Cliff, M.; Lok, S.; Lu, C.; Toivonen, P. M. A. Effect of 1-Methylcyclopropene on the Sensory, Visual, and Analytical Quality of Greenhouse Tomatoes. *Postharvest Biol. Technol.* **2009**, *53*(1), 11–15.

Dhatt A. S.; Mahajan B. V. C. Horticulture Postharvest Technology Harvesting, Handling and Storage of Horticultural Crops. Punjab Horticultural Postharvest Technology Centre, Punjab Agricultural University Campus, Ludhiana, 2007. ttp://nsdl.niscair.res.in/bitstream/123456789/314/4/Revised+Harvesting,+Handling+and+Storage.pdf.

Dhillon, M. K.; Naresh, J. S.; Singh, R.; Sharma, N. K. Evaluation of Bitter Gourd (*Momordica charantia* L.) Genotypes for Resistance to Melon Fruit Fly, *Bactrocera cucurbitae*. *Indian J. Plant Prot.* **2005,** *33*(1), 55–59.

Dhillon, M. K.; Singh, R.; Naresh, J. S.; Sharma, H.C. The melon fruit fly, *Bactrocera cucurbitae*: A review of its biology and management. *J. Insect Sci.* **2005,** *5,* 40.

Duran, J. M.; Retamal, N.; del Hierro, J.; Rodriguez, A. E.; Del Hierro, J. La simulacion de danos de granizo en especies cultivadas. *Agric. Rev. Agropecu.* **1994,** *63*(740), 214–218.

Erickson, A. N.; Markhart, A. H. Flower Developmental Stage and Organ Sensitivity of Bell Pepper (*Capsicum annuum* L) to Elevated Temperature. *Plant Cell Environ.* **2002,** *25,* 123–130.

FAO. Assessment and Collection of Data on Post-harvest Food Grain Losses, FAO Economic and Social Development Paper 13. Rome, 1980.

FAO. Manual for the preparation and sale of fruits and vegetables: from field to market, FAO agricultural services bulletin no. 151. Food and Agriculture Organization of the United Nations, Rome, 2004.

FAO. International scientific symposium 'biodiversity and sustainable diets',final document, 2010. http://www.fao.org/ag/humannutrition/25915-0e8d8dc364ee46865d5841c48976e9980.pdf).

FAO. FAOSTAT statistical database, 2010.

Fogliani, G.; Battilani, P.; Rossi, V. Studio dei processi riparativi dei frutti grandinati. Le lacerazioni su mele. Ponte del Concordato Italiano Grandine, 1985, pp 45–51.

Garcia, C.; Ruiz, M.; Chen, P. *Impact Parameters Related to Bruising in Selected Fruits.* St. Joseph: American Society of Agricultural Engineers, 1988; p 16.

Getinet, H.; Seyoum, T.; Woldetsadik, K. The Effect of Cultivar, Maturity Stage and Storage Environment on Quality of Tomatoes. *J. Food Engin.* **2008,** *87*(40), 467–478.

Ghosal, A.; Chatterjee, M. L; Manna, D. Studies on Some Insecticides with Novel Mode of Action for the Management of Tomato Fruit Borer (*Helicoverpa armigera* Hub.). *J. Crop Weed* **2012,** *8*(2), 126–129.

Goddard, M. S.; Mathews, R. H. Contribution of Fruits and Vegetables to Human Nutrition. *Hortic. Sci.* **1979,** *14*(3), 245–247.

Haard, N. F. Stress Motabolites. In *Postharvest Physiology and Crop Protection;* Lieberman, M., Ed.; Plenum Press: New York, 1983a; p 299.

Haard, N. F. Edible Plant Tissues. In *Principles of Food Science. Part l—Food Chemistry,* 2nd ed.; Fennema, O., Ed.; Marcel Dekker: New York, 1983b.

Hanna, H. Y. Influence of Cultivar, Growing Media, and Cluster Pruning on Greenhouse Tomato Yield and Fruit Quality. *Hortic. Technol.* **2009,** *19*(2), 95–399.

Harvey, J. M. Reduction of Losses in Fresh Market Fruits and Vegetables. *Annu. Rev. Phytopath.* **1978,** *16,* 321–341.

Hazra, P.; Arup Chattopadhyay, K. Karmakar and S. Dutta (2011). Modern Technology of Vegetable Crop Production, New India Publishing Agency, New Delhi, India.

Hazra, P.; Samsul, H. A.; Sikder, D.; Peter, K. V. Breeding Tomato (*Lycopersicon Esculentum* Mill) Resistant to High Temperature Stress. *Int. J. Plant Breed.* **2007,** *1*(1), 31–40.

Hewett, E. W. Perceptions of Supply Chain Management for Perishable Horticultural Crops: An Introduction. *Acta Hortic.* **2003,** *604,* 37–46.

Hodges, R. J.; Buzby, J. C.; Bennett, B. Postharvest Losses and Waste in Developed and Less Developed Countries: Opportunities to Improve Resource Use. *J. Agric. Sci.* **2011,** *149,* 37–45.

Holmes, M.; Koekemoer, J. Wind Reduction Efficiency of Four Types of Windbreaks in the Malelane Area. Inligtingsbull. *Inst. Trop. Subtrop. Gewasse* **1994,** *263,* 16–20.

Hong, K. H.; Kim, Y.S.; Lee, K. K.; Yiem, M. S. An Investigation of Hail Injury at Flowering in Pears. *Korean Soc. Hortic. Sci. Abstr.* **1989,** *7*(1), 150–151.

Hulme, A. C. *The Biochemistry of Fruits and Their Products*, vols. 1 & 2; Academic Press: New York, 1971.

Janse, J. Invloed licht op kwaliteit tomaat- en komkom-mervruchten. *Groenter-en-Fruit* **1984,** *40*(18), 28–31.

Jones, R. A. The Development of Salt-Tolerant Tomatoes: Breeding Strategies. *Acta Hortic.* **1986,** *190,* 101–114.

Kader, A. A. Influence of Preharvest and Postharvest Environment on Nutritional Composition of Fruits and Vegetables. Proc. 1st Int. Symp. Hort. And Human Health ASHS Sump. Ser. **1987,** *1,* 18–32.

Kader A. A. Postharvest Handling. In *The Biology of Horticulture – An Introductory Textbook;* Preece, J. E., Read, P. E., Eds.; Wiley: New York, 1993; pp 353–377.

Kader A. A. Maturity, Ripening, and Quality Relationships of Fruit-Vegetables. *Acta Hortic. (ISHS)* **1995,** *434,* 249–256, Strategies for Market Oriented Greenhouse Production.

Kader, A. A. Increasing Food Availability by Reducing Postharvest Losses of Fresh Produce. *Proceedings of 5th International Postharvest Symposium* Mencarelli, F., Tonutti, P., Eds. *Acta Hortic.* **2005,** *682,* 2169–2175.

Kanahama, K. Various Factors Related to Curved Fruits in Cucumber (1). Cultivation Conditions and the Occurrence of Curved Fruit. *Agric. Hortic.* **1989,** *64,* 47–52.

Kantor, L. S.; Lipton, K.; Manchester, A.; Oliveira, V. Estimating and Addressing America's Food Losses. *Food Rev.* **1997,** *20,* 3–11.

Kitinoja L.; Kader, A. A. Measuring postharvest losses of fresh fruits and vegetables in developing countries. The Postharvest Education Foundation. ISBN 978-1-62027-006-6. PEF White Paper 15–02, 2015.

Kitinoja, L.; Kader, A. A. *Small-scale Postharvest Handling Practices—A Manual for Horticultural Crops,* 4rd ed.; University of California Davis: California, USA, 2003.

Kousik, C. S.; Sanders, D. C.; Ritchie, D. F. Yield of Bell Peppers as Impacted by the Combination of Bacterial Spot and a Single Hail Storm: Will Copper Sprays Help? *Hortic. Tech.* **1994,** *4,* 356–358.

Kumar, D. K.; Basavaraja, H.; Mahajanshetti, S. B. An Economic Analysis of Post-harvest Losses in Vegetables in Karnataka. *Indian J. Agric. Econ.* **2006,** *61*(1), 134–146.

Kumar, N. R.; Pandey, N. K.; Dahiya, P. S.; Rana, R. K.; Arun, P. Postharvest Losses of Potato in West Bengal: an economic analysis. *Potato J.* **2004,** *31*(3/4), 213–216.

Miller, R. A. Harvest and Handling Injury: Physiology Biochemistry and Detection. In Postharvest Physiology and Pathology of Vegetables; Bartz, J. A., Brecht, J. K., Eds.; Marcel Dekker: New York, NY, USA, 2003; pp 177–208.

Moneruzzaman, K. M.; A. B. M. S. Hossain, W.; Sani, M. Saifuddin and M. Alenazi. Effect of Harvesting and Storage Conditions on the Post harvest Quality of Tomato. *Australian J. Crop Sci.* **2009, 3**(2),113–121.

Moretti, C. L.;Sargent, S. A.; Huber, D. J.; Calbo, A. G.; Puschmann, R. Chemical Composition and Physical Properties of Pericarp, Locule, and Placental Tssues of Tomatoes with Internal Bruising. *J. Am. Soc. Hortic. Sci.* **1998,** *123,* 656–660.

Moretti, C. L.; Sargent, S. A.; Huber, D. J. Delayed Ripening does not Alleviate Symptoms of Internal Bruising in Tomato Fruit. *Proc. Flo. State Hort. Soc.* **1999,** *112,* 169–171.

Mujib-ur-Rehman, Khan, N.; Jan. I. Postharvest Losses in Tomato Crop (a Case of Peshawar Valley). *Sarhad J. Agric.* **2007,** *23*(4), 1279–1284.

National Academy of Sciences. Postharvest Food Losses in Developing Countries. National Academy of Sciences; Washington D.C., 1978 P.202.

Nilsson, T. Growth and Carbohydrate Composition of Winter White Cabbage Intended for Long-term Storage. II. Effects of Solar Radiation, Temperature and Degree-days. *J. Hort. Sci.* **1988,** *63*, 431–441.

Oko-Ibom, G. O.; Asiegbu, J. E. Aspects of Tomato Fruit Quality as Influenced by Cultivar and Scheme of Fertilizer Application. *J. Agric.: Food, Environ. Ext.* **2007,** *6*(1), 71–81.

Peirce, L. C. Vegetables: Characteristics, Production and Marketing. John Wiley and Sons: New York, 1987.

Pirovani, M. E.; Guemes, D. R.; Piagnetini, A. M. Predictive Models for Available Chlorine Depletion and Total Microbial Count Reduction During Washing of Fresh Cut Spinach. *J. Food Sci.* **2001,** *66*, 860–864.

Quested, T.; Johnson, H. Household Food and Drink Waste in the UK: Final Report. Wastes and Resources Action Programme (WRAP), 2009.

Rafi, U.; Usmani, M. K.; Akhtar, M. S. Aphids of ornamental plants and winter vegetables and their aphidiine parasitoids (*Hymenoptera: Braconidae*) in Aligarh region, Uttar Pradesh. J. Threatened Taxa 2010, *2*(9), 1162–1164.

Raju, P. S.; Chauhan, O. P.; Bawa, A. S. Postharvest Handling Systems and Storage of Vegetables. In *Handbook of Vegetables and Vegetable processing;* Sinha, N. K., Ed.; Wiley, State Avenue, USA, 2011; pp 185–198.

Raju, S. V. S.; Bar, U. K.; Uma Shankar; Kumar, S. Scenario of infestation and management of eggplant shoot and fruit borer, *Leucinodes orbonalis* Guen.; in India. *Resistant Pest Manage. Newsl.* **2007,** *16*(2), 14–16.

Raymundo, L. C.; Chichester, C. O.; Simpson, K. L. Light-dependent Carotenoid Sysnthesis in the Tomato Fruit. *J. Aric. Food Chem.* **1976,** *24*, 59–64.

Reid, P.; Innes, G. Fruit Tree Pollination Under Nets. Australas. *Beekeeper* **1996,** *98*, 229–231.

Reyes, M. U. Design, Concept and Operation of ASEAN Packinghouse Equipment for Fruits and Vegetables. Postrharvest Horticulture Training and Research Centre, University of Los Banos, College of Agriculture, Laguna, The Philippines, 1988.

Rolle, R. S. Postharvest management of fruit and vegetables in the Asia-Pacific Region. Asian Productivity Organization, Tokyo, Japan and Food and Agriculture Organization of the United Nations, Rome, Italy, 2005.

Roy, S. K. Research achievements (October 1985–March 1991) of Indo-USAID subproject on postharvest technology of fruits and vegetables. Indian Council of Agricultural Research, New Delhi, India, 1993.

Ryall, L. A.; Lipton, W. J. Handling, Transporation and Storage of Fruits and Vegetables, vol. 1 *Vegetables and Melons.* The Avi Publishing Company, INC.: Westport, Connecticut, 1972.

Rylski, L.; Aloni, B.; Karni, L.; Zaidman, Z.; Cockshull, K.; Tuzel, Y.; Gul, A. Flowering, fruit set development and fruit quality under different environmental conditions in tomato and pepper crops. *Acta Hort.* **1994,** *366,* 45–55.

Salunkhe, D. K.; Pao, S. K.; Dull, G. G. Assessment of Nutritive Value, Quality and Stability of Cruciferous Vegetables During Storage and Subsequent to Processing. *CRC Crit. Rev. Food Technol.* **1973,** *5*(1), 1–38.

Sanches, J.; Durigan, J. F.; Durigan, M. F. B. Aplicação de danos mecânicos em abacates e seus efeitos na qualidade dos frutos. *Engenharia Agrícola* **2008,** *28,* 164–175.

Sapers, G. M.; Miller, R. L.; Mattrazoo, A. M. Effectiveness of Sanitizing Agents in Inhibiting *Escherichia coli* in Golden Delicious Apples. *J. Food Sci.* **1999,** *64,* 734–737.

Sapkota, R.; Dahal, K. C.; Thapa, R. B. Damage Assessment and Management of Cucurbit Fruit Flies in Spring-Summer Squash. *J. Ento. Nematol.* **2010,** *2*(1), 7–12.

Sargent, S. A.; Brecht, J. K.; Zoellner, J. J. Sensitivity of Tomatoes at Mature-green and Breaker Ripeness Stages to internal bruising. *J. Amer. Soc. Hort. Sci.* **1992,** *117,* 119–123.

Sarwar, M.; Hamed, M.; Rasool, B.; Yousaf, M.; Hussain, M. Host Preference and Performance of Fruit Flies Bactrocera zonata (Saunders) and Bactrocera cucurbitae (Coquillett) (Diptera: Tephritidae) for Various Fruits and Vegetables. *Int. J. Sci. Res. Environ. Sci.* **2013,** *1*(8), 188–194.

Sato, S.; Peet, M. M.; Thomas, J. F. Determining Critical Pre- and Post-anthesis Periods and Physiological Process in *Lycopersicon esculentum* Mill. Exposed to Moderately Elevated Temperatures. J. Exp. Bot. **2002,** *53,* 1187–1195.

Satpathy, S.; Kumar, A.; Singh, A. K.; Pandey, P. K. Chlorfenapyr: A New Molecule for Diamondback Moth (*Plutella xylostella* L.) Management in Cabbage. *Ann. Plant Prot. Sci.* **2005,** *13,* 88–90.

Shewfelt, R. L. Sources of Variation in the Nutrient Content of Agricultural Commodities from the Farm to the Consumer. *J. Food Qual.* **1990,** *13,* 37–54.

Shivalingaswamy, T. M. S.; Satpathy, S.; Banerjee, M. K. Estimation of Crop Losses due to Insect Pests in Vegetables. In *Resource Management in Plant Protection,* .vol. I; Sarat Babu, B., Vara Prasad, K. S., Anita, K., Rao, R. D., Chakraborty, S. K., Chandukar, P. S., Eds.; Plant Projection association of India: Hyderabad, 2002; pp. 24–31.

Simson, S.P. and M.C. Straus (2010). Post-harvest Technology of Horticultural Crops. Oxford Book Company/Mehra Offset Press, New Delhi, India.

Singh, A. K. (2009). Good Agricultural Practices (Food Safety for Fresh Produce). Studium Press India Pvt. Ltd.; New Delhi, India.

Singh, R.; Kumar, U. Assessment of Nematode Distribution and Yield Losses in Vegetable Crops of western Uttar Pradesh in India. *Int. J. Sci. Res.* **2015,** *4*(5), 2812–2816.

Singh, S K.; Singh, A. K.; Singh, H. M. Relative Resistance of Okra Germplasm to Shoot and Fruit Borer, *Earias vittella* Fabr. Under Field Conditions. *J. Appl. Zoo. Res.* **2007,** *18*(2), 121–123.

Stevens, M. A. Varietal Influence on Nutritional Value. In *Nutritional Qualities of Fresh Fruits and Vegetables;* White, P. L., Selvey, N., Eds.; Futura Publ. Co.: Mt. Kisco, NewYork, 1974; pp 87–110.

Stevens, M. A.; Rudich, J. Genetic Potential for Overcoming Physiological Limitations on Adaptability, Yield, and Quality in Tomato. *Hortic. Sci.* **1978,** *13,* 673–678.

Toivonen, P. M. A. Fruit Maturation and Ripening and Their Relationship to Quality. *Stewart Postharvest Rev.* **2007,** *3*(2), 1–5.

Toivonen, P. M. A. Postharvest Physiology of Vegetables. In *Handbook of Vegetables and Vegetable Processing;* Hui, Y. H., Sinha, N., Ahmed, J., Evranuz E. Ö., Siddiq, M., Eds.; Wiley-Blackwell Publishing: Ames, 2011; 199–220. [ISBN 978-0-8138-1541-1]

Vigneault, C.; Bordini, M. R.; Abrahao, R. F. Embalagem para frutas e hortaliças. In Resfriamento de frutas e hortaliças. Brasília (Eds. L.A.B. Cortez, S.L. Honorio, C.L. Moretti): Embrapa Hortaliças, 2002; pp 96–121.

Visai, C.; Marro, M. Fenomeni di spaccatura e cicatrizzazione nel melo 'Stayman Winesap'. *Not. Ortoflorofrutticoltura* **1986,** *12*(2), 47–53.

Waarts, Y.; Eppink, M.; Oosterkamp, E.; Hiller, S.; Van der Sluis, A.; Timmermans, T. Reducing Food Waste: Obstacles Experienced in Legislation and Regulation. LEI report/LEI Wageningen UR (-059), 2011.

Watada, A. E.; Herner, R. C.; Kader, A. A. Terminology for the Description of Developmental Stages for Horticultural Crops. *Hortic. Sci.* **1984,** *19,* 20–21.

Weinberger, K., Genova II, C.; Acedo, A. Quantifying Postharvest Loss in Vegetables Along the Supply Chain in Vietnam, Cambodia and Laos. *Int. J. Postharv. Technol. Innov.* **2008,** *1,* 288–297.

Weis, E.; Berry, J. A. Plants and High Temperature Stress. *Soc. Expt. Biol.* **1988,** 329–346.

Winsor, G. W. Some Factors Affecting the Quality and Composition of Tomatoes. *Acta Hortic.* **1979,** *93,* 335–341.

Yamaguchi, T.; Blumwald, E. Developing Salt-Tolerant Crop Plants: Challenges and Opportunities. *Trends Plant Sci.* **2005,** *10*(12), 616–619.

CHAPTER 3

MATURITY INDICES IN VEGETABLES—AN OVERVIEW

RAM ASREY[1,*] and KALYAN BARMAN[2]

[1]*Division of Food Science and Postharvest Technology, Indian Agricultural Research Institute, New Delhi 110012, India, *E-mail: ramasrey@iari.res.in*

[2]*Department of Horticulture, Institute of Agricultural Sciences, Banaras Hindu University, Varanasi 221005, Uttar Pradesh, India*

CONTENTS

3.1 INTRODUCTION

Harvesting of vegetables at appropriate stage of maturity is of paramount importance for attaining desirable quality (appearance, texture, flavour and nutritive value) as well as shelf life. Quality implies the degree of excellence endowed with sensory and hidden nutritional attributes (Shewfelt, 1999). The quality of harvested produce develops during its growth when it is attached to plant. Once the produce is harvested, it is detached from its source of water, carbohydrates, nutrient supply, and thus there is no possibility for further improvement in the components that contribute to unique quality attributes of harvested produce. Therefore, after harvesting its quality cannot be improved, only can be maintained at the best, and any mistake made during harvest is reflected and magnified down the line. Therefore, harvesting of horticultural produce at proper stage of maturity is very crucial. Maturity has been divided into two categories—namely physiological and horticultural or commercial. Physiological maturity is the stage when a fruit completes its development process, but it may or may not ripe. Horticultural maturity, which is more pertinent to vegetables, is the stage of development when a plant or plant part possesses the pre-requisites for utilization by the consumers for a particular purpose.

3.2   PRINCIPLES OF HARVEST MATURITY

- Harvested produce should have the peak acceptable quality when it reaches consumers.
- Commodity should have optimum size, shape, appearance and flavour required for the market.
- Produce should not be unacceptable or toxic.
- Harvested produce should have adequate post-harvest life.

3.3   ADVANTAGES OF DETERMINING MATURITY INDICES

- Produce have improved quality (appearance, texture, flavour and nutritive value)
- Delay senescence of harvested produce
- Harvested produce have longer shelf life
- Maximize returns and extend utilization of produce
- Facilitate long-distant marketing

3.4   FACTORS AFFECTING MATURITY

Maturity of vegetables is influenced by several factors including environment, cultural practices, insect-pest infestation and genetic traits of the plant.

3.4.1 ENVIRONMENTAL FACTORS

3.4.1.1 CLIMATE

Temperature during production of vegetables affects its growth, development and maturity. Temperature has a direct influence on metabolism, and thus it indirectly affects cellular structure and other components, which determine maturity (Sams, 1999). It has been found that there exists a positive correlation between temperature and maturity. For example, higher temperature causes turgor change (lower) which leads to early maturity in tomato. Relative humidity also affects the maturity of crop. At the same maximum and minimum temperature, carrots grown under high-humidity condition mature earlier than that are grown at low-humidity condition. Difference in elevation also affects flower-bud differentiation, bloom and maturity.

3.4.1.2 SOIL

Soil texture and structure influence the nutrient availability, heat budget and microbial population in the rhizosphere. Soils with light texture favour more growth than heavy soils under a given set of optimum input conditions. Sandy and sandy loam soils lead to early maturity of vegetable crops. Salinity set early maturity, improved fruit quality in both grafting combinations by increasing firmness, dry matter (DM), acidity and total soluble solids (TSS) contents (Colla et al., 2006).

3.4.2 CANOPY MANAGEMENT

Training and pruning of crops affect crop maturity by harnessing available sunlight, diverting source–sink relationship and avoiding buildup of congenial micro-climate for insects and pests. For example, gherkins, cucumbers and tomatoes mature earlier when trained on bower or trellis. Application of pruning in muskmelon grown under polyhouse condition delayed the fruit maturity compared with un-pruned vines (Barni et al., 2003).

3.4.3   CHARACTERISTICS OF PLANTING MATERIAL

Maturity of vegetables is genetically controlled and varies with the cultivar. Apart from that, rootstocks in grafted seedlings have also been reported to affect maturity. In Korea, about 50% of cucumber is produced from grafted saplings. Lee and Oda (2003) reported that grafted cucurbits and solanaceous vegetables start early bearing and are long lasting in fruiting apart from resistant to low temperature, insects and diseases. Unlike the earlier reports, Georgios et al. (2016) apprehended that earliness of harvest maturity, pronounced in sensitive climacteric scions 'Polynica' and 'Raymond' melons may relate to ethylene-mediated comprehensive acceleration of ripening stressing rootstock-scion synergy. Condition of saplings/seedlings also has effect on crop maturity. Good-quality young sapling/seedling of cauliflower and onion mature early than that of poor quality. Too young and tender saplings delay maturity and cause lower production. Although over-mature saplings tend to grow early as in the case of solanaceous and bulbous vegetable crops, the size of fruits/bulbs drastically reduces.

3.4.4   FLOOR MANAGEMENT

The type of mulching material, its colour and site of application also affects maturity of vegetables. Mulching has been reported to show profound impact on early maturity of winter season vegetables. Antonious et al. (1996) reported that turnip plants grown with blue, green and red mulch triggered late, mid and early maturity, respectively. Tillage has also been reported to affect vegetable maturity. Rapp et al. (2004) reported that maturity of pumpkin was delayed with increasing number of tillage operations. They have mentioned that plants grown at minimum tilled plot show early maturity and low yield due to lower synthesis of gibberellins and cytokinins. Application of herbicides has also been reported to influence maturity of vegetables. For example, application of metribuzin herbicide to the soil before potato emergence resulted in delayed tuber maturity (Mondy and Munshi, 1990). Similarly, application of ciomazone for pre-emergence weed control was associated with early maturity of squash.

3.4.5   WATER MANAGEMENT

Deficiency or excess of moisture also influences the maturity of vegetables. It has been reported that extreme water stress promotes early maturity and reduces yield as well as quality of vegetables, whereas no water stress leads to increase in yield and delay maturity. Cevik et al. (1981) reported that tomato, pepper and other vegetables grown under drip irrigation show early maturity.

3.4.6   NUTRIENT MANAGEMENT

Type of fertilizer, source and concentration also influence vegetable maturity. Among the different fertilizers, nitrogen has utmost effect on crop maturity. Application of excessive nitrogen delays vegetable maturity, whereas optimum phosphorus level in tissue tends to sustain maturity and after harvest helps in maintenance of produce quality. Mineral nutrients such as Fe, Zn, Cu, Mn, Mg, Co, CI and B have non-visible effect, and therefore, it is difficult to trace its impact on vegetable maturity.

3.4.7   GROWTH REGULATORS

Plant-growth regulators are very useful horticultural production tool, which also have effect on vegetable maturity. For example, application of TIBA has been found effective in exerting early maturity in cucurbits. Likewise, ethylene and ethylene-releasing compounds also promote early maturity of crops. On the contrary, gibberellins can be used for delaying crop maturity. For example, maturity of winter season tomato could be enhanced by pre-harvest application of GA_3 (50 ppm) at flower-bud formation stage.

3.5   MATURITY INDICES OF VEGETABLE CROPS

3.5.1   TOMATO (LYCOPERSICON ESCULENTUM MILL.)

Several methods are available to judge the tomato fruit maturity. But the changes in fruit colour are closely linked with most of the desirable chemical constituents of the fruit (Young et al., 1993). They also concluded that regression of fruit colour against fruit soluble solids content, and fructose, glucose and total sugar concentrations are more reliable maturity indicator than days after anthesis, indicating that Hunter 'a' colourimetric values provide a more precise measurement of fruit physiological age. Three maturity stages are generally recognized for tomatoes: mature-green, pink or breaker stage and red-ripe stage (Fig. 3.1). Tomatoes that ripen on the vine are generally considered to be higher in quality and have better flavour than those harvested at earlier stages. Tomatoes harvested at immature green stage are inferior in quality and develop off flavour when ripe on the other hand, fully ripe fruits bruises easily. For long-distant markets, tomatoes should be harvested at mature-green stage. In the case of local or nearby markets, it should be harvested at breaker or ripe stage. At the breaker stage, blossom end of tomatoes turn pinkish or reddish. For canning purpose, tomatoes should be harvested at firm ripe stage. For this, tomatoes should be medium large, smooth, uniform red coloured without green shoulder. Near-infrared spectroscopy is commercially used in Japan for assessing harvest maturity of tomato fruits (Jha and Mutsuoka, 2004).

Mature green stage   **Breaker stage**   **Red ripe stage**

FIGURE 3.1   Maturity and ripening stages of tomato.

3.5.2 *PEPPER* (CAPSICUM SP.)

In pepper, fruit weight, size (diameter and length), colour, duration from anthesis to harvest, TSS, firmness, respiration and ethylene production rates are indicators of maturity. With the onset of maturity, hue angle (colour change from green to red) declines, whereas chroma (colour intensity) and TSS content increases (Tadesse et al., 2002). With advancing maturity, fresh weight (FW) as well as dry weight of sweet pepper increases gradually, reaching a peak after 30–40 days of fruit set and then declines. However, in the case of hot pepper, fresh as well as dry weight increases up to ripening. While optimizing harvesting time for sweet pepper, Rahman et al. (2014) found that desirable physicochemical attributes of sweet pepper var. 'BARI Misti Morich-1' arrived during 33 DAA to 48 DAA. But considering all physical and nutritional character-istics and shelf life along with sensory evaluation scores, fruits of BARI Misti Morich-1 were found suitable to harvest after 45 days of anthesis. At this maturity stage, fruits were crispier, glossy with attractive colour and flavour and contained 58 mg/100 g ascorbic acid.

In both the types of pepper, total carbohydrate and protein content decreases with increasing maturity. Generally, the sweet peppers are harvested prior to seed filling stage, when the flesh attains the firmness between 22 and 28 N and TSS of 6–7 Bx (Molinari et al., 2000). Sweet and hot peppers should be harvested 30 and 36 days after fruit set, respectively, when grown under polyhouse conditions (El-Saeid, 1996). Curry chilli (*Capsicum annum* L.) is a popular crop among horticultural vegetables in southern state of India and Sri Lanka. It is very susceptive to post-harvest losses because of their tender and perishable nature as well as due to harvesting at improper maturity stage. Champa et al. (2005) found the best desirable traits when curry chilli hybrid Hungarian yellow wax was harvested at 17 days after fruit set.

3.5.3 EGGPLANTS (SOLANUM MELONGENA L.)

Eggplants should be harvested at immature stage when the skin becomes bright in colour and shines. At this stage, seeds can be crushed between finger nails (Ranjan and Chakrabarty, 2003). In general, most of the eggplant varieties become mature and produce marketable yields between 15 and 20 days after fruit set (Singh et al., 1990). If the fruits are harvested at over-mature stage, they become pithy and bitter.

3.5.4 CARROT (DAUCUS CAROTA)

Harvest maturity of carrot is judged on the basis of size, shape and colour. It is very difficult to judge the optimum stage of harvest as the maturity stage does not correspond to any physiological and/or biochemical changes. In general, standard varieties are harvested when the roots attain the size of at least 12.25 cm in length and 2.5–4.0 cm in diameter. Carrots are sold in the market with or without tops. Carrots meant for sale with tops should be harvested with bright green turgid leaves and 30–40 cm long roots. If these are sold without tops, the petioles are trimmed, retaining 1 cm green stem from the root. Baby carrots are harvested at an immature stage when it attains the length of about 7.5 cm.

3.5.5   RADISH (RAPHANUS SATIVUS)

Maturity of radish varies from 20 to 50 days after sowing, depending upon cultivars. In general, radish is ready for harvest when it contains 5–6% DM, 3–4% total sugars, 0.6–0.7% crude fibre and 15–20 mg/100 g ascorbic acid. Cultivars such as Pusa Chetki, Pusa Rashmi, Pusa Himani, Punjab Selection and Punjab Pasand take about 40–45 days to reach edible maturity from date of sowing. Although Japanese White and Long White Icicle take about 55–60 days to attain edible root maturity (Langthasa and Barah, 2000).

3.5.6   CABBAGE (BRASSICA OLERACEAE VAR. BOTRYTIS)

Cabbage generally matures 50–120 days after sowing, depending upon cultivars. Firmness or density and weight of cabbage head are the main indicators for its maturity (Everaarts, 1994). A compact and firm head that can be slightly compressed with moderate hand pressure is mature (Fig. 3.2), whereas a very loose head is immature (Fig. 3.3). Apart from that, arrangement of wrapper leaves is also an indicator of cabbage maturity. The wrapper leaves are spread and the head is exposed usually at mature stage. Mature cabbage has a longer post-harvest life than immature cabbage. Maturity at harvest affects the total dietary fibre content in the head and its content of 230–240 g/kg DM is considered to be ready for harvesting cabbage (Wennberg et al., 2002).

FIGURE 3.2 Mature compact cabbage head.

FIGURE 3.3   Immature loose cabbage head.

3.5.7   CAULIFLOWER (BRASSICA OLERACEAE)

In cauliflower, meristem develops in five phases, namely vegetative phase (stem meristem produces the leaves), initiation of future inflorescence (meristem arises in the leaf or bract axil), curd development (meristem formation starts), bract formation and branching initiated from group of meristems, curd maturity (layout of inflorescence is established and the flower differentiation phase starts). Therefore, harvesting of curd should be completed before the onset of fourth phase (Margara and David, 1978). In cauliflower, formation of curd is highly dependent on temperature. So, the transplanting date does not give reliable count for harvesting. According to Booij (1990), most of the cauliflower varieties reach maturity at 17–19 leaf stage. Likewise, Sharma et al. (1997) suggested that curd weight and colour are the good maturity indices for early maturing cauliflower. The symptoms of over-maturity include 'riciness' caused by expanded immature floral parts and browning or spreading of the curds.

3.5.8 CUCUMBER (CUCUMIS SATIVUS)

At harvest maturity, skin colour of cucumber became dark green in colour with wax deposit on the skin (Fig. 3.4). However, in some varieties, the skin colour becomes light green at maturity, and environmental condition during crop growth also affects the fruit colour. Fresh market slicing cucumbers should be at least 15 cm long at maturity and firm to the touch. With the onset of maturity, TSS content increased in cucumber. The chlorophyll content on the peel also increased up to 15–20 days after anthesis. At the initial stage, chlorophyll a content in the peel remains higher than chlorophyll b; however, at the latter stage, it follows a reverse trend (Keynas and Ozelkok, 1991). The abscissic acid and gibberellic acid content increases gradually at the edible maturity of cucumber. Slicing of the cucumber from the centre of fruit is also used to indicate fruit maturity. At proper stage of harvest maturity, a jelly-like material forms in the seed cavity, and the immature seeds are uniformly white in colour. Large, slightly yellow or hard seeds show over-maturity and low fruit quality. The seeds reach physical maturity about 40 days after anthesis (Aroonrungsikul et al., 1994). Cucumbers should not be allowed to turn yellow, as it indicates the sign of over-maturity. Fruit that are too mature have a tough leathery skin and are bitter in flavour.

FIGURE 3.4   Harvest maturity of cucumber.

3.5.9 *MUSKMELON* (CUCUMIS MELO)

In muskmelon, fruits are harvested either at full-slip or half-slip stage. At full-slip stage (45–50 days after anthesis), fruit stems are easily separated from the vine with little pressure or jerk, leaving clean stem cavity (Fig. 3.5). At half-slip stage, it requires more pressure to separate the fruits from the vine. After 50 days of anthesis, there is an increase in the proportion of sucrose in melon fruit (Gao et al., 1999; Pharr and Hubbard, 1994; Hubbard et al., 1991) at maturation, and it was also found associated with a decline in acid invertase activity and an increase in sucrose phosphate synthase enzymes activity (Matsumoto et al., 2010; Kano, 2009). Muskmelons for long-distant markets are harvested at half-slip stage. In addition to this, softening at bloom end, development of aroma, change in colour of inter-netting areas from green to yellow and soluble solids content of at least 13°Bx are the other indices of maturity. In the case of non-slipping melons, maturity can be judged by days from bloom, soluble solids content (at least 13°Bx), softening at bloom end, change in rind colour and development of aroma characteristic for the variety.

FIGURE 3.5   Full-slip stage in muskmelon.

3.5.10 *WATERMELON* (CITRULLUS LANATUS)

Watermelons should be harvested at fully mature stage as the fruits do not develop internal colour or increase in sugar content after being removed from the vine. At maturity, a dull hollow sound is produced from the watermelon when it is tapped with fingers, in contrast to metallic ringing sound produced in immature fruit. The ground spot (the portion of the watermelon touches the soil) changes from pale white to creamy yellow at the appropriate stage of harvest maturity (Fig. 3.6). The ground spot colour can easily be revealed by rolling the fruit gently over to one side, whereas it is still attached to the vine. Another indicator of watermelon maturity is the condition of the tendril (small curly appendage attached to the fruit stem slightly above the fruit). As maturity, the tendril wilts and changes from green colour to a partially desiccated brown colour (Fig. 3.7). Mini-watermelons have shown increased popularity trend since their introduction in 2003 (Walter, 2009). These are generally small in size (weighing 2–4 kg), seedless and having firm flesh, high TSS and lycopene content (Perkins-Veazie et al., 2006). Little is known about external predictors of the internal quality in mini-watermelons. Vinson et al. (2010) established relation between external indicators of maturity such as tendril senescence, ground spot colour, watermelon circumference and/or weight and internal maturity traits by measuring soluble solids, SS: tritratable acidity, TA ratio or pH. In this study, they found that external predictors were most closely linked to fruit pH rather than to SS or SS/TA.

FIGURE 3.6 Development of ground spot on maturity.

FIGURE 3.7 Wilting and drying of tendril at maturity.

3.5.11 PUMPKIN (CUCURBITA MOSCHATA DUCHESNE)

Colour and rind hardness are the important maturity indices for pumpkins. At harvest maturity, the rind develops deep orange colour, and it resists the pressure of finger nail. However, for green use of pumpkin, it may be harvested when the skin is variegated in colour, and the seeds are white and tender (Criollo et al., 1999). After harvesting, curing of pumpkins should be done in a dry place for 10–12 days at warm temperature (80–85°F) to increase their shelf life.

3.5.12 BITTER GOURD (MOMORDICA CHARANTIA)

Bitter gourd follows a sigmoid growth curve and most of the cultivars reach harvest maturity within 8–10 days after fruit set (Pal et al., 2005). Immature pumpkins are deep or light green in colour with sharp-ridged projection on surface and have white flesh (Fig. 3.8). At seed maturity or at over-mature stage, the flesh turns orange in colour with red placenta (Fig. 3.9).

FIGURE 3.8   Bitter gourd at optimum maturity.

FIGURE 3.9   Over-mature bitter gourd fruit.

3.5.13   *PEA* (PISUM SATIVUM L.)

Peas should be harvested when the pods are filled out, which can be judged by touching the pods. At this stage, pods become green in colour and turgid, having chlorophyll *a* content 4800–7300 μg/100 g·FW, lutien content 1200–1900 μg/100 g·FW and beta carotene content 300–490 μg/100 g·FW (Edelenbos et al., 2001). Thermal time requirement of pea from sowing to flowering ranged between 770 and 890 degree days and from sowing to maturity is between 1370 and 1450 degree days depending upon cultivars. Chinese-type peas should be harvested when they reach the desired size, whereas sugar snap-type edible pod peas should be harvested when they are round and firm. The stems should be left on the pods for all types. Everaarts and Sukkel (2000) worked out a relationship between yield and tenderometer readings of smooth- and wrinkled-seeded types of pea cultivars for processing. The results showed a wide variation for tenderometer readings (90–150) between two types of cultivars. At tenderometer readings, lower than 120 the relative yield of wrinkled-seeded cultivars is higher than that of the smooth-seeded type.

3.5.14 OKRA (ABELMOSCHUS ESCULENTUS L. MOENCH)

Increase in size is a good indicator for maturity of okra. The pod length and diameter of okra increases very fast up to 9th day after anthesis. The maximum DM content (12–16%) and minimum toughness (8–9 N) in most of the cultivars are found on the 4th day after anthesis (Ekka et al., 2001). Okra harvested 6th day after anthesis are found more palatable and have longer shelf life with good cooking quality. It should be of medium length (7.1 cm), tender pedicel, bright green in colour and have small seeds. With the advancement of pod development, content of crude fibre increases rapidly, and a delay in harvest of even 1–2 days reduces the marketability of the pods. Likewise, too much tender pods are also not economical as these affect the marketable yield.

3.5.15 POTATO (SOLANUM TUBUROSUM L.)

At maturity of the potato tubers, the haulm starts getting yellow and tips start withering. Tuber skin hardens, and respiration rate, reducing sugar and phenolic contents, declines, whereas DM and protein contents stabilize at the highest (Marwaha, 1990). Peel strength of 2160 cm^{-2} or more found to provide adequate resistance to tuber indicate its maturity (Halderson and Henning, 1993). Parametric measurement also provides clue to determine the tuber maturity. It has been reported that mature tubers take only 1 day to heal about 3.0-mm deep wound, whereas it take 3–4 days to heal the injury in immature tuber (Lulai and Orr, 1995).

3.5.16 ONION (ALLIUM CEPA L.)

Maturity of onion depends upon the variety, end use and growing season. Green vegetable onions are harvested when the bulbs attain the size of 1–2 cm with a plant height of 15–20 cm. Although the dry bulb onions or kharif onions are harvested 100–120 days after transplanting when the top of the plants start drying and leaf colour changes to slight yellow. At this stage, red pigments develop on bulbs, and it attains maximum size (Patil et al., 2003). Rabi onions are harvested 1 week after 60% of top plants fallen over, whereas multiplier onions or small onions are harvested when 50–70% of top plants fallen over. If bulbs are harvested before maturity or at over-mature stage, excessive weight loss take place, and the bulbs become susceptible to sprouting and rotting mainly by *Aspergillus niger, Botrytis, Erwinia* and *Pseudomonas* spp. (Walls and Corgan, 1994). Late harvest in kharif onions results in doubling and bolting. During harvest, leaves should be cut leaving about 2–2.5 cm top above the bulb; this increases DM content during drying. Cutting leaves too close to bulbs increase incidence of decay organisms. Under hot dry conditions, onion harvesting at more than 80% top fall stage followed by fast curing and retaining long neck with bulb reduced weight loss and rot during storage (Eshel et al., 2014).

REFERENCES

Antonious, G. F.; Kasperbauer, M. J.; Byers, M. E. Light Reflected from Colored Mulches to Growing Turnip Leaves Affects Glucosinolate and Sugar Contents of Edible Roots. *Photochem. Photobiol.* **1996,** *64,* 605–610.

Aroonrungsikul, C.; Sukprakarn, S.; Shigenaga, S. GA and ABA in Cucumber During Seed Development and Maturation. *Kasetsart J. (Nat. Sci.)* **1994,** *28,* 314–323.

Barni, V.; Barni, N. A.; Silveira, J. R. P. Melon Plant in Polyethylene Greenhouse: Two Stems are the Best System of Conviction. *Cienc. Rural* **2003,** *33,* 1039–1043.

Booij, R. Cauliflower Curd Initiation and Maturity: Variability Within a Crop. *J. Hort. Sci.* **1990,** *65*(2), 167–175.

Cevik, B.; Kirda, C.; Ding, G. Some Effect of Irrigation Systems on Yield and Quality of Tomato Grown in a Plastic Covered Greenhouse in the South of Turkey. *Acta Hort.* **1981,** *119,* 333–342.

Champa, W. A. H.; Balasooriya, B. M. C.; Fernando, M. D.; Palipane, K. B. Maturity Indices for Harvesting of Curry Chilli (*Capsicum annum* L.) Hybrid: Hungarian Yellow Wax. *Proceedings of the 61st Annual Session, part 1, abstracts, Sri Lanka Association for the Advancement of Science,* 2005.

Colla, G.; Rouphael, Y.; Cardarelli, M.; Massa, D.; Salerno, A. Yield, Fruit Quality and Mineral Composition of Grafted Melon Plants Grown Under Saline Conditions. *J. Hortic. Sci. Biotech.* **2006,** *81,* 146–152.

Criollo, E. H.; Cardozo, C. C. L.; Guevara, C. Determination of the Physiological Maturity and Storage Potential of Seeds of Pumpkin. *Acta Agron.* **1999,** *49,* 24–30.

Edelenbos, M.; Christensen, L. P.; Grevsen, K. HPLC Determination of Chlorophyll and Carotenoid Pigments in Processed Green Pea Cultivars. *J. Agric. Food Chem.* **2001,** *49,* 4768–4774.

Ekka, A. B.; Chakrabarti, A. K.; Pal, R. K. Physico-Chemical Characters and Their Correlation with Pod Development in Okra. *Ann. Agric. Res.* **2001,** *22,* 339–342.

El-Saeid, H. M. Chemical Composition of Sweet and Hot Pepper Fruits Grown Under Plastic House Conditions. *Egypt. J. Hortic.* **1996,** *22,* 11–18.

Eshel, D.; Vinokur, Y.; Rody, V. Fast Curing: A Method to Improve Postharvest Quality of Onions in Hot Climate Harvest. *Postharvest Biol. Technol.* **2014,** *88,* 34–39.

Everaarts, A. P. A Decimal Code Describing the Developmental Stages of Head Cabbage. *Ann Appl Biol.* **1994,** *125,* 207–214.

Everaarts, A. P.; Sukkel, W. Yield and Tenderometer Reading Relationships for Smooth- and Wrinkled-Seeded Processing Pea Cultivars. *Sci. Hortic.* **2000,** *23,* 175–182.

Gao, Z.; Petreikov, M.; Zamski, E.; Schaffer, A. A. Carbohydrate Metabolism During Early Fruit Development of Sweet Melon (*Cucumis melo*). *Physiol. Plant* **1999,** *16,* 207–214.

Georgios, A. S.; Lambros, C. P.; Marios, C. K. Indexing Melon Physiological Decline to Fruit Quality and Vine Morphometric Parameters. *Sci. Hortic.* **2016,** *203,* 207–215.

Halderson, J. L.; Henning, R. C. Measurements for Determining Potato Tuber Maturity. *Am. Potato J.* **1993,** *70,* 131–141.

Hubbard, N. L.; Pharr, D. M.; Huber, S. C. Sucrose Phosphate Synthase and Other Metabolizing Enzymes in Fruits of Various Species. *Physiol. Plant.* **1991,** *82,* 191–196.

Jha, S. N.; Mutsuoka, T. Nondestructive Determination of Acid–Brix Ratio of Tomato Juice Using Near Infrared Spectroscopy. *Int. J. Food Sci. Technol.* **2004,** *39,* 425–430.

Kano, Y. Effects of Mechanically Restricting Melon (*Cucumis melo L.*) Fruit Growth on Cell Size and Sugar Accumulation. *Environ. Control Biol.* **2009,** *47,* 1–12.

Keynas, K.; Ozelkok, S. Variation in Some Pre or Postharvest Characteristics of Cucumber Fruits Harvested at Different Stages of Growth. *Doga Turk Tarim Ve Ormancilik Derg.* **1991,** *15,* 377–383.

Langthasa, S.; Barah, P. Varietal Performance of Radish Under Hill Conditions of Assam. *Ind. J. Hill. Farm.* **2000,** *13,* 85–86.

Lee, J. M.; Oda, M. Grafting of Herbaceous Vegetable and Ornamental Crops. *Hortic. Rev.* **2003,** *28,* 61–124.

Lulai, E. C.; Orr, P. H. Porometric Measurements Indicate Wound Healing Severity and Tuber Maturity Affect the Early Stages of Wound. *Am. Potato J.* **1995,** *72,* 225–241.

Margara, J.; David, C. Morphological Stages in Meristem Development in Cauliflower. *Comptes Rendus Hebdomadarires Des Seances De L'Acedemie Des Sciences Naturelles.* **1978,** *287,* 1369–1372.

Marwaha, R. S. Biochemical Parameters for Processing of Some Indian Potato Cultivars as Influenced by Tuber Maturity. *Ann. Biol.* (Ludhiana) **1990,** *6,* 7–12.

Matsumoto, J.; Goto, H.; Kano, Y. Effects of Nighttime Heating on Cell Size, Acid Invertase Activity, Sucrose Phosphate Synthase Activity, and Sugar Content of Melon Fruit. *J. Am. Soc. Hortic. Sci.* **2010,** *135*(6):501–505.

Molinari, A. F.; Castro, L. R.; Antoniali, S. The Potential for Bell Pepper Harvest Prior to Full Colour Development. *Proc. Fla. State Hortic. Soc.* **2000**, *112*, 143–146.

Mondy, N.; Munshi, C. B. Effect of Herbicide Metribuzin on the Nitrogenous Constituents of Potatoes. *J. Agric. Food Chem.* **1990**, *36*, 636–639.

Pal, R. K.; Behera, T. K.; Sen, N.; Singh, M. Influence of Harvest Maturity on Respiration, Ethylene Evolution, Texture and Nutritional Properties of Bitter Gourd. *J. Food Sci. Tech.* **2005**, *42*, 197–199.

Patil, R. S.; Sood, V.; Garande, V. K.; Masalkar, S. D. Studies on Natural Top Fall in Rangda (Kharif) Onion. *Agric. Sci. Digest.* **2003**, *23*, 47–49.

Perkins-Veazie, P.; Collins, J. K.; Huber, D. J.; Maness, N. Ripening Changes in Mini Watermelon Fruit, In *Proc. Cucurbitaceae 2006;* Holmes, G. J., Ed.; Universal Press: Raleigh, Asheville, NC, 2006; pp 578–584.

Pharr, D. M.; Hubbard, N. L. Melons: Biochemical and Physiological Control of Sugar Accumulation. In *Encyclopedia of agricultural science;* Amtzen, C. J., Ritter, E. M., Ed.; Academic Press: San Diego, CA, 1994; pp 25–37.

Rahman, M. A.; Halim, G. M. A.; Chowdhury, M. G. F.; Rahman, M. M. Changes in Physi-cochemical Attributes of Sweet Pepper (*Capsicum annum* L.) During Fruit Growth and Development. *Bangladesh J. Agric. Res.* **2014**, *39*(2), 373–383.

Ranjan, J. K.; Chakrabarti, A. K. Changes in the Physical Characters of Brinjal (*Solanum melongena* L.) Fruit During Development. *Capsicum Eggplant Newslett.* **2003**, *22*, 139–142.

Rapp, H.; Robin, R.; Chris, H. Reduced Tillage and Herbicide Influence on Weed Suppres-sion and Yield of Pumpkin. *J. Weed Tech.* **2004**, *18*, 953–961.

Sams, C. E. Pre Harvest Factors Affecting Post-Harvest Texture. *Postharvest Biol. Technol.* **1999**, *15*, 249–254.

Sharma, S. R.; Singh, R.; Gill, H. S. Cauliflower (*Brassica oleracea* var. botrytis) Genotypes for May-June Maturity Under North Indian Plains. *Ind. J. Agric. Sci.* **1997**, *67*, 422–423.

Shewfelt, R. L. What is Quality—A Review. *Postharvest Biol. Technol.* **1999**, *15*, 197–200.

Singh, B. P.; Sharma, N. K.; Kalloo, G. Phsico-Chemical Changes with Maturity in Some Promising Varieties of Brinjal (*Solanum melongena* L). *Haryana J. Hortic. Sci.* **1990**, *19*, 318–325.

Tadesse, T.; Hewett, E. W.; Nichols, M. A.; Fisher, K. J. Changes in Physico-Chemical Attributes of Sweet Pepper cv. Domino During Fruit Growth and Development. *Sci. Hortic.* **2002**, *93*, 91–103.

Vinson, E.; Floyd, L.; Wood, M.; Joseph, Kemble, M. Use of external indicators to predict maturity of mini-watermelon fruit. *Hortic. Sci.* **2010**, *45*, 1034–1037.

Walls, M. M.; Corgan, J. N. Post-harvest losses from delayed harvest and storage of onion. *Hortic. Sci.* **1994**, *29*, 802–804.

Walter, S. A. Influence of Plant Density and Cultivar on Mini Triploid Watermelon Yield and Fruit Quality. *Hortic. Technol.* **2009**, *19*, 53–57.

Wennberg, M.; Engqvist, G.; Nyman, M. Effects of Harvest Time and Storage on Dietary Fiber Components in Various Cultivars of White Cabbage. *J. Sci. Food Agric.* **2002**, *82*, 1405–1411.

Young, T. E.; Juvik, J. A.; Sullivan, J. G. Accumulation of the Components of Total Solids in Ripening Fruits of Tomato. *J. Am. Soc. Hortic. Sci.* **1993**, *118*, 286–292.

CHAPTER 4

NON-DESTRUCTIVE QUALITY EVALUATION OF VEGETABLES

ABDEL GAWAD SAAD, PRANITA JAISWAL, AND S. N. JHA*

*Agriculture Structure and Environmental Control Division, Central Institute of Postharvest Engineering and Technology (CIPHET), Ludhiana 141004, Punjab, India, *Email: snjha_ciphet@yahoo.co.in*

CONTENTS

4.1 INTRODUCTION

One of the most valuable groups of food which play a vital role in human health by preventing diseases and repair body via maintaining its alkaline reserve are the vegetables (FSSAI, 2012). Different vegetables are used in different forms such as roots, stems, leaves, fruits and seeds and contribute to a healthy diet (USDA, 2012). Therefore, the main concern is the maintenance of quality of vegetables. However, both quantitative (such as decrease in weight or volume) and qualitative (such as reduced nutrient value and unwanted changes in taste, colour, texture, or cosmetic features of food) losses in vegetables occur between harvest and consumption

(Buzby and Hyman, 2012). Majority of them occur during harvest and post-harvest presses (Kader, 2005; Hodges et al., 2011). Although the qualitative losses, of fresh produce, are difficult to assess than quantitative losses (Kader, 2005; Kitinoja et al., 2011), they significantly affect the overall acceptability of the produce. Quality standards, consumer preferences and purchasing power vary greatly across countries and cultures, and these differences influence marketability and the magnitude of post-harvest losses. In recent years, markets of developed countries have emerged as a major hub of agricultural export for many developing countries. This access to international market has posed many challenges to meet their stringent food safety standards. Therefore, the need of the hour is to develop an effective quality evaluation system for maintaining an acceptable quality level to the end users.

Keeping these things in mind, the objective of post-harvest research round the globe is focussed at (i) understanding the biological and environmental factors responsible for post-harvest losses, (ii) development of suitable post-harvest technology to reduce losses and preserve quality and safety of commodities, (iii) development of rapid, cost effective, user friendly quality evaluation technique. Quality of vegetables is determined by various physicochemical parameters such as colour, shape, size, gloss, firmness, total soluble solids (TSS), pH, dry matter (DM) and acidity, which involve laborious laboratory techniques which are destructive in nature, need trained staff and render the commodity unusable. These problems can be overcome by applying the approach of non-destructive techniques.

Non-destructive techniques can be used for internal quality assessment and sorting of vegetables as well as for measurements of critical selection feature in plant breeding programs. Different non-destructive techniques frequently used for the assessment of vegetables quality are fast, user friendly, cheaper and accurate (Alander et al., 2013; Saldaña et al., 2013). By applying these techniques, it is possible to screen large numbers of diverse samples. The scientific principle of non-destructive technique is to estimate vegetable quality through measuring change in energy, applied on the target (Fig. 4.1).

Various non-destructive techniques such as optics, near infrared (NIR), ultrasonic, X-ray, microwave, acoustic and magnetic resonance/magnetic resonance imaging (MR/MRI) have been applied for quality determination of horticulture produce (Wang et al., 2009; Jha et al., 2010; Lorente et al., 2011).

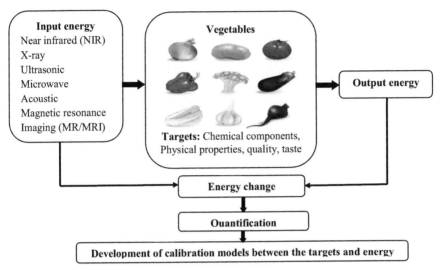

FIGURE 4.1   Scientific principle of non-destructive quality estimation for vegetables.

4.2   SPECTOSCOPY TECHNIQUES

4.2.1   VISUAL SPECTROSCOPY

Chemical components of any food material absorb light energy at specific wavelengths; therefore, some compositional information can be determined from spectra measured by spectrophotometers. In the visible wavelength range, pigments such as chlorophylls, carotenoids, anthocyanins and other coloured compounds are the major light-absorbing component of vegetables (Ignat, 2012). The reflectance properties of any object in the visible region (380–750 nm) are perceived by human eyes as colour, which provide information about the pigment content of the sample (Berns, 2000). When light is absorbed, colours appear, and some part of it is reflected by the target material. If all light is reflected, the object will appear white, and if absorbed, it will appear black. The wavelengths influence the perceived colour which therefore is dependent on both the light source and the absorption by the object (Løkke, 2012).

In some horticultural products such as tomato, vegetable skin colour has been considered as indicator for maturity (Edan et al., 1997). Basically, colour is regarded as one of the appearing property attributed to the spectral distribution of light, which is directly related to the object to which the colour is ascribed as well as the eye of the observer. On the other hand, colour does not appear, in absence of illumination. Hence, a number of factors influenced the radiation and subsequently affect the exact colour that one perceives (Jha, 2010). Jha and Matsuoka (2002) used spectral radiometer to determine the freshness of eggplant based on surface gloss and weight. They established a relationship between gloss index and weight during storage and developed quick and reliable instrumental method for non-destructive estimation of freshness of eggplant. By using this technique, Jha et al. (2002) further developed a freshness index of eggplant.

4.2.2 NEAR-INFRARED SPECTROSCOPY

Near-infrared spectroscopy is one of the most widely studied quality-assessment tools for the last 20 years. It is a rapid, powerful, reliable and non-destructive technique for the measuring qualitative and quantitative properties of biological materials (Jha and Matsuoka, 2004; Teye et al., 2013). For non-destructive measurement of the quality of fruits and vegetables such as soluble solids (Brix), acidity, titratable acidity, water content, dry matter, firmness and so on and for rapid assessment of fibre, protein, fat, ash content and so on this technique is now increasingly used (Jha and Matsuoka, 2004; Bureau et al., 2013).

Near-infrared (NIR) spectroscopy involves use of light in the wavelengths range of 780–2500 nm, and the light penetration depth depends on the wavelength and the sample characteristics. It is up to 4 mm in the 700–900-nm range (Lammertyn et al., 2000; Nicolai et al., 2007). In the NIR region, the absorption is due to overtones and the combination tones of the fundamental infrared (IR) vibration bands of bonds in which the electric dipole moment changes; anharmonic bonds (Thygesen et al., 2003). Information about chemical properties and surface structure reveals the light emitted from the object (sample). The light scattered from fresh produce reveals the microstructure of the tissue, and the light absorption is associated with the presence of chemical components in the sample under study. Therefore, in quality assessment, both phenomena are useful (Nicolai et al., 2007). The fresh agriculture product are generally found to be high in moisture content, hence the NIR spectrum of fresh agriculture product is vastly controlled by water content as water has a high absorption in NIR region of light (Cen and He, 2007). By extracting the relevant information from many overlapping peaks, the spectral pattern information is used to predict the chemical compositions of the sample. So, the pivotal step is to extract the useful information from original spectral data. In addition, automation of NIR measurements can be done after proper calibration. For quantitative analysis of sample constituents, multivariate calibration is required to develop the prediction models. Partial least squares (PLS), principal components regression (PCR), and artificial neural networks are the most used multivariate calibration techniques for the NIR spectroscopy (He et al., 2005). Vis-NIR spectroscopy has been successfully used to develop model for prediction of sensory quality of chicory, soluble solids content and firmness of bell pepper (Francois et al., 2008; Penchaiya et al., 2009). In leafy green vegetables, it has also been used in prediction of chlorophyll content (Xue and Yang, 2009). Slaughter et al. (1996) used NIR spectroscopy to study the soluble solid content of more than 30 varieties of fresh tomatoes. Shao et al. (2007) measured the quality characteristics (soluble solids content, pH and firmness) of tomato "Heatwave", by NIR spectroscopy and found satisfying results.

NIR spectroscopy has been used to measure the nitrate concentrations in vegetables (Shao and He, 2008; Kanda et al., 2010 and Itoh et al., 2011). Shao and He (2008) estimated the strawberry acidity by NIR reflectance spectroscopy, the absorbance data compressed by using wavelength transformation and two models were established to predict strawberry acidity. Matsumoto et al. (2009) developed models to measuring nitrate concentration in the whole body of a lettuce by using absorption spectra in the range of 700–960 nm. Itoh et al. (2011) developed a method for non-destructive measurement of nitrate concentration in Spinach and komatsuna leaves by NIR spectroscopy. They measured absorption spectra of small portion of the leaves, thereafter the nitrate concentrations of the same portion were measured by a liquid chromatography analyser. Finally, PCR or PLS methods with a wavelength selection algorithm were developed to estimate the nitrate concentration.

4.2.3 MICROWAVE DIELECTRIC SPECTROSCOPY

Microwave electromagnetic radiation spectrum stretch from a frequency range 10^8–10^{11} Hz (Chieh, 2012). Microwave dielectric spectroscopy is an emerging technique used in assessment of the internal quality based on dielectric properties of food products (Bohigas et al., 2008). Dielectric properties of all materials are dependent on their molecular structure. Specifically, it depends on the distribution of electric charges, which are either constantly embedded within the molecules or temporarily covers its surfaces. It is also known that the physical and chemical properties of objects are determined by their molecular structure. Therefore, the dielectric properties of various molecules constituting a given material will uniquely identify it. It can successfully diversify physical and chemical properties of a tested material (Fig. 4.3). In agrophysics, the crucial point of the application of the dielectric spectroscopy measurement techniques is the utilization of their advantages for rapid and non-destructive assessment of the quality of the agricultural objects. It may be done by searching for dependencies between the dielectric properties and other physical and chemical properties of tested materials of agricultural origin (Skierucha et al., 2012).

By different microwave measuring sensors, the dielectric properties of food materials in the microwave region can be determined (Kraszewski, 1996). The specific method used depends on the frequency range and the type of target material. By measuring permittivity of watermelon, the water content and soluble solid content of watermelon were evaluated (Nelson et al., 2007). The open-ended probe was used to measure the complex permittivity of watermelon. Soluble solid content and moisture content of watermelon were used as a quality factor for the correlation with the dielectric properties. A high correlation was obtained between the dielectric constant and solid soluble content (Nelson et al., 2007).

The permittivity measurements of honeydew melons and watermelons was used by Nelson and Trabelsi (2008), with an open-ended coaxial-line probe and impedance analyser at frequencies of 10 MHz to 1.8 GHz to provide information about their maturity. Total soluble solid of melons was used as measure of maturity and was correlated with permittivity. Dielectric constant and loss factor correlations with total soluble solid were low, but a high correlation was recorded between the total soluble solid and permittivity from a complex-plane plot of dielectric constant and loss factor, each divided by total soluble solid.

Dielectric spectroscopy can be considered an important non-destructive tool for controlling the freezing process of potato at frequency range of 500 MHz to 20 GHz (Cuibusa et al., 2013). Over a frequency range of 300–3000 MHz at temperatures between 22 and 120°C, the dielectric properties of tomatoes were measured (Penga et al., 2013). To measure the moisture content and tissue density of agricultural materials by predicting heating rates which in turn describe the behaviour of products when exposed to high-frequency or microwave electric fields, the dielectric properties has been used (Venkatesh and Raghavan, 2004).

Nelson (2003) and Nelson et al. (2006) carried out permittivity measurements of cut vegetables (cantaloupe, carrot, cucumber and potato) over a frequency range of 10 MHz to 1.8 GHz and at various temperatures ranging from 5–95°C. The dielectric loss factor was considerably decreased with frequency, whereas slight decrease in dielectric constant was observed with frequency. Similarly, the dielectric loss factor was generally found to be increased with temperature. In fruits and vegetables, if the moisture content is high then the dielectric constant is generally high at a temperature range 5–95°C. McKeown et al. (2012) observed the highest magnitude of permittivity values (dielectric constant and dielectric loss) at low frequencies in carrot. The dielectric constant value of carrot, cantaloupe, potato and finally cucumber was found to be in the decreasing order. Although moisture content did not correlate with the dielectric properties, other factors, such as density, tissue structure, nature of water binding to constituents of vegetables, might have affected the dielectric properties.

4.2.4 X-RAY AND COMPUTERIZED TOMOGRAPHY (CT)

For quickly detecting the strongly attenuating materials, X-ray imaging is an established technique. It has been applied to a number of inspection applications within the agricultural and food industries (Jha and Matsuoka, 2000; Donis-González, 2013).

At present, techniques based on two-dimensional (2D) X-ray and computed tomographic (CT) imaging have been explored and used for internal quality determination of agricultural and food products non-destructively (Abbott, 1999; Haff, 2008). Despite extensive research effort, real-time inspection systems for detection of internal quality of fresh produce are not commercially available due to limitations in useful information when using high-speed systems (Butz et al., 2005). However, new detector technologies, high-performance x-ray tubes, accessibility, real-time imaging, cost diminution and significant reducing in image acquisition time and in-line CT sorting systems are gaining tremendous attraction with the improvement in high-performance computers (Pratx and Xing, 2011).

X-ray is short wave radiation (0.01–10 nm) with high energy $(1.92 \times 10^{-17} - 1.92 \times 10^{-14}J)$ that can easily penetrate matter. X-rays are generated by bombarding electrons on a metallic anode (X-ray tube) (Bushberg et al., 2002). Traditional CT is an imaging modality where an X-ray tube is rotated around an object or objects, and the attenuation is recorded on a detector. Other equipment may contain a rotating stage in front of a fixed x-ray tube and detector (Donis-González et al., 2012). Quenon and De The X-ray method was developed Baerdemaeker (2000) to measure the length of the floral stalk in Belgian endive *Cichorium intybus* L non-destructively. Detection algorithm was developed based on the minimal transmitted intensities along the length. The method is very accurate with an absolute precision of 4.9 mm and allows the study of the influence of storage conditions and time on the internal quality of Belgian endive. They concluded that the X-ray transmission is suitable for a non-destructive measurement of the length of the floral stalk in Belgian endive.

4.3   SOUND WAVES TECHNIQUES

Acoustic sound waves (in the range of human hearing i.e. 20 Hz to 20 KHz) and ultrasonic waves (which are above the range of human hearing i.e. 20 KHz to 1 MHz) are used to evaluate the quality of fresh vegetables non-destructively (Sagartzazu et al., 2008). In acoustic sound, a device is often used to lightly tap or thump the commodity to create a sound wave that pass through the product tissue. As the sound waves pass though the product, their characteristics can be used to indicate the quality attributes of fruit and vegetables during post-harvest processes (Butz et al., 2005).

4.3.1   ACOUSTICS

For non-destructive quality evaluation of fruits and vegetables, acoustic resonance technique is an emerging trend. This technique is based on the response to sound and vibration when the source is gently tapped. It can be used to predict the maturity, internal quality, ripening stage and other similar parameters using the audible frequency range of 20 Hz to 20 kHz. The availability of high-speed data acquisition and processing technology has renewed research interest in the development of impact and sonic response techniques (Vahora et al., 2013). When an acoustic wave reaches to the agricultural products, the reflected or transmitted acoustic wave depends on the acoustic characteristics of the agricultural products. Information can be provided by the reflected or transmitted acoustic wave on the interaction between acoustic wave and agricultural products, and acoustic characteristics such as attenuation coefficient, transmitting velocity, acoustic impedance and natural frequency. Different agricultural products have various acoustic characteristics based on the internal tissue structures (Sugiyama et al., 1994; Trnka et al., 2013).

From last three decades, there has been tremendous development in acoustics technology. It has become a primary method for watermelon sorting and grading (Mizrach et al., 1996). A pendulum hitting device was developed by He et al. (1994) to judge maturity and other internal qualities of watermelons without damage by studying the feature curves of the sound waveform of watermelons. Sugiyama et al. (1994) studied the relationship between the transmission velocity and firmness of muskmelons. In ripened muskmelons, they found that the transmission velocity became lower. Based on the study, they developed an instrument to measure the transmission velocity of sound wave in muskmelons and found that the transmission velocity of sound wave in edible muskmelons ranged from 37 to 50 m/s.

An instrument for measuring the hollow heart and maturity of watermelons (Fig. 4.2) was developed by Applied Vibro-Acoustics (AVA) Company on May 31, 2014. It was based on the theory that everything in the world holds its own special frequency. By using acoustic technology, Lü (2003) and Rao et al. (2004) developed a quality inspecting system. The sound waves were collected via microphones and transformed into electric signal. Thereafter, this electric signal was amplified, followed by filtered via processing circuit, and sampled by a data-acquisition board (Fig. 4.3). A correlation was developed between the transmission velocity and soluble solids content of watermelons, and the best correlation coefficient for different striking positions and growth status of watermelons was found to be 0.81–0.95.

FIGURE 4.2   A portable frequency response measurement instrument.

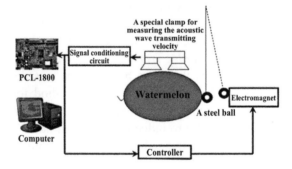

FIGURE 4.3   Diagram of setup the transmitting velocity measurement of watermelon.

Baltazar et al. (2007) used acoustic impact test to study the ripening process of intact tomato. They concluded that the non-destructive acoustic impact technique could detect small physiological changes. Hertog et al. (2004) studied the relation between water loss and firmness measurements in tomatoes during post-harvest period. During post-harvest period, the acoustic stiffness measurement was found to be suitable for determining the softening phenomena of individual pepper samples (Zsom-Muha, 2008). The tests applied on two excitation positions of paprika: top (1) and shoulder part (2) the product showed characteristic frequency peak of the measured acoustic response in these two positions. One dominant frequency peak can be obtained by the excitation on the top of the pepper berry. In contrast to this, by the excitation on the shoulder part, other frequency peak can be seen (probably because of the excitation of other vibration modes), but no significant difference can be observed as acoustic stiffness coefficients on both (top and shoulder part) excitation positions of the pepper berry may be due to the suitability to excite the pepper berry on the top part (Zsom-Muha, 2008).

4.3.2 ULTRASOUND

Ultrasound technology has been known and used in many areas, including medical diagnostics, industrial processes and metal fabrication (Mason et al., 1996). In evaluating and testing the biological and food materials, it has also gained attention and increasing popularity (Jivanuwong, 1998). The advantages of ultrasonic methods include quick measurement and interpretation, penetrating optically opaque materials, accuracy, low cost, freedom from radiation hazards and the ease of online measurement. At high frequencies and low power, it can be used as an analytical and evaluation tool, and at a very high power, it can assist processing (Mason et al., 1996). Ultrasonic vibrations are above the audible frequency range >20 kHz. Ultrasound is generated by a transducer containing a ceramic crystal which is excited by a short electrical pulse that has a typical form of several sine cycles. Through the piezoelectric effect, this electrical energy is converted to a mechanical wave that is propagated as a short sonic pulse at the fundamental frequency of the transducer. This energy is transferred into the material or body under analysis and propagated through it (Krautkramer and Krautkramer, 1990). The ultrasound signal emerging from the test sample is sensed by a piezoelectric element that acts as a receiver, converting any ultrasound impinging on it, back to electrical energy. It was reported by Mizrach et al. (2000) that ultrasonic non-destructively measurement system was relied for the assessment of same transmission parameters which may have quantitative relation with the ripening, maturity, firmness and other internal quality of fruit and vegetables.

4.4 IMAGING ANALYSIS TECHNIQUES

A common way to obtain spatial information of the samples in monochromatic forms or colour images is an imaging system technique. Imaging system is therefore used for colour, shape, size, surface texture evaluation of food products and to detect surface defect in food samples;

however, it cannot identify or detect chemical properties of a food product (Sun, 2009, 2010).

4.4.1 HYPERSPECTRAL IMAGING

Nowadays, hyperspectral imaging system has become a powerful tool and popular for food research. It can capture the spatial data of the whole target at selected wavelengths instead of measuring spectral values at single point (Huang et al., 2014). It is an inherently effective tool because of ability to collect data with both spatial and spectral characteristics of the scanned target (Wang et al., 2013).

Near-infrared hyperspectral imaging system was used by Itoh et al. (2010) to measure the nitrate concentration distribution in a vegetable leaf. The nitrate concentration estimated at each pixel in a leaf image with high accuracy, and the results indicated variation in nitrate concentration inside a leaf. Wang et al. (2009) developed a NIR reflectance hyperspectral imaging system for sour-skin detection in Vidalia onions. The system consisted of an InGaAs video camera, normal lens, liquid crystal tunable filter, frame grabber for acquiring image data and two tungsten halogen lamps as light sources.

One of the very interesting applications of hyperspectral imaging technique is to predict the sugar content distribution in melons (Sun, 2009). Polder et al. (2004) measured the surface distribution of carotenes and chlorophyll in ripening tomatoes at spectral range of 400–700 nm with 1 nm resolution. By using a prototype of hyperspectral transmittance imaging system, Zhang et al. (2013) predicted soluble solid content of tomatoes at spectral range of 720–990 nm. It has also been used to predict the amount of DM, soluble solids content and firmness of onions by using a line-scan hyperspectral imaging system with three sensing modes (reflectance, interactance and transmittance) at spectral range of 400–1000 nm (Wang et al., 2013), TSSs, total chlorophyll, carotenoid and ascorbic acid content during bell pepper maturity at spectral range of 550–850 nm (Itoh et al., 2010).

4.4.2 MACHINE VISION

During recent years, for examination of fruits and vegetables, the machine vision system has been increasingly used, especially for applications in quality inspection and defect sorting applications (Eissa and Abdel Khalik, 2012). Computer vision system is recognized as the integration of devices for non-contact optical sensing, computing and decision processes, which receive and interpret automatically an image of a real scene (Parmar et al., 2011). It includes capturing, processing and analysis of two-dimensional images, with other noting that aims to duplicate the effect of human vision by electronically perceiving and understanding an image. Image processing and image analysis are the core of computer vision with numerous algorithms and methods available to achieve the required classification and measurements (Eissa and Abdel Khalik, 2012).

An automatic strawberry grading system with photoelectric sensors was designed to detect shape and grading of strawberry (Liming and Yanchao, 2010). A machine vision system together with linear discriminate analysis based on colour information of the pixel in dry and wet condition of the object was used to discriminate potato tubers from solid clods (Al-Mallahi et al., 2010). A discrimination rate of 92% for wet condition and 73% for dry condition was successfully achieved (Al-Mallahi et al., 2010). Machine-vision system was developed for guidance of a robot arm to pick the ripe tomato during harvest through acquiring images from tomato plant (Arefi et al., 2011).

4.4.3 MAGNETIC RESONANCE AND MAGNETIC RESONANCE IMAGING

Magnetic resonance imaging (MRI) has become a well-established technique for non-destructive analysis of the internal structure of food. The MRI technique provides a non-destructive method to evaluate both the qualitative and the quantitative properties of biological materials (Chenga et al., 2011). This technique is based on the interaction of certain nuclei, such as carbon and hydrogen, with electromagnetic radiation in the radio frequency range (Slaughter, 2009). It is used to evaluate the property of interest for food processing (as drying), physical tissue damage assessment (as bruising) and others for online sorting process or detection of internal defects (as internal browning) (Defraeye et al., 2013). For internal quality assessment of some fruit and vegetables, magnetic resonance imaging technique has been used as a non-invasive research tool (Mazhar et al., 2013). The physiological change of tomato at different maturity stages has been visualized in MR (Saltveit, 1991; Zhang and McCarthy, 2012). Musse et al. (2009) investigated the change in macroscopic structure and water proton relaxation times during ripening of tomato fruit. Chemical shift imaging technique (nuclear MR spectroscopic method) was employed to investigate spatial–temporal changes in sugar and lycopene contents of tomatoes during ripening, to provide a better conception of the post-harvest ripening process of tomatoes (Chenga et al., 2011). Dedicated MRI has been used to trace the thawing process for boiled and frozen edible vegetables such as green soybeans, broad beans, okra, asparagus and taro. It was measured by the spin-echo method (echo time = 7 ms) with 0.1 or 0.2 s and 1 s repetition times (Koizumia et al., 2006). The pericarp tissue injury in tomatoes was detected by using in-line MRI equipment (Milczarek et al., 2009).

4.5 CONCLUSION

For vegetables quality evaluation, the current review focused on some valuable applications of non-destructive methods. Non-destructive techniques are centre of attraction for its feasibility to predict external and internal quality of vegetables without any loss in structure. Moreover, these techniques provide constitutional variation of the vegetables/vegetable products and their accurate quantification, leading to better characterization and improved quality and safety evaluation results. Considering these advantages, it is expected that non-destructive technology will play more significant role in the field of vegetables/fruits industries in near future.

REFERENCES

Abbott, J. A. Quality Measurement of Fruits and Vegetables. *Postharvest Biol. Technol.* **1999**, *15*, 207–225.

Alander, J. T.; Bochko, V.; Martinkauppi, B.; Saranwong, S.; Mantere, T. A Review of Optical Nondestructive Visual and Near-Infrared Methods for Food Quality and Safety. *Int. J. Spectro.* **2013**, 36.

Al-Mallahi, A.; Kataokab, T.; Okamotob, H.; Shibata, Y. An Image Processing Algorithm for Detecting in-Line Potato Tubers Without Singulation. *Comput. Electr. Agric.* **2010**, *70*, 239–244.

Arefi, A.; Motlagh, A. M.; Mollazade, K.; Teimourlou, R. F. Recognition and Localization of Ripen Tomato Based on Machine Vision. *Austra. J. Crop. Sci.* **2011**, *5*, 1144–1149.

AVA Company. Applied Vibro Acoustic, Japan, 2014. http://www.ava.co.jp.

Baltazar, A.; Espina-Lucero, J.; Ramos-Torres, I.; Gonza'lez-Aguilar, G. Effect of Methyl Jasmonate on Properties of Intact Tomato Fruit Monitored with Destructive and Non-Destructive Tests. *J. Food Eng.* **2007**, *80*, 1086–1095.

Berns, R. S. *Billmeyer and Saltzman Principles of Color Technology*, 3rd ed.; John Wiley Sons: New York, 2000.

Bohigas, X.; Amigo, R.; Tejada, J. Characterization of Sugar Content in Yoghurt by Means of Microwave Spectroscopy. *Food Res. Int.* **2008**, *41*, 104–109.

Bureau, S.; Bertrand, D.; Jaillais, B.; Reling, P.; Gouble, B.; Renard, C. M. G. C.; Dekdouk, B.; Marsh, L. A.; O'Toole, M. D.; Armitage, D. W.; Peyton, A. J.; Alvarez-Garcia, J. FRUITGRADING: Development of a Fruit Sorting Technology Based on Internal Quality Parameters. NIR 2013—16th International Conference on Near Infrared Spectroscopy, la Grande-Motte, France, 2013, 145–148.

Bushberg, J.; Siebert, J.; Leidholdt, E.; Boone, J. *The Essential Physics of Medical Imaging*, 2nd ed.; Lippincott Williams & Wilkins: Philadelphia, 2002.

Butz, P.; Hofmann and, C.; Tauscher, B. Recent Developments in Non-invasive Techniques for Fresh Fruit and Vegetable Internal Quality Analysis. *J Food Sci.* **2005**, *70*, 131–141.

Buzby, J. C.; Hyman, J. Total and Per Capita Value of Food Loss in the United States. *Food Pol.* **2012**, *37*, 561–570.

Cen, H.; He, Y. Theory and Application of Near Infrared Reflectance Spectroscopy in Determination of Food Quality. *Trends Food Sci. Tech.* **2007**, *18*, 72–83.

Chenga, Y.-C.; Wang, T.-T.; Chen, J.-H.; Lin, T.-T. Spatial–Temporal Analyses of Lycopene and Sugar Contents in Tomatoes During Ripening Using Chemical Shift Imaging. *Postharvest Biol. Technol.* **2011**, *62*, 17–25.

Chieh, C. Water Chemistry and Biochemistry. In *Food Biochemistry and Food Processing*, 2nd ed.; Benjamin K. Simpson, Ed.; John Wiley & Sons, 2012, 84–108.

Cuibusa, L.; Castro-Giráldezb, M.; José Fitob, P.; Fabbri, A. Application of Infrared Thermography and Dielectric Spectroscopy for Controlling Freezing Process of Raw Potato Inside Food Symposium. 2013, 9–12.

Defraeye, T.; Lehmann, V.; Gross, D.; Holat, C.; Herremans, E.; Verboven, P.; Verlinden, B.; Nicolai, B. Application of MRI for Tissue Characterisation of 'Braeburn' Apple. *Postharvest Biol. Technol.* **2013**, *75*, 96–105.

Donis-González, I. R. Nondestructive Evaluation of Fresh Chestnut Internal Quality Using x-Ray Computed Tomography (Ct). PhD Thesis, Biosystems Engineering, Michigan State University, 2013.

Donis-González, I. R.; Guyer, D. E.; Pease, A.; Barthel, F. Internal Characteristics Visualization of Fresh Agricultural Products Using Traditional and Ultrafast Electron Beam x-ray Computed Tomography (CT) Imaging. Fifth International Chestnut Symposium: Sheperdstown, WV, USA, 2012.

Edan, Y.; Pasternak, H.; Shmulevich, I.; Rachmani, D.; Guedalia, D.; Grinberg, S.; Fallik, E. Color and Firmness Classification of Fresh Market Tomatoes. *J. Food Sci.* **1997**, *62*, 793–796.

Eissa, A. H.; Abdel Khalik, A. A. Understanding Color Image Processing by Machine Vision for Biological Materials. In *Structure and Function of Food Engineering*; Ayman A. Eissa, Ed.; InTech, 2012, 227–274.

Francois, I. M.; Wins, H.; Buysens, S.; Godts, C.; Van Pee, E.; Nicolai, B.; De Proft, M. Predicting Sensory Attributes of Different Chicory Hybrids Using Physico-Chemical Measurements and Visible/Near Infrared Spectroscopy. *Postharvest Biol. Technol.* **2008,** *49*, 366–373.

FSSAI. Food Safety and Standards Authority of India. Food Safety and Standards authority of India, Government of India. http://fssai.gov.in. (accessed 2 April 2012).

Haff, R. P. Real-Time Correction of Distortion in X-ray Images of Cylindrical or Spherical Objects and its Application to Agricultural Commodities. *Trans. ASABE.* **2008,** *51*, 341–349.

He, D.; Li, Z; Wang, H. On the Characteristics of Sound wave Forms of Watermelons. *Acta Universitatis Agriculturalis Boreali-Occidentalis* **1994,** *22*(3), 105–107.

He, Y.; Zhang, Y.; Xiang, L. G. Study of Application Model on BP Neural Network Optimized by Fuzzy Clustering. *Lect. Notes Comput. Sci.* **2005,** *3789*, 712–720.

Hertog, M. L.; Ben-Arie, R.; Roth, E.; Nicolai, B. M. Humidity and Temperature Effects on Invasive and Non-Invasive Firmness Measurements. *Postharvest Biol. Technol.* **2004,** *33*, 79–91.

Hodges, R. J.; Buzby, J. C.; Bennett, B. Postharvest Losses and Waste in Developed and Less Developed Countries: Opportunities to Improve Resource Use. *J. Agric. Sci.* **2011,** *149*, 37–45.

Huang H.; Liu, L.; Ngadi, M. O. Recent Developments in Hyperspectral Imaging for Assessment of Food Quality and Safety. *Sensors* **2014,** *14*, 7248–7276.

Ignat, T. Non-Destructive Methods for Determination of Quality Attributes of Bell Peppers. PhD Thesis, Department of Physics-Control, Faculty of Food Science, Corvinus University of Budapest, 2012.

Itoh, H.; Tomita, H.; Uno, Y.; Shiraishi, N. Development of Method for Non-destructive Measurement of Nitrate Concentration in Vegetable Leaves by Near-infrared Spectroscopy. International Federation of Automatic Control (IFAC), 2011, 1773–1778.

Itoh, H.; Kanda, S.; Matsuura, H.; Sakai, K.; Sasao, A. Measurement of Nitrate Concentration Distribution in Vegetables by Near-Infrared Hyperspectral Imaging. *Environ. Control Biol.* **2010,** *48*, 31–43.

Jha, S. N. Colour Measurements and Modeling. In *Non-destructive Evaluation of Food Quality: Theory and Practice*; Jha, S.N., Ed.; Springer-Verlag: Berlin Heidelberg, 2010; pp 17–40.

Jha, S. N.; Matsuoka, T. Non-destructive Techniques for Quality Evaluation of Intact Fruits and Vegetables. *Food Sci. Technol. Res.* **2000,** *6*, 248–251.

Jha, S. N.; Matsuoka, T. Development of Freshness Index of Eggplant. *App. Eng. Agr. ASAE* **2002,** *18*, 555–558.

Jha, S. N.; Matsuoka, T.; Miyauchi, K. Surface Gloss and Weight of Eggplant During Storage. *Biosys. Eng.* **2002,** *81,* 407–412.

Jha, S. N.; Matsuoka, T. Nondestructive Determination of Acid Brix Ratio of Tomato Juice Using Near Infrared Spectroscopy. *Int. J. Food Sci. Tech.* **2004,** *39*, 425–430.

Jha, S. N.; Narsaiah, K.; Sharma, A. D.; Singh, M.; Bansal, S.; Kumar, R. Quality Parameters of Mango and Potential of Non-destructive Techniques for Their Measurement—A Review. *J. Food Sci. Technol.* **2010,** *47*, 1–14.

Jivanuwong, S. Nondestructive Detection of Hollow Heart in Potatoes Using Ultrasonics. MSc. Thesis, Biological Systems Engineering, Faculty of Virginia Polytechnic Institute and State University, 1998.

Kader, A. A. Increasing Food Availability by Reducing Postharvest Losses of Fresh Produce. Proc. 5th Int. Postharvest Symp. *Acta Hort. ISHS* **2005,** *682,* 2169–2179.

Kanda, S.; Itoh, H.; Matsuura, H.; Tomoda, S.; Shiraishi, N.; Sakai, K.; Sasao, A. *Development of Hyperspectral Imaging System to Measure Spatial Distribution of Nitrate Concentration in Vegetables.* Proceedings of ISMAB2010 JAPAN, Fukuoka, CD-ROM, 2010.

Kitinoja, L.; Saran, S.; Royb, S. K; Kader, A. A. Postharvest Technology for Developing Countries: Challenges and Opportunities in Research, Outreach and Advocacy. *J. Sci. Food Agric.* **2011,** *91,* 597–603.

Koizumia, M.; Naito, S.; Haishi, T.; Utsuzawa, S.; Ishida, N.; Kano, H. Thawing of Frozen Vegetables Observed by a Small Dedicated MRI for Food Research. *Magn. Reson. Im.* **2006,** *24,* 1111–1119.

Kraszewski, A. *Microwave Aquametry—Electromagnetic Interaction with Water Containing Materials. Piscataway*; IEEE Press: NJ, 1996.

Krautkramer, J.; Krautkramer, H. *Ultrasonic Testing of Materials;* Springer-Verlag: Heidelberg, Germany, 1990.

Lammertyn, J.; Peirs, A.; De Baerdemaeker, J.; Nicolai, B. Light Penetration Properties of NIR Radiation in Fruit with Respect to Non-Destructive Quality Assessment. *Postharvest Biol. Technol.* **2000,** *18,* 121–132.

Liming, X.; Yanchao, Z. Automated Strawberry Grading System Based on Image Processing. *Comput. Electr. Agric.* **2010,** *71,* 32–39.

Løkke, M. M. Postharvest Quality Changes of Leafy Green Vegetables—Assessed by Respiration Rate, Sensory Analysis, Multispectral Imaging, and Chemometrics. PhD thesis, Department of Food Science, Aarhus University, 2012.

Lorente, D.; Aleixos, N.; Gómez-Sanchis, J.; Cubero, S.; García-Navarrete, O. L.; Blasco, J. Recent Advances and Applications of Hyperspectral Imaging for Fruit and Vegetable Quality Assessment. *Food Bioprocess Technol.* **2011,** *5,* 1121–1142.

Lü, F. *Non-Destructive Quality Evaluation of Watermelon Based on its Acoustic Property;* Zhejiang University: Hangzhou, China, 2003.

Mason, T. J.; Paniwnyk, L.; Lorimer, J. P. The Uses of Ultrasound in Food Technology. *Ultrason. Sonochem.* **1996,** *3,* 253–260.

Matsumoto, T.; Itoh, H.; Shirai, Y.; Shiraishi, N.; Uno, Y. *Non-destructive Measurement of Nitrate Concentration in Vegetables by Near Infrared Spectroscopy.* Proceedings of BioRobotics IV, Champaign, CD-ROM, 2009.

Mazhar, M.; Joyce, D.; Cowin, G.; Hofman, P.; Brereton, I.; Collins, R. MRI as a Non-invasive Research Tool for Internal Quality Assessment of 'Hass' Avocado Fruit. *Talking Aocados* **2013,** *23,* 22–25.

McKeown, M.; Trabelsi, S.; Tollner, E.; Nelson, S. Dielectric Spectroscopy Measurements for Moisture Prediction in Vidalia Onions. *J. Food Eng.* **2012,** *111,* 505–510.

Milczarek, R. R.; Saltveit, M. E.; Garvey, T. C.; McCarthy, M. J. Assessment of Tomato Pericarp Mechanical Damage Using Multivariate Analysis of Magnetic Resonance Images. *Postharvest Biol. Technol.* **2009,** *52,* 189–195.

Mizrach, A.; Galili, S.; Gan-mor, S.; Flitsanov, U.; Prigozin, I. Models of Ultrasonic Parameters to Assess Avocado Properties and Shelf Life. *J. Agric. Eng. Res.* **1996,** *65,* 261–267.

Mizrach, A.; Flitsanov, U.; Akerman, M.; Zauberman, G. Monitoring Avocado Softening in Low-Temperature Storage Using Ultrasonic Measurements. *Comput. Electro. Agric.* **2000,** *26*, 199–207.

Musse, M.; Quellec, S.; Cambert, M.; Devaux, M. F.; Lahaye, M.; Mariette, F. Monitoring the Postharvest Ripening of Tomato Fruit Using Quantitative MRI and NMR Relaxometry. *Postharvest Biol. Technol.* **2009,** *53*, 22–35.

Nelson, S. O. Frequency and Temperature-Dependent Permittivities of Fresh Fruits and Vegetables from 0.01 to 1.8 GHz. *Tran. ASAE.* **2003,** *46*, 567–574.

Nelson, S. O.; Trabelsi, S. Dielectric Spectroscopy Measurements on Fruit, Meat, and Grain. *Tran. ASABE,* **2008,** *51*, 1829–1834.

Nelson, S. O.; Trabelsi, S.; Kays, S. J. Dielectric Spectroscopy of Honeydew Melons from 10 MHz to 1.8 GHz for Quality Sensing. *Trans. ASABE.* **2006,** *49*, 1977–1981.

Nelson, S. O.; Guo, W. C.; Trabelsi, S.; Kays, S. J. Dielectric Spectroscopy of Watermelons for Quality Sensing. *Mea. Sci Technol.* **2007,** 1887.

Nicolai, B. M.; Beullens, K.; Bobelyn, E.; Peirs, A.; Saeys, W.; Theron, K.I.; Lammertyn, J. Nondestructive Measurement of Fruit and Vegetable Quality by Means of NIR Spectroscopy: A Review. *Postharvest Biol. Technol.* **2007,** *46*, 99–118.

Parmar, R. R.; Jain, K. R.; Modi, C. K. *Unified Approach in Food Quality Evaluation Using Machine Vision. Advances in Computing and Communications*; Springer: Berlin Heidelberg, 2011; pp 239–248.

Penchaiya, P.; Bobelyn, E.; Verlinden, B. E.; Nicolai, B. M.; Saeys, W. Non-Destructive Measurement of Firmness and Soluble Solids Content in Bell Pepper Using NIR Spectroscopy. *J. Food Eng.* **2009,** *94*, 267–273.

Penga, J.; Tanga, J.; Jiaoa, Y.; Bohnet, S. G.; Barret, D. M. Dielectric Properties of Tomatoes Assisting in the Development of Microwave Pasteurization and Sterilization Processes. *Food Sci. Technol.* **2013,** *54*, 367–376.

Polder G.; van der Heijdena, G. W. A. M.; van der Voeta, H.; Young, I. T. Measuring Surface Distribution of Carotenes and Chlorophyll in Ripening Tomatoes Using Imaging Spectrometry. *Postharvest Biol. Technol.* **2004,** *34*, 117–129.

Pratx, G.; Xing, L. GPU Computing in Medical Physics: A Review. *Med Physics.* **2011,** *38*, 2685.

Quenon, V.; De Baerdemaeker, J. Non-Destructive Method for Internal Quality Determination of Belgian Endive (*Cichorium intybus* L.). *Int. Agrophysics.* **2000,** *14*, 215–220.

Rao, X.; Ying, Y.; Feiling, L.; Jin, B. Development of a Fruit Quality Inspecting System Based on Acoustic Properties. *Tran. CSAM.* **2004,** *35*(2), 69–71.

Sagartzazu, X.; Hervella-Nieto, L.; Pagalday, J. M Review in Sound Absorbing Materials. *Arch. Comput. Method Eng.* **2008,** *15*, 311–342.

SaldañaI, E.; SicheII, R.; Luján, M.; Quevedo, R. Review: Computer Vision Applied to the Inspection and Quality Control of Fruits and Vegetables. *Braz J. Food Technol. Campinas.* **2013,** *16*, 254–272

Saltveit, M. E. Jr. Determining Tomato Fruit Maturity with Non-Destructive in vivo Nuclear Magnetic Resonance Imaging. *Postharvest Biol. Technol.* **1991,** *1*, 153–159.

Shao, Y.; He, Y. Nondestructive Measurement of Acidity of Strawberry Using Vis/NIR Spectroscopy. *Int. J. Food Prop.* **2008,** *11*, 102–111.

Shao, Y.; He, Y.; Gómez, A. H.; Pereir, A. G.; Qiu, Z.; Zhang, Y. Visible/Near Infrared Spectrometric Technique for Nondestructive Assessment of Tomato 'Heatwave' (*Lycopersicum esculentum*) Quality Characteristics. *J. Food Eng.* **2007**, *81*, 672–678.

Slaughter, D. C. Nondestructive Maturity Assessment Methods for Mango: A Review of Literature and Identification of Future Research Needs, pp 1–18, 2009. http://www.mango.org/media/55728/nondestructive_maturity_assessment_methods_for_mango.pdf.

Slaughter, D. C.; Barrett, D.; Boersig, M. Nondestructive Determination of Soluble Solids in Tomatoes Using Near Infrared Spectroscopy. *J. Food Sci.* **1996**, *61*, 695–697.

Skierucha, W.; Wilczek, A.; Szypowska, A. Dielectric Spectroscopy in Agrophysics. *Int. Agrophys.* **2012**, *26*, 187–197.

Sun, D. W. *Hyperspectral Imaging for Food Quality Analysis and Control;* Academic Press/Elsevier: San Diego, California, USA, 2009, 15 Chapters.

Sun, D. W. *Hyperspectral Imaging for Food Quality Analysis and Control.* Academic Press/Elsevier Science: London, 2010; p 496.

Sugiyama, J.; Otobe, K.; Hayashi, S.; Usui, S. Firmness Measurement of Muskmelons by Acoustic Impulse Transmission. *Tran. ASABE* **1994**, *37*, 1234–1241.

Teye, E.; Huang, X.; Afoakwa, N. Review on the Potential use of Near Infrared Spectroscopy (NIRS) for the Measurement of Chemical Residues in Food. *Am. J. Food Sci. Technol.* **2013**, *1*, 1–8.

Thygesen, L. G.; Løkke, M. M.; Micklander, E.; Engelsen, S. B. Vibrational Microspectroscopy of Food. Raman vs. FT-IR. *Trends Food Sci. Tech.* **2003**, *14*, 50–57.

Trnka, J.; Stoklasová, P.; Strnková, J.; Nedomová, Š.; Buchar, J. Vibration Properties of the Ostrich Eggshell at Impact. *ACTA Acta Univ. Agric. Silvic. Mendelianae Brun.* **2013**, *61*, 1873–1880. http://acta.mendelu.cz/61/6/1873/.

USDA (The U.S. Department of Agriculture). *Fresh Fruits and Vegetables Manual*, 2nd ed.; issued 2012. http://www.aphis.usda.gov/permits/.

Vahora, T.; Sinija, V. R.; Alagusundaram, K. Quality Evaluation of Fruits Using Acoustic Resonance Technique: A Review. *J. Food Sci. Technol.* **2013**, *2*, 2278–2249.

Venkatesh, M. S.; Raghavan, G. S. V. An Overview of Microwave Processing and Dielectric Properties of Agri-Food Materials. *Biosys. Eng.* **2004**, *88*, 1–18.

Wang, H.; Li, C.; Wang, M. Quantitative Determination of Onion Internal Quality Using Reflectance, Interactance, and Transmittance Modes of Hyperspectral Imaging. *Tran. ASABE.* **2013**, *56*, 1623–1635.

Wang, W.; Thai, C.; Li, C.; Gitaitis, R.; Tollner, E. W.; Yoon, S.-C. Detection of Sour Skin Diseases in Vidalia Sweet Onions Using Near-Infrared Hyperspectral Imaging. ASABE Annual International Meeting, 2009, Paper No: 096364.

Xue, L. H.; Yang, L.Z. Deriving Leaf Chlorophyll Content of Green-Leafy Vegetables from Hyperspectral Reflectance. *ISPRS J. Photogramm* **2009**, *64*, 97–106.

Zhanga, L.; McCarthy, M. J. Measurement and Evaluation of Tomato Maturity Using Magnetic Resonance Imaging. *Postharvest Biol. Technol.* **2012**, *67*, 37–43.

Zhang, R.; Ying, Y.; Rao, X.; Gao, Y.; Hu, D. Non-Destructive Determination of Soluble Solid Content for Tomato Using Hyperspectral Diffuse Transmittance Imaging. Transactions of the ASABE, Missouri: Kansas City, 2013, Paper number 131595381.

Zsom-Muha, V. Dynamic Methods for Characterization of Horticultural Products. PhD thesis, Corvinus University of Budapest, Department of Physics and Control, Budapest.

CHAPTER 5

BREEDING FOR IMPROVING NUTRITIONAL QUALITIES AND SHELF LIFE IN VEGETABLE CROPS

PRITAM KALIA AND SHRAWAN SINGH

Division of Vegetable Science, Indian Agricultural Research Institute, New Delhi 110012, India

CONTENTS

5.1   INTRODUCTION

Vegetables are rich sources of dietary minerals and vitamins which contribute in various body-building activities and maintenance of good health. They may be edible roots, stems, leaves, fruits or seeds. Each group contributes to diet in its own way. Vegetables-rich diet provides micro-nutrients and health-promoting phytochemicals that alleviate both under-nutrition and obesity. The beneficial health effects are mainly attributed to diverse antioxidant compounds such as vitamins, carotenoids, phenolics, alkaloids, nitrogen-containing compounds, organo-sulphur compounds secondary metabolism and so on. About 3 billion people in the world are malnourished due to imbalanced diets. Under-consumption of vegetables is among the top 10 risk factors leading to micronutrient malnutrition and is associated with the prevalence of chronic diseases. More than 70% of malnourished children live in Asia. At least half of the pre-school children and pregnant women are affected by micronutrient deficiencies. India is home to 194.6 million (out of 795 million worldwide) undernourished people, the highest in the world, according to the 'The State of Food Insecurity in the World'—SOFI 2015 report of the United Nation-Food and Agriculture Organization (UN-FAO), and India missed targets of both Millennium Development Goals (MDGs) and World Food Summit (1996). Arlappa et al. (2011) reported micronutrient deficiencies among rural children of West Bengal and highlighted the need to initiate sustainable long-term interventions for prevention and control of micronutrient deficiencies in children. Besides these, the country has another big challenge of micronutrient malnutrition which is more prevalent in rural and underdeveloped regions. Nearly half of the world's micronutrient-deficient population lives in India (USAID, 2005) which have an estimated economic cost of 0.8–2.4% of India's GDP (Stein and Qaim, 2007).

Deficiency of vitamin A and iron can result in severe anaemia, impaired intellectual development, blindness and even death. Alarming situation of vitamin A deficiency (VAD) is highlighted by the World Health Organization (WHO) estimates (1995–2005) which indicate that 45 (based on night blindness) and 122 (serum retinol concentration <0.70 μmol/l) countries have VAD in preschool-age children. The WHO estimate states that 250 million preschool-age children (>100 countries) and approx. 7.2 million pregnant women in developing countries are vitamin A deficient. The prevalence of global micronutrient deficiencies is depicted in Fig. 5.1. The developing countries have 75% of the world's blind children. Of these, between 250,000 and 500,000 lose their sight each year as a result, and more than half die within 12 months. Around 320,000 children (<16 years) are blind, and this constitutes 1/5th of the world's blind children in country. Cataract is a major cause of blindness (62.6%) and also prevalent

($>15.42\%$) in old age people (>60 years). The prevalence is 3/10,000 in children of affluent societies to 15/10,000 in the poorest communities. In rural India, the prevalence of Bitot's spot (VAD) in pre-school children (objective sign of clinical VAD is 0.8%, higher than the figures recommended by the WHO ($\geq 0.5\%$) (NNMB, 2003). The National Nutrition Monitoring Bureau (NNMB) reported blood VAD (<20 µg/dL) prevalence as 61% in NNMB states which ranges from 52% in Maharashtra to 88% in Madhya Pradesh.

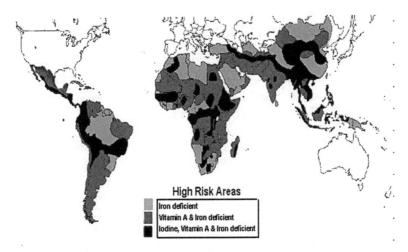

High Risk Areas
Iron deficient
Vitamin A & Iron deficient
Iodine, Vitamin A & Iron deficient

FIGURE 5.1   Adapted from global micronutrient deficiencies.
Source: Sanghvi (1996) modified by Welch (2001).

The World Bank estimates that India is one of the highest ranking countries in the world for the number of children suffering from malnutrition. The prevalence of underweight children in India is among the highest in the world and is nearly double that of sub-Saharan Africa with dire consequences for mobility, mortality, productivity and economic growth. The 2011 Global Hunger Index (GHI) Report ranked India 15th, amongst leading countries with hunger situation. It also places India amongst the three countries where the GHI between 1996 and 2011 went up from 22.9 to 23.7, whereas 78 out of the 81 developing countries studied, including Pakistan, Nepal, Bangladesh, Vietnam, Kenya, Nigeria, Myanmar, Uganda, Zimbabwe and Malawi, succeeded in improving hunger condition. Although India is the second largest producer of vegetables in the world, 20–30% of vegetable is wasted before it reaches the market due to a lack of on-farm primary processing facilities, refrigeration and poor road and marketing infrastructure.

Further, in industrialized countries and rich urban population in developing countries, there is an increasing realization that nutritious and anti-oxidant-rich foods such as fruits and vegetables can play an important role in assuring a healthful life style and alleviate problems related to 'diseases of overabundance' and diet-related chronic diseases, such as some types of obesity, heart disease and certain types of cancer (Steinmetz and Potter, 1996; Bliss, 1999). Hence, improvement of nutritional quality and shelf-life of vegetable crops will be a rewarding activity for plant breeders as we enter the twenty-first century. Although major long-term breeding objectives continue to be increasing yield to meet the food requirement of ever increasing population, to ensure health and nutritional security, it is imperative that nutraceuticals and edible colour-rich varieties with long post-harvest life are to be bred.

5.2   VEGETABLES IN FAO'S SUSTAINABLE DEVELOPMENT GOALS AND UN'S MILLENNIUM DEVELOPMENT GOALS

The contribution of vegetables is magnificent in health and livelihood of millions of people across the globe. Hence, they have special mentions in Sustainable Development Goals of FAO (FAO's SDGs) for their role as source of dietary micronutrients and income. Among them, end poverty (SDP1), Zero hunger (SDP2) and sustainable consumption and production (SDG12) are particularly aimed to reduce poverty and malnutrition including micronutrient deficiencies through sustainable means. It is fact that almost 80% of the world's extreme poor live in rural areas where most are dependent on agriculture. These areas are major hubs of under-nourished but still around 794.6 million people are undernourished and more than 2 billion people suffer from micronutrient malnutrition. Their major share (3/4th) lives in rural or tribal areas in developing countries and, paradoxically, many of them are smallholder farmers. Further, investigations indicate that vegetables are important components of a healthy diet, and their reduced consumption is linked to poor health and increased risk of non-communicable diseases. Annually, around 6.7 million deaths worldwide were attributed to inadequate intake of fruits and vegetables. Although government schemes in form of food supply, food fortification and supplementation and biofortification helped significant population to cross the 'line of undernourishment' but their programmes need additional activities to substantiate the on-going efforts. This may be development of nutrient-rich crop/varieties, increase their availability and accessibility of quality produce, devising proper cooking processes and deworming for bioavailability of dietary minerals.

Further, out of eight MDGs of United Nations, three have direct concern with malnutrition and hunger for them, and nutritious food including vegetables are important tool. These MDGs are eradicating extreme poverty and hunger, reduce child mortality and improve maternal health. Other MDG, that is combat HIV/AIDS, malaria and other diseases are also linked to vegetable sector as antioxidant-rich fruits and vegetables improve the immunity and natural defence system of human body. The common strategies to prevent and control malnutrition in international platforms are food supplementation with micronutrients food (short term), food fortification during cooking/processing steps (medium term) and food biofortification by genetic modifications in plants (long term). All these are in the line of the traditional approach of dietary diversification which remained in food culture for a long period. However, due to shift in food sector, it is essential to highlight the significance of nutrient-rich foods through education and communication, implementing the right to food (and nutrition) in right sprit and ensuring food security and consumption of safe and nutritious food and implementing public health measures for basic services, immunization, sanitation and water supply. These strategies to reduce malnutrition are complementary rather than mutually exclusive. A coherent and multi-sectoral approach including health, food security and agriculture is, therefore, of prime importance wherein the nutrient-rich vegetable varieties could serve much better role.

5.3 NUTRITIVE VALUE OF VEGETABLES

More than 20 mineral elements and more than 40 nutrients are necessary for human health which can be provided by a balanced diet. The currently established human essential nutrients are water, energy, amino acids (histidine, isoleucine, leucine, lysine, methionine, phenylalanine, threonine, tryptophan and valine), essential fatty acids (linoleic and α-linolenic acids), vitamins (ascorbic acid, vitamin A, vitamin D, vitamin E, vitamin

K, thiamine, riboflavin, niacin, vitamin B-6, pantothenic acid, folic acid, biotin and vitamin B-12), minerals (calcium, phosphorus, magnesium and iron), trace minerals (zinc, copper, manganese, iodine, selenium, molybdenum and chromium), electrolytes (sodium, potassium and chloride) and ultratrace minerals. The common deficiencies of micronutrients are anaemia (iron), osteoporosis (calcium), meningomyelocele (folic acid), xerophthalamia (vitamin A), goitre (iodine), beriberi (thiamine), pellagra (niacin), scurvy (vitamin C), ariboflavinosis (riboflavin) and rickets (vitamin D). The sources of dietary micronutrients are food, water and sunlight etc. Role of dietary minerals in human health is given in Table 5.1.

TABLE 5.1   Dietary Minerals and Their Role in Human Health.

Calcium	Bones and teeth health, blood clotting, muscle contraction and nerve transmission, reduce risk of osteoporosis
Chromium	Aids in glucose metabolism and regulates blood sugar
Cobalt	Promotes the formation of red blood cell (RBCs)
Copper	In formation of normal RBCs, connective tissues, acts as a catalyst to store and release iron to help form haemoglobin. Part in central nervous system function
Iodine	Needed by the thyroid hormone to support metabolism
Iron	In RBC formation and function, important for brain function
Magnesium	Activates over 100 enzymes and helps nerves and muscles function
Molybdenum	Contributes to normal growth and development
Phosphorous	Works with Ca for strong bones and teeth, utilization of nutrients
Potassium	Regulates heartbeat, maintains fluid balance and helps muscles contract
Selenium	Essential for antioxidant enzyme, normal growth and development
Sulphur	Needed for muscle protein and hair
Zinc	Digestive enzymes, metabolism, reproduction, wound healing

In foods, cereals contribute carbohydrate, pulses and animal meat source for protein and fruits and vegetables in micronutrients. In vegetarian diets, the role of fruits and vegetables becomes more significant for dietary micronutrients. However, breeding for all these dietary elements is difficult; hence, depending upon economics and prevalence of the incidence and available germ plasm and suitability of breeding strategies in the target crop, the selection of the nutrient(s) for improvement through breeding is done.

TABLE 5.2 Nutritive Value of Important Vegetables (in 100 g). (*Source:* Gopalan et al., 2007; Hazra and Som (1999).)

Vegetables crops	Moisture (g)	Protein (g)	Energy (kcal)	Calcium (mg)	Phosphorus (mg)	Iron (mg)	Carotene (mcg)	Vitamin C (mg)
Vegetable legumes								
Cowpea seeds	13.4	24.1	323	77	414	8.6	12	0
Garden pea	72.9	7.2	93	20	139	1.5	83	9
Broad beans	85.4	4.5	7.2	50	64	1.4	9	12
Cluster beans	81	3.2	10.8	130	57	1.08	198	49
Cowpea beans	85.3	3.5	48	72	59	2.5	564	14
Cowpea pods	91.4	1.7	26	50	28	0.61	132	24
Sword bean	87.2	2.7	44	60	40	2	24	12
Leafy vegetables								
Chenopod leaves	89.6	3.7	30	150	80	4.2	1740	35
Beet leaf	86.4	3.4	46	380	30	16.2	5862	70
Coriander leaves	86.3	3.3	44	184	71	1.42	6918	135
Curry leaves	63.8	6.1	108	830	57	0.93	7560	4
Drumstick leaves	75.9	6.7	92	440	70	0.85	6780	220
Fenugreek leaves	86.1	4.4	49	395	51	1.93	2340	52
Lettuce	93.4	2.1	21	50	28	2.4	990	10
Mustard leaves	89.8	4.0	34	155	26	16.3	2622	33
Spinach	92.1	2.0	26	73	21	1.14	5580	28

TABLE 5.2 (Continued)

Vegetables crops	Moisture (g)	Protein (g)	Energy (kcal)	Calcium (mg)	Phosphorus (mg)	Iron (mg)	Carotene (mcg)	Vitamin C (mg)
Cole vegetables								
Cauliflower	90.8	2.6	4	33	57	1.23	30	56
Cabbage	91.9	1.8	27	39	44	0.8	120	124
Brussels sprouts	85.5	4.7	52	43	82	1.8	126	72
Knol khol	92.7	1.1	21	20	35	1.54	21	85
Broccoli	89.9	3.3	37	80	79	0.8	3500	137
Roots, tubers and bulbs								
Beet root	87.7	1.7	43	18.3	55	1.19	0	10
Carrot	86	0.9	48	80	530	1.03	1890	3
Colocasia	73.1	3.0	97	40	140	0.42	24	0
Onion (big)	86.6	1.2	50	46.9	50	0.6	0	11
Potato	74.7	1.6	97	10	40	0.48	24	17
Radish white	94.4	0.7	17	35	22	0.4	3	15
Sweet potato	68.5	1.2	120	46	50	0.21	6	24
Turnip	91.6	0.5	29	30	40	0.4	0	43
Yam	69.9	1.4	111	35	20	1.19	78	0
Cucurbitaceous vegetables								
Cucumber	96.3	0.4	13	10	25	0.6	0	7
Bottle gourd	96.1	0.2	2.5	20	10	0.46	0	0
Pumpkin (ripe)	86.0	1.4	25	10	30	0.7	2180	2

TABLE 5.2 (*Continued*)

Vegetables crops	Moisture (g)	Protein (g)	Energy (kcal)	Calcium (mg)	Phosphorus (mg)	Iron (mg)	Carotene (mcg)	Vitamin C (mg)
Ash gourd	96.5	0.4	10	30	20	0.8	0	1
Bitter gourd	92.4	1.6	4.2	20	70	0.61	126	88
Ridge gourd	95.2	0.5	17	18	26	0.39	33	5
Snake gourd	94.6	0.5	18	26	20	1.51	96	0
Solanaceous vegetables								
Tomato (ripe)	94.0	1.2	20	48	26	0.4	302	27
Brinjal	92.7	1.4	4	18	47	0.38	74	12
Chilli	85.7	2.9	29	30	80	1.2	292	111
Sweet pepper	92.4	1.3	24	10	30	0.57	427	137
Other vegetables								
Drumstick	86.9	2.5	26	30	110	0.18	110	120
Okra	89.6	1.9	35	66	56	0.35	52	13

The nutritive values of important vegetables are given in Table 5.2. Leafy vegetable nutritive values such as minerals, carotene and vitamin C values and pigmented root vegetable nutritive values except white radish and turnip are given in Table 5.3. Most vegetables are naturally low in fat and calories, whereas none have cholesterol. The protein-rich vegetables (garlic, garden pea, *kakrol* and other vegetable legumes) are essential for body building, mainly tissue, muscles and blood. Potassium-rich vegetables (sweet potatoes, beans, tomato products, beet leaf and spinach) may help one to maintain healthy blood pressure and reduce development of kidney stone. The dietary fibre reduces blood cholesterol levels and lowers risk of heart disease. Fibre is important for proper bowel function. It helps one to reduce constipation and protect from bowel cancer, type 2 diabetes. Vegetables rich in dietary fiber are cowpea, hyacinth bean, drumstick, pointed gourd etc. Further, significant variation is also observed in different vegetables for dietary minerals (Table 5.4) and vitamin profiles (Table 5.5) which suggest for breeding efforts to enrich further to meet adequate share of recommended dietary allowance of dietary minerals of human. However, breeding efforts resulted into some of the cucurbits (pumpkin, cucumber) rich in carotene, radish in anthocyanin, bitter gourd in lycopene and amaranths in anthocyanin.

TABLE 5.3   Vegetables Rich in Dietary Nutrients.

Nutrients	Availability in vegetables
Carbohydrates	Potato, sweet potato, dry beans, yam, tapioca
Protein	Garden pea, lablab, French bean, cowpea, cluster bean, amaranth, broad bean
Vitamin-A	Carrot, beet leaf, amaranth, pumpkin, fenugreek, green pea, paprika
Vitamin-B	Garden pea, lablab
Vitamin-C	Tomato, sweet pepper, green chilli, cauliflower, knol khol, bitter gourd, amaranth, fenugreek, beet leaf, cabbage
Calcium	Fenugreek, kale, beet leaf, amaranth, broccoli, onion, beet root, lablab, cabbage
Potassium	Sweet potato, potato, bitter gourd, radish, lablab, onion, green leafy vegetables
Phosphorus	Garlic, pea, bitter gourd, broccoli, onion, leek, kale, amaranth
Iron	Bitter gourd, amaranth, fenugreek, poi, beet leaf, leek, spinach

TABLE 5.4 Range of Dietary Minerals in Different Vegetable Crops.

Minerals	Brinjal	Cabbage	Cauliflower	Carrot	Bitter gourd
Fe (mg/100 g)	0.33–0.95	0.25–0.43	0.26–0.49	0.33–1.46	0.27–1.17
Zn (mg/100 g)	0.15–0.35	0.14–0.29	0.18–0.34	0.11–0.26	0.13–0.52
Cu (mg/100 g)	0.08–0.20	0.02–0.05	0.03–0.39	0.05–0.34	0.09–4.83
Mn (mg/100 g)	0.13–0.29	0.13–0.22	0.13–0.20	0.11–0.29	0.12–0.23
K (mg/100 g)	227.54–500.32	177.24–267.83	225.61–341.21	168.4–516.6	285.0–546.9
Mg (mg/100 g)	22.23–45.1	16.80–59.14	12.50–18.31	7.12–25.77	30.43–50.70
P (mg/100 g)	28.59–67.40	35.33–46.95	38.70–62.58	19.82–59.16	35.11–56.90
Na (mg/100 g)	2.17–6.46	23.61–43.61	7.67–36.25	35.45–109.86	4.42–11.97
Ca (mg/100 g)	13.99–65.43	37.76–58.47	14.17–23.74	16.90–45.22	10.03–25.03

TABLE 5.5 Range of Vitamins Content in Different Vegetable Crops.

Vitamins	Brinjal	Cabbage	Cauliflower	Carrot	Bitter gourd
Vit B1 (mg/100 g)	0.02–0.05	0.01–0.03	0.01–0.02	0.03–0.08	0.13–0.23
Vit B2 (mg/100 g)	0.03–0.07	0.01–0.04	0.03–0.06	0.03–0.06	0.03–0.08
Vit B3 (mg/100 g)	0.30–0.52	0.24–0.33	0.30–0.55	0.33–0.92	0.20–0.39
Vit B7 (mg/100 g)	2.27–5.18	2.11–6.50	–	4.02–10.90	2.09–32.78
Vit C (mg/100 g)	0.480–6.4058	0.09–0.83	0.009–0.084	6.18–11.54	0.013–0.111
α-Tocopherols (mg/100 g)	0.03–0.08	0.03–0.05	0.01–0.02	0.02–0.15	0.07–0.83
Total carotenoids (mg/100 g)	15.5–605.6	57.07–327.30	20.30–448.91	836.86–11,614.07	53.84–1066.12
β-carotene (mg/100 g)	3.26–11.64	–	–	201.21–6598.20	4.05–54.10

5.4   NUTRIENT-RICH VEGETABLES FOR SMALL-SCALE PRODUCTION SYSTEM

The vegetables are perishable food items and their regular supply in adequate quantity remains a big challenge. Hence, measures to enhance production of vegetables 'in locale' through home gardens, kitchen gardens, container gardening, roof gardening and so on are quite attractive, not only for remote rural areas but also urbanites. Establishment of new home gardens or enriching the existing traditional home gardens with nutrient-rich locally adaptable crops and their superior varieties could serve much better in terms of nutritional security. But this requires identification of region-specific vegetable crops and breeding of varieties rich in dietary nutrients, prolonged harvest period, adaptable to growing situation and high acceptance among the ultimate beneficiaries. The breeding objectives could be rich in dietary nutrients and antioxidants, good in organoleptic scores, tolerance to partial shade, high portion of edible fruits/parts, low gestation period, tolerant to diseases and pest, responsive to organic sources and so on. Thus far, research efforts remained targeted towards development of technologies and varieties for commercial scale, but some varieties are quite fit for small-scale growing in gardens. But efforts are almost required to develop plant types which are ideal for home gardens or other micro-scale production systems.

5.5   BREEDING OBJECTIVES

Vegetable breeding has evolved from yield-oriented traits to stress tolerance and now heading towards futuristic traits such as dietary constituents, climate resilient and fitting to innovative growing situation. In the current chapter, breeding of varieties rich in dietary constituents is discussed, which can contribute in better health and as tool in fight against micronutrient malnutrition. In general, the vegetable breeding for nutrients is centred on high yield and rich in dietary micronutrients which are traditional traits, but there are specific futuristic traits are given in Table 5.1. Although the futuristic traits are affected by consumer preferences and changes with time scale, some of the traits are given in Table 5.6.

TABLE 5.6 Breeding Objectives for Nutrient-Rich Crop Varieties in Vegetable Crops.

Crops	Traditional traits	Futuristic nutritional traits
Carrot	High yield, uniform roots, rich in carotenoids, anthocyanin, lycopene, disease and pest resistance	Rich in β-carotene, lycopene and anthocyanin stable at higher temperatures, high proportion of bioavailable forms of micronutrients (Ca, Fe, Zn) and target carotenoids, mild fragrance
Cauliflower, broccoli	High yield, superior curd traits in different maturity groups	Rich in β-carotene, glucosinolate, anthocyanin, iron and zinc, their stability in different growing temperature regimes, better retention of bioactives during post-harvest life, mild flavour
Radish	High yield, uniform roots, mild to less pungent, rich in minerals and pigments	Rich in folic acid, minerals, anthocyanin, glucosinolates stable in different growing situations
Tomato	Rich in lycopene, total soluble solids (TSS), acidity	High lycopene stable at high temperature, TSS, strong antioxidant activity
Leafy vegetables	High yield, uniformity, rich in minerals and vitamins	Rich in minerals and vitamins, low in anti-nutrients, high bioavailability of micronutrients
Legume vegetables	High yield, protein, vitamins	High level of bioavailable form of proteins, low anti-nutrients (trypsin inhibitor)
Cucurbits	High yield, uniformity	Rich in functional constituents in fruits (cucurbitacins, momordicins, carotenoids minerals etc.), leaves (minerals, vitamin) and seeds (essential fatty acids)

5.6 BREEDING FOR QUALITY

Quality in vegetables is a complex character influenced by both genetic and environmental factors. Breeding for quality has been unsystematic and often empirical, but significant progress has been made in several vegetable crops. Commercial fortification of foods is an intervention familiar to all of us. Minerals and vitamins are added to a particular food vehicle during processing, well after the food has left the farm and before it is distributed through various marketing channels for consumer purchase and consumption. There are two inherent advantages; first and foremost, biofortification would be cost-effective. Once the plants are developed and being grown by farmers, there are no costs year in and year out of buying the fortificants and adding them to the food supply during processing. Second, biofortification would be sustainable.

Conventional breeding in conjunction with molecular biology has bright prospects of developing vegetable varieties high in nutraceuticals, edible colours and bioactive compounds suitable for fresh market as well as fusion food industry. Increasing the health functionality of vegetable crops through breeding and/or genetic modification should create products that deliver greater health benefits than current varieties. Nevertheless, research efforts strive to improve our understanding of crop manipulations for nutrition and in many cases substantial progress have been made by using conventional and molecular tools.

Nutraceuticals and edible colours are natural compounds and are always regulated by several biochemical pathways and controlled by genetic and environmental factors (Table 5.7). From early times, people knowingly or unknowingly selected several vegetable crops for their food purpose. These vegetables contain several nutritional and biochemical compounds (Table 5.8). Many wild or cultivated vegetables are rich in these beneficial compounds. The biochemical pathway and synthesis of the compound are controlled by one or many genes which are scattered in the available or unknown germ plasm of a particular vegetable crop. India is blessed to have diverse agro-climatic region starting from tropical to temperate which makes feasible to grow different kinds of vegetables. In addition, there is plenty of diversity in different vegetable crops which can be exploited for development of special trait varieties. Through conventional breeding, it is possible to develop new vegetable varieties or integrate the favourable genes for nutraceuticals and edible colour into cultivated varieties. Advance in molecular biology and recombinant technology have paved the way for enhancing the pace of special trait variety development using marker-assisted breeding and designing new vegetable crop plants following transgenic approach.

TABLE 5.7 Important Nutraceuticals/Phytochemicals in Plant and Human Health.

Nutraceutical/ phytochemicals	Role in plant system	Role in human health
Phenolics	Signalling molecules, pigments, flavour, defence	Anti-oxidative, anti-inflammatory, antimutagenic, anti-carcinogenic, reduce cardiovascular diseases
Carotenoids	Pigmentation (yellow, orange, red), attract pollinators	Anticancer, anti-cardiovascular, and age related eye diseases, β-carotene: antioxidant, eye health; lycopene: antioxidant, prostate cancer
Anthocyanin	Pigmentation (purple, red)	Antioxidant, anti-inflammatory and anti-carcinogenic activity, cardiovascular disease prevention, obesity control and diabetes alleviation properties
Omega-6-fatty acids (linoleic acid)	Plant defence and metabolic activities	Balanced ratio of omega-3-fatty acids (ex. from fish) and omega-6-fatty acids (ex. from vegetables) is essential for good health
Vitamins	Plant growth and development, quality	Antioxidants, anti-therogenic, anti-carcinogenic, immuno-modulator, prevents colon and breast cancers, some cardiovascular diseases, cataract, arthritis, certain neurological disorders
Organic acids	Precursor for various compounds, protective agents	Taste factor, human metabolism

TABLE 5.8 Vegetable Rich in Nutraceuticals/Bioactive Compounds. (*Source:* Kalia and Saha (2010).)

Vegetables	Nutraceuticals/bioactive compounds
Allium vegetables (garlic, onions, chives, leeks)	Allylsulphides
Cruciferous vegetables (broccoli, cauliflower, cabbage, Brussels sprouts, kale, turnips, kohlrabi)	Indoles/glucosinolates, Sulphaforaphane Isothiocyanates/ thiocyanates, Thiols
Solanaceous vegetables, (tomatoes, peppers)	Lycopene
Umbelliferous vegetables (carrots, celery, parsley, parsnips)	Carotenoids Phthalides Polyacetylenes
Compositae plants (artichoke)	Silymarin
Beans	Flavonoids (isoflavones)

TABLE 5.8   *(Continued)*

Vegetables	Nutraceuticals/bioactive compounds
Carrots, squash, broccoli, sweet potatoes, tomatoes, kale, collards, cantaloupe and pumpkin	Vitamin A (retinol)
Green peppers, broccoli, green leafy vegetables, cabbage and tomatoes	Vitamin C (ascorbic acid)
Green leafy vegetables	Vitamin E

5.7   VEGETABLES AS SOURCE OF FOOD PIGMENTS

Edible colours are natural pigments found in tissue of plants. These colours are the chemical compounds produced by several biochemical pathways which give colours to the food. The Market Research Report from Future Market Insights estimated that the global natural food colour market was to be worth US $ 1144.0 million in 2014 and expected to rise 1697.6 million by 2020 with a compound annual growth rate of 6.8%. The share of natural food colours in global food colour market was estimated around 54.9% in 2014 which is expected to be around 60% by 2020 (Anon., 2015).

The colours change according the growth stage of the plant parts or vegetable product. These include anthocyanins, betalains, carotenoids, chlorophyll and so on. These pigments play important ecological and metabolic functions in the plants (Grotewold, 2006) and are more frequently exploited as the source of nutraceuticals to address a number of human ailments. These pigments have been implicated in regimes to maintain human health, to protect against chronic disease(s), or to restore wellness by repairing tissues after disease has been established. A wide range of bioassays and tests have been forwarded to establish the biological efficacy of natural pigments in human health intervention, including in vitro bioassays, in vivo epidemiological and clinical trials (Lila, 2004).

5.7.1   ANTHOCYANINS

These are natural pigments belonging to the flavonoid family. They are responsible for the blue, purple, red and orange colour of many fruits and vegetables. Anthocyanins are capable of acting on different cells involved in the development of atherosclerosis, one of the leading causes to cardio-vascular dysfunction. Anthocyanins and the aglycone cyanidin were found to inhibit cycloxygenase enzymes, which can be one marker for the initiation stage of carcinogenesis. Both the anthocyanins and cyanidin aglycone from tart cherries have been reported to reduce cell growth of human colon cancer cell lines (Kang et al., 2003). On one hand, they can interfere with glucose absorption and on the other hand, they may have a protective effect on pancreatic cells. The most extensively documented phytomedicinal role of anthocyanin pigments is in improving eyesight, including night vision (Ghosh and Konish, 2007). Anthocyanins exert significant antimicrobial properties, and (in association with other flavonoids) have demonstrated quite effective inhibition of aflatoxin biosynthesis (Norton, 1999).

5.7.2   BETALAINS

Betalains have been widely used as natural colourants for many centuries, but their attractiveness for use as colourants of foods (or drugs and cosmetics) has increased recently due to their reportedly high anti-oxidative, free radical scavenging activities and concerns about the use of various synthetic alternatives.

5.7.3   CAROTENOIDS

Carotenoids are the second most abundant pigments in nature and consist of more than 700 members (Britton, 1998). Carotenoids play an important role in plant reproduction, through their role in attracting pollinators and in seed dispersal, and are essential components of human diets. Carotenoids provide protection to vision and eye function, and against macular degeneration and cataracts. Carotenoids are credited with biological promotion of immune system response. Carotenoids are associated with inhibition of several types of cancers including cervical, oesophageal, pancreatic, lung, prostate, colorectal and stomach.

5.7.4   CHLOROPHYLLS

Chlorophyll is the most important plant pigment and a 'real-life force' that nature uses to explode plants into greenery. Chlorophylls, in contrast, are typically consumed in much higher doses in a diet that incorporates green and leafy vegetables (Delgado-Vargas and Paredes-Lopez, 2003). The anti-mutagenic properties of chlorophylls have been demonstrated in various assays, and clearly, intake of chlorophyll has potential to act as a chemopreventive compound in humans. The colour-rich vegetables are given in Table 5.9.

TABLE 5.9   Edible Colour-Rich Vegetables. (*Source:* Kalia and Saha (2010).)

Colour	Pigments	Vegetables
Red	Lycopene	Tomatoes and watermelon
	Betacyanins	Beet root
Orange	Beta-carotene	Carrots, cantaloupe, pumpkin, sweet potatoes
Blue/purple	Anthocyanins	Eggplant
Yellow	Lutein	Yellow corn
Green	Chlorophyll	Broccoli, kale, spinach, cabbage and asparagus
	Glucosinolates	Broccoli, kale, cabbage and cauliflower
Black	Anthocyanins	Black carrot
Orange–yellow	Flavonoids	Honeydew melon
Yellow–green	Lutein and zeaxanthin	Spinach, baby corn, melon
White-tan	Anthoxanthins	Cauliflower, parsnip, potato, ginger, onions, garlic

5.7.5 PIGMENT RICH VARIETIES

Pigments present in vegetable varieties (Table 5.10) can be extracted and used as natural dyes, edible food colour and also for functional food development.

TABLE 5.10 Edible Colour-Rich Varieties of Vegetable Crops.

Crop	Variety	Predominant edible colour	Associated compound	Salient feature
Carrot	Pusa Rudhira	Red	Lycopene	Lycopene,
	Pusa Meghali, PusaYamdagini, Pusa Nayanjyoti	Orange	β-carotene	β-carotene
	Pusa Asita	Black	Anthocyanin	Anthocyanin
Chenopodium	Pusa Bathua-1	Green	Chlorophyll	Vit A, Vit. C, Fe and Ca
Beet leaf	Pusa Jyoti, Pusa Bharti	Green	Chlorophyll	Vit A, Vit C, Fe, Ca
Tomato	Pusa Uphar, Pusa Hybrid-2, Pusa Rohini	Red	Lycopene	Vit C, lycopene
Vegetable mustard	Pusa Sag-1	Green	Chlorophyll	Vit C, carotene
Amaranths	Pusa Kiran, Pusa Kirti	Green	Chlorophyll	Vit A, Vit C, Ca, Fe
	Pusa Lal Chaulai	Purple	Anthocyanin	Betalains
Bitter gourd	Pusa Hybrid-1, 2	Green	Chlorophyll	Vit A, Vit C, Fe, Ca
	Pusa Rasdar			
Broad bean	Pusa Sumit	Green	Chlorophyll	Protein
French bean	Pusa Parvati	Green	Chlorophyll	Protein
Pumpkin	Pusa Hybrid-1, Pusa Vikas	Orange	β-carotene	Vit A
Radish	Pusa Mridula	Red	Lycopene	Lycopene , Ca, Fe, Mg
	Pusa Jamuni	Purple	Anthocyanin	Anthocyanin, Vit C
	Pusa Gulabi	Purple	Anthocyanin	Anthocyanin, Vit C
Paprika	KTPL-19	Red	Capsanthin	Capsanthin
Beet root	Detroit Dark Red	Purple	Anthocyanin	Betalain, Vit. B & C; beta-carotene, folic acid
Cabbage	Red Cabbage	Purple	Anthocyanin	Anthocyanin, Vit. C, glucosinolates
Broccoli	Palam Vichitra	Purple	Anthocyanin	Anthocyanin, Vit. C, glucosinolates

5.7.6   ANTI-NUTRIENT BOTTLENECKS

Anti-nutrients are important limiting factors in foods which interfere with bioavailability of micronutrients and dietary enzymes from the foods in the human body. These are more common in vegetables as given in Table 5.11. The breeding for low anti-nutrients is difficult, mainly due to cumbersome biochemical estimations and limited availability of phenotypic or DNA markers. But systematic efforts showed success in some food crops such as soybean (Schmidt et al., 2015), rice (Lang et al., 2007) and rhubarb (Libert, 1987). Although many cooking methods such as fermentation, boiling and malting increase the nutritive quality of plant foods through reducing certain anti-nutrients such as phytic acid, polyphenols and oxalic acid, breeding of varieties low in anti-nutrients is the more appropriate way to tackle the problem. This is because of the fact that the energy and substrates diverted to synthesize these compounds could be used in some other physiological process in plant system. Currently, the research is also heading towards use of genetic engineering techniques such as RNA interference (RNAi) approach to eliminate anti-nutrients entirely.

5.8   BREEDING PRINCIPLES AND METHODS

The pre-requisites for breeding of nutrient-rich varieties are (i) breeding must be successful—high nutrient density must be combined with high yields and high profitability; (ii) efficacy must be demonstrated—the micronutrient status of human subjects must be shown to improve when consuming the biofortified varieties as normally eaten; (iii) sufficient nutrients must be retained in processing and cooking and these nutrients must be sufficiently bioavailable; and (iv) the enriched/biofortified varieties must be adopted by farmers and consumed by those suffering from micronutrient malnutrition in significant numbers. Hence, the ultimate impact of any nutrient-rich breeding programme is noticeable when it sustains all these perquisites.

For breeding nutrient-rich varieties in a crop, the exploration of available genetic resources is required for (i) parental genotypes that can be used in crosses, genetic studies, molecular marker development and parent building and (ii) existing varieties, pre-varieties in the release pipeline or finished germ plasm products for 'fast tracking'.

TABLE 5.11   Anti-Nutritional Factors in Vegetables Crops.

Antinutrient	Effect	Dietary source
Phytic acid	Binds minerals K, Mg, Ca, Fe, Zn	Legume vegetables
Trypsin inhibitor	Reduces the activity of the enzyme trypsin and other closely related enzymes that help digest protein	Legume vegetables
Haemagglutinin, for example Lectin	Interfere with cells lining the gastrointestinal tract causing acute symptoms, can bind metals and some vitamins; can be toxic	Legume vegetables
Polyphenolics, tanins	Form complexes with iron, zinc, copper that reduces mineral absorption	Beans
Cyanogens or glycoalkaloids	Inhibit acetyl cholinestrase activity which impair nerve transmission, can damage cell membranes	Cassava, peas, beans
Oxalic acid	Binds calcium to prevent its absorption	Spinach leaf, amaranth, rhubarb, portulaca, colocasia, elephant foot yam
Solanine	Can be toxic, affect gastrointestinal and nervous system	Green parts of potato tubers
Saponins	May irritate the gastrointestinal tract and interfere with nutrient absorption	Soybeans, peas, sugar beets
Goitrogens	Suppress thyroids function	Brassica, allium foods
Cadmium, mercury, lead	May have toxic effects, for example high levels of Hg impair brain development	Contaminated leafy vegetables
Glycosides	Liberate toxic hydrocynic acid with enzymatic action	Tapioca leaves
Dioscorine	Toxic alkaloid	Yam

The pre-breeding and product enhancement is done through combining one of more micronutrients in germ plasm using conventional and molecular marker-based methods. Further, transgressive segregation or heterosis and transgenic approach are also useful approaches for enriching dietary nutrients in vegetable crops. Now-a-days, the use of molecular markers for identification and transfer of quantitative trait loci (QTLs) for nutrients and their use in enrichment programme is not uncommon in vegetable crops. It is to mention that the breeding of plants having lower quality of anti-nutritional factors is an old idea. One can imagine early gatherers and farmers avoiding eating certain plants that made them ill. As research in the nutritional sciences continues to demonstrate the importance to health and well-being of plant-derived nutrients in foods, more breeding is underway to increase the type and amount of nutrients in cereals, legumes, vegetables and fruits. The human diet consists of a diverse array of crops. As specific nutrient levels in many of these crops are often under genetic control, the prospects for improving the nutritional quality of food through breeding are enormous. Breeding method in any crops depends upon the breeding system and genetic architecture resulting from natural selection as well as human selection during the course of cultivation. The genetic architecture or the pattern of inheritance of characters is another important consideration while determining the most appropriate breeding procedure applicable to any particular crop. The choice of breeding method would be largely guided by the nature of gene action and relative magnitude of additive genetic variance, dominance variance and epistasis in a breeding population. The efficient breeding procedure should be effective in manipulation and selection of favourable gene combination, additive genetic variance, exploitation of dominance variance and achieving close relationship between expected genetic gain and realized progress from selection. A line of activities for breeding nutrient-rich varieties is presented in Figure 5.2. However, it is important to the criteria for identifying selectable trait and estimate potential genetic gain in biofortification, or nutrient-enrichment programme are knowledge of existing genetic variation, trait heritability, gene action, association among the traits, available screening techniques and diagnostic tools.

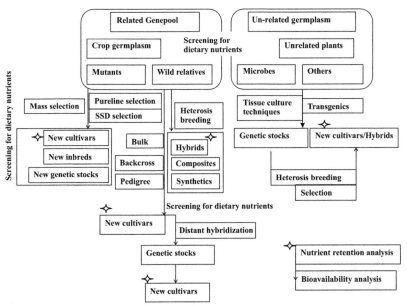

FIGURE 5.2 Breeding programme for nutrient-rich varieties in vegetable crops.

5.8.1 SELF-POLLINATED VEGETABLE CROPS

5.8.1.1 PURE-LINE SELECTION

Pure line is the progeny of single self-fertilized homozygous individuals. In this process, a large number of plants are selected on the basis of phenotypic performance, and their individual progenies are evaluated till homozygosity is attained. The best progeny is developed as a pure-line variety. This procedure may be applied to improve any local variety or exotic collections. In case there is variation with in plants of a progeny, single-plant selection is followed till complete homozygosity is attained.

5.8.1.2 HYBRIDIZATION

It is one of the techniques used to generate variability in segregating progenies, advancing population and selection of desired segregants by using procedures such as pedigree, bulk, single seed descent, backcross and their combinations. In hybridization programme, parents are selected on the basis of phenotypic performance, morphological characteristics and general combining ability (GCA) of the varieties. In general, parents exhibiting high GCA, in which there is involvement of additive genetic variance, are selected in self-pollinated crops. The number of plants selected may be two, three, four or more depending upon the situation. After selecting parents, single cross, three-way cross or double crosses are attempted. Sometimes, F_1s are backcrossed to their respective parents for traits of interest.

5.8.1.3   PEDIGREE METHOD

The current method has been adopted to develop improved germplasm as well a variety for different nutritional quality. In several vegetable crops, improvement has been accomplished through pedigree method and as a result, a large number of varieties have been developed. In this method, single-plant selection is followed up to F_5 or F_6 generations and in advance generation, families are selected on the basis of their phenotypic performance. Pedigree record of selected plants and families are maintained. The F_2 population, which is highly heterozygous, is grown in rows, and single-plant selection is made. Larger the population, there are more chances of getting superior segregants. With the advancement of generation, within-family variation is reduced to increase in homozygosity. The ratio of populations grown and plants selected in F_3 generation may vary from 20:1 to 100:1. It is essential to grow checks after every fifth or sixth row. In F_3 and F_4 generation, there is still within-family variation and therefore, emphasis is laid on within-family selection. As visual selection is applied, the intensity of selection should be considerably high and heritability of characteristics should be taken into account. In this method, selection is more effective for nutraceuticals and edible colours having high heritability.

5.8.1.4   BULK METHOD

This is one of the most economical methods of handling segregating population based on natural selection. In the natural selection, poorly adapted types are eliminated and superior types exist. This method consists of growing large population in each generation and harvesting the seeds in bulk and planting a sample of seeds in the next generations. No selection is made up to F_5 or F_6 generation. In this case, only natural selection operates. In general, there is no selection of plants on the basis of phenotypic performance, but sometimes, superior plants are selected and their seeds are composited. This type of mass selection will help the chance of getting superior plants and their composition of seeds up to F_3, and after that pedigree method is very effective. In F_6, single plants are selected and their progenies are grown in the next generations; superior families are selected on the basis of phenotypic performance, which are further evaluated against standard checks.

5.8.1.5 SINGLE-SEED DESCENT METHOD

It is based on the assumption of additive genetic variance, presumably a rule in self-fertilizing crops. In this method, all the plants in F_2 generation are advanced to the next generations by harvesting and growing on seed from each plant. This process is continued till F_5/F_6 in which superior individual plants are selected and tested in multi-location trials.

5.8.1.6 BACKCROSS METHOD

The current method is applied to transfer the characters controlled by single gene or few genes. It is more convenient to execute this method in self-fertilizing crop. The backcross method is based on the principle of quick and efficient transfer of a gene through repeated backcross and then selection in F_2 and subsequent generation. With continuous backcrossing with the recurrent parent, the backcross population becomes more similar to recurrent parent with additional trait of donor parent.

5.8.2 CROSS-POLLINATED VEGETABLE CROPS

The principle methods for cross-pollinated vegetable crops are introduction, selection (mass selection, line breeding, family breeding, recurrent selection), hybridization and heterosis breeding. The base materials for improvement may be open-pollinated cultivars coming from introduction, population derived through hybridization, synthetics or genetically broad based stocks. The selection methods commonly applied in cross-pollinated vegetable crops are always based on a population concept. Selfing is the most efficient mean of increasing homozygosity. The vegetable crops sensitive to inbreeding depression in descending order are carrot, onion, cabbage and cauliflower (in this, it varies with varieties depending upon proportion of self-compatible and self-incompatible plants occurring in the variety). Therefore, the techniques such as pure-line breeding is never undertaken to develop a variety except as a means of developing inbred lines for producing hybrids in strictly allogamous crops such as cole crops, cucurbits, bulb crops, root crops, leafy vegetables and sweet corn. The selection methods followed are mass selection, recurrent selection and disruptive selection.

5.8.2.1   INTRODUCTION

Introduced materials may be valuable in selecting desirable plants. In the case of non-uniform introductions, the desirable plants can be selected, their progeny increased, purified and later tested against the standard or local varieties for selecting the most promising lines. It is equally important to collect and domesticate the wild relatives of the cultivated species having desirable genes for high nutraceuticals and edible colours. The chances of an introduction depends to a great extent upon the relationship between the agro-climates, particularly temperature and day length, of the donor and receptor areas.

5.8.2.2   MASS SELECTION

In this method, the best individual plants are selected in the population, and their seeds are composited for raising following generation. As the mass selection is made exclusively on the basis of the phenotype of the plants without any progeny testing, the success of selection depends upon the heritability of the characters under selection.

5.8.2.3   LINE BREEDING

In this, superior lines are selected on the basis of progeny test and their seeds composited. Progeny for testing can be developed by various methods such as open pollination, selfing, pair cross, diallel cross or poly cross.

5.8.2.4   FAMILY BREEDING

There is elaborate testing of progenies not only in F_1 generation but also in later 2 or 3 generations and usually more than one cycle of selection. This method is practised in beet and can also be adopted in radish, carrot and cauliflower.

5.8.2.5   RECURRENT SELECTION

The source population for recurrent selection is heterozygous. It may be an open-pollinated variety, single cross or double cross F_1 hybrid, inter-crossed progenies of selected inbreed lines, a synthetic or a composite variety. There are four types of recurrent selection as follows.

In a simple recurrent selection method, the best individual plants are selected and their self-progenies are intercrossed, and further selection are made in the intercrossed progenies. The self-progenies of the selected individuals are again intercrossed next year to complete first cycle of selection. As no test cross is made in this method, it is effective in the improvement of highly heritable characters which can be assessed by simple test or on phenotypic evaluation.

Recurrent selection for GCA and specific combining ability (SCA) is almost similar except for the tester stock used for test cross. The heterozygous-based tester is used in the case of GCA, and narrow-base tester is used for SCA. On the basis of the result of the test crosses of GCA and SCA, the procuring families can be selected for intercrossing and further selection in intercross composite population.

Reciprocal recurrent selection method is useful in simultaneous selection for both GCA and SCA in two heterozygous and preferably genetically unrelated source populations.

5.8.2.6   HETEROSIS BREEDING

The hybrid varieties are developed by exploiting the dominance variance in heterosis breeding. Hybrids are available for commercial cultivation in all the cross-pollinated crops such as cole crops, cucurbits, root crops and bulb crops. Development of F_1 hybrid is very suitable for enhancing nutraceuticals and edible colours. The beta-carotene content in muskmelon has increased manifold in F_1 hybrid (Moon et al., 2002).

5.8.3   ADVANCED TECHNIQUES IN BREEDING

5.8.3.1   MUTATION BREEDING

Out of 3000 mutant varieties developed globally, 776 have been induced for nutritional quality (Jain and Suprasanna, 2007). In a simple way, mutation is a random or directed change in the structure of DNA or the chromosome which often result in a visible or detectable chance in specific or trait. Relatively minor genomic changes (point mutations, single gene insertions) are regularly observed following metabolomic analysis, leading to significant changes in biochemical composition and antioxidants (anthocyanin, lycopene) in the tomato cv. Money maker (Giliberto et al., 2005). In self-pollinated crops, it is well known, whereas in cross-pollinated crops its application is more difficult and identification of the origin of the desirable genotypes is difficult. Both physical (X-ray, UV) and chemical (ethyl ethane sulphonate, ethyl methane sulphonate, 5-bromouracil) mutagens cause macro-mutation (major, mono or oligogenic qualitative mutation) and/or micro-mutation (minor, polygenic or quantitative changes) in the genotypes under treatment. Sapir et al. (2008) reported in tomato that *high pigment-1* (*hp-1*) mutation known to increase flavonoids content in fruits. Spontaneous mutation such as 'orange' cauliflower (Li et al., 2001) and in sweet potato orange mutants are rich in beta-carotene (30–100 ppm) than white fleshed (2 ppm) and some of the common orange fleshed mutant varieties are Nancy Gold and Murff Bush Porto Rico (LaBonte and Don, 2012).

5.8.3.2   POLYPLOIDY BREEDING

Polyploid can be induced because of aberration in cell division. This may occur both in the mitosis as well as in meiosis. This method can be used successfully in vegetable breeding as a means of enhancing nutraceuticals and colours in vegetables. Tetraploids in radish, pumpkin, muskmelon and watermelon are highly productive and have improved quality. Zhang et al. (2010) developed tetraploid muskmelon which is rich in soluble solid, soluble sugar and vitamin C contents and distinctly higher than those in the diploid fruit. The line of activities is given in Figure 5.3. Liu et al. (2010) reported that the range of lycopene content among diploid water-melon fruit was from 33.2 to 54.8 mg/kg, and in triploid it was ranged from 41.2 to 61.8 mg/kg. The range of lycopene content among tetraploid was from 38.1 to 59.8 mg/kg. They also reported that the lycopene content in the triploid and tetraploid was higher than that of diploid. However, ploidy did not affect lycopene content in variety Fan Zu No. 2. Marzougui et al. (2009) induced polyploidy in *Trigonella foenum-graecum* L. using a 0.5% colchicine solution and reported that the autotetraploid cultivar has larger leaf area and larger productivity compared with the diploids concerning seed number, pod number and branch number. Its leaves are richer in potassium, sodium, calcium and phosphorus.

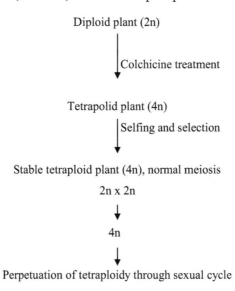

FIGURE 5.3   Process for development of tetraploids.

5.8.3.3   *BIOTECHNOLOGICAL APPROACHES*

5.8.3.3.1   **Marker-Assisted Selection**

Molecular markers such as random amplified polymorphic DNA, inter-simple sequence repeats (ISSR), microsatellites or SSR, sequence-characterized amplified regions (SCAR), cleaved amplified polymorphic sequences, sequence tag sites, expressed sequence tags, single nucleotide polymorphisms and diversity arrays technology are used to study linkage with gene responsible for high nutraceuticals and edible colours using mapping population. In marker-assisted selection, a marker (morphological, biochemical or one based on DNA/RNA variation) is used for indirect selection of a trait of interest. The mapping populations such as near isogenic lines (NILs) and recombinant inbreed lines (RILs) are used to identify the molecular marker linked to genes.

5.8.3.3.2   Development of Mapping Population

The mapping populations are usually created from F_1 lines that are derived from two parents that show differing phenotypes for a target trait. The commonly used mapping populations are F_2 population, F_3 population, NILs, RILs and doubled haploids (DH). The F_2 mapping population is easy to develop by selfing F_1 cross from two contrasting parents of the target trait(s). The NILs are developed by crossing the source line (non-recurrent parent) with the recipient line (recurrent parent). The progenies are again crossed with the recurrent parent followed by phenotypic selection. The positive lines are forwarded for backcrossing. Generally, six to eight backcrosses are made to recover recurrent parent genome. For development of RILs, the parents are crossed to produce F_1 population. The F_1 is selfed to produce F_2 population, and the F_2 population is selfed or intermitted among different F_2s and subsequent generations. This is followed till F_7/F_8 generation to produce RILs. In RILs, generally the population size of 50–250 is better for high-resolution mapping. The DH population is developed by use of microspore culture of F_1 individuals. Haploid plantlets produced from embryos usually undergo spontaneous doubling producing DH lines. This can be achieved using colchicine; colchicine binds to tubulin and prevents microtubule polymerization, acting as a mitotic poison, pretty potent secondary metabolite for its source plant, *Autumn crocus*. DH lines are advantageous as they are homozygous or fixed lines and as such can be replicated in trials allowing better estimates of within and between line variations. Moreover, they can be produced in a relatively short period (1–2 years). Besides these, advanced intercross lines and advanced backcross lines or backcross inbred lines are also useful in understanding the genetics of target traits. The TILLING (T_{arget} I_{nduced} L_{ocal} L_{esions} I_n G_{enomes}) populations are also common as an exploratory tool for trait improvement. However, these are not strictly a mapping population. TILLING populations are mutagenized populations that allow for the identification of point mutations within genes of interest following a screening procedure.

5.8.3.3.3   Nutrient Enrichment Using Markers

Zhang et al. (2008) found SCAR markers linked to *or* gene inducing beta-carotene accumulation in Chinese cabbage. Ripley and Roslinsky (2005) identified an ISSR Marker for 2-propenyl glucosinolate content in *Brassica*. The eye-appealing orange cauliflower was first discovered in Bradford Marsh, Ontario, Canada in 1970. The orange cauliflower results from a spontaneous mutation of a single dominant gene designated as *Or* for orange gene (Dickson et al., 1988). This *Or* mutant was originally found in white curded autumn crop cv. Extra Early snowball. This trait is absent in Indian cauliflower in which large population is suffering from carotene deficiency. Hence, we at Indian Agricultural Research Institute designed and led the biofortification of Indian cauliflower with β-carotene enhancing native *Or* gene following marker-assisted breeding (Fig. 5.4).

5.8.3.3.4   Anti-nutrient Elimination Using Markers

Peters et al. (2011) used RNAi approach to reduce pathogenesis-related protein-10 (PR10), allergenic potency in carrot. They silenced genes in transgenic carrots strongly decreased PR10 protein compared with untransformed controls. The decrease of the allergenic potential in Dau c 1-silenced plants was sufficient to cause a reduced allergenic reactivity in patients with carrot allergy, as determined with skin prick tests. Although simultaneous silencing of multiple allergens will be required to design hypoallergenic carrots for the market, but the study demonstrates the feasibility of creating low-allergenic food by using RNAi.

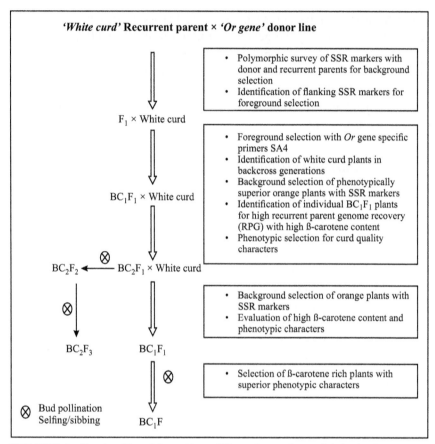

FIGURE 5.4   Carotenoid enrichment in cauliflower by marker-assisted selection sources.

The marker-assisted breeding is also being used to reduce anti-nutritional factors in vegetable crops as reported in faba bean (*Vicia faba*). Faba bean contains condensed tannins that reduce the value of the inherently high-protein levels of the crop. Gutierrez et al. (2007) reported that these tannins can be removed by the activity of two genes, zt-1 and zt-2, which are pleiotropic for white-flowered plants. A SCAR marker linked to the zt-2 gene was identified, one of the genes controlling absence of tannins, that is associated with increased protein levels and reduced fibre content of faba bean seeds, which would facilitate the development of tannin-free faba cultivars.

In plants, most of the phosphorus is in the form of phytic acid (inositol hexaphosphate), which cannot be digested by non-ruminants such as humans, pigs, fish and chickens. In addition, it affects bioavailability of minerals and enzymes. However, reducing phytic acid has resulted in a concomitant reduction in seed and plant performance; it compromises germination, emergence, stress tolerance and yield, but several interesting biotechnological approaches have been suggested to remedy this problem, including embryo-specific silencing of an ATP binding cassette transporter responsible for phytic acid accumulation (Shi et al., 2007) and the engineering of high-phytase seeds (Raboy, 2009). Biotechnological solutions might ultimately help one to avoid phytic acid accumulation in seeds, but scientists have already demonstrated that mutagenesis was sufficient to develop a low-phytic acid mutant with yield performance comparable with its wild counterpart (Campion et al., 2009).

5.8.3.3.5 Quantitative Trait Loci Analysis

The QTL analysis is the study of the alleles that occur in a locus and the phenotypes (physical forms or traits) that they produce. Most traits of interest are governed by more than one gene; defining and studying the entire locus of genes related to a trait gives hope of understanding the effect the genotype of an individual. The advent of molecular maps and the derived QTL mapping technology has provided strong evidence that despite the inferior phenotype, exotic germ plasm is likely to contain QTL that can increase the quality of elite breeding lines. These results have motivated the development of a new molecular breeding strategy, referred to as advanced backcross QTL (AB-QTL) method which integrates QTL analysis with variety development, by simultaneously identifying and transferring favourable QTL alleles from un adapted to cultivated germ plasm. Bin 3-C has previously been described as harbouring a single gene mutation *r* yellow-flesh in tomato (Fray and Grierson, 1993).

5.8.3.3.6 Advanced Backcross Quantitative Trait Loci Analysis

The AB-QTL strategy has so far been tested in tomato and pepper. The most extensive experiments have been conducted in tomato, in which populations involving crosses with five wild *Lycopersicon* species have been genotyped and field tested in a number of locations around the world for numerous traits important for the tomato-processing industry. Through the application of marker and phenotypic analysis of segregating generations of the cultivated tomato and wild *Lycopersicon* species, it had been possible to reveal QTL that improve fruit colour. QTL that improve fruit colour originating from red-fruited (*L. pimpinellifolium*) and green-fruited (*L. peruvianum*, *L. Hirsutum*, *L. parviflorum*) wild relatives had been detected in segregating populations of crosses of these species and the cultivated tomato. The QTL associated with carotenoids and tomato fruit colour using introgression populations of *L. pennellii*, *L. peruvianum* and *L. hirsutum* have been described by Bernacchi et al. (1998).

5.8.3.3.7   *Introgression Line Libraries*

An introgression line (IL) library consists of a series of lines harbouring a single homozygous donor segment introgressed into a uniform, cultivated background. The key advantage of IL libraries is in reducing the complexity of polygenic traits by separating them into a set of monogenic loci. IL libraries also provide perfect starting material for this purpose: each line containing a locus of interest can be backcrossed to the recurrent parent (and, if necessary, selfed) to create a large segregating population. This population can be used to identify recombinants within the introgression segment using flanking markers. IL libraries consist of homozygous 'immortal' lines and therefore can be phenotype repeatedly and used for the simultaneous mapping of many traits. IL libraries contain homogenous genetic backgrounds, only differing from one another by the introgressed donor segment. A tomato IL population that combines single chromosomal segments introgressed from the wild, green-fruited species *Lycopersicon pennellii* in the background of the domesticated tomato; *Lycopersicon esculentum* was used to identify QTL for nutritional and antioxidant contents. The concentration of ascorbic acid, total phenolics, lycopene, beta-carotene and the total antioxidant capacity of the water-soluble fraction were measured in the ripe fruits. Utilizing the developed ILs between *L. esculentum* and *L. Pennellii*, Liu et al. (2003) applied the candidate gene approach to link sequences that have known functional roles in carotenoid biosynthesis to QTLs that are responsible for the variation of the tomato red fruit colour.

5.8.4   TRANSGENIC APPROACH

Three genes, encoding phytoene synthase (*CrtB*), phytoene desaturase (*CrtI*) and lycopene beta-cyclase (*CrtY*) from *Erwinia* have been introduced in potato to produce beta carotene. Romer et al. (2000) developed transgenic tomato to enhance the carotenoid content and profile of tomato fruit, we have produced transgenic lines containing a bacterial carotenoid gene (*CrtI*) encoding the enzyme phytoene desaturase which converts phytoene into lycopene. Gerjets and Sandmann (2006) developed genetically engineered potato for the production of commercially important keto-carotenoids including astaxanthin (3, 3'-dihydroxy 4, 4'-diketo-β-carotene).

Lu et al. (2006) suggested that transgenic cauliflower with *Or* transgenesis associated with a cellular process that triggers the differentiation of proplastids or other non-coloured plastids into chromoplasts for carotenoids accumulation and also reported *Or* gene can be used as a novel genetic tool to induce carotenoid accumulation in a major staple food crop. The post-transcriptional gene silencing (PTGS) is also used for change in nutrient biosynthesis pathways which improve the nutrient contents.

5.8.4.1   QUALITY IMPROVEMENT

Tomatoes have been altered in attempts to improve their flavour or nutritional content. In 2000, the concentration of pro-vitamin A was increased by adding a bacterial gene encoding phytoene desaturase, although the total amount of carotenoids remained equal. When the snapdragon (*Antirrhinum*) genes were used, the fruits had similar anthocyanin concentrations to blackberries and blueberries, and when fed to cancer susceptible mice, extended their life span. The RNAi-mediated suppression of De-Ethiolated 1 expression under fruit-specific promoters has recently shown to improve carotenoid and flavonoid levels in tomato fruits with minimal effects on plant growth (Williams et al., 2004).

One of the drawbacks in inducing the synthesis of ketocarotenoids in transgenic plants is the accompanying decrease in β-carotene level, especially in transgenic plants overexpressing only a β-carotene ketolase. An attempt to overcome this drawback isolated a β-carotene ketolase (*HpBkt*) complementary DNA (cDNA) from *Haematococcus pluvialis* and generated transgenic carrot plants overexpressing *HpBkt* cDNA under the control of the *ibAGP1* promoter and its transit peptide sequence. Over expression of *HpBkt* caused an increase in the transcript levels of the endogenous carotenogenic genes, including phytoene synthase 1 (*PSY1*), phytoene synthase 2 (*PSY2*), lycopene β-cyclase 1 (*LCYB1*) and β-carotene hydroxylase 1 (*CHXB1*), which resulted in elevated β-carotene levels in the *HpBkt*-transgenic plants (range 1.3–2.5-fold) compared with the wild-type plants. Thus, *HpBkt*-overexpressing carrot plants under the control of the *ibAGP1* promoter and its transit peptide are capable of both newly synthesizing keto-carotenoids and enhancing their β-carotene level.

Further, Morris et al. (2008) adopted transgenic approach to increase the levels of Ca in the roots using a modified calcium/proton antiporter (known as short cation exchanger 1) to increase Ca transport into vacuoles. They also showed that consumption of Ca-fortified carrots results in 41% increase in Ca absorption. These studies highlight the potential of increasing plant nutrient content through expression of a high-capacity transporter and illustrate the importance of demonstrating that the fortified nutrient is bioavailable. Similar efforts are suggested for biofortification of important dietary micronutrients.

Another big health challenge is folate deficiency which causes neural tube defects and other human diseases. Folates are synthesized from pteridine, *p*-aminobenzoate (PABA) and glutamate precursor. Díaz de la Garza et al. (2004, 2007) engineered tomatoes by fruit- specific over-expression of guanosine triphosphate cyclohydrolase I that catalyzes first step of pteridine synthesis, and aminodeoxychorismate synthase that catalyzes the first step of PABA synthesis. Vine-ripened fruits contained on average 25-fold more folate than controls by combining PABA- and pteridine overproduction traits through crossbreeding of transgenic tomato plants. This research programme showed way for biofortification of folate in other vegetables for millions of people deprived of adequate folate in their regular diets, however, adequate attention is required on safety issues.

5.8.4.2   IMPROVEMENT OF TASTE

When geraniol synthase from lemon basil (*Ocimum basilicum*) was expressed in tomato fruits under a fruit-specific promoter, 60% of untrained taste testers preferred the taste and smell of the transgenic tomatoes. The fruits contained around half the amount of lycopene, reducing the health benefits of eating them.

5.9 BREEDING FOR INCREASING SHELF LIFE

Shelf life is a major problem for any produce in the world, especially in India due to non-availability of controlled-environment storage. High levels of ethylene are produced at the beginning of the ripening stage in climacteric fruit vegetables. Increased firmness through regulation to decrease fruit softening enzymes has met with limited success thus far. Control of ethylene content appears to be more promising. Ethylene controls ripening and rotting in many vegetables. By reducing ethylene concentration, it is possible to arrest development of the vegetables and permits the fruit vegetables to remain on the plants until ripe.

Tomatoes have been used as a model organism to study the fruit ripening of climacteric fruit. To understand the mechanisms involved in the process of ripening, scientists have genetically engineered tomatoes. In 1994, the 'Flavr savr' became the first commercially grown genetically engineered food to be granted a license for human consumption. A second copy of the tomato gene *polygalacturonase* was inserted into the tomato genome in the antisense direction. The polygalacturonase enzyme degrades pectin, a component of the tomato cell wall, causing the fruit to soften. When the antisense gene is expressed, it interferes with the production of the polyga-lacturonase enzyme, delaying the ripening process. The 'Flavr savr' failed to achieve commercial success and was withdrawn from the market in 1997. Similar technology, but using a truncated version of the polygalac-turonase gene, was used to make a tomato paste. DNA Plant Technology (DNAP), Agritope and Monsanto developed tomatoes that delayed ripening by preventing the production of ethylene, a hormone that triggers ripening of fruit. All three tomatoes inhibited ethylene production by reducing the amount of 1-aminocyclopropane-1-carboxylic acid (ACC), the precursor to ethylene. DNAP's tomato, called Endless Summer, inserted a truncated version of the *ACC synthase* gene into the tomato that interfered with the endogenous *ACC synthase*. Monsanto's tomato was engineered with the *ACC deaminase* gene from the soil bacterium *Pseudomonas chlororaphis* that lowered ethylene levels by breaking down ACC. Agritope introduced an *S*-adenosylmethionine hydrolase encoding gene derived from the *E. coli* bacteriophage T3, which reduced the levels of *S*-adenosylmethionine, a precursor to ACC. Endless Summer was briefly tested in the market-place, but patent arguments forced its withdrawal.

Meli et al. (2010) delayed the ripening of tomatoes by silencing two genes encoding N-glycoprotein modifying enzymes, α-mannosidase and β-d-*N*-acetylhexosaminidase. The fruits produced were not visibly damaged after being stored at room temperature for 45 days, whereas unmodified tomatoes had gone rotten.

Broccoli is a very perishable crop. Ethylene plays an important role in in the yellowing of broccoli as chlorophyll loss is associated with increase in floret ethylene synthesis. Research on transgenic aimed at increasing shelf life of ethylene sensitive broccoli. Antisense version of two regulatory genes in the ethylene biosynthesis path way ACC oxidase and ACC synthase have been used to produce transgenic broccoli which reduce ethylene synthesis. However, reduction in ethylene synthesis increases shelf life for 1 or 2 days.

Post-harvest yellowing in broccoli is known to result from chlorophyll degradation, with chlorophyllase being the first enzyme to degrade chlorophyll. In broccoli, three putative chlorophyllase genes (*BoCLH1*, *BoCLH2* and *BoCLH3*) were cloned using degenerate primers from the conserved regions of known chlorophyllases. Among these three genes, only *BoCLH1* is transcribed during the course of broccoli post-harvest senescence. No individual antisense *BoCLH2* or *BoCLH3* transformants showed significant slowing of post-harvest yellowing. The results suggested that genes other than the *BoCLH*-Chlases obtained in the present study might also be essential in the yellowing process.

5.10   RNA INTERFERENCE IN VEGETABLES

The discovery of RNAi and its regulatory potentials has opened new vista for crop improvement (Jagtap et al., 2011). RNAi technology is precise, efficient, stable and better than antisense technology and even the chances of acceptance of products is more in public. The RNA silencing is a novel gene regulatory mechanism that limits the transcript level by either suppressing transcription (transcriptional gene silencing) or by activating a sequence-specific RNA degradation process (PTGS/RNAi) (Agrawal et al., 2003). This technology has been employed successfully to alter the gene expressions for improved quality traits by increasing antioxidants in tomatoes (Niggeweg et al., 2004) or suppressing over expression of negative traits such as sinapate esters in canola (Husken et al., 2005) and alpha-linolenic acid in soybean (Flores et al., 2008). The role of RNAi technology in extending the shelf-life of tomato by blocking the expression of ACC oxidase gene (Xiong et al., 2005) and suppression of two ripening specific N-glycoprotein modifying enzymes, α-mannosidase and β-D-N-acytl hexosaminidase (Meli et al., 2010) is well established. The role RNAi in regulation of reproductive behaviour (male sterility and fertility) is under investigations because in changing climate the traditional sterility systems are getting affected which is affecting hybrid seed production and ultimately crop productivity. Peter et al. (2011) used RNAi technology to develop Dau c 1.01 and Dau c 1.02-silenced transgenic carrot plants that show reduced allergenicity to patients with carrot allergy. McCormick et al. (2004) developed virus (*Carrot virus Y*) resistant carrot genotypes using RNAi technology. Moreno et al. (2013) also demonstrated that a pronounced reduction in storage root thickness and colour of carrots was obtained in *DcLcyb1* transgenic silenced lines. Negligible attempts were made to utilize this novel technology for improvement of vegetable crops in general and carrot in particular in India. However, the RNAi technology has great promise in improvement of carrot for specific traits such as beta-carotene in tropical carrot, late bolting in tropical carrots, diseases and pests and in male sterility for hybrid seed production.

5.11   CONCLUSION

Consumer interest in whole foods with enhanced nutritional qualities is at an all-time high, and more consumers are choosing foods on the basis of their health characteristics that describe health benefits. Biofortification could make an impact in relatively remote rural areas where food staples do not enter the marketing system or where processing facilities are relatively small, numerous and widely dispersed. Growers and traders would be interested in longer shelf life for perishable crops so that they can make more money. In this respect, breeding programmes for improving the content of nutrients and shelf life in vegetables are becoming more important for public and private breeders. Improvement programmes always must consider more than one trait. The traits may not be given equal emphasis, however. Certain traits may be more economically important than others, traits for which consumers are willing to pay a premium. Alternatively, some traits may receive more attention because they are easier to improve as a result of higher heritability or rapid and inexpensive testing procedures.

REFERENCES

Agrawal, N.; Dasaradhi, P. V. N.; Mohmmed, A.; Malhotra, P.; Bhatnagar, R. K.; Mukherjee, S. K. RNA Interference: Biology, Mechanism, and Applications. *Microbiol Mol Biol Rev.* **2003**, *67*, 657–685.

Anon. Natural Food Colours Market: Global Industry Analysis and Opportunity Assessment 2014–2020, 2015. http://www.futuremarketinsights.com/. (accessed Jan, 2015).

Arlappa, N.; Laxmaiah, A.; Balakrishna, N.; Harikumar, R.; Kodavanti, M. R. Micronutrient Deficiency Disorders Among the Rural Children of West Bengal, India. *Ann Hum Biol.* **2011**, *38*(3), 281–289.

Bernacchi, D., Beck-Bunn, T.; Eshed, Y.; Lopez, J.; Petiard, V.; Uhlig, J.; Zamir, D.; Tanksley, S. Identification of QTLs for Traits of Agronomic Importance from *Lycopersicon-hirsutum. Theor. Appl. Genet. 1998,* 97, 381–397.

Bliss, F. A. Nutritional Improvement of Horticultural Crops Through Plant Breeding. *Hortic. Sci.* **1999**, *34*(7), 1163–1167.

Britton, G. Overview of Carotenoid Biosynthesis. In *Carotenoids*; Britton, G., Ed.; Birkhauser: Basel, Switzerland, 1998; pp 13–147.

Campion, B.; Sparvoli, F.; Doria, E.; Tagliabue, G.; Galasso, I.; Fileppi, M.; Bollini, R.; Nielsen, E. Isolation and Characterisation of an LPA (low phytic acid) Mutant in Common Bean (*Phaseolus vulgaris* L.). *Theor. Appl. Genet.* **2009**, *118*, 1211–1221.

Delgado-Vargas, F.; Paredes-Lopez, O. *Natural Colorants for Food and Nutraceutical Uses.* CRC Press: Florida, USA, 2003.

Diaz de la Garza, R. I.; Quinlivan, E. P.; Klaus, S. M. J.; Basset, G. J. C; Gregory, J. F; Hanson, A. D. Folate Biofortification in Tomatoes by Engineering the Pteridine Branch of Folate Synthesis. *Proc. Natl. Acad. Sci. U. S. A.* **2004**, *101*, 13720–13725.

Diaz de la Garza, R. I.; Gregory, J. F.; Hanson, A. D. Folate Biofortification of Tomato Fruit. *Proc. Natl. Acad. Sci. U. S. A.* **2007**, *104*, 4218–4222.

Dickson, M. H.; Lee, C. Y.; Blamble, A. E. Orange-Curd High Carotene Cauliflower Inbreds, NY 156, NY 163, and NY 165. *Hortic. Sci.* **1988**, *23*, 778–779.

Flores, T.; Karpova, O.; Su, X.; Zheng, P.; Bilyeu, K.; Sleper, D. A.; Nguyen, H. T.; Zhang, Z. J. Silencing of the Gm*FAD3* gene by siRNA Leads to Low a-linolenic Acids (18:3) of *fad3*-Mutant Phenotype in Soybean *Glycine max* (Merr.). *Transgenic Res.* **2008**, *17*, 839–850.

Fray, R. G.; Grierson, D. Identification and Genetic-Analysis of Normal and Mutant Phytoene Synthase Genes of Tomato by Sequencing, Complementation and Co-Suppression. *Plant Mol. Biol.* **1993**, *22*, 589–602.

Gerjets, T.; Sandmann, G. Keto-Carotenoid Formation in Transgenic Potato. *J. Exp. Bot.* **2006**, *57*(14), 3639–3645.

Ghosh, D.; Konish, T. Anthocyanins and Anthocyanin-Rich Extracts: Role in Diabetes and Eye Function. *Asia Pac. J. Clin. Nutr.* **2007**, *16*(2), 200–208.

Giliberto, L.; Perrotta, G.; Pallara, P.; Weller, J. L.; Fraser, P. D.; Bramley, P. M. Manipulation of the Blue Light Photoreceptor Cryptochrome 2 in Tomato Affects Vegetative Development, Flowering Time and Fruit Antioxidant Content. *Plant Physiol.* **2005,** *137,* 199–208.

Gopalan, C.; Sastri, B. V. R.; Balasubramanian, S. C. Nutritive Value of Indian Foods. National Institute of Nutrition: Hyderabad, India, 2007.

Grotewold, E. The Genetics and Biochemistry of Floral Pigments. *Annu. Rev. Plant Biol.* **2006,** *57,* 761–780.

Gutierrez, N.; Avila, C.; Rodriguez-Suarez, C.; Moreno, M.; Torres, A. Development of SCAR Markers Linked to a Gene Controlling Absence of Tannins in Faba Bean. *Mol. Breed.* **2007,** *19,* 305–314.

Hazra, P.; Som, M. G. *Technology for Vegetable Production and Improvement;* Naya Prokash: Kolkata, India, 1999.

Hüsken, A., Baumert, A.; Strack, D.; Becker, H. C; Möllers, C.; Milkowski, C. Reduction of Sinapate Ester Content in Transgenic Oilseed Rape (*Brassica napus* L.) by dsRNAi-based Suppression of BnSGT1 Gene Expression. *Mol. Breed.* **2005,** *16,* 127–138.

Jagtap, U. B.; Gurav, R. G.; Bapat, V. A. Role of RNA Interference in Plant Improvement. *Naturwissenschaften* **2011,** *98,* 473–492.

Jain, S. M.; Suprasanna, P. Induced Mutations for Enhancing Nutrition and Food Production. *Geneconserve* **2007,** *40,* 201–215.

Kalia, P.; Saha, P. Breeding Principle for Enhancing Nutraceuticals and Edible Colours in Vegetable Crops. In *Designing Nutraceutical and Food Colorant Rich Vegetable Crop Plants: Conventional and Molecular Approach;* Kalia, P, Behera, T. K. Eds.; Division of Vegetable Science, Indian Agricultural Research Institute: New Delhi, 2010.

Kang, S.; Seeram, N.; Nair, M.; Bourquin, L. Tart Cherry Anthocyanins Inhibit Tumor Development in Apc (Min) Mice and Reduce Proliferation of Human Colon Cancer Cells. *Cancer Lett.* **2003,** *194,* 13–19.

LaBonte; Don, R. *Sweet Potato, Lists 1-26 Combined. Vegetable Cultivar Descriptions for North America. Department of Horticulture, Louisiana State University, USA,* 2012.

Lang N. T.; Nguyet, T. A.; Phang, N.V.; Buu, B. C. Breeding for Low Phytic Acid Mutants in Rice (*Oryzasativa* L.). *Omonrice* **2007,** *15,* 29–35.

Li, L.; Paolillo, D. J.; Parthasarathy, M. V.; DiMuzio, E. M.; Garvin, D. F. A Novel Gene Mutation That Confers Abnormal Patterns of Beta-Carotene Accumulation in Cauliflower (*Brassica oleracea* var. *botrytis*). *Plant J.* **2001,** *26,* 59–67.

Libert, B. Breeding Alow-Oxalate Rhubarb (*Rheum* sp. L.). *J. Hortic. Sci.* **1987,** *62*(4), 523–529.

Lila, M. A. Plant Pigments and Human Health. In *Plant Pigments and Their Manipulation*; Davis, K., Eds.; Ann Plant Rev. 2004, Vol. 14, pp 248–271.

Liu, W.; Zhao, S.; Cheng, Z.; Wan, X.; Yan, Z.; King, S. R. Lycopene and Citrulline Contents in Watermelon (*Citrulluslanatus*) Fruit with Different Ploidy and Changes During Fruit Development. *Acta Hortic.* **2010,** *871,* 543–547.

Lu, S.; Eck, J. V.; Zhou, X; Lopez, A. B; Halloran, D. M.; Cosman, K. M. The Cauliflower *Or* Gene Encodes a DnaJ Cysteine-Rich Domain-Containing Protein That Mediates High Levels of ß-Carotene Accumulation. *Plant Cell* **2006,** *18,* 3594–3605.

Marzougui, N.; Boubaya, A.; Elfalleh, W.; Ferchichi, A.; Beji, M. Induction of Polyploidy in *Trigonella foenum-graecum* L.: Morphological and Chemical Comparison Between Diploids and Induced Autotetraploids. *Acta Bot Gallica.* **2009,** *156,* 379–389.

McCormick, N. L.; Ford, R.; Taylor, P. W. J.; Rodoni, B. *The Development of Virus Resistant Carrot Genotypes Using RNAi Technology.* International Crop Science Congress: Brisbane, Australia, 2004.

Meli, V. S.; Ghosh, S.;Prabha, T. N.; Chakraborty, N.; Chakraborty, S.; Datta, A. A. Enhancement of Fruit Shelf Life by Suppressing N-Glycan Processing Enzymes. *Proc. Natl. Acad. Sci. USA.* **2010,** *107*(6), 2413–2418.

Moon, S. S.; Verma, V. K.; Munshi, A. D. Gene Action of Quality Traits in Muskmelon (*Cucumis melo* L.). *Veg. Sci.* **2002,** *29*(2), 134–136.

Moreno, J. C.; Pizarro, L.; Fuentes, P.; Handford, M.; Cifuentes, V.; Stange, C. Levels of Lycopene β-Cyclase1 Modulate Carotenoid Gene Expression and Accumulation in *Daucus carota.* *PLoS One* **2013,** *8*(3), e58144. DOI: 10.1371/journal.pone.0058144.

Morris, J.; Hawthorne, K. M.; Hotze, T.; Abrams, S. A.; Hirschi, K. D. Nutritional Impact of Elevated Calcium Transport Activity in Carrots. *Proc. Natl. Acad. Sci. U. S. A.* **2008,** *105,* 1431–1435.

Niggeweg, R.; Michael, A. J.; Martin, C. Engineering Plants with Increased Levels of the Antioxidant Chlorogenic Acid. *Nat. Biotechnol.* **2004,** *22,* 746–754.

NNMB (National Nutrition Monitoring Bureau). *Prevalence of Micronutrient Deficiencies;* Report no. 22. National Institute of Nutrition (Indian Council of Medical Research): Hyderabad, India, 2003.

Norton, R. A. Inhibition of Aflatoxin B1 Biosynthesis in *Aspergillusfavus* by Anthocyanidins and Related Flavonoids. *J. Agric. Food Chem.* **1999,** *47,* 1230–1235.

Peters, S.; Imani, J.; Mahler, V.; Foetisch, K.; Kaul, S.; Paulus, K. E.; Scheurer, S.; Vieths, S.; Kogel, K. H. Dau c 1.01 and Dau c 1.02-silenced Transgenic Carrot Plants Show Reduced Allergenicity to Patients with Carrot Allergy. *Transgenic Res.* **2011,** *20*(3), 547–56.

Raboy, V. Approaches and Challenges to Engineering Seed Phytate and Total Phosphorus. *Plant Sci.* **2009,** *177,* 281–296.

Ripley, V. L.; Roslinsky, V. Identification of an *ISSR Marker for 2-Propenyl Glucosinolate Content in Brassica juncea* and Conversion to a SCAR Marker. *Mol. Breed. 2005,* *16*(1), 57–66.

Romer, S.; Fraser, P. D.; Kiano, J. W.; Shipton, C. A.; Misawa, N.; Schuch, W.; Bramley, P. M. Elevation of the Pro-Vitamin A Content of Transgenic Tomato Plants. *Nat Biotechnol.* **2000, 18,** 666–669.

Sanghvi, T. G. Economic Rationale for Investing in Micronutrient Programs. A Policy Brief Based on New Analyses, Office of Nutrition, Bureau for Research and Development, United States Agency for International Development: Washington, D.C., 1996; pp 1–12.

Sapir, M.; Shamir, M. O.; Ovadia, R.; Reuveni, M.; Evenor, D.; Tadmor, Y. Molecular Aspects of *Anthocyanin Fruit* Tomato in Relation to *high pigment-1. J Hered.* **2008,** *99*(3), 292–303.

Schmidt, M.; Hymowitz, A. T.; Herman, E. M. Breeding and Characterization of Soybean Triple Null; a Stack of Recessive Alleles of Kunitz Trypsin Inhibitor, Soybean Agglutinin, and P34 Allergen Nulls. *Plant Breed.* **2015,** *134,* 310–315.

Shi, J.; Wang, H.; Schellin, K.; Li, B.; Faller, M.; Stoop, J. M.; Meeley, R. B.; Ert, D. S.; Ranch, J. P.; Glassman, K. Embryo-Specific Silencing of a Transporter Reduces Phytic Acid Content of Maize and Soybean Seeds. *Nat Biotechnol.* **2007,** *25,* 930–937.

Stein, A. J.; Qaim, M. The Human and Economic Cost of Hidden Hunger. *Food Nutr Bull.* **2007,** *28,* 125–134.

Steinmetz, K. A.; Potter. J. D. Vegetables, Fruit, and Cancer Prevention: A Review. *J Am. Dietet. Assn.* **1996,** *96,* 1027–1039.

USAID (The United States Agency for International Development). USAID'S OMNI Micronutrient Fact Sheets. USAID: India, 2005.

Welch, R. M. Micronutrients, Agriculture and Nutrition; Linkages for Improved Health and Well Being. In Perspectives on the micronutrient nutrition of crops; Singh, K., Mori, S., Welch, R. M., Eds.; Scientific Publishers: Jodhpur, India, 2001; 247–289.

Williams, M.; Clark, G.; Sathasivan, K.; Islam, A. S. RNA Interference and its Application in Crop Improvement. *Plant Tissue Cult. Biotechnol.* **2004,** 1–18. DOI: 10.1.1.98.1069.

Xiong, A, S.; Yao, Q. H.; Peng, R. H.; Li, X.; Han, P. L.; Fan, H. Q. Different Effects on ACC Oxidase Gene Silencing Triggered by RNA Interference in Transgenic Tomato. *Plant Cell Rep.* **2005,** *23,* 639–646

Zhang, F.; Wang, G.; Mang, M.; Liu, X.; Zhao, X. Identification of SCAR Markers Linked to *or*, a Gene Inducing Beta-Carotene Accumulation in Chinese Cabbage. *Euphytica,* **2008,** *164*(2), 463–471.

Zhang, W.; Xiucun, H.; Leyuan, M.; Zhao, C.; Yu, X. Tetraploid Muskmelon Alters Morphological Characteristics and Improves Fruit Quality. *Sci Hort.* **2010,** *125*(3), 396–400.

CHAPTER 6

APPLICATIONS IN POST-HARVEST MANAGEMENT OF VEGETABLE CROPS

ACHUIT K. SINGH[1], AVINASH CHANDRA RAI[1], ASHUTOSH RAI[1], and MAJOR SINGH[1*]

[1]Division of Crop Improvement, ICAR—Indian Institute of Vegetable Research, Varanasi 221305, UP, India, E-mail: achuits@gmail.com

[2]Division of Crop Improvement, ICAR—Indian Institute of Vegetable Research, Varanasi 221305, UP, India, *E-mail: singhvns@gmail.com

CONTENTS

6.1   INTRODUCTION

In order to fulfil the basic requirement and earning point of view, the world is shifting from a rural population to high concentration in the urban area at the moment. This situation seems dangerous, and urban areas have been shifting very rapidly to rural agricultural lands. Vegetable farms are far away from areas of food consumption. This situation creates critical condition to the vegetable producer who faces problems to bring the right food to the right place with a minimum of losses. Post-harvest problems can do substantial losses, and its volume depends on the country, crop and technology. Most of the developing countries are situated in tropical regions where high temperature, water shortage and humidity create many problems. Moreover, the developing countries are also lacking behind in advance technology in comparison with countries such as the United States of America, Japan, Australia and European countries, which are applying technically advanced technologies to minimize losses. Sighting the current scenario, changes in climate such as water scarcity, increase in population and rapid increase of food demand make the researchers to emphasize not only to improve the nutritional quality and food quantity but also to enhance the shelf life of vegetables by using the biotechnological tools (Abano and Buah, 2014). The current chapter focuses an insight into biotechnological approach to improve the nutritional quality and shelf life of vegetables as a part of post-harvest management.

Vegetables are one of the important components of the diet of a human being (Table 6.1). There are several reports in literature stating that vegetables have a wide range of factors such as nutritional value, flavour, colour, texture, processing qualities and shelf life (Bapat et al., 2010; Vadivambal and Jayas, 2007). Therefore, before manipulation of vegetable yield and quality, it is important to know the fundamental biological processes about what are the things which influence fruit set, maturation, stages of fruit ripening. In this respect, biotechnology has played a significant and crucial role which revealed the constraints surrounding the extensive reproductive phase in some vegetables that have extended juvenile periods, the complex reproductive stage, greater heterozygosity, inter and intra incompatibility and sterility of breeding of vegetable plants for improvement (Bapat et al., 2010). The technique such as genetic modification is used in the vegetables to enable plants to tolerate the biotic and abiotic stresses, and plant resistances to pests and diseases, which may improve nutritional contents and extend the shelf life of the vegetable products.

TABLE 6.1   Vitamins Supplied by Vegetables and Root Crops.

Vitamin	Source	Characterized by
A (retinol)	From carotene in dark green leaves, tomatoes, and carrots	(Bureau and Bushway, 1986)
B1 (thiamine)	Pulses, green vegetables	(Ifon and Bassir, 1980)
B 2 (riboflavin)	Green leafy vegetables	(Ismail and Fun, 2003)
B 6 (pyridoxin)	Green leafy vegetables	(Ifon and Bassir, 1980)
PP niacin (nicotinic acid)	Pulses, peanuts	(Ismail and Fun, 2003)
Folate (folic acid)	Dark green leaves, broccoli, spinach, beets, cabbage, lettuce	(Scholl and Johnson, 2000)
C (ascorbic acid)	Dark green leaves, spinach, cauliflower, sweet pepper	(Oboh and Akindahunsi, 2004)
K	Tomato, spinach, turnip, lettuce	(Klein and Perry, 1982)
E	Green leafy vegetable	(Stahl and Sies, 1997)

After maturity or complete of their life span, the plants start a process of decay, and it is expected that the death of the plants/organism will finally lead to conversion in its organic matter. This process is even more rapid and fast in harvested tissue of a particular organ instead of the whole plant. Overall, these processes are called senescence. In general, senescence is the final stage in the life of plants, and it was thought that it has an essentially disordered nature in which cell components break down without any particular order. However, recent advances in science revealed that the process of senescence is a very well-programmed developmental stage, involving highly coordinated cellular events that require the consistent action of many genes.

Basic knowledge of senescence has greatly improved in the last few years, but we are still far from having a good understanding of the underlying biochemical and molecular nature of senescence. There are two kinds of senescence process in nature: one comes after the end of the useful life of an organ, and the second is an environmentally induced senescence. For example, in a natural process, fruits will develop and ripen to attract predators and ensure seed dispersal. A natural senescence process will eventually end up with the spoilage of the fruit, when the seeds contained in the fruit are not viable anymore. In another natural process, old leaves shrivel and fall, whereas new ones develop to keep the photosynthetic capacity viable of the plant. These events are pre-programmed in the genetic code of the plants. Harvested plant tissue for commercialization causes a series of stresses in the detached tissue that will extensively trigger senescence, although these natural and induced senescence processes do not always follow the same mechanisms. To increase the useful life of a particular foodstuff in a very long form is the main objective of post-harvest technology. Due to diverse nature of food stuff of horticultural crops, complex variety of plant tissues is being commercialized such as fruit, leaves, flowers, roots and tubers. Although senescence has followed some common pathways and features, due to diverse nature of tissue and organs of vegetable crops, they have specific characteristics, and therefore, there is the need to study them separately. The huge complexities/diversities of horticultural crops and their tissues and/or organs make them to not use for a single post-harvest treatment. Treatments of leafy vegetables are different from the fruity vegetables and even vegetable fruits have a wide variety of post-harvest problems depending on the fruit size shape, type and commercial value of the fruit variety.

It is important to know the nature of the crop and the problem faced with those crops before establishing the biotechnological approaches, then a fruitful and possible biotechnological approach should be attempted. Further, it is also important to emphasize that the same problem can be solved with different approaches. Metabolic-engineered biotechnological approaches can target internal processes, but they are not restricted to endogenous genes. The present form of genetic engineering allows us to overcome the species barriers, and therefore, genes that would not normally be accessible by traditional breeding can now be inserted into the targeted plant species. Using of physical instruments, such as humidity controller, refrigeration, irradiation, and so on, has proven useful in controlling post-harvest losses. In rural areas, due to the lack of infrastructure and specialized skilled, the implementation of many existing and new technologies is quite difficult in many developing countries. In countries, such as India and others, crops are grown on small farms, which indicated that a single grower is not able to afford the economic cost of setting up treatment of plants and their specialized packaging requirements. Regional centres, either governmental or private, can alleviate the situation to some extent, but they require a level of organization that is commonly missing or rarely exists.

6.2 BIOTECHNOLOGICAL APPROACHES IN POST-HARVEST: AN OVERVIEW

Due to the short lifespan of many vegetables, very quick and extensive actions need to be taken to extend their lifespan or delaying the senescence after harvesting the crop. The rate of delay in senescence in vegetable crops is not an easy task either by conventional or biotechnological approaches. The post-harvest-induced senescence creates complex changes in the biochemical and cellular processes of the cell components, and due to this complexity and lack of knowledge of the cellular processes, it is very difficult to develop a new biotechnological technique. Hormone physiology-based techniques are used to alter the senescence, for example, by enhancing cytokinin production or blocking ethylene production or perception. For extending the post-harvest life of leafy vegetables, there is a high need to focus on the minute observation of these events that occur in regular leaves during senescence. It has been reported that cytokinins can delay leaf senescence and at that time, there is a drop in endogenous ethylene levels (Van Staden et al. 1988). Smart et al. (1991) demonstrated that the overproduction of a bacterial enzyme, IPT, catalyses the rate-limiting reaction in the biosynthesis of cytokinins under the control of strong constitutive promoter, CaMV35S; result indicates that there are high levels of cytokinins and delayed leaf senescence in transgenic plants. However, these transgenic plants showed developmental abnormalities apart from delay senescence. An ingenious solution to the use of cytokinins to delay senescence has been provided by Gan and Amasino (1997) who placed the IPT gene under the control of a senescence-specific promoter (SAG12). In this system, the onset of senescence triggers the SAG12 promoter, leading to the production of cytokinins and preventing cytokinin from accumulating to a level that would interfere with other aspects of plant development. Transgenic tobacco plants which are obtained in this way contained leaves with extremely delayed senescence that maintained high levels of photosynthetic activity with no other developmental abnormalities. This type of practices can be applied to leafy vegetables.

Similarly, another approach through which we can reduce the ripening of vegetable fruits and delay the fruiting senescence of the vegetables is by controlling ethylene production in the fruits of vegetables. Ethylene formation is the result of biosynthetic pathway in which first Methionine is converted to S-adenosyl-L-methionine (SAM) by the enzyme methionine adenosyltransferase, after that SAM is converted to 1-aminocyclopropane-1-carboxylic acid (ACC) by the enzyme ACC synthase (ACC-S), and in the last step, ACC is converted to ethylene by the enzyme ACC oxidase (Imaseki, 1991). Genes encoding the enzymes involved in biosynthesis (ACC synthase and ACC oxidase) have been identified, isolated and further cloned into a suitable vector. This technique has been blocking the expression of those genes, which ultimately block or reduce the biosynthesis of ethylene. Alternatively, other enzymes have been identified in microorganisms that metabolize the intermediates in ethylene biosynthesis. These include ACC deaminase and SAM hydrolase, an enzyme capable of degrading ACC and SAM, respectively. The insights of pathway study of ethylene production and their concentration in the fruits are complicated and require several techniques that developed over past the two decades. These technologies include gene integration into plants, selection of transformed cells, regeneration of plants from genetically transformed cells and ultimately identification of targeted gene expression. Ethylene often has a reverse effect to cytokinins in promoting senescence, but it is not essentially used for the regulation of the process in vegetative plant tissues. Transgenic tomato plants with reduced levels of ethylene production have shown interrupted leaf senescence (Klee 1993; Good et al., 1994; John et al., 1995). Transgenic *Arabidopsis* plants with the *etr1-1* dominant mutation that make them insensitive to ethylene have also shown delayed leave senescence (Grbic and Bleecker 1995). Likewise, a natural pigment anthocyanin, present in purple tomato fruit, can significantly extend shelf life by slowing down the ripening process and also reducing or controlling the infection of *Botrytis cinerea*, one of the most important post-harvest pathogens (Zhang et al., 2013). Reduced susceptibility to *B. cinerea* is dependent specifically on the accumulation of anthocyanins; this may alter the spreading of the ROS burst during infection. This increases antioxidant activity of purple tomato fruits and slows the processes of overripening. Enhancing the levels of natural antioxidants such as anthocyanin in tomato facilitates a novel strategy for extending shelf life both by genetic engineering and conventional breeding (Zhang et al., 2013).

In most of the cases, it has been observed that the delay of ripening can be achieved only at the time of onset. But once the process starts, it goes in normal patterns. It is evident from available data that ethylene activates the senescence-related genes only if the tissue is ready to senescence. Ethylene has already been known for its involvement and role in both onset and regulation process of senescence in carnations (Brandt and Woodson et al., 1992). The antisense copies of tomato ACO gene have been earlier used to produce transgenic broccoli (Henzi et al., 1999a; Henzi et al., 1999b; Henzi et al., 2000). Transgenic lines displayed a noticeable upsurge in ethylene production in the early phase of post-harvest with levels three times higher than control samples; nevertheless, 74 h after harvest ethylene production in controls markedly increased, whereas the transgenic lines showed reduced ethylene levels. In addition to the ethylene, the metabolic parameters of immature vegetables after harvest undergo significant changes with loss of proteins and lipids and accumulation of free amino acids and ammonia that ultimately lead to tissue breakdown.

Abiotic stress such as improper soil nutrition, pre/post-harvest moisture, temperature and so on also induces some of the post-harvest problem. For example, the inappropriate temperature and humidity combinations provide a favourable environment to pathogens. To reduce such types of decay, antibiosis have been used to control bacteria and fungi. This antibiosis is produced by cloning antibiotic genes or other generalized antibiotics compounds. Nikkomycin, an antibiotic produced by *Streptomyces tendae*, has been shown to be effective against fungi (Kinzel, 1989). Similarly, chitinase shows significant effect over fungi, whose cell wall contains chitin as a key structural unit. Due to the deleterious effect of Chitinase, they have been proved equally effective on insect and nematodes. On the basis of these, exogenous production of chitinase may lead to develop such type of plants which may prevent fungal mycelia development. Lund et al. (1989) demonstrated transgenic tobacco plants expressing cloned bacterial chitinase gene that are effective against some fungal pathogens but produce some developmental deformities in transgenic plants. Further, in another experiment, transgenic tobacco plants having chitinase specific promoters

178 Advances in Postharvest Technologies of Vegetable Crops

which control overproduction of chitinase showed resistant against fungal pathogens without any deformities in transgenic plants (Ahmad et al., 2012; Irving and Kuc, 1990; Linthorst et al., 1990). A bean chitinase gene, constitutively expressed in transgenic tobacco, was found resistant towards *Rhizactonia solani* (Kühn) (Brogue et al., 1991). Other inducible antimicrobial compounds normally produced by plants might also be exploited for bioengineering (Bowles, 1990). High peroxidase activity has been implicated in phytopathogen resistance (Vanitha and Umesha, 2008; Mohan and Kolattukudy, 1990). Moreover, the interest may be hydrolyses, proteinases and other pathogen-defence-related compounds induced in response to pathogen attack.

Cloning antibiosis proteins and mobilizing antibiotic genes are well studied, and their integration into plants plays an important role in regulating pathogens genes. In another approach, the virulence of pathogens was stopped by bioengineering plants with genes which do not directly kill the pathogens but destroy their ability to cause disease. Plant cell wall degrading enzymes play a vital role in facilitating bacterial and fungal pathogens to penetrate into fruits and vegetable causing extensive postharvest deterioration. To clone a plant with specific resistance to all of these enzymes would be difficult, whereas many *Erwinia Spp.* have multiple, genetically independent pectolytic enzymes. However, Chatterjee et al. (1994) detailed the discovery of regulatory genes that are necessary for cell export of several major bacterial cell wall-degrading enzymes produced by the post-harvest pathogen *Erwinia carotovora* (Murata et al., 1991).

Bt toxins were placed as a model system proven to be effective towards definite insect pest due to activity of crystal protein of *Bacillus thurangiensis* (Berliner). Many transgenic crop plants having Bt toxin gene have already been produced which are insect resistant (Delannay et al., 1989; Grierson and Covey, 1988). In the same way, the process of ripening controlling genes in plants may be a way to prevent post-harvest losses. As this involves manipulating the plant's own internal biological clock, using antisense RNA is an appropriate bioengineering method. Plant enzymes and hormones involved in ripening are the most logical targets for genetic manipulation. Giovannoni et al. (1989) demonstrated that tomato polygalacturonase (PG) had a major role in degrading cell wall polyuronides, but little effect on fruit softening or colour development. This study indicated that PG repression might be useful in increasing fruit shelf life. Bioengineering this enzyme has been accomplished with tomatoes (Kramer, 1990; Kramer et al., 1990). Antisense RNA to the mRNA of the tomato's own PG was introduced into the plant. The resulting produce has a longer shelf life, and neither flavour nor texture has been compromised. As the tissue does not soften at the same rate as normal tissue, the fruit should have increased resistance to post-harvest pathogen attack. Transgenic tomatoes with an antisense gene to ethylene, a hormone involved in ripening, have also been developed (Hamilton et al., 1990). The tomato studies offer a model system that should be readily applicable to other crops.

6.3 NEED OF BIOTECHNOLOGY IN POST-HARVEST VEGETABLE PRODUCTION

A number of challenges, such as increase in population, water scarcity, changing global climate, high post-harvest decays and short shelf life of vegetables, need to be addressed by help of biotechnological applications in the vegetable production. As the world is facing great water scarcity and drought conditions, whereas most of the vegetables by their own internal properties require more water, this condition will cope with the biotechnological approaches where biotechnologists have produced the high-yielding vegetable crops with huge water scarcity (Rai et al., 2012, 2013). World population is expected to rise to 10 billion by 2050 (Avano and Buah, 2014). Freshwater, vital for agricultural productivity, is becoming lower to alarming environmental change could increase temperature, drought and uncertainty. By the use of biotechnological approaches, it will be needed to enhance existing technologies that will be used to develop new crop varieties to meet the challenges of global climate changes. In this way, biotechnology has been successfully used to develop insect and herbicide resistance in a limited number of crops. In future, the insights pathway study of crop plants through transcriptomics data analysis may explore the actual metabolism of crop plants will be altered to produce new/improved varieties or species that are tolerant to environmental stresses (Shukla et al., 2015). In addition, the nutritional value of staple food such as rice will be enhanced. In future, the focus will also be on harvesting sunlight to provide an alternative of fossil fuels and reduce the emission of CO_2. Some of these new crop plants are already in field trials and will be available to farmers in the near future (Gellatly and Dennis, 2011).

Post-harvest decay is the major challenge of vegetables throughout the world, and degree of post-harvest loss through decay is well documented. In developed countries also, it is estimated that about 20–25% of the harvested vegetables are decayed by pathogens during post-harvest handling (Sharma et al., 2009; Singh and Sharma, 2007, Droby, 2005; Zhu, 2006; El-Ghaouth et al., 2004). This situation is much more serious in the developing countries, where post-harvest decays are always above 35%, due to poor storage, processing and proper transportation facilities (Abano and Sam-Amoah, 2011). The use of harmful synthetic fungicides such as benomyl and iprodione to check the post-harvest diseases of vegetables is well documented in the scientific literature (Fan et al., 2001; El-Ghaouth et al., 2004; Korsten, 2006; Zhu, 2006; Zhang et al., 2007; Singh and Sharma, 2007).

The continuous use of synthetic fungicides and the food and vegetable products has seriously affected health and environments, alarming the policy-makers and consumers to obtaining greener technologies and food products from the food industries as well as the scientific community. In the past two decades, to reduce post-harvest decays in vegetables, microbial antagonists (yeast, fungi and bacteria) have been used with limited success (Roberts, 1990; Droby et al., 1992; Janisiewicz and Korsten, 2002; Droby, 2005; Korsten, 2006; Zhang et al., 2007; Sharma et al., 2009). However, the advance biotechnological approaches can be applied to develop vegetables with improved quality and shelf life. The molecular, physiological and biochemical traits of the plants is directly related to post-harvest storage of vegetables to maintain the quality and shelf life of post-harvest products. These traits are genetically determined and can be manipulated using genetic breeding and/or biotechnology. It has been revealed in published research results that the manipulated genes have potential and can be used to develop quality post-harvest crop plants. The application of this biotechnological knowledge should not only lead to major improvements in the post-harvest storage of fresh fruits and vegetables but as well improve the human food supply. In most vegetables, the appearance and eating quality deteriorate rapidly due to post-harvest diseases. Moreover, the poor post-harvest characteristics such as deficient flavour, short shelf life, rapid softening, fast and easy spoilage, sensitivity to low temperatures and susceptibility to pathogens are major reasons to decrease in profitability for the domestic market, as well as bottle neck for expansion of existing and new export markets.

There are basically two targets to be achieved for post-harvest management of vegetables, which are as follows: (i) extension of shelf life and (ii) resistance to pathogen attack. There is involvement of large number of biochemical pathway for the ripening process which results in marked changes in the texture, taste and colour. A large number of genes involved at the molecular level, and they are tightly regulated to induce the right changes at the right time in a highly coordinated process. In general, vegetables are classified as climacteric or non-climacteric depending upon their patterns of respiration and ethylene synthesis during ripening. In the suitable climatic conditions, fruits have an increase in the respiration rate during their early stage in the ripening process completed by autocatalytic ethylene production, whereas during unfavourable climatic conditions, vegetables show a different respiration pattern and display a lack of autocatalytic ethylene synthesis. Most of the attempts made to improve the shelf life of vegetable were found to focus on the general appearance or the firmness of the membrane and cell wall and the rate of ripening (ethylene production or perception). In these approaches, the indigenous gene, which are involved in ripening process, were down-regulated aiming gene silencing.

6.4   RIPENING RATE AND PERISHABILITY IN VEGETABLES

In favourable and unfavourable climatic conditions, vegetable fruits ripening and softening is the major cause which contributes towards the perishability of vegetable fruit crops. Vegetables such as tomato take about a few days after which it is considered inedible due to overripening. Due to perishability, fruits spoil, which includes excessive softening, changes in taste, aroma and skin colour. This extreme condition brings significant losses to both farmers and consumers in a parallel way. The physiological, biochemical and molecular changes associated with ripening is an irreversible process, and once started, it cannot be stopped, whereas ripening in vegetables can be extended through several external procedures (Prasanna et al., 2007; Martínez-Romero et al., 2007).

Due to short generation time and better understanding of genetics of tomato, its species can be extensively used due to their long existing history of physiological, biochemical and molecular research in genetics for fruits, vegetable development and ripening process. Besides this, Bapat et al. (2010) reported that successful and stable transformation system, huge availability of expression sequence tag (EST) resources, microarrays and completed genome sequencing effort, availability of large number of germplasms, in well-characterized centres in the world (AVRDC, USDA, TGRC), well-characterized mutants and high-density genetic maps have assisted in the understanding of development and ripening in vegetable to a large extent. The earlier researchers were focused on ethylene synthesis, whereas due to advances in molecular biology approaches, scientist have focused on ripening metamorphosed for identification and study of gene sequences, recently, in the DNA of organism to demonstrate insights into ripening control of ethylene, ripening-related signal transduction systems and downstream metabolic networks (Theologis 1992) and manipulation of cell wall and proteins structure (Rose et al., 2004). In tomato, when a single gene determines two or more apparently unrelated character-istics of the same organism in ripening, mutations can be added to the understanding of ripening in fresh vegetables. The single gene that affects multiple characters (Pleiotropic) in tomatoes includes colourless non-ripening (Cnr), ripening-inhibitor (rin), never-ripe (Nr), green-ripe (Gr) and high-pigment (hp-1 and hp-2).

Characterization of various ripening-related mutations in tomato have been possible for the use of molecular techniques such as positional cloning, genetic mapping of mutant loci and candidate genes characterize in details the various tomato ripening mutations. The Cnr and rin are two muta-tions identified by recessive and dominant, respectively, which effectively block the ripening process. This was responsible for failure to produce high ethylene or reply to exogenous ethylene during ripening (Vrebalov et al., 2002; Manning et al., 2006). The Nr mutation evident an ethylene receptor gene, and Gr has been found to encode a novel factor of ethylene signalling. With positional cloning, GR gene was cloned which underlined a dominant ripening mutation (Barry and Giovannoni, 2006). Although the biochemical nature of the GR remains unclear, protein sequence indi-cates its membrane localization and possible copper-binding activities. Although studies are carried out on different mutants, changes in gene sequence and their characters through mutations also facilitates to develop transgenic tomato fruit with different genes. This process has provided a better insight of fruit ripening and genes expression. ACO and ACS are a multigene family of five and nine members in tomato, whose expres-sions are differentially regulated during fruit development and ripening

(Abano and Buah, 2014). The antisense copy of one member of ACS gene family with its untranslated region in tomato was used to develop transformed plants. Such type of transgenic tomatoes did not ripe without exogenous treatment of ethylene and reduced 99.5% ethylene production. In another attempt, anti-ACS containing transgenic tomato plants showed a 30% decrease in ethylene production by fruits (Abano and Buah, 2014) (Table 6.2).

TABLE 6.2 Classification Vegetables According to Principal Causes of Post-Harvest Losses and Poor Quality in Order of Export.

Group	Examples	Principal causes of post-harvest losses and poor quality (in order of export)	Shelf life at ambient conditions (days)
Root vegetables	Carrots	Mechanical injuries	20–60
	Beets	Improper curing	20–60
	Garlic	Water loss (shrivelling)	30–120
Leafy vegetables	Lettuce	Water loss (wilting)	2–3
	Chard	Loss of green colour (yellowing)	2–3
	Spinach	Mechanical injuries	3–4
	Cabbage	Relatively high respiration rates	5–15
	Green onions	Decay	3–7
Flower vegetables	Artichokes	Mechanical injuries	3–7
	Broccoli	Yellowing and other discolourations	5–15
	Cauliflower	Abscission of florets and Decay	5–12
Immature-fruit vegetables	Cucumbers	Over-maturity at harvest	5–8
	Squash	Water loss (shrivelling)	10–15
	Eggplant	Bruising and other mechanical injuries	5–15
	Peppers	Chilling injury, breaking of tips, mechanical injuries	14–21
	Okra	Decay	4–7
	Snap beans	Decay	5–15
Mature-fruit vegetables	Tomato	Bruising	7–21
	Melons	Over-ripeness and excessive softening at harvest	10–20

6.5 BIOTECHNOLOGICAL APPROACHES APPLIED TO VEGETABLES FOR POST-HARVEST MANAGEMENT

The increasing world food demands need to use an emerging technology to fulfil this; plant tissue culture has provided a great platform to improve both the agriculture and industries. This technique has made significant contributions to the improvement of agricultural sciences, and today they build up an essential tool in modern agriculture. Without the previous example, biotechnology has been introduced into agricultural practices. Tissue culture practices facilitate the production and propagation of genetically uniform and disease-free plant material. For promoting soma clonal variation, in vitro cell and tissue culture techniques are a useful tool. Such variations in the form of genetic variability could be used as a source of variability to obtain new stable genotypes. Presently, interventions of biotech approaches by using tissue culture techniques as regeneration, mass micropropagation and gene transfer studies in vegetable species have been encouraging extensively. In vitro cultures of mature and/or immature zygotic embryos (embryo rescue) are also applied to recover plants obtained from interspecies crosses that do not produce fertile seeds. Such type of practices will give the wild crops of a wild cross with cultivated lines and may produce multiple biotic and abiotic stress-resistant plants (Prasanna, 2015; Shah et al., 2015). The introduction of genetic engineering can make possible a number of improved/new crop varieties having high-yield potential and resistance to biotic and abiotic stresses.

Recent reports have been showed that the recombinant DNA technology has been provided to be important to give flexibility to farmers for marketing their produce and ensuring consumer needs for quality vegetables (Bapat et al., 2010). Genetically modified grapes were developed for modified auxin production for fungal and virus resistance as well as vegetable quality and appearance (DeFrancesco, 2008).

Fresh vegetables have a short post-harvest life and due to physiological activity, mechanical injury and decay, they are inclined to post-harvest losses. For reducing the post-harvest losses, low temperature and refrigerated storage techniques are widely used which creates delaying in senescence in vegetables ultimately maintaining their post-harvest quality. Due to refrigerated storage, tropical and subtropical crop plant species have initiated a set of physiological changes which are commonly known as the chilling injury that revelry affect the quality of vegetable products and ultimately lower the economic value of the products. Membrane damage and reactive oxygen species (ROS) are the key chilling injury factors which affect sensitive horticultural produce. The ability to accumulate heat-shock proteins (HSP) of certain plant species is responsible for chilling injury tolerance. HSPs plays various housekeeping activities providing resistance towards chilling tolerance due to their chaperon activity but also due to their capability to function as ROS scavenger, membrane stability and wonderful massagers for coordinating cell antioxidant systems. Due to these synergistic actions, HSPs are also important for balancing osmotic adjustments, which is an important feature for chilling tolerance. Due to defence action of these small HSPs, it can be presumed to use during the low temperature storage, by transforming the horticultural crops using biotechnological approaches to induce the production and action of HSP to express and stabilize chilling tolerance-related proteins during their post-harvest cold storage (Sevillano et al., 2009, Aghdam et al., 2013). The molecular weight of stress response heat shock protein ranges from 15 to 115 kDa. These are the five families of HSP which have been identified, namely HSP70, chaperonins (HSP60), HSP90, HSP100. Along with cytoplasm and nucleus, HSPs are also spread widely in cell compartments such as mitochondria, chloroplast and endoplasmic reticulum (Wang et al., 2004; Timperio et al., 2008).

Post-harvest treatments include physical treatments, (ultraviolet C light and heat), chemical treatment (methyl salicylate, methyl jasmonate), those involved in controlling the atmosphere surrounding the produce by regulating carbon dioxide (CO_2) and oxygen (O_2) concentrations. Accumulation of HSP has enhanced, due to these treatments, which effectively describes their protective role against chilling injury. Although the application to apply these chilling injury tolerance technologies is readily available, performances depend greatly on capacity limitations and energy expenses of the facilities required to set up above application on an industrial scale. At molecular level, a specific family of heat-shock transcription factors (HSTFs) regulated the expression of genes those encoding HSPs, which are capable to 'sense' a change in temperature and then activate HSP gene expression by binding to cis-acting elements, situated in the TATA-box-proximal 5'-flanking regions, which promotes genes encoding HSP. The alternating units of 5'-nGAAn-3' and efficient HSTF binding requires, at least, three units of this consensus motif (Schöffl et al., 1998; Larkindale et al., 2005). HSTFs are present in normal conditions in cell as a dormant monomer, unable to both DNA binding as well as transcriptional activity. These transcription factors (HSTFs) are translocated to the nucleus and then converted to trimeric form, which is capable in heat-shock consensus element-binding activity and transcriptional activation during the stress (Sorger and Nelson, 1989; Al-Whaibi, 2011).

The hot-air treatment (38°C for 48 h) to tomato provides acquired tolerance toward low-temperature (2°C) storage (Sabehat et al., 1996), whereas in another experiment, the hot-air treatment (38°C for 3 days) provides chilling injury tolerance due to accumulation of HSP 17 under low temperature storage at 2°C (Lurie et al., 1996). Small heat shock proteins (HSPs), such as tom66 and tom111, get induced after hot air treatment of (38°C for 3 days), and they continue their expression level during subsequent cold storage (2°C) which is associated with chilling injury resistance (Sabehat et al., 1998a, b). It has been observed that the expressions of heat shock transcription factors are induced higher in heat than chilling stress (Swindell et al., 2007). The higher induction of HSTFs under high temperature positively correlates them with level of expressions of HSPs, concluding that these HSTFs are the transcriptional regulators of HSPs (Borges and Ramos, 2005). The *Arabidopsis* HSTF AtHSFA1b has a crucial role providing chilling tolerance when initiated by high-temperature treatment: transgenic tomato overexpressing AtHSFA1b not only shows higher tolerance to heat stress but also to low temperatures without the need for a previous induction of heat treatment to induce it (Li et al., 2003). Further, it has been also observed that genetically engineered tomato plants over expressing the AtHSFA1b (*Arabidopsis* HSTF) activate sHSP biosynthesis as well as the expression of genes coding for APX (Li et al., 2003). The results of this set of processes are responsible for the improved cold tolerance of transgenic tomato (Li et al., 2003). Saltveit and Hepler (2004) and Pressman et al. (2006) suggested that fruits and vegetables treated with low temperature should increase their tolerance to heat and vice versa; heat treatment enhances chilling tolerance of sensitive fruits and vegetables, and HSTFs play a crucial role in this condition of enhanced cold tolerance. It has been suggested by Waters et al. (1996) that the key response of fruits and vegetables to heat treatment at the cellular level was an induction of gene expression and protein accumulation of HSPs. Heat treatment also enhanced APX activity conducting increasing chilling injury tolerance in rice seedlings (Sato et al., 2001). Thus, manipulation of HSP gene expressions opens great possibilities for its biotechnological application with the goal of mitigating post-harvest chilling injury. In the post-harvest life of vegetables, understanding the regulatory roles of HSP would create new eventualities for implying biotechnological tools to improve tolerance to chilling temperatures and to extend the period of refrigerated post-harvest storage of vegetables which are chilling injury susceptible.

6.6   APPROACHES FOR MANAGEMENT OF NUTRITIONAL QUALITY OF VEGETABLES

The desirable property in the fresh vegetables such as nutritional value, flavour, colour, texture, processing qualities and shelf life has been reported by many revisions in the wide range of determinants (Vadivambal and Jayas, 2007; Bapat et al., 2010). Studies found that tomato plants transformed with yeast SAMDC gene under the control of E8 promoter exhibited improvement in tomato lycopene content, better fruit juice quality and vine life without affecting fruit colour and softness (Bapat et al., 2010). In another report, over-expression of Nr (wild-type) gene, in tomato, along with constitutive 35S promoter produced plants that showed reduction in ethylene sensitivity (Ciardi et al., 2000). Ethylene receptors belong to a multigene family, except in the case of LeETR4 antisense reduction in expression of individual receptors did not show a major effect on ethylene sensitivity possibly due to the redundancy of individual receptor genes. Along with CaMV35s promoter, LeETR4 transformed antisense plants showed constitutive ethylene expression and were severally affected (Tieman et al., 2001). Acceleration in fruit softening cause exclusive fruit spoilage which needs to be checked. A rin transformed transgenic plants accumulate reduced amount of which indicate to developing antisense PG transgenic under the control of E8 promoter. Although transgenic fruits were produced with PG, enzyme activity was 60% wild type; however, a target of reduced softening of fruit was achieved.

However, constitutive expression of an antisense PG transgene driven by the CAMV 35S promoter yielded transgenic fruits, showed down-regulated accumulation of PG mRNA, retaining only 0.5–1% of wild-type levels of PG enzyme activity without affecting overall fruit ripening and softening of fruits (Rose et al., 2003; Saladié et al., 2007), whereas transgenic developed with antisense suppression of pectinesterase (PE) under CaMV35S promoter produced fruits with reduced PE activity and suppression in the rate of softening during ripening (Phan et al., 2007). In tomato, a large and divergent multigene family encodes EGases (cellulases) consisting of minimum eight members. The highly divergent *EGases LeCel1* and *LeCel2* mRNA accumulation were suppressed individually by constitutive expression of antisense transgenes (Rose et al., 2003). Without affecting the expression of other family members of E Gasses and fruit softening, in both of the cases, mRNA accumulation was decreased by 99% in fruit pericarp compared with wild type. Galactosidases in tomato are encoded by seven members (*TBG1–7*) multigene family. These members show differential expression patterns during fruit development (Smith and Gross, 2000). Transgenic plants have been developed to reduce the softening process using the members of this family. A short gene-specific region sense suppressed *TBG1* cDNA reduced TBG1 mRNA abundance to 10% of wild-type levels in ripe fruit but did not reduce total exo-galactanase activity as well as cell wall galactose content or fruit softening (Carey et al., 2001). Transgenic tomatoes of antisense beta-galactosidase 4 (*TBG4*) and 7 (*TBG7*) cDNAs along with CaMV35S promoter exhibited modulated fruit firmness in comparison with control fruit (Moctezuma et al., 2003). Ethylene response factors (*ERFs*) play an important role in regulating ethylene-induced ripening in fruits. ERFs are transcriptional regulators belonging to a multigene family. These mediate ethylene-dependent gene expressions by binding to the GCC motif in the ethylene-regulated promoter up and down-regulated expression of these individual *ERFs* in tomato has proved their role in plant development and ripening. The sense/antisense *LeERF1* transgenic tomato under the constitutive control of CaMV35 promoter was developed. Typical ethylene triple response on the etiolated seedling was observed during the overexpression of *LeERF1* in tomato plants, whereas in another report, antisense *LeERF1* fruits showed longer shelf life compared with wild-type tomato (Li et al., 2007). Over-expression of the Sl-ERF2 gene in transgenic tomato lines resulted in premature seed germination and indicative of increased ethylene sensitivity which enhanced hook formation of dark-grown seedlings (Pirrello et al., 2006). The expression of the *Mannanase 2* gene was up-regulated in *Sl-ERF2*-over-expressing seeds, indicating that *Sl-ERF2* induced the mannanase 2 gene which ultimately stimulated seed germination. Previously studied unreported tomato fruit cultivar named as Delayed Fruit Deterioration (DFD) undergoes normal ripening but remains firm and shows no loss of probity for at least six months (Rose et al., 2003). Ripening

DFD fruit significantly showed minimal water loss by transpiration and high cellular turgor, whereas expression of genes associated with wall segregation was similar with other cultivars (Saladié et al., 2007). Based on the biomechanical analysis, this group has introduced a model in which softening of tomato fruit is directly affected by cuticle providing physical support and regulating fruit water status. Although there is no candidate, gene/genes are identified for this trait till now, but once identified, it would be of much interest for biotechnological ends. At present, new and important components of complex gene-regulatory networks which controlling plant development are known as micro RNAs (miRNAs) (Jones-Rhoades et al., 2006). Though miRNAs and their targets have been identified in the number of plant species, not much work has been carried out in relation to their involvement in fruit development and ripening. Recently, scientists (Yin et al., 2008; Zhang et al., 2008) discriminated a set of miRNAs and their goals from tomatoes that were conserved with the phase change from vegetative to generative growth. In addition, a recent technology of high-throughput pyrosequencing has evident miRNAs s targeting genes that are involved in vegetable fruit ripening (Moxon et al., 2008; Shukla et al., 2008).

In another report, however, partial fragmentations (1038 bp cDNA) of melon invertase expressed in antisense orientation under the control of constitutive CaMV35S promoter were observed (Yu et al., 2008). The transgenic melon fruits were 60% reduced in size and reported increased sucrose and acidity invertase levels, with degraded chloroplast as a result of decreased photosynthetic rate compared with control fruits.

6.7   IMPACT OF BIOTECHNOLOGICAL APPROACHES ON THE SHELF LIFE OF VEGETABLES

In developing countries, on account, almost 50% of the post-harvest losses of vegetable produce has been observed due to excessive softening; the world second vegetable producing country, India, faces 35–40% of vegetable losses every year. Ripening of fruits intensifies their softening, which quickly damaged during the shipping and handling processes. The softening that accompanies ripening of fruits exacerbates damage during shipping and handling processes. This situation creates a direct impact on taste, consumer acceptability, shelf life and post-harvest diseases of the harvested products, which ultimately play a major role to reduce the cost/income of the producer. Biotechnological applications such as genetic engineering have a great potential to engineer the fruits which delayed ripening characters. In 1994 Calgene, United States of America had got first approval for commercial sale of a genetically engineered transgenic tomato product 'Flavr-savr' tomato which had delayed ripening trait; fruits with improved shelf life which were suitable for food processing can be developed (using down-regulation/modification of ethylene metabolism and manipulating cell wall metabolism).

In the current approach to prevent fruit, through the exploration of anti-sense RNA technology, ripening is to inhibit ethylene production. This antisense gene is produced by inverting the orientation of the protein-coding region of a gene related to its promoter region. This gene-produced mRNA has the same sequence as the antisense strand of the normal gene, and this mRNA is called antisense RNA. The transcription of the two genes yields antisense and sense RNA transcripts respectively which are complementary to each other and therefore pair to form double stranded RNA molecules, when sense and antisense gene are present in the same nucleus. The duplex formed inactivates the mRNA, and no further protein can be formed. Antisense RNA technology has been used to inhibit the expression of both ethylene biosynthesis (*ACC synthase* and *ACC oxidase*); gene and fruits from these transgenic plants were found to be more resistant to overripening and desiccating than control fruits.

It has been reported that the shelf life of transgenic fruits was last up to 60 days at room temperature without any change in their hardness and colour (Rose et al., 2003). RNAi technique, conferring delayed ripening in tomato has been reported using ACO gene. In another report antisense, transgenic lines of tomato have also been produced by altering ethylene biosynthesis with an *anti-ACO* gene (Nath et al., 2006). Researchers also reported that transgenic tomato fruits had an extended shelf life of at least 120 days (Xiong et al., 2005).

6.8 CHALLENGES ASSOCIATED WITH COMMERCIALIZATION OF POST-HARVEST VEGETABLES DEVELOPED THROUGH BIOTECHNOLOGICAL APPROACHES

Various studies revealed that even though biotechnological approaches are being seen by the scientific community as a universal cure to recent increased demands of nutritional security, the technology is more of a scientific jargon than a commercially viable entity. The probable cause is the dilemma and uncertainties which remain up to today regarding the consumption of genetically engineered vegetables. The cessation has created hurdle with consumption of genetically reshaped vegetables in many countries and some continents mainly due to the complexities surrounding its use. Very few biotech vegetables have been introduced, after the first biotech crop genetically modified tomato introduced and commercialized for processing as a consumer tomato paste (Anthony and Ferroni, 2012). There are also few reported cases with potential benefits for farmers in developing countries including virus-resistant papaya in China, which is now commercially grown, and, at present, the high-profile case of Bt eggplant is in progress in India (Choudhary and Gaur, 2009). Brinjal fruits are highly susceptible to fruit and shoot borer insects; to reduce the economic loss and yield of the crop, extensive use of pesticides has been applied to the crop which is ultimately harmful to the consumers. In India, GEAC (Genetic Engineering Appraisal Committee) recommended India, the Indian commercial release of Bt brinjal (Event EE1, developed by ICAR-IIVR) in 2010; but unfortunately, no authorization was given by the Ministry of Environment and Forestry (Jayaraman, 2010). Moreover, with brinjal, a wide array of transgenic vegetables such as tomato, broccoli, cabbage and okra are also developed in India (James et al., 2009).

Since 1996, 183 events in 24 crop species, mainly for non-consumable crops having more than 900 registrations, have been approved and granted (CERA. GM Crop Database; ISAAA GM Approval Database) (James et al., 2009). Very less number of registrations had been awarded since 1990 for commercialization of various vegetables. Small farmers can be significantly benefited by approving suitably developed and stewarded high-value vegetables, because of the relatively high prices of these crops on the market. A study to understand factors responsible for the lack of traits for commercialization have been carried out by Miller and Bradford (2010). The study revealed that among 300 research papers published during last decades, describing approximately 250 unique transgenic events, 20% of the papers were from India and China. Many input traits such as insect resistance, herbicide tolerance, output traits such as yield, post-harvest quality, biotic, abiotic stresses, traits where modifications to compositions of oil, starch, protein and nutrients had been addressed by various researchers. The primary results of the study concluded that the traits were not introduce to markets, not due to poor performance or lack of interest but because of regulatory affairs.

A review has inferred that opinion and debate on acceptance of transgenic agricultural biotechnology remain centralized both 'for and against' and is often not aligned with ad mount review and balanced; empirically grounded evaluation of socio-economic and community benefits, human safety, environmental considerations such as non-target safety, biodiversity, gene flow and associated risks (Anthony and Ferroni, 2012). At a time when biotech crops have been grown extensively in the Americas and Asia for more than 13 years, the preventive principle conquers in many countries even for the traits included in these crops. For example, in the European Union (EU), GM vegetables are not allowed on the market till now, and none of the GM plants currently authorized in the EU are purporting for direct consumption (GMO Compass 2006). In the other case, GM tomatoes are lurking in grocery stores in the United States of America and never received authorization in the EU. Although there are some controversies in Europe when meant reports in EU members indicate that some countries such as Finland, Germany and Greece have strongly opposed the commercialization of GM crops including vegetables, Spain and United Kingdom do not fundamentally oppose cultivating GM crops, but they have taken some precautionary action.

Although lots of debates are available since last 25 years of transgenic crops in literature, we are neither directly privileged nor have dismissed the genetically modified crops. Apart from government's policy on regulations of genetically modified crops, there are few key factors which influence the future availability of biotech vegetables. Developing countries are managing capacity and the ability of technology providers. Excellence in stewardship (2009) reveals that stewardship biotechnology of vegetables includes not just management of biosafety and adherence with regulatory authorities' requirements but also product probity and integrity along the whole product life cycle right from initial research ideas to the withdrawal of crop varieties. In this situation, there is urgent need of national and international harmonized solutions. As the private sector will keep trying to release the supporting technology to the public sector to assist farmers in developing countries, especially for food crops such as vegetables that could cross national borders or enter international trade channels, this situation makes cautions and altering the governments. Additionally, little impact has been realized to date with vegetables because of development timescales for molecular breeding and development and regulatory costs and political considerations facing biotech crops in many countries.

6.9 NOVEL APPROACHES FOR COMMERCIALIZATION OF GENETICALLY MODIFIED VEGETABLES

Although there are huge number of biotechnological tools offering potential to excel the development and introduction of improved varieties and possibility for greater genetic diversity, there is huge opportunity to improve and to introduce biotechnological applications for the full benefits, which are yet to be established. In order to reduce yield gap, the development and adoption of technology-based solutions have many constrains that are to be overcome. The new paradigm shift proposed to include both public and private entities to integrate a broader trust and creation of propagation material system and markets backed by extension and other services for betterment of poor farmers. There is need of transformational change in research, product development and the delivery of seed embodied technology to farmers by commitment to increase funding of agricultural R&D. Public–private partnerships-related policies and operational level to enable formation of such type of bonds should be promoted. The technologies having mild chemical fungicides, herbal anti-microbial extracts and physical means, such as hot water treatment, irradiation with ultraviolet light, microwave and infrared treatment in the post-harvest biocontrol process, have reduced the loss of post-harvest products. Further, in genetical improvement in the expression of key recombinant DNA genes and/or combining genes from different agents in the mass production, formulation and storage, or in response to exposure and contact with parent plant tissue after application will also help to improve the shelf life of crop plants. Use of genetically modified organisms as biocontrol agents to enhance the post-harvest quality and shelf life of fruits and vegetables and the research towards discovering new DNA genes instead of the ones currently used in practices, only a small portion of the earth microflora has been identified and characterized.

6.10 FUTURE TRENDS

In the current chapter, discussion on post-harvest interventions through biotechnological approaches insights the available promising methods for crop improvement. The examples which are provided in the chapter are simple and are also provided with proof of evidences. The feasibility of gene inactivation through antisense techniques, endogenous enhancement of plant hormone levels and ethylene biosynthesis inhibition by gene knockout are the examples to improve the shelf life of vegetable crops for post-harvest management (Hamilton, 1990; Oeller et al., 1991; Grbic and Bleecker, 1995). The role of ethylene biosynthesis and their mode of action give insights to perform basic research over the ethylene along with development of improved varieties. Use of hydrolases in fruit softening and senescence in leaf using up/down concentration of cytokines also played an important role to extend shelf life of leafy and fruity vegetables. Although the various experiments using biotechnological application extensively improve the basic research over post-harvest technology, there is need of lot of refinement, and these are only achieved though expensive basic research. There are several types of lacunas in development of the genetically engineered, commercially viable vegetable crops, which can provide better post-harvest characters. The harvested tissue has also very rapidly changed their biochemical, molecular and cellular activities, and we have a challenge to reduce these activities so that we can retain fresh tissues and improved commercial value of fruit/leafy vegetable crops. As changes in cellular processes may change the metabolic activity of the crop, the current race to adopt new technology by the private and public sector (especially private sector) needs to pay attention to adopt new techniques and product. In this way, might be we are at the risk of such type of products which have potentially adverse agronomic characteristics. To avoid such type of situations, we need to pay full attention to release or commercialize these products. We can recover from such conditions when government takes possible action to reduce the fund over the basic research and pay attention to the applied research in their universities and research institutions, which might full recover the cost of investment. When we improve our fundamental research, we discover new genes which may be used for betterment and advancement of post-harvest technologies. It is known that pathogens are the most serious problem, which reduce the shelf life of vegetable and fruits/leaves. The discovery of new genes is urgently needed to insure resistant to various post-harvest pathogens. Although there is need of specific promoters which regulate and allow the precise expression of a gene in a tissue specific manner, again there is need of inducible promoters which may act according to their situation and applicable artificially turn on and off, whenever required. Form the above discussion, it can be inferred that the most reliable techniques are the gene knockout. Researchers mutate the genes to check their expression for various traits. A lot of gene knock out models are available which are enabling us to obtain promising results. There is also a need of new, promising, reliable and faster transformation techniques. Further, there is next generation, in which there is urgent need of crop improvement to benefit directly to the consumers rather that produces. Although longer shelf life is an urgent need targets, without research to the targeted pathways insight of the crop plant, it will always be a complex objective. We have required a long shelf life vegetable fruit, which ripen much more slowly, whereas slow ripening should not be only an objective because as we know that majority of the fruits are destroyed due to action of pathogens; so there should be priority to develop such techniques which also protects the fruits for pathogens. Further, there should not be accepted vegetable fruits on the cost of loss of ideal level of proteins, vitamins and aroma. There are developed genetically engineered vegetable crop varieties which have enhanced nutritional quality, including such components that are normally not present in traditional cultivars. In future, biotechnology is emerging as a powerful tool for crop improvement. Although we are in its initial stages, we are having the potential to apply these techniques to enhance our agronomical and nutritional characteristics of crops. Most of the consumers, who do not have enough knowledge over biotechnological applications, can only see the tip of the iceberg, but the duty of a biotechnologist is to have patience to develop safe and improved crops which can serve human beings.

6.11 CONCLUSION

In the present chapter, we discuss biotechnological applications to enhance nutritional quality and storability of vegetables using the biotechnological approach. It was marked that developed biotechnological approaches have the potential to enhance the yield, quality and shelf life of pre- and post-harvest intervention of vegetables to meet the demands of consumers. However, the developing countries and most of the European country are yet to adopt biotech approaches for vegetables, and it has more of academic jargon than a commercial reality. This is due to the uncertainty of release of the products and dilemma of the policy-makers. Many stakeholders in the industry, private sectors, agriculturalists, policy-makers, biotechnologists, scientists, extension agents, farmers and the general public must be engaged in making policy, seed embodiments and products development in order to confirm that the current debates and complexities surrounding the registration and the commercialization of genetically modified vegetables are adequately addressed. The government has thought much more in emphasizing manner to adopt to meet the gap between demand and production. The fruits of knowledge can be availed if there are the total devolution by all stakeholders regarding increased and sustained funding, increase agricultural R&D and less cost and time for registration and commercialization of new traits.

REFERENCES

Abano, E. E.; Sam-Amoah, L. K. Effects of Different Pretreatments on Drying Characteristics of Banana Slices. *APRN J. Eng. App. Sci.* **2011**, *6*, 121–129.

Abano, E. E.; Buah, J. N. Biotechnological Approaches to Improve Nutritional Quality and Shelf Life of Fruits and Vegetables. *Int. J. Eng. Technol.* **2014**, *4*, 11.

Aghdam, M. S.; Laura, S.; Francisco, B. F.; Samad, B. Heat Shock Proteins as Biochemical Markers for Postharvest Chilling Stress in Fruits and Vegetables. *Scientia Hortic.* **2013**, *160*, 54–64.

Ahmad, P.; Muhammad, A.; Muhammad, Y.; Xiangyang, H.; Kumar, A.; Akram, N. A.; Al-Qurainy, F. Role of Transgenic Plants in Agriculture and Biopharming. *Biotechnol. Adv.* **2012**, *30*(3), 524–540.

Al-Whaibi; Mohamed, H. Plant Heat-Shock Proteins: A Mini Review. *J. King Saud Univ. Sci.* **2011**, *23*(2), 139–150.

Anthony, V. M.; Ferroni, M. Agricultural Biotechnology and Smallholder Farmers in Developing Countries. *Curr. Opin. Biotechnol.* **2012**, *23*(2), 278–285.

Bapat, V. A.; Trivedi, P. K.; Ghosh, A.; Sane, V. A.; Ganapathi, T. R.; Nath, P. Ripening of Fleshy Fruit: Molecular Insight and the Role of Ethylene. *Biotechnol. Adv.* **2010**, *28*(1), 94–107.

Barry, C. S.; Giovannoni, J. J. Ripening in the Tomato Green-Ripe Mutant is Inhibited by Ectopic Expression of a Protein that Disrupts Ethylene Signaling. *Proc. Natl. Acad. Sci.* **2006**, *103*(20), 7923–7928.

Borges, J. C.; Ramos, C. H. I. Protein Folding Assisted by Chaperones. *Protein Peptide Lett.* **2005**, *12*(3), 257–261.

Bowles, D. J. Defense-Related Proteins in Higher Plants. *Ann. Rev. Biochem.* **1990**, *59*(1), 873–907.

Brandt A. S.; Woodson, W. R. Variation in Flower Senescence and Ethylene Biosynthesis Among Carnations. *Hortic. Sci.* **1992**, *27*(10), 1100–1102.

Brogue, K.; Chet, I.; Holliday, M.; Cressman, R.; Biddle, P.; Knowlton, S.; Mauvais, C. J.; Broglie, R. Transgenic Plants with Enhanced Resistance to the Fungal Pathogen Rhizoctonia Solani. *Science* **1991**, *254*(5035), 1194–1197.

Bureau, J. L.; Bushway, R. J. HPLC Determination of Carotenoids in Fruits and Vegetables in the United States. *J. Food Sci.* **1986**, *51*(1), 128–130.

Carey, A. T.; Smith, D. L.; Harrison, E.; Bird, C. R.; Gross, K. C.; Seymour, G. B.; Tucker, G. A. Down-Regulation of a Ripening-Related β-Galactosidase Gene (TBG1) in Transgenic Tomato Fruits. *J. Exp. Bot.* **2001**, *52*(357), 663–668.

Chatterjee, A.; Murata, H.; McEvoy, J. L.; Chatterjee, A. Global Regulation of Pectinases and Other Degradative Enzymes in Erwinia Carotovora Subsp. Carotovora, the Incitant of Postarvest Decay in Vegetables. *Hortic. Sci.* **1994**, *29*(7), 754–758.

Choudhary, B.; Gaur, K. *The Development and Regulation of Bt Brinjal in India (Eggplant/ Aubergine).* International Service for the Acquisition of Agri-biotech Applications, 2009.

Ciardi, J. A.; Tieman, D. M.; Lund, S. T.; Jones, J. B.; Stall, R. E.; Klee, H. J. Response to *Xanthomonas campestris* pv. vesicatoria in Tomato Involves Regulation of Ethylene Receptor Gene Expression. *Plant Physiol.* **2000**, *123*(1): 81–92.

DeFrancesco, L. Vintage Genetic Engineering. *Nat. Biotechnol.* **2008**, *26*(3), 261–263.

Delannay, X.; LaVallee, B. J.; Proksch, R. K.; Fuchs, R. L.; Sims, S. R.; Greenplate, J. T.; Marrone, P. G. Field Performance of Transgenic Tomato Plants Expressing the *Bacillus thuringiensis* var. Kurstaki Insect Control Protein. *Nat. Biotechnol.* **1989**, *7*(12), 1265–1269.

Droby, S. Improving Quality and Safety of Fresh Fruits and Vegetables After Harvest by the Use of Biocontrol Agents and Natural Materials. In *I International Symposium on Natural Preservatives in Food Systems* 2005, *709*, pp. 45–52.

Droby, S.; Chalutz, E.; Wilson, C. L.; Wisniewski, M. E. Biological Control of Postharvest Diseases: A Promising Alternative to the use of Synthetic Fungicides. *Phytoparasitica* **1992**, *20*(1), S149–S153.

El Ghaouth, A.; Wilson, C.; Wisniewski, M. Biologically-Based Alternatives to Synthetic Fungicides for the Control of Postharvest Diseases of Fruit and Vegetables. In *Diseases of Fruits and Vegetables:* Vol. II; Naqvi, S. A. M. H., Ed.; Springer: Netherlands, 2004; pp 511–535.

Fan, Q.; Tian, S. P. Postharvest Biological Control of Grey Mold and Blue Mold on Apple by Cryptococcus Albidus (Saito) Skinner. *Postharvest Biol. Technol.* **2001**, *21*(3), 341–350.

Gan, S.; Amasino, R. M. Making Sense of Senescence (Molecular Genetic Regulation and Manipulation of Leaf Senescence). *Plant. Physiol.* **1997**, *113*(2), 313.

Gellatly, K.; Dennis, D. T. *Plant Biotechnology and GMOs. Comprehensive Biotechnology,* 2nd ed.; Vol. 4; Murray Moo-Young, Ed.; Elsevier: Spain, 2011; pp 9–22.

Giovannoni, J. J.; DellaPenna, D.; Bennett A. B.; Fischer, R. L. Expression of a Chimeric Polygalacturonase Gene in Transgenic Rin (ripening inhibitor) Tomato Fruit Results in Polyuronide Degradation but not Fruit Softening. *Plant. Cell* **1989**, *1*(1), 53–63.

Good, X.; Kellogg, J. A.; Wagoner, W.; Langhoff, D.; Matsumura, W.; Bestwick, R. K. Reduced Ethylene Synthesis by Transgenic Tomatoes Expressing S-adenosylmethionine Hydrolase. *Plant. Mol. Biol.* **1994**, *26*(3), 781–790.

Grbić V.; Bleecker, A. B. Ethylene Regulates the Timing of Leaf Senescence in Arabidopsis. *Plant J.* **1995**, *8*(4), 595–602.

Grierson, D.; Covey, S. N. Structure and Expression of Nuclear Genes. In *Plant Molecular Biology;* Springer: US, 1988; pp 22–46.

Hamilton, A. J.; Lycett, G. W.; Grierson, D. Antisense Gene that Inhibits Synthesis of the Hormone Ethylene in Transgenic Plants. *Nature* **1990**, *346*, 284–287.

Henzi, M. X.; McNeil, D. L.; Christey, M. C.; Lill, R. E. A Tomato Antisense 1-Aminocyclopropane-1-Carboxylic Acid Oxidase Gene Causes Reduced Ethylene Production in Transgenic Broccoli. *Funct. Plant Biol.* **1999a**, *26*(2), 179–183.

Henzi, M. X.; Christey, M. C.; McNeil, D. L.; Davies, K. M. Agrobacterium Rhizogenes-Mediated Transformation of Broccoli (*Brassica oleracea* L. var. italica) with an Antisense 1-Aminocyclopropane-1-Carboxylic Acid Oxidase Gene. *Plant Sci.* **1999b**, *143*(1), 55–62.

Henzi, M. X.; Christey, M. C.; McNeil, D. L. Morphological Characterisation and Agronomic Evaluation of Transgenic Broccoli (*Brassica oleracea* L. var. italica) Containing an Antisense ACC Oxidase Gene. *Euphytica* **2000**, *113*(1), 9–18.

Ifon, E. T.; Bassir, O. The Nutritive Value of Some Nigerian Leafy Green Vegetables—Part 2: The Distribution of Protein, Carbohydrates (including ethanol-soluble simple sugars), Crude Fat, Fibre and Ash. *Food Chem.* **1980**, *5*(3), 231–235.

Imaseki, H. The Biochemistry of Ethylene Biosynthesis. *Plant Hormone Ethylene* **1991**, 1–20.

Irving, H. R.; Kuć, J. A. Local and Systemic Induction of Peroxidase, Chitinase and Resistance in Cucumber Plants by K2HPO4. *Physiol. Mol. Plant Pathol.* **1990**, *37*(5), 355–366.

Ismail, A.; Fun, C. S. Determination of Vitamin C, β-Carotene and Riboflavin Contents in Five Green Vegetables Organically and Conventionally Grown. *Malays. J. Nutr.* **2003**, *9*(1), 31–39.

James, C. Brief 41: Global Status of Commercialized Biotech/GM Crops: 2009. *ISAAA Brief*, International Service for the Acquisition of Agri-biotech Applications 290: Ithaca, NY, 2009.

Janisiewicz, W. J.; Korsten, L. Biological Control of Postharvest diseases of Fruits. *Ann. Rev. Phytopathol.* **2002**, *40*(1), 411–441.

Jayaraman, K. Bt Brinjal Splits Indian Cabinet. *Nat. Biotechnol.* **2010**, *28*(4), 296–296.

John, I.; Drake, R.; Farrell, A.; Cooper, W.; Lee, P.; Horton, P.; Grierson, D. Delayed Leaf Senescence in Ethylene-Deficient ACC-Oxidase Antisense Tomato Plants: Molecular and Physiological Analysis. *Plant J.* **1995**, *7*(3), 483–490.

Jones-Rhoades, W; Matthew, D.; Bartel, P.; Bartel, B. MicroRNAs and Their Regulatory Roles in Plants. *Annu. Rev. Plant Biol.* **2006**, *57*, 19–53.

Kinzel, B. Nikkomycin: Antibiotic for Plants. *Agricultural Research-US Department of Agriculture, Agricultural Research Service (USA)*, 1989.

Klee, H. J. Ripening physiology of Fruit from Transgenic Tomato (*Lycopersicon esculentum*) Plants with Reduced Ethylene Synthesis. *Plant Physiol.* **1993**, *102*(3), 911–916.

Klein, B. P.; Perry, A. K. Ascorbic Acid and Vitamin A Activity in Selected Vegetables from Different Geographical Areas of the United States. *J. Food Sci.* **1982**, *47*(3), 941–945.

Korsten, L. Advances in Control of Postharvest Diseases in Tropical Fresh Produce. *Int. J. Postharvest Technol. Innovat.* **2006**, *1*(1), 48–61.

Kramer, M. Genetically Engineered Plant Foods: Tomatoes. In *Report 2, Agricultural Biotechnology, Food Safety and Nutritional Quality for the Consumer*; MacDonald J. F.

Ed.; National Agriculture Biotechnology Council, Union Press: Binghamton, New York, 1990; pp 127–130.

Kramer, M.; Sanders, R. A.; Sheehy, R. E.; Melis, M.; Kuehn, M.; Hiatt, W. R. Field Evaluation of Tomatoes with Reduced Polygalacturonase by Antisense RNA, In *Hortic biotechnol.*; Bennett A. B., O'Neill, S. D., Eds.; Wiley-Liss: New York, 1990; pp 347–355.

Larkindale, J.; Hall, J. D.; Knight, M. R.; Vierling, E. Heat Stress Phenotypes of Arabidopsis Mutants Implicate Multiple Signaling Pathways in the Acquisition of Thermotolerance. *Plant Physiol.* **2005**, *138*(2), 882–897.

Li, H. Y.; Chang, C. S.; Lu, L. S.; Liu, C. A.; Chan, M. T.; Charng, Y. Y. Over-Expression of Arabidopsis Thaliana Heat Shock Factor Gene (AtHsfA1b) Enhances Chilling Tolerance in Transgenic Tomato. *Botanical Bull. Academia Sinica.* **2003**, 44.

Li, Y.; Zhu, B.; Xu, W.; Zhu, H.; Chen, A.; Xie, Y.; Shao, Y.; Luo, Y. LeERF1 Positively Modulated Ethylene Triple Response on Etiolated Seedling, Plant Development and Fruit Ripening and Softening in Tomato. *Plant Cell Rep.* **2007**, *26*(11), 1999–2008.

Linthorst, H. J. M.; Loon, L. C.; Memelink, J.; Bol, J. F. Characterization of cDNA Clones for a Virus-Inducible, Glycine-Rich Protein from Petunia. *Plant Mol. Biol.* **1990**, *15*(3), 521–523.

Lund, P.; Lee, R. Y.; Dunsmuir, P. Bacterial Chitinase is Modified and Secreted in Transgenic Tobacco. *Plant Physiol.* **1989**, *91*(1), 130–135.

Lurie, S.; Handros, A.; Fallik, E.; Shapira, R. Reversible Inhibition of Tomato Fruit Gene Expression at High Temperature (effects on tomato fruit ripening). *Plant Physiol.* **1996**, *110*(4), 1207–1214.

Manning, K.; Tör, M.; Poole, M.; Hong, Y.; Thompson, A. J.; King, G. J.; Giovannoni, J. J.; Seymour, G. B. A Naturally Occurring Epigenetic Mutation in a Gene Encoding an SBP-Box Transcription Factor Inhibits Tomato Fruit Ripening. *Nat. Genetics* **2006**, *38*(8), 948–952.

Martínez-Romero, D.; Bailén, G.; Serrano, M.; Guillén, F.; Valverde, J. M.; Zapata, P.; Castillo, S.; Valero, D. Tools to Maintain Postharvest Fruit and Vegetable Quality Through the Inhibition of Ethylene Action: A Review. *Crit. Rev. Food Sci. Nutr.* **2007**, *47*(6), 543–560.

Miller, J. K.; Bradford, K. J. The Regulatory Bottleneck for Biotech Specialty Crops. *Nat. Biotechnol.* **2010**, *28*(10), 1012–1014.

Moctezuma, E.; Smith, D. L.; Gross, K. C. Antisense Suppression of a β-Galactosidase Gene (TB G6) in Tomato Increases Fruit Cracking. *J. Exp. Botany* **2003**, *54*(390), 2025–2033.

Mohan, R.; Kolattukudy, P. E. Differential Activation of Expression of a Suberization-Associated Anionic Peroxidase Gene in Near-Isogenic Resistant and Susceptible Tomato Lines by Elicitors of Verticillium Albo-Atratrum. *Plant Physiol.* **1990**, 92(1), 276–280.

Moxon, S.; Jing, R.; Szittya, G.; Schwach, F.; Pilcher, R. L. R.; Moulton, V.; Dalmay, T. Deep Sequencing of Tomato Short RNAs Identifies MicroRNAs Targeting Genes Involved in Fruit Ripening. *Genome Res.* **2008**, *18*(10), 1602–1609.

Murata, H.; McEvoy, J. L.; Chatterjee, A.; Collmer, A.; Chatterjee, A. K. Molecular Cloning of an Aep A Gene That Activates Production of Extracellular Pectolytic, Cellulolytic, and Proteolytic Enzymes in Erwinia Carotovora Subsp. Carotovora. *Mol. Plant Microbe Int.* **1991**, (4), 239–246.

Nath, P.; Trivedi, P. K.; Sane, V. A.; Sane, A. P. Role of Ethylene in Fruit Ripening. In *Ethylene Action in Plants;* Nafees, A., Khan, Eds.; Springer Berlin: Heidelberg, 2006; pp 151–184.

Oboh, G.; Akindahunsi, A. A. Change in the Ascorbic Acid, Total Phenol and Antioxidant Activity of Sun-Dried Commonly Consumed Green Leafy Vegetables in Nigeria. *Nutr. Health* **2004**, *18*(1), 29–36.

Oeller, P. W.; Lu, M. W.; Taylor, L. P.; Pike, D. A.; Theologis, A. Reversible Inhibition of Tomato Fruit Senescence by Antisense RNA. *Science* **1991**, *254*(5030), 437–439.

Phan, T. D.; Bo, W.; West, G.; Lycett, G. W.; Tucker, G. A. Silencing of the Major Salt-Dependent Isoform of Pectinesterase in Tomato Alters Fruit Softening. *Plant Physiol.* **2007**, *144*(4), 1960–1967.

Pirrello, J.; Miranda, F. J.; Sanchez-Ballesta, M. T.; Tournier, B.; Ahmad, Q. K.; Regad, F.; Latché, A.; Pech, J. C.; Bouzayen, M. Sl-ERF2, a Tomato Ethylene Response Factor Involved in Ethylene Response and Seed Germination. *Plant Cell Physiol.* **2006**, *47*(9), 1195–1205.

Prasanna, H. C.; Kashyap, S. P.; Krishna, R.; Sinha, D. P.; Reddy, S.; Malathi, V. G. Marker Assisted Selection of Ty-2 and Ty-3 Carrying Tomato Lines and Their Implications in Breeding Tomato Leaf Curl Disease Resistant Hybrids. *Euphytica* **2015**, *204*(2), 407–418.

Prasanna, V.; Prabha, T. N.; Tharanathan, R. N. Fruit Ripening Phenomena–an Overview. *Crit. Rev. Food Sci. Nutr.* **2007**, *47*(1), 1–19.

Pressman, E.; Shaked, R.; Firon, N. Exposing Pepper Plants to High day Temperatures Prevents the Adverse Low Night Temperature Symptoms. *Physiol. Plant* **2006**, *126*(4), 618–626.

Rai, A. C.; Singh, M.; Shah, K. Engineering Drought Tolerant Tomato Plants Over-Expressing BcZAT12 Gene Encoding a C_2H_2 Zinc Finger Transcription Factor. *Phytochemistry* **2013**, *85*, 44–50.

Rai, A. C.; Singh, M.; Shah, K. Effect of Water Withdrawal on Formation of Free Radical, Proline Accumulation and Activities of Antioxidant Enzymes in ZAT12-Transformed Transgenic Tomato Plants. *Plant. Physiol. Biochem.* **2012**, *61*, 108–114.

Roberts, R. G. Postharvest Biological Control of Gray Mold of Apple by *Cryptococcus laurentii. Phytopathology* **1990**, *80*(6), 526–530.

Rose, J. K. C.; Saladié, M.; Catalá, C. The Plot Thickens: New Perspectives of Primary Cell Wall Modification. *Curr. Opin. Plant Biol.* **2004**, *7*(3), 296–301.

Rose, J. K. C.; Catalá, C.; Gonzalez-Carranza, C. Z. H.; Roberts, J. A. Plant Cell Wall Disassembly. In *Plant Cell Wall*; Rose J. K. C, Ed.; Blackwell Publishing Ltd.: Oxford, 2003, Vol. 8; pp 264–324.

Sabehat, A.; Weiss, D.; Lurie, S. The Correlation Between Heat-Shock Protein Accumulation and Persistence and Chilling Tolerance in Tomato Fruit. *Plant. Physiol.* **1996**, *110*(2), 531–537.

Sabehat, A.; Weiss, D.; Lurie, S. Heat-Shock Proteins and Cross-Tolerance in Plants. *Physiol. Plant.* **1998a**, *103*(3), 437–441.

Sabehat A, Lurie, S.; Weiss, D. Expression of Small Heat-Shock Proteins at Low Temperatures a Possible Role in Protecting Against Chilling Injuries. *Plant Physiol.* **1998b,** *117*(2), 651–658.

Saladié, M.; Matas, A. J.; Isaacson, T.; Jenks, M. A.; Goodwin, S. M.; Niklas, K. J.; Xiaolin, R. A Reevaluation of the Key Factors that Influence Tomato Fruit Softening and Integrity. *Plant Physiol.* **2007,** *144*(2), 1012–1028.

Saltveit, M. E.; Hepler, P. K. Effect of Heat Shock on the Chilling Sensitivity of Trichomes and Petioles of African Violet (*Saintpaulia ionantha*). *Physiol. Plant.* **2004**, *121*(1), 35–43.

Sato Y.; Murakami, T.; Funatsuki, H.; Matsuba, S.; Saruyama, H.; Tanida, M. Heat Shock-Mediated APX Gene Expression and Protection Against Chilling Injury in Rice Seedlings. *J. Exp. Botany* **2001,** *52*(354), 145–151.

Schöffl, F.; Prändl, R.; Reindl, A. Regulation of the Heat-Shock Response. *Plant. Physiol.* **1998,** *117*(4), 1135–1141.

Scholl, T. O.; Johnson, W. G. Folic Acid: Influence on the Outcome of Pregnancy. *Am. J. Clin. Nutr.* **2000,** *71*(5), 1295–1303.

Sevillano, L.; Sanchez-Ballesta, M. T.; Romojaro, F.; Flores, F. B. Physiological, Hormonal and Molecular Mechanisms Regulating Chilling Injury in Horticultural Species. Postharvest Technologies Applied to Reduce its Impact. *J. Sci. Food Agric.* **2009,** *89*(4), 555–573.

Shah, K.; Singh, M.; Rai, A. C. Bioactive Compounds of Tomato Fruits from Transgenic Plants Tolerant to Drought. *LWT–Food Sci. Technol.* **2015,** *61*(2), 609–614.

Sharma, R. R.; Singh, D.; Singh, R. Biological Control of Postharvest Diseases of Fruits and Vegetables by Microbial Antagonists: A Review. *Biol. Control* **2009,** *50*(3), 205–221.

Shukla, A.; Singh, V. K.; Bharadwaj, D. R.; Kumar, R.; Rai, A.; Rai, A. K.; Mugasimangalam, R.; Parameswaran, S.; Singh, M.; Naik, P. S. *De Novo* Assembly of Bitter Gourd Transcriptomes: Gene Expression and Sequence Variations in Gynoecious and Monoecious Lines. *PLoS One* **2015,** *10*(6), e0128331.

Shukla, L. I.; Chinnusamy, V.; Sunkar, R. The Role of MicroRNAs and Other Endogenous Small RNAs in Plant Stress Responses. *Biochimica et Biophysica Acta (BBA)-Gene Regulatory Mechanisms* **2008,** *1779*(11), 743–748.

Singh, D.; Sharma, R. R. *Postharvest Diseases of Fruit and Vegetables and Their Management;* Daya Publishing House: New Delhi, India, 2007.

Smart C. M.; Scofield, S. R.; Bevan, M. W.; Dyer, T. A. Delayed Leaf Senescence in Tobacco Plants Transformed with tmr, a Gene for Cytokinin Production in Agrobacterium. *Plant. Cell* **1991,** *3*(7), 647–656.

Smith, D. L.; Gross, K. C. A Family of at Least Seven β-Galactosidase Genes is Expressed During Tomato Fruit Development. *Plant. Physiol.* **2000,** *123*(3), 1173–1184.

Sorger, P. K.; Nelson, H. C. M. Trimerization of a Yeast Transcriptional Activator Via a Coiled-Coil Motif. *Cell* **1989,** *59*(5), 807–813.

Stahl, W.; Sies, H. Antioxidant Defense: Vitamins E and C and Carotenoids. *Diabetes* **1997,** *46*(S2), S14–S18.

Swindell, W. R.; Huebner, M.; Weber, A. P. Transcriptional Profiling of Arabidopsis Heat Shock Proteins and Transcription Factors Reveals Extensive Overlap Between Heat and Non-Heat Stress Response Pathways. *BMC Genomics* **2007,** *8*(1), 1.

Theologis, A.; Zarembinski, T. I.; Oeller, P. W.; Liang, X.; Abel, S. Modification of Fruit Ripening by Suppressing Gene Expression. *Plant. Physiol.* **1992,** *100*(2), 549.

Tieman, D. M.; Ciardi, J. A.; Taylor, M. G.; Klee, H. J. Members of the Tomato LeEIL (EIN3-like) Gene Family are Functionally Redundant and Regulate Ethylene Responses Throughout Plant Development. *Plant. J.* **2001,** *26*(1), 47–58.

Timperio, A. M.; Egidi, M. G.; Zolla, L. Proteomics Applied on Plant Abiotic Stresses: Role of Heat Shock Proteins (HSP). *J. Proteomics* **2008,** *71*(4), 391–411.

Vadivambal, R.; Jayas, D. S. Changes in Quality of Microwave-Treated Agricultural Products—A Review. *Bio. Eng.* **2007,** *98*(1), 1–16.

Van Staden, J. Cytokinins and Auxins in Carnation Senescence as Related to Chemical Treatments. In *IV International Symposium on Postharvest Physiology of Ornamental Plants* 1988, *261*, 69–80.

Vanitha, S. C.; Umesha, S. Variations in Defense Related Enzyme Activities in Tomato During the Infection with Bacterial wilt Pathogen. *J. Plant Interact.* **2008,** *3*(4), 245–253.

Vrebalov, J.; Ruezinsky, D.; Padmanabhan, V.; White, R.; Medrano, D.; Drake, R.; Schuch, W.; Giovannoni, J. A MADS-box Gene Necessary for Fruit Ripening at the Tomato Ripening-Inhibitor (rin) Locus. *Science* **2002,** *296*(5566), 343–346.

Wang, W.; Vinocur, B.; Shoseyov, O.; Altman, A. Role of Plant Heat-Shock Proteins and Molecular Chaperones in the Abiotic Stress Response. *Trends Plant. Sci.* **2004,** *9*(5), 244–252.

Waters, E. R.; Lee, G. J.; Vierling, E. Evolution, Structure and Function of the Small Heat Shock Proteins in Plants. *J. Exp. Botany* **1996,** *47*(3), 325–338.

Xiong, A. S.; Yao, Q. H.; Peng, R. H.; Li, X.; Han, P. L.; Fan, H. Q. Different Effects on ACC Oxidase Gene Silencing Triggered by RNA Interference in Transgenic Tomato. *Plant. Cell Rep.* **2005,** *23*(9), 639–646.

Yin, J. Q.; Zhao, R. C.; Morris, K. V. Profiling MicroRNA Expression with Microarrays. *Trends Biotechnol.* **2008,** *26*(2), 70–76.

Yu, X., Wang, X.; Zhang, W.; Qian, T.; Tang, G.; Guo, Y.; Zheng, C. Antisense Suppression of an Acid Invertase Gene (MAI1) in Muskmelon Alters Plant Growth and Fruit Development. *J. Exp. Botany* **2008,** *59*(11), 2969–2977.

Zhang, H.; Zheng, X.; Yu, T. Biological Control of Postharvest Diseases of Peach with *Cryptococcus laurentii. Food Control.* **2007,** *18*(4), 287–291.

Zhang, J.; Zeng, R.; Chen, J.; Liu, X.; Liao, Q. Identification of Conserved MicroRNAs and Their Targets From *Solanum lycopersicum* Mill. *Gene* **2008,** *423*(1), 1–7.

Zhang, Y.; Butelli, E.; Stefano, R. D.; Schoonbeek, H.; Magusin, A.; Pagliarani, C.; Wellner, N. Anthocyanins Double the Shelf Life of Tomatoes by Delaying Overripening and Reducing Susceptibility to Gray Mold. *Curr. Biol.* **2013,** *23*(12), 1094–1100.

Zhu, S.; Noureddine, B.; Norio, S. Non-Chemical Approaches to Decay Control in Postharvest Fruits. *Adv. Postharvest Technol. Hortic. Crops.* **2006,** 297–313.

CHAPTER 7

EDIBLE COATING: A SAFE WAY TO ENHANCE SHELF LIFE OF VEGETABLE CROPS

SUDHIR SINGH, RUCHI MISHRA, and B. SINGH

Indian Institute of Vegetable Research, Post Bag No. 1, Jakhini (Shahanshapur), Varanasi 221305, UP, India

CONTENTS

7.1   INTRODUCTION

Fruits and vegetables play a significant role in human nutrition, not only to provide essential nutrients, but also to provide compounds related to health promotion and disease prevention. Being the cheapest sources of vitamins and minerals, vegetables are the high-valued sources of nutrition to the poor family. Many lifestyle diseases, such as cardiovascular, cancerous, renal and gastric diseases, are controlled by vegetables to a great extent.

In India, vegetables are largely produced by small and marginal farmers with small-scale subsistence-level farming systems. Except potatoes and onions, most of the vegetables are not stored for long duration due to non-availability of simple and cost-effective storage systems. Large quantities of vegetables perish due the negligible number of cold chain management facilities of horticultural produce, leading to huge post-harvest losses that ultimately affect the livelihood of poor farmer's family.

Research on edible coating has been intense in recent years with greater combinations due to higher qualitative and quantitative losses of fruits and vegetables. Edible coatings are made of edible materials that are used to cover the surface with fancy attractive coating, providing a semipermeable barrier to gases and water vapour (Baldwin et al., 1999).

Edible coating provides partial barrier to moisture and gas, especially carbon dioxide and oxygen exchange. Coating on the surface of vegetables improves mechanical handling properties (Mellenthin et al., 1982), thus retaining volatile flavour compounds (Nisperos-Carriedo et al., 1990) and carrying food additives containing anti-microbial agents, anti-oxidants and so on (Kester and Fennema, 1988). The basic purpose of edible coatings for fruits and vegetables is enhancing the natural barrier properties or replacing with a new barrier in which handling and washing has partially removed or altered it.

7.2   HISTORY OF EDIBLE COATINGS

In twelfth and thirteenth centuries, the use of wax edible coating has been started to prevent desiccation on citrus fruits. Water loss from the surface of fruits and vegetables was slowed down by the use of wax. Since 1950s, the application of carnauba wax and oil in water emulsions have been practiced (Kaplan, 1986). There has been much advancement on the use of edible coatings from a variety of polysaccharides, proteins and lipids alone or in mixture to produce composite films (Kester and Fenemma, 1988; Ukai et al., 1975). Lipid films made of acetylated monoglycerides waxes were the most successful coatings. The coatings were basically used to block moisture transfer and reduce surface abrasion during handling of fruits and vegetables (Hardenburg, 1967). Earlier reported wax emulsions generally consisted colloidal suspensions of waxes dispersed in water by means of soap. The acid portion of a soap was one of higher molecular weight fatty acids (oleic or linoleic), and the basic portion was ammonia, sodium or potassium hydroxide or more commonly triethanolamine. Portions of shellac, gum, resin or mineral oil were sometimes added to the dispersed phase of the emulsion which was paraffin, carnauba or some wax. As carnauba alone imparted a hard, brittle and high-lustre quality, and paraffin provided an efficient moisture barrier, carnauba wax was often used in combinations of paraffin wax (Claypool, 1940).

7.3 TECHNICAL DEVELOPMENT OF EDIBLE FILMS AND COATINGS

Providing a semipermeable barrier against gases and vapour is the main purpose of employing edible films and coatings in addition other purposes such as texture enhancing, anti-microbials, anti-oxidants and others. Successful application of edible films or coatings as barriers for fruits and vegetables mainly depends on developing film or coating that can provide appropriate internal gas composition for a specific fruit/vegetable (Park 1999).

Thus, we need to completely understand the physiology of the fruit/ vegetable being coated and have a clear idea of the function of the film or coating relative to the product to develop the appropriate film or coating.

7.3.1 LIST OF ADDITIVES INCORPORATED INTO EDIBLE FILMS AND COATINGS FOR FRUITS AND VEGETABLES

Sl. no.	Purposes	Additives
1.	Anti-browning compounds	Sodium sulphite and sodium chloride
2.	Anti-microbial agents	Potassium sorbate and sodium benzoate
3.	Texture enhancers	Calcium chloride, starch and sodium alginate
4.	Nutrients	Vitamins E, A and C
5.	Aroma precursors	Linoleic acid
6.	Probiotics	*Bacillus lactis* and *Lactococcus lactis sub sp. Lactis*

7.4 NUTRITIONAL QUALITY

Nutritional quality of fruits and vegetables can be affected by edible films and coatings in terms of carriers of nutrients, and they can produce abiotic stress that could modify the metabolism of fruits and vegetables during storage. Han et al. (2004) obtained higher amounts of vitamin and calcium on strawberries coated with chitosan containing calcium and vitamin E in the formulation due to diffusion of nutrients into the fruit. Romanazi et al. (2002) observed an increased phenylalanine ammonia-lyase activity, a key enzyme for synthesis of phenolic compounds on chitosan-coated grapes. Edible coatings have also been used as carriers of probiotics, *Bacillus lactis*, which was maintained for 10 days on fresh-cut fruits under refrigeration when applied on alginate and gellan-based edible coating (Tapia et al., 2007).

7.5   TYPES OF EDIBLE COATING

Proteins, lipids and polysaccharides are commonly the main constituents of edible films and coatings. Among the commonly used protein-based edible coating, wheat gluten, corn zein, soy proteins, rice proteins, egg albumin and milk proteins are generally used as coating in vegetables (Perez-Gago et al., 2005; Falcao-Rodrigues et al., 2007). Polysaccharide-based edible coatings, such as alginate, pectin, cellulose and derivatives, starch and sucrose polyesters, have been used for extending the shelf life of fruits and vegetables (Rhim 2004; Rojas-Grau et al., 2007). To extend the shelf life of perishables, lipids as well as resins are also used (Perez-Gago et al., 2006).

7.5.1   EDIBLE COATING FROM PROTEINS

The film-forming ability of several proteinaceous substances has been utilized in industrial applications for a long time. A number of proteins, both from plant and animal origin, have received attention for production of films and coating. These proteins are corn zein, wheat gluten, soy protein, peanut protein, keratin, collagen, gelatin, casein and whey proteins (Gennadios et al., 2002).

7.5.1.1   CORN ZEIN

Corn grains have protein contents in the range of 7–11% (Wall and Paulis, 1978). Zein is the prolamin fraction of corn. About 80% of zein is soluble in 95% ethanol and is characterized as α-zein, and the remaining 20%

insoluble fraction in ethanol is characterized as β-zein (Turner et al., 1965). α-Zein consists of monomers and a series of disulphide-linked oligomers of varying molecular weight, whereas β-zein consists of higher molecular weight oligomers (Paulis, 1981).

Zein does not dissolve in water as well as in anhydrous alcohols. Aqueous aliphatic alcohol solutions are solvent systems used for zein. In general, suitable solvents for zein are either mixtures of water with an organic compound, such as alcohols, acetone and acetonylacetone, or mixtures of two anhydrous organic compounds, such as alcohols and chlorinated hydrocarbons or glycols (Pomes, 1971).

7.5.1.2   WHEAT GLUTEN

Wheat proteins account for 8–15% of the dry weight of wheat kernels (Kasarda et al., 1976). About 70% of the total protein is contained in the wheat endosperm, the major source of wheat flour (Kasarda et al., 1971). Prolamins and glutelins are the major protein fractions among the four major fractionations of proteins such as albumins, globulins, prolamins and glutelins, which are generally referred as gliadin and glutelin accounting for 85% of the wheat flour protein content (Kasarda et al., 1976). Various industrial methods for production of vital wheat gluten from wheat flours are based on hydrating the flour and washing away the starch milk while applying mechanical mixing. Alkali treatment of flour to dissolve gluten and separate it from alkali insoluble starch is also used. Centrifugation of dispersed wheat flour slurries is another alternative (Fellers, 1973).

Gliadins are low-molecular-weight proteins in wheat gluten complex known for their solubility in 70% ethanol. It distinguishes into four gliadin groups viz. α-, β-, γ- and ω- gliadins. Extensive hydrogen bonding among the gliadin polypeptide chains is promoted by the terminal amide group of glutamine. Hydrophobic interactions among non-polar amino-acid residues as well as the very small number of ionizable basic and acidic groups are responsible for the limited solubility of gliadin in neutral water (Krull and Wall, 1969).

Glutenins are high-molecular-weight proteins in the wheat gluten complex that are primarily responsible for dough viscoelasticity. It is believed that cross-linking of polypeptide sub-units results in formation of glutenins. Glutenin molecules are linear, and polypeptide chains are not highly cross-linked through disulphide bonds. All bonds of this type in the glutenin structure are linked through inter-molecular and intra-molecular disulphide bonds as well as with non-covalent bonds (Khan and Bushuk, 1979).

7.5.1.3   SOY PROTEIN

The protein content of soybeans (38–44%) is much higher than that of cereal grains (8–15%) (Snyder and Kwon, 1987). Most of the protein in soybean can be classified as globulin. Along with a number of other functional properties such as cohesiveness, adhesiveness, water and fat absorption, emulsification, dough and fibre formation, texturizing capability and whippability, the film-forming ability of soy protein has been noted (Wolf and Cowan, 1975). Edible films from soybeans have been traditionally produced on the surface of heated soymilk. Protein is the major component of these films, but significant amounts of lipids and carbohydrates are also incorporated. These films are referred as multi-component films (Gennadios et al., 2002).

7.5.1.4   WHEY PROTEINS

Accounting for 20% of total milk proteins, whey proteins are characterized by their solubility at pH 4.6 (Brunner, 1977). Whey proteins contain five principal protein types such as α-lactalbumin, β-lactoglobulin, bovine serum albumin, immunoglobulins and proteose peptone. When appropriately processed, whey proteins produce transparent, flavourless and flexible edible film. Whey protein isolate films are produced by heat treating 8–12% solutions of whey proteins at temperatures between 75 and 100°C. Heat is necessary for formation of inter-molecular disulphide bonds required to produce intact films. But heating solutions caused brittle films which can be improved by addition of food-grade plasticizer (McHugh et al., 1994).

7.5.2   EDIBLE COATING FROM LIPIDS AND RESINS

Edible coating in the form of lipids and resins are extensively used to prevent weight loss, slow down aerobic respiration and improve appearance by providing gloss. Lipids and resins are useful in reducing surface abrasion during handling operations, while at the same time, sealing tiny scratches that may nevertheless occur on the surface of fruits and vegetables (Hernandez, 2002).

7.5.2.1   PARAFFIN WAX

From the wax distillate fraction of crude petroleum, paraffin wax is derived. Synthetic paraffin wax consists of a mixture of solid hydrocarbons, resulting from catalytic polymerization of ethylene. Both natural and synthetic paraffins are refined to meet Food and Drug Administration (FDA) specifications for ultraviolet absorbance. Paraffin waxes are permitted for use as protective coating for raw fruits and vegetables (Bennett, 1975).

7.5.2.2   CARNAUBA WAX

An exudate of palm tree leaves from Brazilian origin of Tree of Life is carnauba wax, *Copernica cerifera*. It has the highest melting point and specific gravity of commonly found natural waxes and is added to other waxes to increase melting point, hardness, toughness and lustre. Carnauba wax is generally considered as generally recognise as safe (GRAS) substance and is permitted for use in coatings for fresh fruits and vegetables, chewing gums, confectionery products and in sauces with no limitations (Hernandez, 1991). Carnauba wax emulsion is formed by melting and emulsifying the wax with 7–9% oleic acid as emulsifier followed by addition of hot water and caustic soda solution in hot condition for stable solution. The emulsion is neutralized by addition of acetic acid and is subsequently diluted suitable for coating on the surface of vegetables (Srivastava, 2011).

7.5.2.2.1   Application

As compared with 5–6 days in fully control carrot samples during storage at 26–32°C, the shelf life of carnauba-wax-coated carrot sample at 5.3% wax level in polypropylene pouches was increased to 15 days, whereas the acceptable overall acceptability was 25 days at 15°C as compared with 8–10 days in fully control carrot samples during storage (Singh and Singh,

2015). In an another studies Koley et al., (2009) carnauba wax entend the shelf life of pointed gourd for 10 days under cold storage conditions (8–10°C and 85% RH). Similarly, the shelf life of carnauba-wax-coated capsicum followed by packaging in polypropylene pouches was extended to 20–22 days at 15°C as compared with 5–6 days in fully control capsicum. The decrease in ascorbic acid of carnauba-wax-treated capsicum was 22–24% after 20 days of refrigerated storage as compared with 33–35% after decrease after 5 days of storage in fully control capsicum samples (Srivastava, 2011).

7.5.2.3   BEESWAX

Beeswax is also known as white wax secreted by honey bees for comb building. The wax is harvested by centrifuging honey from wax combs and then melted with hot water, steam or solar drying. The wax is subsequently refined with diatomaceous earth and activated carbon, and finally bleached with permagnates or bichromates. The wax is very plastic at room temperature but becomes brittle at colder temperature (Tulloch, 1970).

7.5.2.4   CANDELILLA WAX

Candelilla wax, an exudate of the Candelilla plant (*Euphorbia cerifera, E. antisphylitica, Pedilanthus parvonis, Praecoxanthus aphyllus*), is a red-like plant that grows mostly in Mexico and southern Texas. The wax is recovered by immersing the plant in boiling water, after which the wax is skimmed off the surface, refined and bleached. Its degree of hardness is between beeswax and carnauba and is generally regarded as GRAS (Hernandez, 2002).

7.5.2.5   SHELLAC RESINS

Shellac resin, mostly produced in central India, is a secretion by the insect *Lacifer lacca*. This resin is composed of a complex mixture of aliphatic alicyclic hydroxy acid polymers, such as aleuritic and shelloic acids (Griffin, 1979). This resin is soluble in alcohols and alkaline solutions. It is also compatible with most waxes, resulting in improved moisture barrier properties and increased gloss on the coated surface. It is permitted as an indirect food additive in food coatings and adhesives. The shellac coating emulsion is prepared by dissolving dewaxed and bleached shellac in alkaline aqueous medium with the addition of polyvinyl alcohol as binding and coating agent. Binder and defoaming agent is added to make stable emulsion after adding hot water, triethanol amine as surfactant and oleic acid as lubricant. Sodium alginate or carboxy methyl cellulose as thickener is added for effective and stable coating on the surface of fruits and vegetables (Chitravathi et al., 2014).

7.5.2.5.1 *Application*

The shelf life of shellac-coated capsicum is increased to 30–35 days at refrigerated storage temperature of 10°C as against 5–6 days of fully control capsicum during refrigerated storage. There had been 28–30% decrease in ascorbic acid, 33–35% in green colour and 28–30% in hardness during shellac-coated capsicum after 30–35 days of refrigerated storage (Singh and Singh, 2015). Similarly, the shelf life of shellac-coated pointed gourd is increased to 15 days at refrigerated storage as against 6–8 days of control pointed gourd during refrigerated storage. There has also been significant increase in shellac-coated eggplant during ambient storage. As compared with 2–3 days of control eggplant during ambient storage, the shelf life of shellac-coated eggplant is increased to 7 days. There has been increase in physiological loss in weight from 6.0 to 25.7% and increase in total solids from 8.5 to 11.9% after 7 days of ambient storage. There had been 35–40% decrease in texture and 50–55% in anti-oxidant activity (Singh and Singh, 2015).

7.5.2.6   WOOD ROSIN

Either as an exudate or as tall oil, rosins are obtained from the oleoresins of pine trees, a by-product from the wood pulp industry. After distillation of volatiles from the crude resin, Rosin is the residue left. Wood resin is approximately 90% abietic acid and its isomers, and 10% dehydroabietic acid. Wood resins can be modified by hydrogenation, polymerization, isomerization and decarboxylation to make it less susceptible to oxidation and discolouration and to improve its thermoplasticity. In coatings for citrus and other fruits, rosins and its derivatives are widely utilized (Sward, 1972).

7.5.3   EDIBLE COATINGS AND FILMS BASED ON POLYSACCHARIDES

The development of coatings from water-soluble polysaccharides has brought a surge of new types of coatings for extending the shelf life of fruits and vegetables because of selective of these polymers to oxygen and carbon dioxide. Polysaccharide-based coatings can be utilized to modify the atmosphere, thereby reducing fruits and vegetables respiration (Nisperos-Carriedo and Baldwin, 1988).

7.5.3.1   CELLULOSE AND DERIVATIVES

Cellulose is a polysaccharide compound of linear chains of (1→4)-B-D-glucopyranosyl units. Due to high level of intra-molecular hydrogen bonding in the cellulose polymer, native cellulose is insoluble in water. Changing the level methoxyl, hydroxypropyl and carboxy methyl substitution affects a number of physical and chemical properties such as water holding capacity, sensitivity of electrolytes and other solutes, dissolution temperatures, gelation properties and solubility in non-aqueous systems. It is possible to hydrate the polymer by placing along the chain substituents that interfere with the formation of crystalline unit cell. The etherification process involves reacting cellulose with aqueous caustic, then with methyl chloride, propylene oxide or sodium monochloroacetate to yield methylcellulose, hydroxymethylcellulose, hydroxypropyl cellulose and sodium carboxymethyl cellulose (Nisperos-Carriedo, 2002). Changing the level of methoxyl, hydroxypropyl and carboxymethyl substitution affects a number of physical and chemical properties such as water holding capacity, sensitivity to electrolytes and other solutes, gelation properties and solubility of non-aqueous systems (Guilbert, 1986).

7.5.3.2   STARCHES AND DERIVATIVES

Starch is the most common food hydrocolloid (Whistler and Paschall, 1967). Starch provides wide range of functional properties in natural and modified forms and partly because of its low cost relative to alternatives. Starches can be derived from tubers, stems and cereals.

7.5.3.2.1   Modified Starch

Modification of native starch by disruption of hydrogen bonding through reduction of molecular weight or chemical substitution leads to lower gelatinization temperatures and reduced tendency of retrogradation. Modification by acid treatment shortens the chain length of the amylose fraction and converts some of the branched amylopectin into linear amylose units. The ultimate result is to significantly lower the hot paste viscosity, which allows high-solid pastes to be prepared. Especially in the cases where food processing demands stability to high shear and low pH, cross-linking of starch polymers stabilizes the granules and ensures that they do not over swell or rupture. Pre-gelatinized starches are produced by instantaneously cooking/drying starch suspensions on steam heated rollers, puffing/extruding and then spray drying the products. The dry products swell or dissolve in cold water and, therefore, can be used by food processors for the formulations of instant products (Nisperos-Carriedo, 2002). Water-soluble transparent films have been produced from hydroxypropylated amylomaize starch having an apparent amylose content of 71% (Roth and Mehltretter, 1967). A very low permeability to oxygen is exhibited by the film. Hydroxypropylation reduced the dry tensile strength of amylomaize starch film but increased bursting strength and elongation considerably. The hydroxylpropyl derivative may be compounded with other ingredients to improve coating pliability, speed setting upon cooling and/or drying to control rate or resolubility (Jokay et al., 1967). To retard the development of oxidative rancidity during storage, coating formulations from this type of starches have been developed for candies, prunes, raisins, dates, figs, nuts and beans (Scheick et al., 1970).

7.5.3.3   PECTINS

Pectins are complex group of structural polysaccharides which occur widely in citrus peel and apple pomace. Pectic substances are polymers composed mainly $(1{\rightarrow}4)$-a-D-galactopyranosyluronic acid units. One aspect of difference among the pectic substances is their content of methyl esters or degree of esterification, which have the vital effects on solubility and gelation properties. High-methoxyl pectins require a minimum amount of soluble solids (55–80%) and pH within a narrow range (pH 2.8–3.7) to form gels. Low-methoxyl pectins require the presence of a controlled amount of calcium and do not require sugar or acid. As a coating agent for certain foods, the use of low-methoxyl pectinate has been proposed as it is edible and gives an attractive, non-sticky surface to covered foods (Swenson et al., 1953).

7.5.3.4   CARRAGEENANS

Carrageenan is extracted from several species of red seaweeds, mainly *Chondrus crispus*, with water and small amount of alkali, then filtered and recovered by alcohol precipitation. Carrageenan is a complex mixture of several polysaccharides. The three principal carrageenan fractions are kappa, iota and lambda that differ in sulphate ester and 3,6-anhydro-α-D-galactopyranosyl content (Glicksman, 1984). Carrageenan fraction of kappa contains lowest number of sulphate groups and highest concentrations of 3,6-anhydro-α-D-galactopyranosyl units. However, iota-carrageenan fraction differs from kappa-carrageenan with an additional sulphate group at the position 2. Gelation of kappa and iota-carrageenan fraction occurs with both monovalent and divalent ions (Nisperos-Carriedo, 2002).

In food systems, carrageenans are used mainly in gel formation, in stabilizing suspensions and emulsions and for gelation and structural viscosity of milk-based products. Carrageenan gels can be used as food coatings (Meyer et al., 1959). Enhanced stability against growth of surface microorganisms in an intermediate moisture cheese analogue was obtained by enrolling it in a carrageenan–agarose gel matrix that contains sorbic acid (Torres et al., 1985). A carrageenan-based coating applied on cut grape fruit halves resulted in less shrinkage, leakage or deterioration of taste after 2 weeks of storage at 40°F (Bryan, 1972).

7.5.3.5   ALGINATES

Alginates are salts of alginic acid. They occur mainly as the major structural polysaccharides of brown seaweeds known as Phaeophyceae. Alginic acid is considered to be a linear $(1{\rightarrow}4)$ linked polyuronic acid containing three types of block structures of poly-β-D-mannopyranosyluronic acid (M) blocks, poly-a-L-gulopyranosyluronic acid (G) blocks and MG blocks containing both polyuronic acids (McDowell, 1973). These block regions determine the shape of the polymer which in turn governs how effectively the chains associate during gel formation. The block contents of the alginate are responsible for the different gel strengths of products derived from different seaweeds. Gels prepared with alginates are rich in L-gulopyranosyluronic acid tend to be stronger, more brittle and less elastic than those prepared with alginates rich in D-mannopyranosyluronic acid. Alginates in solution are compatible with a wide variety of materials, including other thickeners, synthetic resins, organic solvents, enzymes, surfacatants, plasticizers and alkali metal salts (Kelco, 1987).

7.5.3.6   CHITOSAN

Chitin is β-1,4-linked linear polymer of 2-acetamido-2-deoxy-D-glucopyranosyl residues (BeMiller, 1965). It occurs as the major organic skeletal substance of invertebrates and as a cell wall constituent of fungi and green algae. Chitin does not exist alone but is closely associated with calcium carbonate and/or protein and other organic substances, making the compound stable to most reagents. Fusion of chitin with alkalies gives the product chitosan, a heterogenous substance in various stages of deacetylation and depolymerization (Nisperos-Carriedo, 2002).

Chotosan is non-toxic (Arai et al., 1968) and its applications include coatings, flocculating agents and ingredients for foods and feeds. It can form a semipermeable coating which can modify the internal atmosphere, thereby delaying ripening and decreasing transpiration rates in fruits and vegetables (Arai et al., 1968). Theoretically, it is an ideal coating material based on many inherent good qualities for shelf life extension in fruits and vegetables. It can inhibit the growth of fungi and phytopathogens and can induce chitinase, a defence enzyme (El Ghaouth et al., 1991; Hirano and Nagao, 1989; Kendra and Hadwiger, 1984; Mauch et al., 1984). Carolan et al. (1991) have developed a method for preparing a chitosan derivative with a wide variety of agricultural and industrial applications. N,O-carboxy methyl chitosan (NOCC) was prepared from the reaction of chitosan with monochloroacetic acid under alkaline conditions. NOCC is water soluble, biodegradable and forms selectively permeable non-toxic films. Nutri-Save, an NOCC-based coating (Nova Chem, Halifax, Nova Scotia, Canada), is reported to show positive effect on extending the shelf life of fruits and vegetables (El Ghaouth et al., 1991; Meheriuk, 1990; Elson, 1985). High permeabilities for oxygen and carbon dioxide are shown by chitosan-acetic acid complex membrane, and synthetic polymers can modify the permeation behaviour of the chitosan membrane for two gases (Bai et al., 1988).

7.5.3.7   POTENTIAL ADVANTAGES OF POLYSACCHARIDE-BASED COATINGS

The attractive non-greasy and low-calorie polysaccharide films and coatings can be used to extend the shelf life of fruits and vegetables by preventing dehydration and surface browning. Its wider application has been recently popularized during last few years due to their ability to modify the internal atmosphere. The outcome of numerous studies has reported the occurrence of off-flavour as a result of over modification of the internal atmospheres in wax or oil-coated fruits and vegetables. The ability of water-soluble polysaccharides to reduce oxygen and increase the level of carbon dioxide levels in internal atmosphere reduces respiration rates, thereby extending the shelf life of fruits and vegetables in a manner similar to controlled atmospheric storage (Nisperos-Carriedo, 2002).

Sufficient evidence is available to establish the beneficial effects of polysaccharide-based edible coating on fruits and vegetables, among which are (i) improved retention of flavour, acids, sugars, texture and colour, (ii) increased stability during shipping storage, (iii) improved appearance and (iv) reduced spoilage (Nisperos-Carriedo, 2002).

Polysaccharide-based edible coating enables the ruled-out possibilities of creating anaerobic conditions as well as retention of original flavour and aroma of fruits and vegetables. Further desirable qualities in terms of glossy appearance as well as prevention of moisture loss can be improved with the use of functional ingredients. Food additives, such as resins and rosins, plasticizers, surfactants, oils, waxes and emulsifiers, can be added to make the edible coating permeable (Nisperos-Carriedo et al., 1992).

7.5.4   EDIBLE COATINGS FOR MINIMALLY PROCESSED VEGETABLES

The market demand globally is increasing for minimally processed vegetables. Minimally processed products remain biologically and physiologically active as the tissues are living and respiring with a shifting of cellular processes and interactions in response to the tissue damage inflicted by the operations. There have been many intricate problems as a result of physiological changes. The causes are related to disruption of cell tissues and breakdown of cell membrane, leading to leakage of ions, loss of components, alteration in flux potential and loss of turgor (Davies, 1987).

Edible coatings can act as protective cover on the cut surface of minimally processed fruits and vegetables, thus preventing further deterioration. The most well-known example is the coating of fresh fruits and vegetables with waxes or waxy-water emulsions. The idea of coating cut pieces of fruits and vegetables sounds attractive. The coating should be planned with the idea of (i) forming an efficient barrier properties to moisture loss, (ii) having a selective permeability of gases, (iii) controlling migration of water-soluble solutes to retain the natural colour pigments and nutrients and (iv) incorporating additives such as colourings, flavours or preservatives that impart specific functions and properties (Kester and Fennema, 1986).

The binding of cut surfaces in fruits and vegetables with continuous release of exudate presents a considerable technical problem towards the coating in fruits and vegetables. One solution implies towards some means of setting the coating material by forming a tight matrix. An emulsion mixture of casein and acetylated monoglyceride will form a coagulum by adjusting the pH to isoelectric point of 4.6 (Krochta et al., 1988). The lipid molecules are presumably trapped within the matrix of casein coagulum. Further researches have highlighted the importance of ionic cross-linking for binding purposes. In a caseinate/acetylated monoglyceride/ alginate emulsion, the association of polyglucuronic acid and calcium ions provides junction zones which cross-link the alginate polymers into a three-dimensional network containing casein, with the acetylated monoglyceride dispersed in the interstices. These types of binding technique using alginate has the additional advantage of forming cross-links between the coating and the endemic pectin on the cut surface of fruits and vegetables. In most experiments, ascorbate solutions containing calcium are used, both for the purpose of cross-linking and for preventing the cut surface from enzymatic browning (Wong et al., 2002).

Krochta et al. (1988) initiated the application of emulsion coating consisting of sodium caseinate and acetylated monoglyceride using isoelectric pH for coagulation. Recently, milk-protein-based coating has been recommended for minimally processed vegetables. These are primarily casinate-lipid coatings. Sodium caseinate/stearic acid mixtures, tested on the coating of peeled carrots during storage at 2.5°C, 70% RH and an air flow of 20 cc/min, were shown to eliminate the formation of white blush and decrease the rate of dehydration. Optimization of formulations estimated by response surface methodology suggested that coating composition of 1.4–1.6% sodium caseinate 0.1–0.2% stearic acid increased water vapour resistance by 84% and retarded the formation of white blush that are usually formed at cut carrot surfaces. Further experiments towards coating with caseinate/acetylated monoglyceride resulted in 75% reduction of moisture loss in uncut celery sticks stored at 2.5°C and 70% RH. Respiration and ethylene production rates showed no changes compared with uncoated samples. However, coating containing one component did not give satisfactory results, whereas emulsions of mixed components seemed to perform best. Results indicate that the optimal composition ratio of caseinate–acetylated-monoglyceride–alginate was 10:15:0.5%. The emulsified product was viscous. Lowering the concentration of any one component caused a drastic change in moisture retention. In general, proteins or polysaccharides are mostly hydrophilic and cannot be used as moisture barriers on cut surfaces with high surface water activity. Edible coating containing protein or polysaccharides alone provides no detectable protection against water loss from cut surfaces of fruits and vegetables (Wong et al., 1992).

7.5.5   METHODS OF EDIBLE COATING APPLICATION ON VEGETABLES

7.5.5.1   DIP APPLICATION

Dip application is an industrial process of application in which coating was accomplished by submerging the commodities into the tank of emulsion. Due to superior shelf life and gloss of appearance, consumers developed a preference of dip application; as a result, tomatoes and other vegetables were being dipped for increasing the shelf life (Platenius, 1939).

Dipping fruits and vegetables into a tub or tank of the coating material is adequate usually for small quantities of fruits and vegetables. The produce is washed, dried and then immersed in the dip tank. Time of immersion is not important. But complete wetting of fruits and vegetables is imperative for good coverage (Long, 1964). The commodity is then either conveyed to a drier in which water is removed or allowed to dry under ambient conditions. The dip tanks can be equipped with porous basket which can be lifted to strain and remove debris. In addition, to avoid dilution of the resin solution or emulsion coating, vegetables entered in the dip tank must be completely dry.

7.5.5.2   FOAM APPLICATION

Foam application is used for larger application of vegetables and the application is used with the help of applicator. A foaming agent is added to the coating or compressed air (less than 5 psi) blown into the applicator tank (Long and Leggo, 1959). The agitated foam is then dropped onto the commodity as it moves over the rollers. Cloth flaps or brushes distribute the emulsion over the commodity. Excess coating is removed with squeegees positioned below the rollers and then often recirculated. The foamed emulsion contains very little water and thereby facilitates the drying process. As extensive tumbling action is necessary to break the foam and distribute the coating, even distribution of coating is difficult to achieve (Grant and Burns, 2002).

7.5.5.3   SPRAY APPLICATION

Spray application is the most acceptable conventional method for applying most coatings to fruits and vegetables. Low-pressure spray applicator used in past delivered coating in excess. Often, recovery wells were utilized and excess coating was recirculated. There have been concerns of dilution and contamination during spray application. Later high-pressure spray applicators delivering coating of 60–80 psi became available which used much less coating material and yielded better coverage, thus requiring no recovery wells and recirculation. However, nozzle size was critical as small nozzles often plugged and large nozzles delivered too much wax. Air-atomizing systems connected to metering pumps have also been used. Air is delivered to the nozzle head at pressure usually at 5 psi or below, and coating is delivered with a metering pump at less than 40 psi. Usually the coating is sprayed onto the vegetables which pass under a set of fixed or mobile nozzles. Vegetables travel over a slowly rotating bed of brushes. The waxer brush bed is one of the most important pieces of equipment for coating purposes. Straight-cut brushes on the brush bed are most effectively used with round or slightly elliptical fruits and vegetables, whereas spiral-cut or tumble-trim brush designs are used with small, flat, irregularly shaped produce which requires more tumbling action for good coating coverage. The bed is typically composed of 12 or 14 brushes, with four to six brushes positioned after the applicator. Too many brushes after the applicator can remove the wax on the commodity surface. All brushes are recommended to be made of 50% mixture of horse hair and polyethylene bristles (Grant and Burns, 2002).

7.5.5.4   DRIP APPLICATION

The most commercial method used today to apply coatings to fruits and vegetables is the drip-application method. Different sizes of emitters are available that will deliver a variety of large droplet sizes. Usually mounted on a dual-bank manifold, emitters are spaced 1-in. apart along the manifold and width of the bed. A metering pump is used to deliver the coating at pressure not higher than 40 psi. It is most economical method of coating on the surface of fruits and vegetables. In view of large droplet sizes, good coverage is achieved when there has been adequate tumbling action over several brushes which are saturated with the coating (Grant and Burns, 2002).

7.5.6   KEY SUCCESS ISSUES RELATED TO COATING ON FRUITS AND VEGETABLES

All coating equipment, especially applicator, should be cleaned after each operation thoroughly with hot water. Failure to implement a routine clean-up procedure will result in pump breakdown and clogging of nozzles. Buildup of hardened coating due to non-uniform cleaning would result in commodity injury and subsequent decay. The washed fruits and vegetables should be air dried to remove the excess water from the surface. Excess adhering water serves to dilute the coating as a result less coating is applied and emulsification action is less. Other benefits such as dilution of fungicides are forfeited for effective post-harvest management in fruits and vegetables (Grant and Burns, 2002).

Brush wear should be routinely monitored. Incomplete coating coverage can often be avoided by ensuring that clean fruits and vegetable surfaces are used for coating. Soils and debris present on fruits and vegetables should be effectively removed by brushes or rollers. Coating materials should be dried as much possible before packaging. Some coating formulations remain sticky after drying and fruits and vegetables must be handled carefully to avoid the creation of missed spots on the finished products. Various post-coating operations such as inadequate air flow, temperature, underfill and overfill of drier significantly affect the coating quality of fruits and vegetables (Grant and Burns, 2002).

The use of coatings, although increasingly under scrutiny, has become an important component of fresh fruits and vegetable industry. The ability to control desiccation, incorporate fungicides for decay control and to control aspects of product physiology serves to lengthen the marketing period with extended shelf life. Certain fruits and vegetables are transported to cater the distant market with extended shelf life. Higher natural quality demand can be fulfilled with suitable combinations of edible coating. The increasing trend of global marketing demands the greater importance of coating in the large usage of edible coatings (Grant and Burns, 2002).

7.5.7   CHALLENGES AND FUTURE THRUSTS OF EDIBLE COATINGS

Many types of edible coatings have been developed to enhance the shelf life of fruits and vegetables. There is a need to identify the relationship between the internal atmosphere caused by the coating and the velocity of physiological biochemical processes related to ripening such as respiration, tissue softening, metabolic reactions, production of metabolites and secondary compounds generated during storage. Many research workers are interested to have very long extended shelf life without understanding the physiology of respiration processes in terms of minimum oxygen concentration which would prevent the onset of anaerobic respiration. Therefore, it is very pertinent to know the optimum oxygen concentration at which the consumption is minimized without promoting the development of anaerobic respiration in various fruits and vegetables. There is a need of availability of suitable composition of edible coating suitable for leafy vegetables as immediate wilting takes place in leafy vegetables immediately after harvest.

7.5.7.1   SALIENT CHALLENGES ARE LISTED BELOW

- Characterization of physiochemical properties edible coating solutions such as composition, concentration, solubility, viscosity density, surface tension on changes in pH, temperature and time.
- Characteristics of edible coatings in terms of gas and vapour permeability, mechanical and sensory changes on fruits and vegetables, thickness, solubility, digestibility and so on.
- Studies on metabolic changes occurring within coated fruits and vegetables with respect to partial modified atmosphere.
- Monitoring of gaseous composition with the coating thickness and composition of edible material during storage to the defined fruits and vegetables.
- Determination of optimal coating methods for quicker application and quicker handling for obtaining high-quality products at compatible price.
- Studies on wider consumer acceptability of coatings.
- Studies on impact of edible coatings on final cost of fruits and vegetables.

- Easy removal of edible coating from the surface of fruits and vegetables while washing with water and no residual left over.
- No detrimental effect on the quality of fruits and vegetables.

REFERENCES

Arai, L; Kinumakai, Y.; Fujita, T. Toxicity of Chitosan. *Bull. Tokai Reg. Fish. Res. Lab.* **1968,** *56,* 89.

Bai, R.; Huang, M.; Jiang, Y. Selective Permeabilities of Chitosan-Acetic Acid Complex Membrane and Chitosan Polymer Complex Membranes for Oxygen and Carbon Dioxide. *Polym. Bull.* **1988,** *20,* 83–88.

Baldwin, E. A.; Burns, J. K.; Kazokas, W.; Brecht, J. K.; Hagenmaier, R. D.; Bender, R. J.; Pesis, E. Effect of 2 Edible Coatings with Different Permeability Characteristics on Mango (*Mangifera indica* L.) Ripening During Storage. *Postharvest Bio. Technol.* **1999,** *17,* 215–226.

BeMiller, J. N. Chitosan. In *Methods in Carbohydrate Chemistry. General Polysaccharides,* Vol. V.; Whistler R. L., Ed.; Academic Press: New York, 1965; pp 103–106.

Bennett, H. *Industrial Waxes, Vol. 1;* Chemical Publ. Co.: New York, NY, 1975.

Brunner, J. R. Milk Proteins. In *Food Proteins;* Whitaker, J. R., Tanenbaum S. R., Eds.; AVI Publishers, Ins; Wesport CT, 1977; pp 175–208.

Bryan, D. S. Prepared Citrus Fruit Halves and Method of Making the Same. U.S. Patent 3,707,383, 1972.

Carolan, C.; Blair, H. S.; Allen, S. J. Chitosan Derivative Keeps Apples Fresh. *Postharvest News Info.* **1991,** *2,* 75.

Chitravathi, K.; Chauhan, O. P.; Raju, P. S. Postharvest Shelf Life Extension of Green Chillies (*Capsicum annum* L.) Using Shellac Based Edible Surface Coatings. *Postharvest Biol. Technol.* **2014,** *92,* 146–148.

Claypool, L. L. The Waxing of Deciduous Fruits. *Am. Soc. Hortic. Sci. Proc.* **1940,** *37,* 443–447.

Davies, E. Plant Response to Wounding. In *The Biochemistry of Plants. Vol. 12;* Davies, D. D., Stumpf, Conn, E. E., Ed.; Academic Press: California, USA, 1987; pp 243–264.

Elson, C. W.; Hayes, E. R.; Lidster, P. D. Development of the Differentially Permeable Fruit Coatings Nutri Save for the Modified Atmospheric Storage of Fruit. In *Proceedings of the fourth National Controlled Atmospheric Research Conf.* Blankenship S. M., Ed.; Dept of Hort Sci: North Carolina State Univ., Rpt. 1985; Vol. 126, 248.

Fellers, D. A. Fractionation of Wheat into Major Components. In *Industrial Users of Cereals;* Pomeranz, Y. Ed.; American Association for Cereal Chemists: St. Paul, MN, 1973; pp 207–228.

Griffin, W. C. Emulsions. Kirk Othmer Encyclopaedia of Chemical Technol 3rd ed. 1979; 8, 913–916.

Gennadios, A.; McHugh, T. H.; Weller, C. L.; Krochta, J. M. Edible Coatings and Films Based on Proteins. In *Edible Coatings and Films to Improve Food Quality;* Krochta, J. M., Baldwin, E. A., Nisperos-Carriedo M., Ed.; CRC Press: New York, USA, 2002; pp 201–277.

Ghaouth, A. E. L.; Arul, J.; Ponnampalan, R.; Boulet, M. Chitosan Coating Effect on Storability and Quality of Fresh Strawberries. *J. Food Sci.* **1991**, *56,* 1618.

Glicksman, M. *Food Hydrocolloids,Vol.3;* CRC Press: Boca Raton, FL, 1984.

Grant L. A.; Jackie, B. Application of Coatings. In *Edible Coatings and Films to Improve Food Quality;* Krochta J. M., Baldwin E. A., Nisperos-Carriedo, M., Ed.; CRC Press: New York, USA, 2002; pp 189–200.

Guilbert, S. Technology and Application of Edible Protective Films. In *Food Packaging and Preservation Theory and Practices;* Mathlouthi, M., Ed.; Elsevier Applied Science Publishing Co.: London, UK, 1986, p 371.

Han, C.; Zhao Y.; Leonard, S. W.; Traber, M. G. Edible Coating to Improve Storability and Enhance Nutritional Value of Fresh and Frozen Strawberries (*Fragaria ananassa*) and Raspberries (*Rubus ideaus*). *Postharvest Biol. Technol.* **2004,** *33,* 67–68.

Hardenburg, R. E. *Wax and Related Coatings for Horticultural Products. A Bibliography. Agri. Res. Bull.;* US Department of Agriculture.: Washington, DC., 1967; 51–15.

Hernandez, E. Edible Coatings from Lipids and Resins. In *Edible Coatings and Films to Improve Food Quality.* Krochta J. M., Baldwin E. A., Nisperos-Carriedo M., Ed.; CRC press: New York, USA, 2002; pp 279–303.

Hernandez, E.; Baker, R. A. Candelilla Wax Emulsion, Preparations and Stability. *J. Food Sci.* **1991,** *56,* 1382–1383.

Hirano, S.; Nagao, N. Effect of Chitosan, Pectic Acid, Lysozyme and Chitinaseon the Growth of Several Phytopathogens. *Agric. Bio. Chem.* **1989,** *53,* 3065.

Jokay, L.; Nelson, G. E.; Powell, E. L. Development of Edible Amylaceous Coatings for Foods. *Food Tech.* **1967,** *21,* 12–14.

Kaplan, H. J. Washing, Waxing and Colour Adding. In *Fresh Citrus Fruits;* Wardowdki W. F., Nagy S., Grierson W. Eds.; AVI Publishing Co.: Westport CT, 1986; p 379.

Kasarda, D. D.; Bernardin, J. E.; Nimmo, C. C. Wheat Proteins. In *Advances in Cereal Science and Technology,* Vol. 1; Pomeranz Y., Ed.; American Association Cereal Chemists: Paul St. MN., 1976, pp 158–236.

Kasarda, D. D.; Nimmo, C. C.; Kohler, G. O. Proteins and the Amino Acid Composition and Wheat Fractions. In *Wheat Chemistry and Technology, 2nd Eds.* Pomeranz Y., Ed.; American Association Cereal Chemists: Paul St MN., 1971; pp 227–299.

Kelco. *Alginate Products for Scientific Water Control;* Kelco Products Bull: San Diego, CA, 1987.

Kendra D. F.; Hadwiger, L. A. Characterization of the Smallest Chitosan Oligomer that is Maximally Antifungal to *Fusarium solani* and Elicits Pisatin Formation in *Pisum sativum. Exp. Mycol.* **1984,** *8,* 276.

Kester, J. J.; Fennema, O. R. Edible Films and Coatings: A Review. *Food Tech.* **1986,** *40,* 47–59.

Kester, J. J.; Fennema, O. R. Edible Films and Coatings: A Review. *Food Technol.* **1988,** *42,* 47–49.

Khan, L.; Bushuk, W. Studies of Glutenin. Comparison by Sodium Dodecyl Sulphate-Polyacrylamide Gel Electrophoresis of Unreduced and Reduced Glutenin from Various Isolation and Purification Procedures. *Cereal. Chem.* **1979,** *56,* 63–73.

Koley, T. K.; Asrey, R.; Pal, R. K.; Samuel. D. V. K. Shelf-life extension in pointed gourd (Trichosanthes dioica Roxb.) by post-harvest application of sodium hypochlorite, potassium metabisulphite and carnauba wax. *J. Food Sci Technol.* **2009**, 46, 581–584.

Krochta, J. M.; Hudson, J. S.; Camirand, W. M.; Pavlath, A. E. Edible Films for Light Processed Fruits and Vegetables. Paper presented at International Winter Meeting of the American Society of Agricultural Engineering, Chicago, IL, 1988.

Krull, L. H.; Wall, J. S. Relationship of Amino Acid Composition and Wheat Protein Properties. *Baker's Dig.* **1969,** *43,* 30–34.

Long, J. K.; Leggo, D. Waxing Citrus Fruits. *CSIRO Food Preserv. Quart.* **1959,** *19,* 32–37.

Long, W. G. Better Handling of Florida's Fresh Citrus Fruit. *Fla. Agri. Exp. Sta. Bull.* **1964,** *681,* 25–28.

Mauch, F.; Hadwiger, L. A.; Boller, T. Ethylene: Symptoms, not Signal for the Induction of Chitinase and β-1,3-Glucanase in Pea Pods by Pathogens and Elicitors. *Plant Physiol.* **1984,** *76,* 607.

McDowell, R. H. *Properties of Alginates;* Alginate Industrial Limited: London, UK, 1973.

McHugh, T. H.; Aujard, J. F.; Krochta, J. M. Plasticized Whey Proteins Edible Films: Water Vapour Permeability Properties. *J. Food Sci.* **1994,** *59,* 416–419.

Meheriuk, M. Skin Colour in Newton Apples Treated with Calcium Nitrate, Urea, Diphenylamine and a Film Coating. *Hortic. Sci.* **1990,** *25,* 775–776.

Mellenthin, W. M.; Chen, P. M.; Borgic, D. M. Inline Application of Porous Wax Coating Materials to Reduce Friction Discolouration of Bartlett and Danjou Pears. *Hortic. Sci.* **1982,** *17,* 215–217.

Meyer, R. C.; Winter, A. R.; Weiser, H. H. Edible Protectice Coatings for Extending the Shelf Life of Poultry. *Food Tech.* **1959,** *13,* 146.

Nisperos-Carriedo, M. O. Edible Coatings and Films Based on Polysaccharides. In *Edible Coatings and Films to Improve Food Quality;* Krochta J. M., Baldwin E. A., Nisperos-Carriedo M., Eds.; CRC Press: New York, USA, 2002; pp 305–335.

Nisperos-Carriedo, M. O.; Baldwin, E. A. Effect of Two Types of Edible Films on Tomato Fruit Ripening. *Proc. Florida State Hortic. Soc.* **1988,** *101,* 217–220.

Nisperos-Carriedo, M. O.; Baldwin, E. A.; Shaw, P. E. Development of Edible Coating for Extending Postharvest Life Selected Fruits and Vegetables. *Proc. Florida State Hortic. Soc.* **1992,** *104,* 122–125.

Nisperos-Carriedo, M. O.; Shaw, P. E.; Baldwin, E. A. Changes in Volatile Flavour Components of Pineapple Orange Juice as Influenced by the Application of Lipid and Composite Film. *J. Agric. Food Chem.* **1990,** *38,* 1382–1387.

Park, H. J. Development of Advanced Edible Coatings for Fruits. *Trends Food Sci. Technol.* **1999,** *10,* 254–260.

Paulis, J. W. Disulfide Structures of Zein Proteins from Corn Endosperm. *Cereal Chem.* **1981,** *58,* 542–546.

Perez-Gago, M. B.; Serra, M. Del Rio, M. A. Colour Change of Fresh Cut Apples Coated with Whey Protein Concentrate Based Edible Coatings. *Postharvest Biol. Technol.* **2006,** *39,* 84–92.

Perez-Gago, M. B.; Serra, M.; Alonso, M.; Mateos, M.; Del Rio, M. A. Effect of Whey Protein and Hydroxylpropyl Methylcellulose Based Edible Composite Coatings on Colour Change of Fresh Cut Apples. *Postharvest Biol. Technol.* **2005,** *36,* 77–85.

Platenius, H. Wax Emulsion for Vegetables. *New York Agric. Exp. Sta. Bull.* **1939,** *723.*

Pomes, A. F. Zein. *Encyclopaedia of Polymer Science and Technology: Plastics, Resins, Rubbers, Fibers,* Vol. 15, Mark H. F., Gaylord N. G., Bikales N. M., Eds.; Inter science Publishers: New York, 1971; pp 125–132.

Rhim, J. W. Physical and Mechanical Properties of Water Resistant Sodium Alginate Films. *Lebensmittel Wissenchaft Technol.* **2004,** *37,* 323–330.

Rojas-Grau, M. A.; Tapia, M. S.; Rodri Guezb, F. J.; Carmonac, A. J.; Martin-Belloso, O. Alginate and Gellan Based Edible Coatings as Carriers of Antibrowning Agents Applied on Fresh Fuji Apples. *Food Hydrocolloids* **2007,** *21,* 118–127.

Romanazi, G.; Nigro, F.; Ippolito, A.; Di Venere, D.; Salerno, M. Effects of Pre- and Post-harvest Chitosan Treatments to Control Storage Grey Mold of Table Grapes. *J. Food Sci.* **2002,** *67,* 1862–1867.

Roth, W. B.; Mehltretter, C. L. Some Properties of Hydroxpropylated Amylomaize Starch Films. *Food Tech.* **1967,** *21,* 72–74.

Scheick K. A.; Jokay, L.; Nelson, G. E. U.S. Patent 3,527,646, September 8, 1970.

Singh, S.; Singh, B. Value Addition of Vegetable Crops. *IIVR Technical Bull.* **2015,** *65,* 1–61.

Snyder, H. E.; Kwon, T. W. Soybean Utilization, Van Nos-trand Reinhold Company Inc.: New York, 1987.

Srivastava, B. Studies on Carnauba Wax Coating on Shelf Life of Capsicum (*Capsicum annum* L.). MSc. Dissertation, Allahabad University, Allahabad, 2011.

Sward, G. G. Natural Resins. *Am. Soc. Test. Mat.* **1972,** 77–91.

Swenson, H. A.; Miers, J. C.; Schultz, T. H.; Owens, H. S. Pectinate and Pectate Coatings. II. Application to Nuts and Fruit Products. *Food Tech.* **1953,** *7,* 232–235.

Tapia, M.; Rojas-Grati, M.; Rodriguez, F.; Ramirez, J.; Carmona, A.; Maratin-Belloso, O. Alginate and Gellan Based Edible Films for Probiotic Coatings on Fresh Cut Fruits. *J. Food Sci.* **2007,** *72,* E190–E196.

Torres, J. A.; Bouzas, J. O.; Karel, M. Microbial Stabilization of Intermediate Moisture Food Surfaces. II. Control of Surface pH. *J. Food Process. Preserv.* **1985,** *9,* 93.

Tulloch, A. P. The Composition of Beeswax and Other Waxes Secreted by Insects. *Lipids* **1970,** *5,* 247–258.

Turner, J. E.; Boundy, J. A.; Dimler, R. J. Zein: a Heterogeneous Protein Containing Disul-phide-Linked Aggregates. *Cereal Chem.* **1965,** *42,* 452–461.

Ukai, N. T.; Tsutsumi, T.; Marakami, K. U.S. Patent 3,997,674, February 25, 1975.

Wall, J. S.; Paulis, J. W. Corn and Sorghum Grain Proteins. In *Advances in Cereal Science and Technology,* Vol. 2; Pomeranz Y., Ed.; American Association of Cereal Chemists: St Paul MN, 1978; pp 135–219.

Whistler, R. L.; Paschall, E. F. *Starch: Chemistry and Technology,* Vol. 2; Academic Press: New York, 1967.

Wolf, W. J.; Cowan, J. C. *Soybeans as a Food Source;* CRC Press, Inc.: Cleveland OH, 1975.

Wong, D. W. S.; Pavlath, A. E.; Tillin, S. J. Edible Double Layer Coating for Slightly Processed Fruits and Vegetables. Symposium on Edeible Coatings for Food, American Chem Soc, Washington DC, 1992, August 23–28, 1992.

Wong, D. W. S; Camirand, W. M.; Pavlath, A. E. Development of Edible Coatings for Minimally Processed Fruits and Vegetables. In *Edible Coatings and Films to Improve Food Quality;* Krochta, J. M., Baldwin, E. A., Nisperos-Carriedo, M., Ed.; CRC press: New York, USA, 2002; pp 65–88.

CHAPTER 8

STRATEGIES FOR LOW-TEMPERATURE STORAGE OF VEGETABLES

R K PAL[1], TANMAY KUMAR KOLEY[2], AND SUDHIR SINGH[2]

[1]National Research Centre on Pomegranate, Kegaon, Solapur 413255, Maharashtra, India, E-mail: rkrishnapal@gmail.com

[2]Indian Institute of Vegetable Research, Varanasi 221305, Uttar Pradesh, India

CONTENTS

8.1 INTRODUCTION

Ever since the civilization of mankind, efforts have been directed towards accumulating and storing foods when they are in plenty to meet needs during the days of scarcity. Although in the case of food grains not much problem was faced due to nature's noble way of reducing the moisture level as the grains mature, in the case of vegetables, long-term storage in their fresh form was not possible (until development of modern methods) primarily due to their high degree of perishability due to high moisture content of these commodities at the time of harvest.

Storage of vegetables in their fresh form prolongs their usefulness and in some cases improves their quality. It also checks market glut, enables in orderly marketing and increases financial gain to the producers by preserving the quality of produce. Horticultural produce such as fruits, vegetables and cut flowers are living, and they continue respiration even after being separated from the parent plant. Therefore, the aim of storage is to control various physiological processes, namely respiration, transpiration and other metabolic activities, to keep the produce in maximum usable form.

Post-harvest temperature surrounding the produce is the single-most important factor which maintains the quality of vegetables. Storage in refrigerated atmosphere retards the ageing process due to ripening, softening and colour changes. Refrigeration also reduces the undesirable metabolic changes and respiratory heat production. Temperature control throughout the supply chain from harvest to utilization has been found to be the most important factor in maintaining the product quality. Close control of temperature is essential to optimize the storage life. One of the most important functions of refrigeration is to control the respiratory rate of the produce.

In India, significant development on the growth of cold storage industry has been observed in the recent years with more than 6000 units with a total installed capacity of 255 lakh metric tonne in the year 2008. However, percentage share of vegetables stored cold store is comparatively less. About 85% of the cold stores are occupied with potato. The major problems and constraints faced by the cold storage industries in India are as follows:

1. Unplanned growth in number and capacity of cold storages in some pockets
2. Dependence on single commodity for storage
3. Lack of diversification
4. Low or no capacity utilization during the off season
5. High cost of operation due to inefficient technology, poor insulation and mismatch of plant machinery with capacity
6. Lack of working capital finance
7. Lack of awareness about the processed produce that can be stored
8. Irregular power supply with high power tariff

In any cold chain management system, cold stores form the core activity for its success. In the beginning of twentieth century, cold stores were first established in India. The units were designed for storage of potato. They were mainly located in areas such as UP, West Bengal, Punjab, Bihar and so forth. However, around 1960 in Maharashtra, the idea of multi-product, multi-chamber cold stores were introduced. The cold storage sector is undergoing a major transformation through establishment of cold chain system focusing on reduction in post-harvest losses and investment opportunities for food-processing industries. Developments of energy-efficient cold stores are given priority by several agencies. Every part of a cold chain renders itself amenable for improvement with the advent of newer materials/equipment. As a result, all of them are witnessing changes with respect to type of construction, insulation, refrigeration equipment, type of controls and so forth.

Different techniques have been developed to keep the stored vegetables fresh for longer period. However, usually long-term storage of vegetables is expensive and requires a high level of technical knowledge. Recommended storage conditions and approximate length of storage period for commercial storage of vegetables are mentioned in Table 8.1.

TABLE 8.1 Recommended Storage Conditions and Approximate Length of Storage Period for Commercial Storage of Vegetables.

Vegetable	Temperature (°C)	Relative humidity (%)	Approximate length of storage
Asparagus	0	85–90	3–4 weeks
Beans, Lima			
Unshelled	0	85–90	2–4 weeks
	4.5	85–90	10 days
Shelled	0	85–90	15 days
	4.5	85–90	4 days
Beans, snap	0–4.5	85–90	2–4 weeks
Beets			
Topped	0	95–98	1–3 months
Bunch	0	85–90	10–14 days
Broccoli, Italian	0–1.6	90–95	7–10 days
Brussels sprouts	0–1.6	90–95	3–4 weeks
Cabbage	0	90–95	3–4 months

TABLE 8.1   *(Continued)*

Vegetable	Temperature (°C)	Relative humidity (%)	Approximate length of storage
Carrots			
Topped	0	95–98	4–5 months
Bunch	0	85–90	10–14 days
Cauliflower	0	85–90	2–3 weeks
Celery	0	90–95	2–4 months
Sweet corn	0	85–90	4–8 days
Cucumbers	7–10	85–95	10–14 days
Brinjal	7–10	85–90	10 days
Endive.	0	90–95	2–3 weeks
Garlic (dry)	0	70–75	6–8 months
Horse–radish	0	95–98	10–12 months
Jerusalem artichoke	0	90–95	2–5 months
Kohlrabi.	0	95–98	2–4 weeks
Leeks (green)	0	85–90	1–3 months
Lettuce	0	90–95	2–3 weeks
Muskmelons	0–1	75–78	7–10 days
Honey Dew melons	2	75–85	2–4 weeks
Casaba and Persian melons	2–4	75–85	4–6 weeks
Okra	10	85–95	2 weeks
Onions	0	70–75	6–8 months
Onion sets	0	70–75	6–8 months
Parsnips	0	90–95	2–4 months
Peas (green)	0	85–90	1–2 weeks
Peppers (sweet)	0	85–90	4–6 weeks
Potatoes	3.3–10	85–90	5–8 months
Pumpkin	10–13	70–75	2–6 months
Pointed gourd	10–13	90–95	8 days
Radishes	0	95–98	2–4 months
Rhubarb	0	90–95	2–3 weeks
Rutabagas	0	95–98	2–4 months
Salsify	0	95–98	2–4 months
Spinach	0	90–95	10–14 days

TABLE 8.1 *(Continued)*

Vegetable	Temperature (°C)	Relative humidity (%)	Approximate length of storage
Squashes			
Summer	4–10	85–95	2–3 weeks
Winter	10–13	70–75	4–6 months
Sweet potatoes	10–13	80–85	4–6 months
Tomatoes			
Ripe	4–10	80–85	7–10 days
Mature green	13	80–85	3–5 weeks
Turnips	0	95–98	4–5 months
Watermelons	2–4	75–85	2–3 weeks

8.2 PRE-STORAGE TREATMENTS

Pre-cooling is the process of rapid removal of field heat/respiratory heat usually practiced for fresh fruits, vegetables and flowers immediately after harvest but before shipment or storage or processing. This is the first step of good temperature management. The primary advantages of pre-cooling are (i) inhibition of the growth of decays causing organisms, (ii) restriction of the enzyme activity, (iii) reduction of water loss, (iv) reduction in rate of respiration and C_2H_4 liberation and (v) rapid wound healing. The production and action of ethylene from harvested fruits, vegetables and flowers are temperature dependent. Harvested produce kept at 25°C with 30% relative humidity (RH) shows a tendency of 36 times more water loss as compared with that stored at 0°C with 90% RH (Barman et al., 2011). Hence pre-cooling serves as an essential practice in any successful cool chain management of horticultural produce. The speed of cooling depends upon the following factors: (i) product accessibility to the refrigerating medium, (ii) temperature difference exists between the product and refrigerating medium, (iii) velocity of refrigerating medium and (iv) type of cooling medium. To predict the end point of pre-cooling, it is essential to know the half-cooling time. The time required to reduce the temperature difference between commodities and the coolant by one-half in known as half-cooling time. This is independent of the initial temperature of the commodity. The time of cooling is depending on several factors like the shape, density and surface area/volume ratio of the vegetables and the type of packaging. Some products do not tolerate exposure to water. In general, more packaging results slower cooling rate. When the packaging is expected to be exposed to water, waxing on paper-based packaging is recommended (Brosnan and Sun, 2001).

There are two most important factors in pre-cooling like temperature and time. Vegetable must be pre-cooled as soon as possible. The process of pre-cooling follows a logarithmic function. A rapid cooling is observed at initial stage and then the process gradually slows down. It is always advised to pre-cool the vegetables to 7/8th of their recommended storage temperature, because it is difficult to remove all the field heat during the process of pre-cooling. The remaining 1/8th of the heat can be gradually removed with less energy cost by placing the vegetable in the storage. When several vegetables are handled, the cooling system must be compatible to all of them. Highly perishable vegetables like asparagus, broccoli, spinach, sweet corn and so forth have a high respiratory rate, need fast cooling and require high refrigeration capacity and fast pre-cooling methods. There are basically four methods used for horticultural commodities. These are (i) room/air cooling, (ii) water/hydro cooling, (iii) forced air-cooling, (iv) vacuum cooling and (v) package icing.

Before storage of the commodities, different pre-treatments have been standardized to keep them fresh or enhance their shelf life. For example, treatment of potatoes and onions either with maleic hydrazide (2000–2500 ppm) (Kleinkopf et al., 2003) or irradiation with Gamma rays (0.15 KGy) inhibit sprouting (Kader, 1986). This technology is now commonly used in commercial storage of potatoes and onions. Irradiation treatment has been quite promising and effective technology currently available to overcome quarantine restrictions in international trade. Vapour heat treatment and hot water treatment are very simple and useful which are becoming popular pre-treatments before storage or transportation of some fruits, especially mango.

8.2.1　CONTROL OF SPOILAGE OF VEGETABLES

The spoilage of vegetables in storage houses are caused by various kinds of micro-organisms. Among them, bacteria and fungi are most predominant. The inoculums of bacteria are generally proliferated by direct contact of vegetables with contaminated surfaces of packaging used in various stages of transport and storage chain. Sometimes contaminated water used for pre-treatment or cooling provides sufficient quantity of inoculums to spoil them. Although most of the bacteria are saprophytic in nature, some pathogenic bacteria also have been observed in specific vegetable like *Listeria monocyte* genes (Likotrafiti et al., 2003). Vegetables are mostly spoiled by bacteria due to their high pH (4.5–7.0), whereas many fruits particularly the acidic fruits (pH < 4.5) inhibit the growth of bacteria but encourage fungal spoilage. An RH less than 94–95% is considered low enough for growth of bacteria. Majority of the soft rot causing bacteria have their optimum temperature for growth around 30°C (Table 8.2). The following are the major bacterial species that infect the fruits and vegetables.

TABLE 8.2   Spoilage Causing Spoilage Pathogen in Fruits and Vegetables.

Sl. no.	Commodity	Causal organism
1	Potato, carrot and so forth.	*Erwiniacartovora*
		Bacillus subtilis
		Clostridium sp.
2	Onion	*Pseudomonas capacia*
3	Cabbage, Lettuce	*Pseudomonas cicorii*
4	Cauliflower, cabbage, tomato, bean, radish	*Xanthomonas* spp.

Bacteria like *Erwinia* and *Xanthomonas* are mainly disseminated by leafhoppers.

Control of bacterial diseases: Chlorine compounds added to water as chorine gas or as hypochlorite solution could be used to kill bacteria rapidly. In commercial practice, 2% Na-hypochlorite or 50–100 mg of chlorine or 1–1.5 g/l chlorine dioxide are frequently used to control the bacteria. Low temperature storage is considered to be the best method for controlling the bacterial diseases.

Three major fungal organisms are mostly associated with post-harvest diseases of vegetables. These are (i) *Fusarium* spp. cause rots in tinda, bean, cowpea, chilli, tomato, cauliflower, potato, cucumber, pointed gourd and pumpkin. (ii) Geotrichum diseases in tomato, pointed gourd, watermelon and muskmelon caused by *Geotrichumcandidum*. As a thumb rule, the foods containing lactic acid are attacked by this fungus. (iii) Pithium diseases caused by *Pithium* spp. that attack okra and cucurbits. Table 8.3 shows the time temperature combinations for control of the fungal organisms.

TABLE 8.3   Spoilage Causing Fungal Pathogen in Fruits and Vegetables.

Organism	Temperature (°C)	Time/Duration
Botryodiplodia. Gloeosporium, Colletotrichum, Rhizopus	43–49	20
Colletotrichum, Diplodia, Aspergillus, Bortyodiolodia, Rhizopus	47–55	10–20
Monilia, Rhizopus	49–84	1.5–3.5
Alternaria, Geotrichum	48–57	5–10

8.2.2   ON-FARM STORAGE TECHNIQUES

Evaporative cooling (EC) technology is one of the most efficient and economical technology for reducing the temperature and increasing the RH surrounding the perishable produce like vegetables and thus, enhanced their storage life. It is a simple and cheap method for vegetables when they are intended for short-term storage. It does not depend on conventional energy sources. During summer months, the system can maintain 20–25°C temperature and 90–95% RH as compared with ambient conditions of 23–45°C and 30–75.5% RH under north Indian conditions. The shelf life of commodities is extended by 20–80% in EC storage as compared with the storage at ambient condition. The freshness is well-maintained due to the high humidity of the environment. EC stored tomatoes develop better colour and ripe uniformly (Thiagu et al., 2007).

Pusa Zero Energy Cool Chamber is an on-farm cuboidal storage chamber developed at IARI, New Delhi that works on the principle of EC. It can be constructed easily anywhere with locally available materials like brick, sand, bamboo, *khaskhas* or straw, gunny bags and so forth and its operation needs a steady source of water. It consists of a double-walled bricks having cavity in between. The cavity is filled up with fine riverbed sand on all four sides. The porous bricks allow water for seepage. During seepage, the water evaporated from the walls and sand matrix and thus temperature of the cool chamber get reduced. The process of water seepage through inner wall provides necessary moisture in the enclosure, and thus, it increases the RH of the chamber (Roy and Pal, 1993).

Commercial version of Pusa zero energy cool chamber was also developed having dimension $3.66 \times 3.66 \times 3.66$ m. It can store about 6–8 tons of fruits or vegetables. It is also a double brick-wall and sand-filled cuboidal structure. An overhead tank is constructed which continuously supplies water to the walls of the chamber. The floor is made of wooden planks and bottom of the chamber is provided with air ducts. Roof has an exhaust fan at the centre. This runs for a definite time with the help of a sequential timer.

The major benefit of Pusa zero energy cool chamber are (i) extension of growing period of button mushrooms with 24% higher yield than conventional growing, (ii) orderly marketing of potatoes, (iii) quality assurance through better appeal and high retention of essential vitamins in fruits and vegetables, (iv) uniform ripening of tomatoes and banana even in peak summer months. The commercial size (6–8 MT) chamber could be successfully utilized for onion storage during rainy season after withholding the water supply.

8.3 CONTROLLED ATMOSPHERE STORAGE OF VEGETABLES

The effect of gas on harvested fruits is known since ancient times. In earlier days, fruits were taken to temples for improving ripening. The scientific explanation of this phenomenon could be attributed to the volatiles released on burning of incense containing hydrocarbon gases causing ripening of fruits. Earliest scientifically documented evidence on post-harvest physiology was recorded by J.E. Bernard in 1819 in France who noticed that harvested fruits absorb O_2 and liberate CO_2. Atmosphere devoid of O_2 caused no ripening in peach, prunes and apricot for several days but ripening continued when they were placed back in air. A commercial cold storage built in 1856 by B. Nice, and ice was used to maintain a temperature of 1°C. After a decade he experimented with modifying the cold store with CO_2 and O_2 gases inside. It was claimed that 4000 bushels of apples were kept in good condition in his store for 11 months under low O_2 level (such that the flame would not burn). In 1907, Foulton J. observed the increase in fruit damage by large accumulation of CO_2 in the storage atmosphere. In 1915, R.W. Thatcher on their experiments with apples in sealed boxes containing various levels of gases concluded that CO_2 greatly inhibited ripening. The first scientific evidence on the effect of CO_2 on respiratory rates was established by Kidd and West (1917) in seeds. In

1927, Kidd and West carried out comprehensive studies on the effect of CO_2 and O_2 on storage of fruits and vegetables. In the early 1940, the term of 'gas storage' was replaced by controlled atmosphere (CA) storage.

Atmosphere at ambient conditions comprises maximum of nitrogen (78.08%) followed by oxygen (20.98%) and minute quantities of carbon dioxide (0.03%). Any deviation from this normal atmosphere composition, for example elevated level of CO_2 reduced level of O_2, N_2 or any other combination, is known as 'Modified atmosphere'. When this deviated normal atmosphere is precisely kept under control then it is termed 'Controlled atmosphere'. This control can be done in package (CA packaging) or in the storage chamber CA storage. Similar is the case for modified atmosphere storage (MA storage) and MA packaging (MAP). In general, O_2 below 8% and CO_2 above 1% are used in CA storage. Under CA storage, the atmosphere is changed from the existing ambient atmosphere, and gaseous composition is maintained throughout the storage. Atmospheric modification is a supplementary practice to temperature management in preserving quality and safety of fresh fruits, vegetables, ornamentals and their products throughout postharvest handling

Essentiality of CA/MA technology should be justified only if (i) the commodities are having high market value, (ii) it significantly enhances storage life, (iii) it retains significantly better quality, and (iv) it fetches better price compared with conventional cool stored produce. Retardation of ripening, reduction in decay and prevention of specific disorders and maintenance of product texture are some of the potential advantages of CA/MA storage. However, initiation or aggravation of certain physiological disorders, namely black heart in potato and brown stain in lettuce take place in CA storage if appropriate gaseous regimes are not maintained (Kader et al., 1989). A concentration of less than 2% O_2 or more than 5% CO_2 in the storage/package atmosphere results in irregular ripening of banana, tomato and pear fruits. Too low O_2 or too high CO_2 can increase the susceptibility to decay causing organisms. Off-flavour development in fruits stimulation of sprouting and retardation of periderm formation in potatoes are some of the ill effects of improper CA condition.

CA storage and preservation of packaged horticultural produce is a widely utilized technology for fresh fruits, vegetables and flowers. The utilization of inert gases or vacuum can allow unique applications that control microorganisms as well as maintain product colour and freshness. It should be kept in mind that CA and MA are the supplementary practices to low temperature storage. The quality retention is mainly due to reduction in the respiratory and metabolic activities and check in ethylene liberation by fresh horticultural produce during storage.

8.3.1 BIOLOGICAL BASIS OF CONTROLLED ATMOSPHERE EFFECTS

Our atmosphere is composed around 78% nitrogen, 21.5 oxygen, 0.93% argon, 0.093% carbon dioxide and traces of other gases. Atmospheric storage of vegetables reduces the oxygen concentration and increases the level of carbon dioxide and subsequently it creates a condition inside the cabinet so that respiratory activity is reduced as a result senescence stage in vegetables is delayed, and subsequently, the shelf life of vegetables is increased. Normally the shelf life of vegetables under controlled storage environment is extended to 2–4 times than the usual storage. Long-term storage of vegetables actually involves inhibiting the ripening and ageing processes, thus retaining flavour and quality. Controlled atmospheric storage retains the quality and freshness for longer time without the use of any chemical additives.

Fruits and vegetables are considered living entity, and as a result fruits and vegetables tolerate the reduced levels of O_2 and increased levels of CO_2 to the limited range. However, respiratory activity and ethylene production is altered after adjusting the gaseous composition beyond the limited range of O_2 and CO_2 level. This condition leads to stress condition. Stress environment causes physiological disorders and increased susceptibility to decay. Various factors such as exposure to higher storage temperature than the recommended storage temperature, and RH can play synergistic effect in stress condition (Pal and Buescher, 1993).

Normally, there has been aerobic respiration during the development of fruits and vegetables. However, there has been shift from aerobic to anaerobic respiration after the harvest and during storage under controlled atmospheric storage. The change in respiration pattern is dependent on maturity condition, ripening status, temperature of storage and stress exposure of oxygen and carbon dioxide concentration. However, plants have the capability to tolerate the level of reduced oxygen and elevated carbon dioxide level at certain limit for limited period. Plants can tolerate low level of oxygen ($<1\% \ O_2$) and elevated carbon dioxide ($>10\% \ CO_2$) for a very short period of time. Climacteric fruits have lower tolerance limit as compared with non-climacteric fruits in terms of recovering the stress. The extent of stress recovery is governed by period of stress and maturity levels of fruits and vegetables (Kader, 2001).

The activity of 1-aminocyclopropane-1-carboxylate (ACC) synthase enzyme responsible for ethylene biosynthesis is increased at low O_2 concentration, whereas it is reduced at high CO_2 concentration (Gorny and Kader, 1996). The optimum storage conditions retain the quality attributes of vegetables for longer time. It caused lesser degradation of chlorophyll in vegetables; therefore, the natural green is retained for longer time. It also caused increased biosynthesis of carotenoids; thus, yellow and orange is maintained for longer time. The optimum environment storage conditions helped in biosynthesis and oxidation of phenolic compounds which resulted in brown colour development in vegetables (Zagory et al., 1989). Furthermore, the enzymatic activity responsible for cell wall degrading enzyme is slowed down significantly during atmospheric storage as a result pericarp of vegetables retained toughening for longer time during storage. Various natural biochemical activities are also altered during controlled atmospheric storage. Vegetables during storage at low-O_2 and high-CO_2 storage result in slower degradation of acidity; conversion of starch to sugar is also reduced. Furthermore, the sugar inter-conversion rate and biosynthesis of flavour volatiles are also affected. Nutritional profile is also maintained for longer time during controlled atmospheric storage. Vitamins especially vitamin C retention was reported for longer time during controlled atmospheric storage as a result nutritional quality is maintained for longer time (Kader, 2001).

Vegetables normally experience stress with change the gaseous conditions during storage. The stress situation leads to reduction in cytoplasmic pH, ATP and pyruvate dehydrogenase activity, whereas pyruvate decarboxylase, alcohol dehydrogenase and lactate dehydrogenase activities are activated. Various hazardous chemicals such as acetaldehyde, ethanol, ethyl acetate and lactate are accumulated in plant tissues due to alteration in enzymatic activities during stress conditions beyond the tolerance limit. However, the accumulation of hazardous chemicals varied upon cultivars, maturity and ripening stage, storage temperature and period time and in some cases ethylene concentrations (Kader, 1995).

Nitrogen gas is used to vary the gaseous composition of reduced O_2 and increased CO_2 level. However, in certain experiments, the diffusivity of O_2, CO_2 and C_2H_4 is increased upon replacement of nitrogen with argon and helium. However, the experiment could not be further continued due to costs factors.

Increased O_2 concentration upto 80% level is reported to trigger the ethylene production which would result in rapid change in colour from green to yellow in non-climacteric fruits and rapid increase in ripening due to increased respiration rate, ethylene production and onset of physiological symptoms. However, higher O_2 concentration beyond 80% level leads to infestation of post-harvest pathogens due to oxygen toxicity (Kader, 2016).

8.3.2   BENEFICIAL EFFECTS OF CONTROLLED ATMOSPHERE

- The rate of biochemical changes is reduced. Slow respiration rate caused decreased softening of tissues. The storage of vegetables in CA storage resulted in delayed ethylene production as a result senescence process is delayed. Reduction of sensitivity to ethylene action at O_2 levels < 8% and/or CO_2 levels > 1%.
- Occurrence of physiological order symptoms during CA storage in certain vegetables under CA storage has direct or indirect effect on post-harvest pathogens (bacteria and fungi) and consequently caused spoilage. The storage of vegetables with carbon dioxide at 10–15% CO_2 significantly inhibited the development of Botrytis rot in certain vegetables.
- Low O_2 (< 1%) and/or elevated CO_2 (40–60%) can be successful for controlling insect during storage of dried products from vegetables.

8.3.3   DETRIMENTAL EFFECTS OF CONTROLLED ATMOSPHERE

- Increased physiological disorder symptoms such as brown stain of lettuce and chilling injury of some commodities.
- The reduced level of O_2 (below 2%) and higher levels of CO_2 (above 5%) resulted in uneven ripening in tomatoes.
- CA storage sometimes can encourage anaerobic respiration and fermentative metabolism during storage of vegetables with very low O_2 and very high CO_2 concentrations.
- There have been increased chances for decay of vegetables during short storage of vegetables due to physiological injury during storage with very low oxygen and very high carbon dioxide concentration.
- CA storage may lead to sprouting in potato and carrots.

8.3.4 COMMERCIAL APPLICATION OF CONTROLLED ATMOSPHERE STORAGE

CA storage systems are used commercially for the long-term storage of fresh horticultural crops. Commercial use of CA storage is the maximum on apples and pears worldwide. Recent research has shown the potential advantages of this method in short-term (a few days) and medium-term (a few weeks) storage of certain types of produce. Optimizing storage conditions requires facilities that allow the temperature and the composition of gases in storage rooms to be controlled precisely. Each product reacts in different ways to different concentrations of gases.

Various types of CA systems are used in commercial storage. These include generating nitrogen by separation from compressed air using molecular sieve beds or membrane systems. Various types of CA storage system include ultra-low O_2 (1.0–1.5%) storage, low ethylene (<μl/l) CA storage; rapid CA storage (rapid establishment of optimal levels of O_2 and CO_2), and programmed or sequential CA storage (e.g. storage in 1% O_2 for 2–6 weeks followed by storage in 2–3% O_2 for the remainder of the storage period) and so forth (Kader, 2016). Recent reports of short-term CA exposure techniques indicate a great promise of simulated effect of continuous CA-storage system with particular reference to delay in disease development, delaying the senescence and quality assurance of CA-insensitive climacteric fruits (Pal, 1997). Other developments include use of atmospheric modification during transport and distribution, improved technologies of establishing, monitoring and maintaining CA, using edible coatings or polymeric films with appropriate gas permeability to create a desired atmospheric composition around the commodity.

8.4 MODIFIED ATMOSPHERE PACKAGING

MAP is defined as alteration of gaseous environment during respiration process by the addition and removal of gases in the closed environment of vegetables to manipulate the levels of O_2, CO_2, N_2 and C_2H_4. Vegetables respire in sealed package during storage with respiration process resulting in onset of passive MAP. Even passive MAP storage extends the shelf life of vegetables due to change in respiratory activity, as a result bio-chemical changes are affected during storage. The packaging materials also transmit O_2, CO_2 and water vapour which further modify the gaseous environment.

Active MAP is developed by addition or removal of gases to attain the gaseous composition different from initial composition. The modification of gases during MAP storage controls the normal physiological processes and reduces the microbial infestation to increase the shelf life and maintain the quality of vegetables for longer time.

MAP was first introduced in 1927 for increasing the shelf life of apples during storage with change in environment of reduced oxygen and increased carbon dioxide. The process gained popularity during subsequent years. In 1939, the MAP storage was used for long transport of fruits and beef carcasses in ships, and shelf life was increased up to 100% (Davies, 1995). The storage under MAP was introduced commercially for retail packs until the early 1970s in European markets. MAP techniques are not restricted to fruits and vegetables, but it widely followed in variety of fresh and cooked foods including meat, fish, poultry, pasta, dairy, coffee, tea and bakery products (Philips, 1996).

8.4.1 ADVANTAGES AND DISADVANTAGES OF MODIFIED ATMOSPHERE PACKAGING

Lower O_2 and higher CO_2 concentration during MAP storage reduces respiration rate, delayed ethylene production and ultimately delayed ripening and reduced textural softening. Other notable biochemical changes such as chlorophyll degradation, enzymatic browning and loss of vitamins can be minimized to greater length of time. CA/MAP technique is cumbersome storage process for extending the shelf life of vegetables as respiration rate varies from one vegetable to another vegetable. Therefore, selection of suitable packaging film, thickness and gaseous composition in terms of oxygen and carbon dioxide vary for all types of vegetables. Therefore, selection of packaging film and thickness depends on respiration rate. Based on the respiration rate, the gaseous composition of oxygen and carbon dioxide is decided for extending the shelf life of vegetables. MAP technique is effective only when vegetables are stored above the chilling injury of vegetables.

8.4.2 RELATIVE TOLERANCE TO LOW O_2 AND ELEVATED CO_2 CONCENTRATION

The benefits associated with CA and MAP vary on type of cultivar, maturity of vegetables at time of harvest, initial quality, type and thickness of packaging materials, gaseous composition, temperature of storage and duration of exposure. Vegetables also suffer stress conditions during adjusting the gaseous composition below or above the tolerance limits than the optimum level. The stress condition reflects physiological disorders and detrimental effects in vegetables during storage.

Tables 8.4 and 8.5 include classifications of vegetables according to their relative tolerance to low O_2 or elevated CO_2 concentrations when kept at or near their optimum storage temperature and RH (Table 8.1) (Kader et al., 1989).

TABLE 8.4 Classification of Vegetables According to Their Tolerance to Low O_2 Concentrations. (*Source:* Kader et al., 1989.)

Minimum O_2 concentration tolerated (%)	Commodities
1.0	Broccoli, mushroom, garlic, onion
2.0	Cantaloupe, green bean, celery, lettuce, cabbage, cauliflower, Brussels sprouts
3.0	Tomato, pepper, cucumber, artichoke
5.0	Green pea, asparagus, potato, sweet potato

TABLE 8.5 Classification of Fresh Vegetables According to Their Tolerance to Elevated CO_2 Concentrations. (*Source:* Modified from Kader et al., 1989; Saltveit, 1997.)

Maximum CO_2 concentration tolerated (%)	Commodities
1	Onion
2	Tomato, sweet pepper, lettuce, Chinese cabbage, artichoke, sweet potato
5	Pea, chilli, brinjal, cauliflower, Brussels sprouts, radish, carrot, cucumber, bell pepper, potato
7	French bean
10	Okra, asparagus, Brussels sprout, cabbage, celery, leek, dry onion, garlic, sweet corn
15	Spinach, kale, broccoli
20	Cantaloupe, mushroom

8.4.3 METHODS OF CREATING MODIFIED ATMOSPHERE CONDITIONS

MA can be created either *passively* by the respiration of commodity also called *commodity generated* MA or *active* MA generation.

8.4.3.1 PASSIVE MODIFIED ATMOSPHERE

Passive MA is developed placing vegetables in hermetically sealed package as a result of respiration process, so that oxygen concentration decreases and carbon dioxide concentration builds up. However, appropriate passive atmosphere is developed with matching of film permeability and change of gaseous composition. This atmosphere must be established rapidly and without danger of the creation of anoxic conditions due to high level of carbon dioxide.

8.4.3.2 ACTIVE MODIFIED ATMOSPHERE

Active MA is practised to avoid the too low level of oxygen and too high level of carbon dioxide to vegetables during storage. This can be done by using partial vacuum and replacing the package atmosphere with gas mixture. The gaseous composition is further adjusted through the use of absorbers or adsorbers in the package to scavenge these gases.

8.4.4 PACKAGING MATERIALS FOR MODIFIED ATMOSPHERE PACKAGING

Most suitable packaging materials for MAP storage are made from one or more of four polymers such low density polyethylene, polyvinyl chloride, polyethylene terephthalate, polyethylene and polypropylene films. The choice of suitable packaging material aims at increasing the shelf life along with protection the natural physical and chemical attributes of vegetables. Horticultural commodities are preferred in polymeric films with specific gas permeabilities and storage at low temperature. However, other characteristics such as water vapour transmission rate, mechanical properties, types of package, transparency, sealing reliability and microwaveability are also considered for selection of packaging materials (Day, 1993).

8.4.5   EFFECTS OF MODIFIED ATMOSPHERE PACKAGING ON MICROBIAL GROWTH

Inherent microorganisms present on vegetables spoil the vegetables during storage. MAP reduces the growth rate of inherent microorganisms as a result the vegetables remain organoleptically acceptable for longer time. MAP also can delay senescence in vegetables, therefore reduces the susceptibility of pathogenic microorganisms (Yam and Lee, 1995).

8.4.5.1   SPOILAGE ORGANISMS

The commonly known spoilage microorganisms of vegetables are *Pseudommonas* sp., *Erwinia herbicola, Flavobacterium, Xanthomonas, Enterobacter agglomerans* lactic acid bacteria such as *Leuconostoc mesenteroids* and *Lactobacillus* sp. Although the growth of spoilage microorganisms depends on storage conditions, cabbage during storage at 7 and 14°C spoil at the same rate. However, the reduction of microbial load was significant during storage of cabbage at 7°C (King et al., 1976). Similar findings have been reported with shredded chicory salads and shredded carrots in which mesophilic counts decreased with low temperature. The growth rate of food-borne pathogens is decreased during storage at low temperature as well as low temperature storage increases the solubility of carbon dioxide in liquid phase surrounding the vegetables (Carlin et al., 1989). Similarly, the storage of broccoli with 10% carbon dioxide and 5% oxygen caused inhibition of aerobic spoilage bacteria at 1°C (Barth et al., 1993). The growth of lactic acid bacteria is favoured with decreased oxygen and increased carbon dioxide concentration. Lactic acid bacteria sensitive to lettuce, chicory leaves and carrots are spoiled under MAP. There has been negligible effect of MAP on yeasts. The growth molds are affected during storage under MAP, as molds are aerobic and growth inhibition of molds is reflected at 10% carbon dioxide concentration during MAP (Molin, 2000).

8.4.5.2   PATHOGENIC ORGANISMS

Many pathogenic organisms such as Clostridium botulinum, Yersinia enterocolitica, Listeria monocytogenes, Aeromonas hydrophyla, Salmonela spp., Clostridium perfinges, Bacillus cereus grow at 5–12°C and spoil the vegetables. It is observed that some of the clostridia and campylobacter species may be able to survive better in MA as compared with normal atmosphere. Certain pathogenic microorganisms such as *L. monocytogenes, Y. enterocolitica* and *A. hydrophila* are capable of growing at extremely low temperature in certain modified temperature (Farber, 1991). MAP stored vegetables are vulnerable from safety point of view as MAP inhibits the growth of spoilage organisms, whereas the growth of pathogenic are encouraged.

Certain enteric pathogenic organisms such as *Salmonella, Shigella* and *E. coli* are concerns about their behaviour under MAP storage. Under high oxygen concentration and moderate carbon dioxide concentration, the growth of *S. typhimurium, S. enteritidis, S. typhimurium, L. monocytogenes* and *pathogenic E. coli* were unaffected (Francis et al., 1999).

REFERENCES

Barman, K.; Asrey, R.; Pal, R. K. Putrescine and Carnauba Wax Pretreatments Alleviate Chilling Injury, Enhance Shelf Life and Preserve Pomegranate Fruit Quality During Cold Storage. *Sci. Hortic.* **2011,** *130*(4), 795–800.

Barth, M. M.; Kerbel, E. L.; Broussard, S.; Schmidt, S. J. Modified Atmosphere Packaging Protects Market Quality in Broccoli Spears Under Ambient Storage. *J. Food Sci.* **1993,** *58,* 1070–1072.

Brosnan, T.; Sun, D. W. Precooling Techniques and Applications for Horticultural Products—A Review. *Int. J. Refrig.* **2001,** *24*(2), 154–170.

Carlin, F.; Nguyen-the, C.; Cudennec, P.; Reich, M. Microbiological Spoilage of Fresh Ready-to-Use Grated Carrots. *Sci. Alim.* **1989,** *9,* 371.

Davies, A. R. *Advances in Modified Atmosphere Packaging, New Methods of Food Preservation*; G. W. Glaskow: UK, 1995.

Day, B. P. F. Fruit and Vegetables. In *Principles and Applications of Modified Atmosphere Packaging;* Parry, R. T., Ed.; Blackie: Glaskow, UK, 1993; pp 114–133.

Farber, J. M. Microbiological Aspects of Modified Atmosphere Packaging—A Review. *J. Food Prot.* **1991,** *54,* 58–70.

Francis, G. A.; Thomas, C.; O'Beirne, D. The Microbiological Safety of Minimally Processed Vegetables—A Review. *Int. J. Food Sci. Technol.* **1999,** *34,* 1–22.

Gorny, J. R.; Kader, A. A. Controlled-Atmosphere Suppression of ACC Synthase and ACC Oxidase in Golden Delicious' Apples During Long-Term Cold Storage. *J. Am. Soc. Hortic. Sci.* **1996,** *121*(4), 751–755.

Kader, A. A. Potential Application of Ionizing Radition in Post-Harvest Handling of Fresh Fruits and Vegetable. *Food Technol.* **1986,** *40*(6), 117–121.

Kader, A. A. Regulation of Fruit Physiology by Controlled/Modified Atmosphere. *Acta. Hortic.* **1995,** *398*, 59–70.

Kader, A. A. Physiology of CA Treated Produce. In VIII International Controlled Atmosphere Res Conference. *Acta. Hortic.* **2001,** *600,* 349–354.

Kader, A. A. Controlled Atmosphere Storage. *The Commercial Storage of Fruits, Vegetables, and Florist and Nursery Stocks;* Gross, Kenneth C., Chien Yi Wang, Mikal Saltveit, Eds.; Washington, DC, USA, 2016; pp 66, 22–25.

Kader, A. A.; Zagory, D.; Kerbeland, E. L.; Wan, E. L. Modified Atmosphere Packaging of Fruits and Vegetables. *Crit. Rev. Food. Sci. Nutr.* **1989,** *28*(1): 1–30.

Kidd, F.; West, C. The Controlling Influence of Carbon Dioxide, IV. On the Production of Secondary Dormancy in Seeds of Brassica Alba Following Treatment with Carbon Dioxide, and the Relation of this Phenomenon to the Question of Stimuli in Growth Processes. *Ann. Bot.* **1917,** *31*(123/124), 457–487.

King, A. D.; Michener, H. D.; Bayne, H. G.; Mihara, K. L. Microbial Studies on Shelf Life of Cabbage and Coleslaw. *Appl. Environ. Microbiol.* **1976,** *31,* 404–407.

Kleinkopf, G. E.; Nathanand, X. A. O.; Olsen, N. L. Sprout Inhibition in Storage Current Status, New Chemistries and Natural Compounds. *Am. J. Potato Res.* **2003,** *80*(5), 317–327.

Likotrafiti, E.; Manderson, K.; Tuohy, K.; Gibson, G. R.; Rastall, R. A. Screening of Probiotic Strains Isolated from the Elderly for Antimicrobial Activity Against Gastrointestinal Pathogens. In VTT Symposium **2003,** *226,* 52.

Molin, G. Modified Atmospheres. *The Microbiological Safety and Quality of Foods*; Lund, B. M., Baird-Parker, D. C., Gould, G. W., Eds; Springer: Gaithersburg, MA, 2000; pp 214–234.

Pal, R. K. Short Term Exposure to Controlled Atmosphere on Shelf Life and Quality of Mango. 84th Indian Science Congress Delhi University. Jan 3–8, 1997.

Pal, R. K.; Buescher, R. W. Respiration and Ethylene Evolution of Certain Fruits and Vegetables in Response to Carbon-dioxide in Controlled-Atmosphere Storage. *J. Food Sci. Technol.* **1993,** *30*(1), 29–32.

Philips, C. A. Modified Atmosphere Packaging and its Effects on the Microbiological Quality and Safety of Produce—A Review. *Int. J. Food Sci. Technol.* **1996,** *31,* 463–479.

Roy, S. K.; Pal, R. K. A Low-Cost Cool can Chamber an Innovative Technology for Developing Countries. In ACIAR Proceeding. *Aus. Centre Int. Agric. Res.* **1993,** 393–393.

Saltveit, M. E. A Summary of CA and MA Requirements and Recommendations for Harvested Vegetables, In: Proc. 7th Intl. Controlled Atmosphere Res. Conf. vol. 4. Vegetables and ornamentals. Saltveit, M. E., Ed.; Univ. Calif. Postharvest Hort. Ser. 18. 1997; pp 98–117.

Thatcher, R. W. Enzymes of Apples and Their Relation to the Ripening Process. *J. Agric. Res.* **1915,** *5*(3), 103–105.

Thiagu, R.; Nagin, C.; Habibunnisa, E. A.; Prasad, A. B.; Raman, K. V. R. Effect of Evaporative Cooling Storage on Ripening and Quality of Tomato. *J. Food Qual.* **2007,** *14*(2), 127–144.

Yam, K. L.; Lee, D. S. Design of MAP for Fresh Produce. IN *Active Food Packaging;* Rooney, M. L. Ed.; Chapman and Hall Inc.: USA, 1995; p 55.

Zagory, D.; Kader, A. A.; Wash, D. C. Quality Maintenance in Fresh Fruits and Vegetables by Controlled Atmospheres. In *Quality Factors of Fruits and Vegetables, Chemistry and Technology*; Joseph J. J, Ed.; American Chemical society: Washington D.C., 1989; pp 174–188.

CHAPTER 9

ACTIVE AND SMART PACKAGING TECHNIQUES IN VEGETABLES

D. V. SUDHAKAR RAO[1], ANURADHA SRIVASTAVA[2], AND RANJITHA K.[1]

[1]Division of Post-Harvest Technology, Indian Institute of Horticultural ResearchHessaraghatta Lake Post, Bengaluru, India

[2]ICAR Research Complex for Eastern Region (ICAR-RCER), Patna, India

CONTENTS

9.1   INTRODUCTION

Consumer demand for fresh vegetables is on the unprecedented rise due to the growing population and improved public awareness on their unique nutritional qualities pertaining to antioxidants, vitamins and functional properties. Moreover, improved economic growth and employment has increased number of 'rich in money-poor in time' households; for whom, the ready-to-use vegetables are a boon. Fresh and fresh-cut produce are highly perishable being living tissues and undergo rapid senescence resulting in

major post-harvest losses and limited consumer availability. The microbiological spoilage or their mutual synergistic action in the warm temperatures
aggravate the spoilage apart from the physiological degradation, and even
pose food safety risks to the public. Unlike processed products, in which
rigorous processing parameters take care of the spoilage, the shelf life of
fresh and fresh-like produce is enhanced mainly by resorting to simple
post-harvest treatments and suitable packaging technologies. Scientific
interventions have taken the packaging technology from the traditionally
used gunny bags to a long way, through modified atmosphere packaging
(MAP), with the current research being focussed on smart packaging technologies to evaluate real-time quality of the produce in a non-invasive and
non-destructive manner. For packing any living tissue, a good packaging
material should ensure the maintenance of a suitable in-pack modified
atmosphere (O_2, CO_2 and N_2), and this is an invariable prerequisite for packaging fresh vegetables too. Modified atmosphere packagingof fresh vegeta-
blesin total is beyond the preview of the current chapter; the use of active,
intelligent or smart systems as supplementary to MAP are described below.

9.2   ACTIVE PACKAGING SYSTEMS

Active packaging refers to the incorporation of certain additives into packaging systems (whether loose within the pack, attached inside packaging
materials or incorporated within the packaging materials themselves) with
the aim of maintaining or extending produce quality and shelflife. When
it performs some desired role in food preservation other than providing
an inert barrier to external conditions, packaging may be termed active
(Hutton, 2003). Active packaging has been defined as packaging, which
'changes the condition of the packed food to extend shelflife or to improve
safety or sensory properties, while maintaining the quality' (Ahvenainen,
2003). Examples of active-packaging systems for fresh vegetables are
incorporation of ethylene absorbers, oxygen scavengers, moisture regulators, taint removal systems, ethanol and carbon dioxide emitters, and
antimicrobial-releasing systems in the packages.

9.2.1   ETHYLENE CONTROL

Ethylene is known as ripening hormone, andit accelerates senescence and
softening, increases chlorophyll degradation and reduces shelf life of fresh

and minimally processed vegetables (De Kruijfetal., 2002). Ethylene scavenging systems are commonly based on potassium permanganate or finely dispersed minerals (zeolite, active carbon, pumice etc.) to absorb ethylene and are generallyplaced as sachet inside the fresh produce package. The mineral scavengersnot only absorb ethylene but also alter the permeability of the film so that ethylene and CO_2 diffuse more rapidly and O_2 enters more readily than through flexible films. In climacteric produce like tomato, cucumber and other crops suchas carrots, potatoes and asparagus, ethylene absorbers find vast application (De Kruijf et al., 2002). Shelf life of many vegetables can be extended by the application of low concentrations of 1-methylcyclopropene (1-MCP), an ethylene action inhibitor (Blankenship and Dole, 2003). 1-MCP in the form of a white powder is encapsulated in a molecular encapsulating agent such as alpha-cyclodextrin, and the controlled release of the 1-MCP is facilitated through addition of water to this complex (Daly and Kourelis, 2000).

9.2.2   OXYGEN SCAVENGERS

To remove the oxygen by means of a chemical reaction, oxygen scavengers are easily oxidizable substances included in the packaging system. The substance is usually contained in sachets made of a material highly permeable to air but it can also be included in the plastic film matrix. The most common substances used are iron powder and ascorbic acid. The scavengers may be of self-reaction type or moisture dependent. The reaction only takes place after moisture has been absorbed from the food product, in the latter case, whereas in the first case, the reaction starts as soon as the scavenger is exposed to air (Charles et al., 2006).

9.2.3   CARBON DIOXIDE GENERATORS

The growth of aerobic microorganisms is delayed by the high level of carbon dioxide in the headspace of packaged fruits and vegetables, and this reduces the respiration and senescence processes (Chakraverty, 2001). When carbon dioxide is present in excess amounts, the use of a carbon dioxide emitting package should not induce anaerobic metabolism. Therefore, the film permeability and the respiration rate should be taken into consideration. Most carbon dioxide emission processes are activated by moisture that usually comes from packaged foods (Ozdemir and Floros, 2004). Therefore, this activation mechanism may have limited applications with intermediate moisture foods but may work well with high-moisture foods such as minimally processed vegetables and fruits. This technology uses the reaction of sodium bicarbonate and hydrating agents such as water with acidulates to produce carbon dioxide (Ozdemir and Floros, 2004).

9.2.4 TEMPERATURE COMPENSATING FILMS

Temperature strongly affects respiration rate of fresh produce, whereas the gas permeability of traditional films changes negligibly due to temperature fluctuations. This results in rapid accumulation of carbon dioxide and depletion of oxygen in the package during temperature-abused storage environments. The introduction of a new array of films called temperature-compensating films are used to overcome these problems, whose permeability changes significantly with change in temperature. The activation energy of permeability in temperature-compensating films matches more closely with the respiration rate activation energy of produce. Therefore, changes in the respiration rate caused by temperature fluctuation tend to be compensated by changes in the gas transmission rate of film, avoiding anoxic conditions (Yam and Lee, 1995).

Side-chain crystallizable polymers are used by Landec Corporation (USA) in its Intelimer-polymer technology. They consist of an acrylic backbone with fatty acid-based side chains of various lengths. In general, they remain in a crystalline state, but side-chains of different lengths melt at different temperatures, altering the polymers' physical properties in response to specific, predetermined temperature variation. The polymer abruptly becomes amorphous and more permeable to oxygen when the temperature rises above the set temperature. This change is physically reversible, thus the polymer recrystallizes and the permeability returns to its originally set level if there is a decrease in temperature (http://landec.com/technology/platforms/intelimer-polymers/). The Intellipac membrane is an adhesive patch coated with Intelimer-polymer that is positioned over a die-cut hole on a plastic bag, allowing the produce to 'breathe'.

9.2.5   HUMIDITY AND CONDENSATION CONTROL

Spoilage of packed fresh and fresh-cut vegetables is accelerated by the condensation of water vapour in the package released during the process of respiration and transpiration (De Kruijf et al., 2002; Powers and Calvo, 2003). The control of excessive water can be achieved by application of drip-absorbent sheets which enclose a layer of super absorbent polymer such as polyacrylate salts or cellulose fibres. Another application to control excess moisture in packaged food is to use humectants such as propylene glycol which could be placed between two plastic films. Moisture regulators can prevent the growth of yeast and bacteria in high a_w foods such as minimally processed vegetables (De Kruijf et al., 2002). Shirazi and Cameron (1992) proved the usefulness of sodium chloride in prevention of surface mould development due to its humidity-controlling capacity in packed tomatoes. A multilayer package wall comprising a layer of moisture-absorbent material was developed by Patterson and Joyce (1993), such as polyvinyl alcohol, or a cellulosic fibre-based material like paper, sandwiched between an water vapour and liquid water impermeable outer layer, such as polyethylene, and an inner layer that is hydrophobic but permeable to water vapour, which is spot welded to the moisture absorbent layer. This structure prevents condensation by allowing the absorbent layer to take up water vapour when the relative humidity increases as a result of reduced temperature and prevents dehydration of the produce by releasing water vapour into the headspace of the package when the relative humidity decreases in response to increasing temperature. This approach has undergone further recent improvement (Gibbard and Symons, 2005) and forms the basis of Commonwealth Scientific and Industrial Research Organisation's Moisture Control Technology, which has been reported (CSIRO, 2006) to provide substantial extension to the storage life of cauliflower in trial shipments from Australia to Singapore.

9.2.6   ODOUR CONTROL

To remove odours associated with the volatile compounds such as aldehydes, amines and sulfides accumulating inside a package as a result of food degradation, a number of active packaging technologies have been developed (Brody et al., 2001). However, some produce, such as onions and potatoes, have a naturally pungent odour. In these cases, a primary concern is the prevention of cross-contamination during storage and transportation of mixed loads containing these commodities. Morris (1999) designed an odour-proof package for the storage and transportation of durian in which the package is comprised of an odour-impermeable plastic, such as polyethylene terephthalate or polyethylene of a suitable thickness, to prevent transmission of odour, together with a port to allow for the passage of respiratory gases. Over this port is affixed a sachet containing an odour-absorbing material made from a mixture of charcoal and about 10% w/w nickel.

9.2.7   MICROBIAL CONTROL

Fresh and minimally processed vegetables get contaminated due to exposure to unhygienic environment during cultivation, harvesting, transportation, packaging and processing operations (Erdoğrul and Şener, 2005). Although freshly harvested fruits and vegetables contain mixed initial flora of coliforms especially *Escherichia coli,* lactic acid bacteria, *Pseudomonas* and *Erwinia*, the primary causes of the spoilage of fresh-cut fruits and vegetables are yeasts, moulds and *Pseudomonas*, especially when stored aerobically under refrigeration (May and Fickak, 2003).

To improve safety and shelf life of food products, antimicrobial packaging is a promising form of active packaging. The three basic categories of antimicrobial packaging systems include incorporation of antimicrobial substances into a sachet attached to the package from which the volatile bioactive substance is released during further storage, direct incorporation of antimicrobial agent into the packaging film and coating of packaging with a matrix that acts as a carrier for the antimicrobial agent (Appendini, 2002). Non-volatile antimicrobial substances must contact the surface of the food, so that the antimicrobial substances can diffuse to the surface of the food product. Therefore, diffusion of incorporated antimicrobial agents from the packaging material to the surface of the food is crucial in exerting the antimicrobial activity (Cooksey, 2000). The use of multilayer films or blending antimicrobial substances into packaging materials, in which only one layer is impregnated with antimicrobial substances, improved microbial stability of fresh as well as fresh-cut produce (Ozdemir and

Floros, 2004; Rojas-Grau et al., 2007). Nanocomposite films containing antimicrobial agents with improved mechanical, thermal, optical and physicochemical properties (immobilized films) are studied for more effective delivery of the agents (Han, 2004; Rhim and Ng, 2007; Tunç and Duman, 2011). Recently, the demand for natural preservative/antimicrobial agents is more than ever. Many natural antimicrobial agents such as bacteriocins, essential oils and enzymes have been effectively incorporated into biodegradable films . Improvements in fundamental characteristics of food-packaging materials such as strength, barrier properties, antimicrobial properties and stability to heat and cold are being achieved using nanocomposite materials (Lagaron et al., 2005). Significant shelf life prolongation up to more than two months was achieved using silver-montmorillonite active packaging of fresh-cut carrots.

The two major applications of antimicrobial packaging in vegetables are ethanol- and carbon dioxide-releasing systems. In high concentrations, ethanol acts against vegetative cells of microorganisms, and it also has a preserving action in low concentrations. The use of ethanol-generating sachets or strips avoids the ethanol spraying directly onto the product surface prior to packaging. Carbon dioxide has a bacteriostatic effect on certain microorganisms, extending the lag phase and decreasing the growth rate during the logarithmic growth phase (Smith et al., 1995). Allylisothiocyanate in cyclodextrin matrix in inner liners of plastic films was practically tested by American Air Liquideand its Japanese partner against various pathogenic bacteria. Sulfur dioxide-releasing pads were commercialized by QuimicaOsku (Chile), and from a mixture of organic acid and calcium sulphite, sulfur dioxide-releasing plastic films were developed by Food Science, Australia. Chlorine dioxide-releasing films are commercially available for the disinfection of packaged fruits and vegetables (Scully and Horsham, 2007; Han and Floros, 2007).

9.2.8 COMMERCIAL PRODUCTS FOR ACTIVE PACKAGING OF VEGETABLES

Several active packaging materials based on the above requirements have been developed, patented and commercialized. These are summarized in the table below.

Product name	Manufacturer	Principle
Ethylene control		
Ethylene control®	Ethylene Control, Inc. (USA)	$KMnO_4$ is immobilized in an inert substrate such as alumina or silica gel
Air Repair®	DeltaTrak, Inc	Activated aluminium beads impregnated with potassium permanganate
SendoMate®	Mitsubishi, Japan	Activated carbon based on a palladium catalyst
Hatofresh System®	Honshu Paper, Japan	Activated carbon impregnated with bromine-type inorganic chemicals
Neupalon®	Sekisui Jushi (Japan)	
Evert-Fresh Bags®	Evert fresh corporation, USA	Activated earth based
Oxygen scavenger		
Ageless®	Mitsubishi, Japan	Iron powder and ascorbic acid based
Carbon dioxide generator		
Verifrais™	SARL Codimer (France)	Sodium bicarbonate and ascorbate based
Temperature compensating packs		
Intellipack®	Landec Corporation, USA	Adhesive patch coated with Intelimer-polymer
Antimicrobial packs		
Ethicap®	Freund Corporation, Japan	Silicon dioxide powder (35%) containing adsorbed alcohol (35%) and water (10%)
Negamold®	Freund Corporation, Japan	Carbon dioxide emitter

9.3   SMART OR INTELLIGENT PACKAGING

Intelligent or smart packaging is defined as a packaging system which monitors the condition of packaged food to give information about the quality of the packaged food during transport and storage (Ahvenainen, 2003). Although distinctly different from the concept of active packaging, features of intelligent packaging can be used to monitor the effectiveness and integrity of active packaging systems. Smart packaging informs the consumer a plethora of food pack attributes such as product safety/quality, tamper evidence and pack integrity, traceability/anti-theft devices and product authenticity. Ordinarily, an external or internal indicator for the active product history and quality determination is included in it (Krumhar and Karel, 1992; Hurme et al., 1994; Barnetson, 1995). Thus it is an interactive packaging system carrying out intelligent functions such as detecting, sensing and recording, tracing and communicating to inform consumers about the quality and safety of the food inside the package and warn about possible problems. Intelligent packaging can be consideredan extension of the communication function of traditional packaging and communicates information to the consumer based on its ability to sense, detect, or record external or internal changes in the product's environment (Restuccia et al., 2010). The indicator systems of interest for smart packaging of fresh fruits and vegetables are the gas, time-temperature, microbial load, freshness and ripeness.

9.3.1 GAS INDICATORS

To monitor the changes in the gas composition of the package, gas indicators or sensors in the form of a package label or printed on packaging films are used (Yam et al., 2005). Gas indicators especially oxygen and carbon dioxide indicators could be useful in modified atmosphere packages of vegetables. Most of the commercial internal oxygen and carbon dioxide indicators change their colour due to chemical or enzymatic reaction (Krumhar and Karel, 1992). These indicators are in contact with gas atmosphere and work on the basis of the gas change in the package. To monitor maturity stage of vegetables in the package, oxygen and carbon dioxide indicators could also be used. Another application of the gas indicators is to determine package leakage which not only causes change in internal atmosphere but also microbial contamination from the environment. To determine improper sealing and quality deterioration of modified atmosphere packages, gas indicators were used (Ahvenainen et al., 1997). Oxygen indicators react with oxygen coming out from the package through leakages (leakage indicator), or they are intended to be used with oxygen absorbers to verify that all oxygen has been absorbed from the package headspace (e.g. Yoshikawa et al., 1979; Perlman and Linschitz, 1985; Krumhar and Karel, 1992). For a leakage indicator, atwo-component system has been developed and patented, which both indicates leakage and absorbs residual oxygen (Ahvenainen and Hurme, 1997). This oxygen indicator was designed specifically for leak detection of modified atmosphere packages.

Carbon dioxide indicators are used to monitor the modified carbon dioxide level in (MAP systems. In general, carbon dioxide is used in MAP, together with inert nitrogen, because of its bacteriostatic effect. For respiring produce, the MAP conditions must be optimized for each fresh vegetable or fruit type. In high CO_2 packages, a leak will result in a lowered CO_2 level and an increased O_2 level. These changes can be monitored by leakage indicators (Smolander et al., 1997).

Both the above indicator types can be used successfully to confirm that the product has been properly packaged. However, they have their deficiencies in distribution. Most of the oxygen indicators are too sensitive to oxygen from a gas packaging point of view, and they are also reversible in the colour change reaction. So, in the first case, the indicators may react to the residual oxygen entrapped in the gas packaging process or in the second case, they may indicate that there is no oxygen in a package (i.e. a package should be intact), even though the product is spoiled by microbial growth which has consumed oxygen entrapped through leakage (Ahvenainen et al., 1995). Reversible colour change of the carbon dioxide indicators may in the worst case result in false information for the consumer: if the packed product is microbiologically spoiled, for example due to leakage, the carbon dioxide produced by microbial metabolism may still keep the headspace carbon dioxide at a level that indicates good product quality. The colour change of the indicators should, therefore, be irreversible.

9.3.1.1   COMMERCIAL EXAMPLES

Ageless eye: This indicator is marketed by Mitsubishi Gas Chemical Co., Japan. Ageless eye® can be includedwith oxygen scavengers in active packages; it turns pink near anaerobic environments (0–1% or less) and blue if oxygen exists (0–5% or more) (Abe, 1990).

9.3.2   TIME AND TEMPERATURE INDICATORS

The deterioration of fruits and vegetables by delaying senescence or alienating chilling injury is slowed down by precise control of temperature (Atkin and Tjoelker, 2003). A time–temperature integrator (TTI) is used as an indicator of product safety and quality (Taoukis, 2008). TTI's are devices that show an irreversible change in a physical characteristic, usually colour or shape, in response to temperature history. The TTI's are expected to mimic the change of a certain quality parameter of the food product undergoing the same exposure to temperature. The TTI's presently in the market have working mechanisms based on different principles: biological, chemical and physical. For the first type, the change in biological activity, such as microorganisms, spores or enzymes is the basic working principle. The others are based on a purely chemical or physical response towards time and temperature, such as an acid–base reaction, melting, polymerization and so on (Scetar et al., 2010). These indicators are already commercialized for products which require cold chain such as modified atmosphere (MA) products. However, to make sure that proper handling during transportation and storage is done, the application of TTIs should be extended to all MA products. Taking into account the low cost of the TTI tags, their wide usage is suggested to monitor the various steps of the real distribution chain of fresh produce (Scetar et al., 2010; Schilthuizen, 1999).

9.3.2.1   COMMERCIAL PRODUCTS

9.3.2.1.1   Fresh-Check®

An example of a TTI is the Fresh-Check® indicator (LifeLines Technology, USA) that has been developed for consumer use. It is a full history indicator whose working mechanism is based on the colour change of a polymer formulated from diacetylene monomers. It consists of a small circle of polymer surrounded by a printed reference ring. The inside polymer circle darkens upon accumulated temperature exposure. If the polymer centre is darker than the outer ring, the consumer is advised not to consume the product, regardless of the use-by date. The indicators are activated by temperature and are stored deep-frozen before use. The system has been validated for a number of vegetables such as lettuce and other chilled, fresh vegetables.

9.3.2.1.2   Vitsab® Indicator

The Vitsab TTI indicator (Vitsab Sweden AB, Sweden) is based on an enzymatic reaction causing a pH change in the reaction mixture. The device consists of a bubble-like dot containing two compartments: one for the enzyme solution, lipase plus a pH indicating dye compound and the other for the substrate, consisting primarily of triglycerides. By application of pressure on the plastic bubble, the dot is activated at the beginning of the monitoring period, which breaks the seal between compartments. The ingredients are mixed and as the reaction proceeds, a pH change results in a colour change. The dot, initially green in colour, becomes progressively yellow as product approaches the end of shelflife. The reaction is irreversible and will proceed faster as temperature is increased and slower as temperature is reduced.

9.3.2.1.3 3MMonitorMark®

ATTI manufactured by Introtech, Netherlands, functions on the basis of melting of a coloured fatty acid ester at a predetermined temperature and its migration from a reservoir along a carrier. This diffusion can be observed through transparent windows.

9.3.3 MICROBIOLOGICAL SPOILAGE INDICATORS AND PATHOGEN SENSORS

A biosensor, which can inform the user of the growth of microorganisms or even a specific microorganism in the package, can be used as microbial growth indicators. The quality of packaged foods is determined by a spoilage indicator by measuring the specific by-products formed through the deterioration process in the food and therefore inside the package. Most of the spoilage indicators are based on the detection of volatile compounds such as CO_2, diacetyl, amines, ammonia and hydrogen sulfide produced during the ageing of foods (De Jong et al., 2005; De Kruijf et al., 2002). Microbial detector systems are based on depositing, on the bar code, a plastic layer loaded with specific antibodies of pathogenic microorganisms such as *Salmonella* or *Listeria* whose presence can be detected when the bar code is read. In recent years, nanoparticles such as gold and silver nanoparticles have been used in different immunoassay readout techniques suitable for smart packaging, on the account of their long-term stability and biocompatibility with antibodies, proteins and nucleic acids (Omidfar et al., 2013). Nanosensors could be used to detect chemicals, pathogens and toxins in foods (Liu et al., 2007).

9.3.4   FRESHNESS INDICATORS

By gas and volatile material measurements in the headspace, freshness of vegetables can be determined. This indicator detects volatile metabolites generated by the produce such as oxygen, carbon dioxide, diacetyl, amines, ethanol and hydrogen sulfide (Brody et al., 2001; Smolander, 2003; Han and Floros 2007; Poças et al., 2008). Ethanol concentration in package headspace can be measured by an enzymatic reaction based on chromogenic substrates from which the colour indicates the degree of fermentation (Smolander 2008). Bromothymol blue and methyl red are chromogenic indicators of fruit and vegetable fermentation. They react with carbon dioxide produced by fermentation, and the colour density is used to determine the degree of fermentation (Smolander, 2008). As a large amount of carbon dioxide is produced by respiration, quality indicators based on carbon dioxide production have limited applications in fresh fruits and vegetables, which masks the amounts produced by microbial metabolism (Smolander, 2008). Indicators based on aroma measurements have also been developed to monitor the degree of fruit fermentation.

9.3.4.1   COMMERCIAL PRODUCTS

9.3.4.1.1   RipeSense®

RipeSense® is a polymeric freshness indicator that was developed by Ripesense Ltd. It is designed to sense the aroma compounds given off by the fruit or vegetable. The device signals ripeness by label visual cue, colour change for fruit that does not change external colour during ripening. A customer can choose the fruit with the required ripening stage by matching the colour of the sensor with a reference colour standard (Pocas et al., 2008).

9.3.5   RADIO FREQUENCY IDENTIFICATION

This technology uses radio waves for product identification and traceability. It includesincorporating a radio frequency identification (RFID) tag into the package, from which a proper sensor is used to collect data about the item's status. The data stored in tags are activated by the sensor, which is then transmitted to a reader for decoding and processing by a computer system (Yam et al., 2005; Brody et al., 2008). The data are about the product identification (e.g. description of the label content) and its history (e.g. how long the product took to move through the supply chain, its temperature, pressure, humidity and gas leakage) and can be collected at any point during processing and distribution. The information obtained from data analysis can be used for judgment of the produce status such as traceability in the case of outbreaks of food-borne infections. A new cold-chain monitoring service system using the RFID system was developed to monitor fresh produce. This system, which is called X-Track™, is comprised of an RFID label connected to TTI that continuously monitors and stores data about produce temperature and time of exposure. The data are then uploaded via an RFID reader to secure base servers from which the customer can retrieve them anywhere and anytime (Poç‚as et al., 2008).

9.4   APPLICATION OF NANOTECHNOLOGY IN SMART AND ACTIVE PACKAGING

In the near future, among emerging active and smart packaging technologies, nanotechnology applications in food packages are predicted to make up a significant portion of the food packaging market, although not yet widely widespread (Ray et al., 2006; Weiss et al., 2006). Improvements in fundamental characteristics of food-packaging materials such as strength, barrier properties, antimicrobial properties and stability to heat and cold are being achieved by utilizing nano composite materials (Lagaron et al., 2005). Moreover, the incorporation of active components that can deliver functional attributes beyond those of conventional active packaging, and the sensing and signalling of relevant information are some of the other basic applications of nanotechnology in food packaging. To improve traceability and monitoring of the condition of food during transport and storage, nanoscale technologies are also developed. Nanofood packaging materials may extend food life, improve food safety, alert consumers that food is contaminated or spoiled, repair tear in packaging and even release preservatives to extend the life of the food in the package. Nanotechnology applications in the food industry can be utilized to detect bacteria in packaging, or produce stronger flavours and colour quality, and safety by increasing the barrier properties. To food safety, nano-structured materials will prevent the invasion of bacteria and microorganisms. Embedded nano-sensors in the packaging will alert the consumer if a food has gone bad (Yadollahi et al., 2010).

Nanocomposites can improve mechanical strength, reduce weight, increase heat resistance and improve barrier against oxygen, carbon dioxide, ultraviolet radiation, moisture and volatiles of food package materials. Polymer nanocomposites are thermoplastic polymers that have nanoscale inclusions, 2–8% by weight. Nanoscale inclusions consist of nanoclays, carbon nanoparticles, nanoscale metals and oxides, and polymeric resins. In addition, nanocomposites could also be characterized by an antimicrobial activity (Sletmoen et al., 2008). To give information of enzymes produced in the breakdown of food molecules, packaging containing nanosensors are coming to food stores and this makethem unsafe for human consumption. The packages could also be used to let air and other enzymes out but not in, thus increasing shelf life, as well as the reduction of man-made preservatives in our foods. The degradation of ripening gas is another important potential application of nanoparticles in food packaging, such as ethylene (Maneerat et al., 2006). Nano-Ag with little dimension, quanta and large external area effect can have more effective antibacterial activity than Ag^+. Moreover, nano-Ag has the function of absorbing and decomposing ethylene (Hu and Fu, 2003). Chitosan-based nanocomposite films, especially silver-containing ones, showed a promising range of antimicrobial activity (Rhim et al., 2006). Researchers in the Netherlands are going one step further to develop intelligent packaging that will release a preservative if the food within begins to spoil. This 'release on command' preservative packaging operates by using a bio-switch developed through nanotechnology.

9.4.1 COMMERCIAL PRODUCTS

Durethan from Bayer Polymers (Pittsburg, USA) is a nanocomposite film enriched withan enormous number of silicate nanoparticles that reduce entry of oxygen and other gases and the exit of moisture, thus preventing food from spoiling. Nanotechnology has the potential to influence the packaging sector by retarding oxidation and controlling moisture migration, microbial growth, respiration rates and volatile flavours and aromas (Brody et al., 2008).

9.5 FUTURE PROSPECTS

It is likely that the trend of integrating active and smart packaging technologies into the packaging material itself will continue, eventually culminating in the development of truly interactive package that responds directly to the needs of the produce. The feasibility of such intelligent packaging is approaching rapidly with recent advances in the development of flexible electronic components, such as nanocomposites and biosensors. Another relatively unexplored area is the incorporation of smart packaging technologies into edible coating that has the potential to grow in the future.

REFERENCES

Abe, Y. Active Packaging—A Japanese Perspective. *Conference Proceedings International Conference On Modified Atmosphere Packaging,* 15–17thOct 1991. Alveston Manor Hotel, Stratford upon Avon, UK. (Gloucestershire: The Campden Food and Drink Research Association), Part I: 18 pp. 1990

Ahvenainen, R.; Hurme, E.; Randell, K.; Eilamo, M. The Effect of Leakage on the Quality of Gas-Packed Foodstuffs andthe Leak Detection. *VTT Research Notes 1683,* (Espoo: Technical Research Centre of Finland), 84 pp. + app. 14 pp 1995.

Ahvenainen, R.; Hurme, E. Active and Smart Packaging for Meeting Consumer Demands for Quality and Safety. *Food Addit. Contam.* **1997,** *14,* 753–763.

Ahvenainen, R. *Novel Food Packaging Techniques;* Woolhead Pub.: Cambridge, England, 2003.

Appendini, P.; Hotchkiss, J. H. Review of Antimicrobial Food Packaging. *Innovative Food Sci. Emerg. Technol.* **2002,** *3,* 113–126.

Atkin, O. K.; Tjoelker, M. G. Thermal Acclimation and the Dynamic Response of Plant Respiration to Temperature. *Trends Plant Sci.* **2003,** *8*(7), 343–350.

Barnetson, A. In *Intelligent Packaging of Foods,* Conference Proceedings Modified Atmosphere Packaging (MAP) and Related Technologies, Campden & Chorleywood Food Research Association: Chipping Campden, 1995; p 18.

Blankenship, S. M.; Dole, J. M. 1-Methylcyclopropene: A Review. *Postharvest Biol. Technol.* **2003,** *28,* 1–25.

Brody, A.; Strupinsky, E.; Kline, L. *Active Packaging For Food Applications;* Brody, A., Strupinsky, E., Kline, L., Eds.; CRC press: New York, USA, 2001; pp 1–211.

Brody, A. L.; Bettybugusu, J. H. H.; Clairekoelsch, S.; Mchugh, T. H. Scientific Status Summary: Innovative Food Packaging Solutions. *J. Food Sci.* **2008,** *73*(8), R107–116.

Chakraverty, A. *Postharvest Technology;* Scientific Publishers: Enfield, NH, 2001.

Charles, F.; Sanchez, J.; Gontard, N. Absorption Kinetics of Oxygen and Carbon Dioxide Scavengers as Part of Active Modified Atmosphere Packaging. *J. Food Eng.* **2006,** *72,* 1–7.

Cooksey, K. Utilization of Antimicrobial Packaging Films for Inhibition of Selected Microorganisms. In *Food Packaging:Testing Methods and Applications;* Risch, S., Ed.; ACS: Washington D. C, 2000; pp 17–25.

CSIRO Plant Industry Communication Group. http://www.csiro.au/files/files/p2if.pdf (accessed June 2006).

Daly, J.; Kourelis, B. Synthesis Methods, Complexes and Delivery Methods for the Safeand Convenient Storage, Transport and Application of Compounds for Inhibiting the Ethylene Response in plants. U.S. Patent 6,017,849, Jan 25, 2000.

De Jong, A. R.; Boumans, H.; Slaghek, T.; Van Veen, J.; Rijk, R.; Van Zandvoort, M. Activeand Intelligent Packaging for Food: is it the Future? *Food Addit. Contam.* **2005,** *22,* 975–979.

De Kruijf, N.; Van Beest, M.; Rijk, R.; Sipilainen-Malm, T.; Losada, P. P.; De Meulenaer, B. Active and Intelligent Packaging: Applications and Regulatory Aspects. *Food Addit. Contam.* **2002,** *19,* 144.

Erdoğrul, Ö.; Şener, H. The Contamination of Various Fruit and Vegetables with *Enterobiusvermicularis, Ascaris*eggs, *Entamoebahistolyca*cysts and *Giardia* Cysts. *Food Control***2005,** *16,* 559–562.

Gibbard, M. R.; Symons, P. J. International Patent Application WO 05/053955, 2005.

Han, J. H. Mass Transfer Modeling in Closed Systems for Food Packaging Particulate Foods and Controlled Release Technology. *Food Sci. Biotechnol.* **2004,** *13*(6), 700–706.

Han, J. H.; Floros, J. D. Active Packaging. In *Advances in Thermal and Non-thermal Food Preservation;* Tewari, G.; Juneja, V. K., Eds.; Blackwell Professional: Ames, Ia, 2007; pp 167–183.

Hu, A. W.; Fu, Z. H. Nano Technology and its Application in Packaging and Packaging Machinery. *Packag. Eng.* **2003,** *24,* 22–24.

Hurme, E.; Vaari, A.; Ahvenainen, R. Active and Smart Packaging of Foods. In *Minimal Processing of Foods, VTT Symposium142;* Ahvenainen, R.; Mattila-Sandholm, T.; Ohlsson, T., Eds.; Technical Research Centre of Finland: Espoo, 1994; pp 149–172.

Hutton, T. *Food Packaging: An Introduction. Key Topics in Food Science and Technology (no. 7);* Campden & Chorleywood Food research Association Group: Chipping Campden, Glousestershire, UK, 2003; p 108.

http://landec.com/technology/platforms/intelimer-polymers.

Krumhar, K. C.; Karel, M. Visual Indicator System. U.S. Patent 5,096,813, 1992.

Lagaron, J. M.; Cava, D.; Cabedo, L.; Gavara, R.; Gimenez, E. Increasing Packaged Food Quality and Safety: (II) Nanocomposites. *Food Addit. Contam.* **2005,** *22,* 994–998.

Liu, Y.; Chakrabartty, S.; Alocilja, E. Fundamental Building Blocks for Molecular BiowireBased Forward Error-Correcting Biosensors. *Nanotechnology* **2007,** *18,* 1–6.

Maneerat, C.; Yasuyoshi, H. Y. Antifungal Activity of TiO_2 Photocatalysis Against *Penicillium expansum* in vitro and in Fruit Tests. *Int. J. Food Microbiol.* **2006,** *107*(2), 99–103.

May, B. K.; Fickak, A. The Efficacy of Chlorinated Water Treatments in Minimizing Yeast and Mold Growth in Fresh and Semi-Dried Tomatoes. *Drying Technol.* **2003,** *21*(6), 1127–1135.

Morris, S. Odour-Proof Package, WO 99/25625, 1999.

Omidfar, K.; Khors, F.; Azizi, M. D. New Analytical Applications of Gold Nanoparticles as Label in Antibody Based Sensors. *Biosens. Bioelectron.* **2013,** *43,* 336–347.

Ozdemir, M.; Floros, J. D. Active Food Packaging Technologies. *Crit. Rev. Food Sci.* **2004,** *44*(3), 185–193.

Patterson, B.; Joyce, D. C. A Package Allowing Cooling and Preservation of Horticultural Produce Without Condensation and Desiccants. PCT application PCT/AU9,300,398, 1993.

Perlman, D.; Linschitz, H. Oxygen Indicator for Packaging. U.S. Patent 4,526,752, 1985.

Powers, T.; Calvo, W. J. Moisture Regulation. In *Novel Food Packaging Techniques;* Ahvenainen, R. Ed.; Woodhead Publishing Limited: England, 2003; pp 172–185.

Restuccia, D.; Spizzirri, U. G.; Parisi, O. I.; Cirillo, G.; Curcio, M.; Iemma, F.; Puoci, F.; Vinci, G.; Picci, N. New EU Regulation Aspects and Global Market of Active and Intelligent Packaging for Food Industry Applications. *Food Control* **2010,** *21,* 1425–1435.

Ray, S.; Easteal, A.; Quek, S. Y.; Chen, X. D. The Potential Use of Polymer Clay Nano Composites in Food Packaging. *Int. J. Food Eng.* **2006,** *2,* 1–11.

Rhim, J. W.; Hong, S. I.; Park, H. M.; Ng, P. K. Preparation and characterization of chitosan-based nanocomposite films with antimicrobial activity. *J. Agric. Food Chem.* **2006,** *54*(16), 5814–5822.

Rhim, J. W.; Ng, P. K. W. Natural Biopolymer-Based Nanocomposite Films for Packaging Applications. *Crit. Rev. Food Sci. Nutr.* **2007,** *47,* 411–433.

Rojas-Grau, M. A.; Grasa-Guillem, R.; Martin-Belloso, O. Quality Changes in Fresh-Cut Fuji Apple as Affected by Ripeness Stage, Anti-Browning Agents, and Storage Atmosphere. *J. Food Sci.* **2007,** *72,* 36–43.

Scetar, M.; Kurek, M.; Galic, K. Trends in Fruit and Vegetable Packaging—A Review. *Croat. J. Food Technol. Biotechnol. Nutr.* **2010,** *5*(3–4), 69–86.

Schilthuizen, S. F. Communication with Your Packaging: Possibilities for Intelligent Functions and Identification Methods in Packaging. *Packag. Technol. Sci.* **1999,** *12,* 225–228.

Scully, A. D.; Horsham, M. A. Active Packaging for Fruits and Vegetables. In *Intelligent and Active Packaging for Fruits and Vegetables;* Wilson, C. L., Ed.; Taylor & Francis: Boca Raton, FL, 2007, pp 57–71.

Shirazi, A.; Cameron, A. C. Controlling Relative Humidity in Modified Atmosphere Packages of Tomato Fruit. *Hortic. Sci.* **1992,** *27,* 336–339.

Sletmoen, M.; Maurstad, G.; Stokke, B. T. Potentials of Bionanotechnology in the Study and Manufacturing of Self-Assembled Biopolymer Complexes and Gels. *Food Hydrocolloids.* **2008,** *22*(1), 2–11.

Smith, J. P.; Hoshino, J.; Abe, Y. Interactive Packaging Involving Sachet Technology. In *Active Food Packaging;* Rooney, M. L., Ed.; Blackie Academic Professional: London, 1995; pp 143–173.

Smolander, M.; Hurme, E.; Ahvenainen, R. Leak Indicators for Modified-Atmosphere Packages. *Trends Food Sci. Technol.* **1997,** *4,* 101–106.

Smolander, M. The Use of Freshness Indicators in Packaging. In *Novel Food Packaging Technologies;* Ahvenainen, R., Ed.; Woodhead: Cambridge, UK, 2003; pp 127–143.

Smolander, M. Freshness Indicators for Food Packaging. In *Smart Packaging Technologies for Fast Moving Consumer Goods;* Kerry, J., Butler, P., Eds.; John Wiley: Hoboken, NJ, 2008; pp 111–127.

Taoukis, P. S. Application of Time-Temperature Integrators for Monitoring and Management of Perishable Product Quality in the Cold Chain. In *Smart Packaging Technologies for Fast Moving Consumer Goods;* John Wiley: Hoboken, NJ, 2008, pp 61–74.

Tunç, S.; Duman, O. Preparation of Active Antimicrobial Methyl Cellulose/Carvacrol/Montmorillonite Nanocomposite Films and Investigation of Carvacrol Release. *LWT—Food Sci. Technol.* **2011,** *44,* 465–472.

Weiss, J.; Takhistov, P.; Mc Clements, J. Functional Materials in Food Nanotechnology. *J. Food Sci.* **2006,** *71,* 107–116.

Yadollahi, A.; Arzani, K.; Khoshghalb, H. The Role of Nanotechnology in Horticultural Crops Post-Harvest Management. *Acta Hortic.* **2010,** *875,* 49–56. DOI: 10.17660/ActaHortic.2010.875.4.

Yam, K. L.; Lee, D. S. Design of Modified Atmosphere Packaging for Fresh Produce. In *Active Food Packaging;* Rooney, M. L., Ed.; Champman & Hall, 1995; pp 55–72. ISBN: 0751401919.

Yam, K. L.; Takhistov, P. T.; Miltz, J. Intelligent Packaging: Concepts and Applications. *J. Food Sci.* **2005,** 70(1), R1–10.

Yoshikawa, Y.; Nawata, T.; Goto, M.; Fujii, Y. Oxygen indicator. U.S. Patent 4,169,811, 1979.

CHAPTER 10

ANTI-NUTRITIONAL COMPOUNDS IN VEGETABLES

JAGDISH SINGH[2], G. K. SRIVASTAVA[1], RAJANI KANAUJIA[1], and SHWETA PAL[1]

[1]*Division of Basic Sciences, ICAR-Indian Institute of Pulses Research, Kanpur 208024, UP, India, E-mail:jagdish1959@gmail.com*

[2]*Principal Scientist (Biochemistry) & Head of Division*

CONTENTS

10.1 INTRODUCTION

Vegetables are nutrient dense, low in energy and are a good source of minerals and vitamins dietary fibre and a range of phytochemicals including carotenoids. Vegetable production and consumption are a potent mechanism for small-scale, disadvantaged farmers to obtain the required nutrients in their diets and to generate much-needed income through trade. Diversifying diets with vegetables is a cheaper and more sustainable way to supply a range of nutrients to the body and combat malnutrition and associated health problems.

Vegetables are rich source of protein, dietary fibre, complex carbohydrates, resistant starch and a number of vitamins and minerals, namely folate, potassium, selenium and zinc. In addition to the macronutrients, they contain a wide variety of anti-nutritional compounds (ANCs) such as toxic amino acids, saponins, cyanogenic glycosides, tannins, phytic acid, oxalates, goitrogens such as glucosinolates, lectins (phytohaemagglutinins), protease inhibitors, chlorogenic acid and amylase inhibitors (Champ, 2002). These ANCs are known to interfere with metabolic processes such as growth, and the bioavailability of nutrients are negatively influenced (Abara, 2003). Phytate and oxalates have the ability to form chelates with di- and trivalent metallic ions such as Cd, Mg, Zn and Fe to form poorly soluble compounds that are not readily absorbed from the gastrointestinal tract, thus decreasing their bioavailability.

The presence of ANCs in crop plants is often the result of an evolutionary adaptation which enables the plant to survive and complete its life cycle under natural conditions. Indeed, due to their anti-nutritional or even toxic properties, various seed components have been shown to play a protective role against insects, fungi, predators and a number of stress conditions. For example, hydrolase inhibitors have proven to act as protective agents against insect attack. However, a reappraisal on their roles and effects is currently taking place as recent researches have associated the consumption of vegetables with a decreased risk for a variety of chronic degenerative diseases such as cancer, obesity, diabetes and cardiovascular diseases (Anderson and Major, 2002).

ANCs do not appear to be equally distributed in all vegetables, and their physiological effects are diverse. Some of these compounds are important in plant-defence mechanisms against predators or environmental conditions. Others are reserve compounds, accumulated in seeds as energy stores in readiness for germination. These non-nutritive bioactive compounds earlier considered as anti-nutrients because of their activity to reduce protein digestibility and mineral bioavailability have recently been shown to also have health-protective effects (Mathers, 2002). Phytic acid exhibits antioxidant activity and protects DNA damage (Phillippy, 2003), phenolic compounds have antioxidant and other important physiological and biological properties (Yeh and Yen, 2003), saponins have hypocholesterolaemic effect and anticancer activity (Shi et al., 2004).

10.2 MAJOR ANTI-NUTRIENT COMPOUNDS IN SOLANACEOUS VEGETABLES

The Solanaceae include a number of commonly collected or cultivated species. The most economically important genus of the family is *Solanum*, which contains the potato (*S. tuberosum*), the tomato (*S. lycopersicum*), and the eggplant or aubergine (*S. melongena*). Another important genus, *Capsicum*, produces both chilli peppers and bell peppers. Solanaceae are known for having a diverse range of alkaloids. These alkaloids can be desirable, toxic or both. The tropanes are the most well known of the alkaloids found in the Solanaceae. Tomato (*Solanum lycopersicum*) accumulates a variety of secondary metabolites including phenolic compounds, phytoalexins, protease inhibitors and glycoalkaloids. These metabolites protect against adverse effects of hosts of predators including fungi, bacteria, viruses and insects. Glycoalkaloids are N-containing secondary plant metabolites found in numerous solanaceous plant species including eggplant, potato and tomato. α-Tomatine, a glycoalkaloid consisted of the aglycontomatidine and a tetrasaccharide side chain composed of xylose, galactose, and two glucose units and occurs naturally in tomato. The glycoalkaloid tomatine actually consists of a mixture of two glycoalkaloids, α-tomatine and dehydrotomatine. Both compounds are present in all parts of the tomato plant. The glycoalkaloids content depends on cultivar, fruit ripening stage and agricultural practices. Immature green tomatoes contain up to 500 mg of α-tomatine/kg of fresh fruit weight. The compound is largely degraded as the tomato ripens until, at maturity, it reaches levels in red tomatoes of 5 mg/kg of fresh fruit weight. These metabolites are also reported to have a variety of pharmacological and nutritional properties in animals and humans. Several studies have shown that tomatine molecule possessed antibiotic properties against a variety of fungi and the human pathogens *Escherichia coli* and *Staphylococcus aureus*. α-Tomatine and its hydrolysis products have also been associated with a variety of effects on human health, including toxicity, cholesterol lowering, enhanced immune responses as cancer chemotherapy agents, and protection against pathogenic fungi and other microorganisms (Friedman, 2002). Besides tomatine, there is another toxic ingredient called atropine in the tomato plant. Atropine is a toxic alkaloid present in belladonna and some other members of the nightshade family. This drug depresses the action of the vagus nerve, a nerve which controls a number of functions, including heart rate. It also interferes with the action of acetylcholine, causing the muscles of the body to relax. Atropine typically dilates the pupils and elevates the heart rate. It can also cause dizziness, nausea and a variety of neurological symptoms, because it crosses the blood–brain barrier.

In potatoes, there is solanidine. This steroid alkaloid is the nucleus (i.e. aglycone) for two important glycoalkaloids, solanine and chaconine. Other plants in the *Solanum* family including tomato also contain solanum-type glycoalkaloids. Solanine is a toxic glycoalkaloid with a bitter taste, it has the formula $C_{45}H_{73}NO_{15}$. It is formed by the alkaloid solanidine with a carbohydrate side chain. It is found in leaves, fruit and tubers of various Solanaceae such as the potato and tomato. Its production is thought to be an adaptive defence strategy against herbivores. Substance intoxication from solanine is characterized by gastrointestinal disorders (diarrhoea, vomiting, abdominal pain) and neurological disorders (headache). The median lethal dose is between 2 and 5 mg per kg of body weight. Symptoms become manifest 8 to 12 h after ingestion. The amount of these glycoalkaloids in potatoes, for example, varies significantly depending of environmental conditions during their cultivation, the length of storage and the variety. The average glycoalkaloid concentration is 0.075 mg/g of potaton (Zeiger, 1998). Solanine has occasionally been responsible for poisonings in people who ate berries from species such as *Solanumnigrum* or *Solanumdulcamara*, or green potatoes. The concentration of these glycoalkaloids is highest in potato sprouts and green potato skins, and tomato vines and green tomatoes. Care should be taken to prevent the exposure of potatoes to sunlight. These alkaloids are not destroyed by cooking or drying at high temperatures. New potato varieties should not be introduced unless they contain less than 20 mg glycoalkaloids/100 g. These solanum-type glycoalkaloids cause: a bitter taste above 14 mg/100 g, and a burning sensation to mouth and throat above 20 mg/100 g. Solanum-type glycoalkaloids are gastrointestinal tract irritants causing inflamed intestinal mucosa, ulceration, haemorrhage, stomach pains, constipation or diarrhoea and also act as cholinesterase inhibitors and thus affect the nervous system causing apathy, drowsiness, salivation, laboured breathing, trembling, ataxia, muscle weakness, convulsions, involuntary urination, paralysis, loss of consciousness, coma and death due to respiratory paralysis.

10.3 MAJOR ANTI-NUTRIENT COMPOUNDS IN CUCURBITS

Cucurbits contain chemical compounds known as terpenoids that are also responsible for flavour and scent. The terpenoids responsible for bitterness in cucumbers are called cucurbitacins. Cucurbitacin is a triterpenoid which have diverse pharmacological and biological activities. These substances are present in the roots, stems, leaves and fruit. Specifically, two genes account for the bitter taste or lack of a dominant one that makes the cucumber bitter and a recessive one that inhibits the formation of cucurbitacins, thus suppressing the bitterness. Nevertheless, the accumulation of cucurbitacins is generally not heavy enough to make the fruit bitter, or if it is, the concentration is at the stem and just under the skin. Bitterness in cucumbers also appears to be controlled by the presence of an enzyme known as elaterase. Elaterase converts cucurbitacins to non-bitter compounds. Scientists believe that the elaterase activity operates independently of the genes that control bitterness, when the level of bitterness in cucumbers varies from year to year; it is due in part to environmental effects that either stimulate or suppress the elaterase activity.

Bitter gourd (Momordica charantia L.) is widely used as a functional food to prevent and treat diabetes and associated complications (Ray et al., 2010). In addition to the hypoglycaemic effect, the antitumor activity of crude bitter gourd extract has also been reported in various types of cancer cells in vitro and in vivo (Jilka et al., 1983). The eported chemical constituents of bitter gourd include glycosides, saponins, alkaloids, fixed oils, triterpenes, polypeptides and steroids. Recent studies have demonstrated that wild bitter gourd extracts exhibited multiple pharmacological activities associated with anti-inflammation and anti-diabetics, including those of suppressing inflammatory response in macrophages, overcoming insulin resistance in skeletal muscle in fructose-fed rats, enhancing insulin signalling in skeletal muscle in high-fat-diet-fed mice, and up-regulating mRNA expression of peroxisome proliferator-activated receptor (PPAR) α, PPARγ and their target genes in mice. Together, these activities might account for the ability of wild bitter gourd to improve metabolic syndrome in humans. Bioactive constituents that contribute to this pharmacological effect, however, remain undefined. Jing-Ru Weng et al. (2013) reported that 3β,7β-dihydroxy-25-methoxycucurbita-5,23-diene-19-al, a cucurbitane-type triterpene (Fig. 10.1) isolated from wild bitter gourd, induced apoptotic death in breast cancer cells through PPAR γ activation.

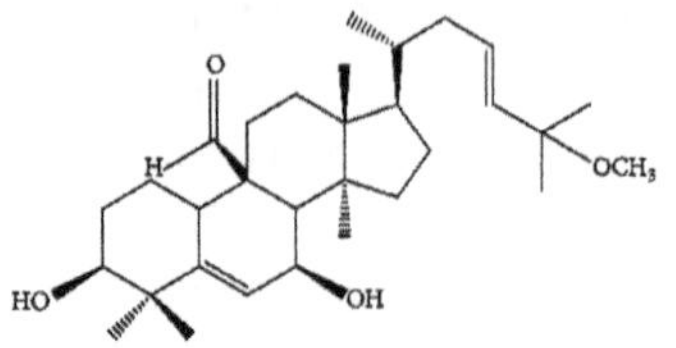

FIGURE 10.1   3β, 7β-dihydroxy-25-methoxycucurbita-5, 23-diene-19-al.

One chemical from bitter gourd has clinically demonstrated the ability to inhibit the enzyme guanylate cyclase that is thought to be linked to the cause of psoriasis and also necessary for the growth of leukaemia and cancer cells. In addition, a protein found in bitter gourd, momordin, has clinically demonstrated anticancerous activity against Hodgkin's lymphoma in animals. Other proteins in the plant, α- and β-momorcharin and *cucurbitacins* B, have been tested for possible anticancerous effects. Charantin, an anti-diabetic compound, is a typical cucurbitane-type triterpenoid in *M. charantia* and is a potential and promising substance for the treatment of diabetes. Cucurbitane glycosides exhibited a number of biologic effects beneficial to diabetes and obesity. They stimulate GLUT4 translocation to the cell membrane—an essential step for inducible glucose entry into cells. This was associated with increased activity of *adenosine monophosphate*-activated protein kinase, a key pathway mediating glucose uptake and fatty acid oxidation (Tan et al., 2008).

10.4   MAJOR ANTI-NUTRIENT COMPOUNDS IN CRUCIFEROUS VEGETABLES

Cruciferous or Brassica vegetables come from plants in the family Cruciferae or, alternately, Brassicaceae. Many commonly consumed cruciferous vegetables come from the Brassica genus, including broccoli, Brussels sprouts, cabbage, cauliflower, collard greens, kale, kohlrabi, mustard, rutabaga, turnips, bokchoy and Chinese cabbage. Cruciferous vegetables are unique as they are rich sources of glucosinolates, sulphur-containing compounds that impart a pungent aroma and spicy taste. The hydrolysis (breakdown) of glucosinolates by an enzyme called myrosinase results in the formation of biologically active compounds, such as indoles and isothiocyanates. Myrosinase is physically separated from glucosinolates in intact plant cells.

However, when cruciferous vegetables are chopped or chewed, myrosinase comes in contact with glucosinolates and catalyses their hydrolysis.

High intakes of cruciferous vegetables have been associated with lower risk of lung and colorectal cancer in some epidemiological studies. Although glucosinolate hydrolysis products may alter the metabolism or activity of sex hormones in ways that could inhibit the development of hormone-sensitive cancers, evidence of an inverse association between cruciferous vegetable intake and breast or prostate cancer in humans is limited and inconsistent.

Very high intakes of cruciferous vegetables, such as cabbage and turnips, have been found to cause hypothyroidism (insufficient thyroid hormone) in animals. Two mechanisms have been identified to explain this effect. The hydrolysis of some glucosinolates found in cruciferous vegetables (e.g. progoitrin) may yield a compound known as goitrin, which has been found to interfere with thyroid hormone synthesis. The hydrolysis of another class of glucosinolates, known as indoleglucosinolates, results in the release of thiocyanate ions, which can compete with iodine for uptake by the thyroid gland.

10.4.1 GLUCOSINOLATES AND ISOTHIOCYANATES

Cruciferous vegetables, such as bokchoy, broccoli, Brussels sprouts, cabbage, cauliflower, horseradish, kale, kohlrabi, mustard, radish, rutabaga, turnip and watercress, are rich sources of glucosinolate precursors of isothiocyanates. Glucosinolates are water-soluble compounds that may be leached into cooking water. Boiling cruciferous vegetables from 9–15 min resulted in 18–59% decreases in the total glucosinolate content of cruciferous vegetables (McNaughton and Marks, 2003). Cooking methods that use less water, such as steaming or microwaving, may reduce glucosinolate losses. However, some cooking practices, including boiling, steaming and microwaving at high power (750–900 W), may inactivate myrosinase, the enzyme that catalyses glucosinolate hydrolysis. Even in the absence of plant myrosinase activity, the myrosinase activity of human intestinal bacteria results in some glucosinolate hydrolysis. However, several studies in humans have found that inactivation of myrosinase in cruciferous vegetables substantially decreases the bioavailability of isothiocyanates.

Unlike some other phytochemicals, glucosinolates are present in relatively high concentrations in commonly consumed portions of cruciferous vegetables. Total glucosinolate contents of selected cruciferous vegetables are presented in Table 10.1.

TABLE 10.1   Glucosinolate Content of Selected Cruciferous Vegetables.

Food (raw)	Serving (g)	Total Glucosinolates (g)
Brussels sprouts	44	0.104
Garden cress	25	0.098
Mustard greens	28	0.079
Turnip	65	0.060
Cabbage, savoy	45	0.035
Kale	67	0.067
Watercress	34	0.032
Kohlrabi	67	0.031
Cabbage, red	45	0.029
Broccoli	44	0.027
Horseradish	15	0.024
Cauliflower	50	0.022
Bok choy	35	0.019

Isothiocyanates are derived from the hydrolysis (breakdown) of glucosinolates—sulphur-containing compounds found in cruciferous vegetables. Isothiocyanates, such as sulphoraphane (SFN), may help prevent cancer by promoting the elimination of potential carcinogens from the body and by enhancing the transcription of tumour suppressor proteins. Cruciferous vegetables contain a variety of glucosinolates, each of which forms a different isothiocyanate when hydrolysed. For example, broccoli is a good source of glucoraphanin, the glucosinolate precursor of SFN and sinigrin, the glucosinolate precursor of allylisothiocyanate. Watercress is a rich source of gluconasturtiin, the precursor of phenethylisothiocyanate, whereas garden cress is rich in glucotropaeolin, the precursor of benzyl isothiocyanate. Table 10.2 lists vegetables that are relatively good sources of some of the isothiocyanates that are currently being studied for their cancer-preventive properties.

TABLE 10.2 Food Sources of Selected Isothiocyanates and Their Glucosinolate Precursors.

Glucosinolate	Isothiocyanate	Food sources
Sinigrin	Allyl isothiocyanate	Broccoli, Brussels sprouts, cabbage, horseradish, mustard, radish
Glucotropaeolin	Benzyl isothiocyanate	Cabbage, garden cress, Indian cress
Gluconasturtiin	Phenethyl-isothiocyanate	Watercress
Glucoraphanin	Sulphoraphane	Broccoli, Brussels sprouts, cabbage

No serious adverse effects of isothiocyanates in humans have been reported. The majority of animal studies have found that isothiocyanates inhibited the development of cancer when given prior to the chemical carcinogen (pre-initiation).

10.5 MAJOR ANTI-NUTRIENT COMPOUNDS IN LEGUME VEGETABLES

10.5.1 PROTEASE INHIBITORS

Protease inhibitors are the most commonly encountered class of antinutritional factors of plant origin. The protein inhibitors of hydrolases present in legume vegetables are active against proteases, amylases, lipases, glycosidases and phosphatases. Protease inhibitors have the ability to inhibit the activity of proteolytic enzymes within the gastrointestinal tract of animals (Liener and Kakade, 1980). Protease inhibitors isolated from legume vegetables are generally classified into two families, referred to as Kunitz and Bowman-Birk on the basis of their molecular weights and cystine contents. Kunitz type inhibitors have a molecular mass of ~20 kDa, with two disulphide bridges, and act specifically against trypsin. Bowman-Birk type inhibitors, with a molecular mass of 8–10 kDa, have seven disulphide bridges, and inhibit trypsin and chymotrypsin simultaneously at independent binding sites. In common beans, lima bean and cowpea, protease inhibitors have been characterized as members of the Bowman-Birk family. These inhibitors have been reported to be partly responsible for the growth-retarding property of raw legumes. The retardation has been attributed to inhibition of protein digestion. Trypsin inhibitors have been implicated in reducing protein digestibility and in pancreatic hypertrophy

(Liener, 1976); however, their effect is usually manifested only if the seed or the flour are consumed uncooked, as heat denaturation, as the consequence of legume seed cooking, normally inactivates these proteins. Once inactivated, the protein inhibitors may even play a positive nutritional role, due to their high content of sulphur-containing amino acids relative to the majority of the seed proteins.

10.5.2   AMYLASE INHIBITORS

Amylase inhibitors are also known as starch blockers because they contain substances that prevent dietary starches from being absorbed by the body. Starches are complex carbohydrates that cannot be absorbed unless they are first broken down by the digestive enzyme amylase and other secondary enzymes (Choudhury et al., 1996). Beans are the second largest group of seeds after cereals reported as natural sources of α-amylase inhibitors. The content of α-amylase inhibitors differs greatly among legumes, with the highest amount found in dry bean. These inhibitors have been found to be active over a pH range of 4.5–9.5 and are heat labile. Amylase inhibitors inhibit bovine pancreatic amylase but fail to inhibit bacterial, fungal and endogenous amylase.

10.5.3   LECTINS (PHYTOHAEMAGGLUTININ)

Seed lectins are another family of protein ANCs. Lectins are ubiquitous (glyco)-proteins which exhibit specific and reversible carbohydrate binding activities. The specificity of a lectin is defined in term of the monosaccharide or simple oligosaccharide which inhibits the lectin-induced cell agglutination reaction. However, many, but not all, lectins have haemagglutinating activity. Their structures are diverse and their specificity is restricted to carbohydrates. The toxicity of lectins is characterized by growth inhibition in experimental animals, and by diarrhoea, nausea, bloating and vomiting when injected in humans. Lectins have the capability to directly bind to the intestinal mucosa (Almeida et al., 1991), interacting with the enterocytes and interfering with the absorption and transportation of nutrients (particularly carbohydrates) during digestion (Santiago et al., 1993) and causing epithelial lesions within the intestine (Oliveira et al., 1989). Although the biological role of lectin in legume seeds is still controversial, there is evidence that lectins can be defence proteins against potential plant enemies. As the majority of plant lectins exhibit specificity against carbohydrates of animal origin, one can reasonably argue that plants use this type of protein as a defence against phytophagous invertebrates and herbivorous animals. Many lectins are non-toxic, such as those from pea and faba bean. *Viciafaba* agglutinin, a lectin present in broad bean, aggregated and stimulated the morphological differentiation and reduced the malignant phenotype of colon cancer cells (Jordinson et al., 1999). Heat processing can reduce the toxicity of lectins, but low temperature or insufficient cooking may not completely eliminate their toxicity.

10.5.4 PHYTIC ACID

Phytic acid (myo-inositol hexaphosphate or $InsP_6$) occurs naturally throughout the plant kingdom and is present in considerable quantities within many of the major legumes and oilseeds. It has been reported that about 62–73% and 46–73% of the total phosphorus within cereal grains and legume seeds respectively is in form of organically bound phytin phosphorus. The major part of the phosphorus contained within phytic acid is largely unavailable to animals due to the absence of the enzyme phytase within the digestive tract of monogastric animals. Phytic acid acts as a strong chelator, forming protein and mineral-phytic acid complexes, the net result being reduced protein and mineral bioavailability. Phytic acid is reported to chelate metal ions such as calcium, magnesium, zinc, copper, iron and molybdenum to form insoluble complexes that are not readily absorbed from gastrointestinal tract. Phytic acid also inhibits the action of gastrointestinal tyrosinase, trypsin, pepsin, lipase and α-amylase. Erdman (1979) stated that the greatest effect of phytic acid on human nutrition is its reduction of zinc bioavailability.

In vivo and in vitro studies have demonstrated that inositol hexaphosphate ($InsP_6$, phytic acid) exhibits significant anticancer (preventive as well as therapeutic) properties. It reduces cell proliferation and increases differentiation of malignant cells with possible reversion to the normal phenotype and is involved in host defence mechanism, and tumour abrogation (Shamsuddin, 2002). Phytic acids delay postprandial glucose absorption, reduce the bioavailability of toxic heavy metal such as cadmium and lead and exhibit antioxidant activity by chelating iron and copper (Minihane and Rimbach, 2002). Dietary and endogenous phyticacid has protective effect against cancer and heart disease and may be responsible for the cancer-protective effects of high-fibre foods. The anti-carcinogenic properties of phytic acid may result from numerous factors, including its ability to chelate metal ions; this depends on the phytate retaining its integrity in the colon. *Myo-inositol* has been evaluated for its ability to improve the mental health of patients with various psychiatric disorders (Einat and Belmaker, 2001). In addition to *myo-inositol,* smaller amounts of *epi-* and *scyllo-inositol* are present in human brains. *Myo-inositol* and InsP$_6$ have synergistic or additive effects in inhibiting the development of cancer. In mice, dietary *myo-inositol* has been shown to be effective in preventing cancer of the lung, fore stomach, liver, colon, mammary gland, prostate and skin.

10.5.5 PHENOLIC ACIDS

The major polyphenolic compounds of legume vegetables consist mainly of tannins, phenolic acids and flavonoids. The seed colour is mainly due to the presence of polyphenolic compounds, namely flavonoids such as flavonol glycosides, anthocyanins and condensed tannins (proanthocyanidins). Until recently, phenolic compounds were regarded as non-nutritive compounds, and it was reported that excessive content of polyphenols, in particular tannins, may have adverse consequences because it inhibits the bioavailability of iron (South and Miller, 1998) and blocks digestive enzymes in the gastrointestinal tract. Phenolic compounds can also limit the bioavailability of proteins with which they form insoluble complexes in the gastrointestinal tract (Shahidi and Naczk, 2004). Later on, the significance of phenolic compounds was gradually recognized and several researches have now reported that phenolics offer many health benefits and are vital in human nutrition. High correlations between phenolic compositions and antioxidant activities of legume extracts were observed. Among all phenolic compounds detected, sinapic acid was the predominant phenolic acid, and (+)-catechin and (−)-epicatechin were the predominant flavonoids.

10.5.6 TANNINS

Tannins are water soluble phenolic compounds with a molecular weight greater than 500 Da. Tannins form complexes with proteins, making them insoluble and inactivating their enzymatic activity. They may also bind to other macromolecules such as starch, causing a reduction in the nutritional value of food. There are two different groups of tannins: hydrolysable tannins and condensed tannins. Tannins may form a less digestive complex with dietary proteins and may bind and inhibit the endogenous protein, such as digestive enzymes (Kumar and Singh, 1984). Tannin-protein complexes involve both hydrogen bonding and hydrophobic interactions. The precipitation of the protein-tannin complex depends upon pH, ionic strength and molecular size of tannins. Tannins have been found to interfere with digestion by displaying anti-trypsin and anti-amylase activity. Tannins also have the ability to complex with vitamin B12 (Liener, 1980). Other adverse nutritional effects of tannins have been reported to include intestinal damage, interference with iron absorption and the possibility of tannins producing a carcinogenic effect (Butler, 1989).

10.5.7 SAPONINS

Saponins are a heterogeneous group of naturally occurring foam-producing triterpene or steroidal glycosides that occur in a wide range of plants, including pulses and oil seeds. It has been reported that saponins can affect animal performance and metabolism in a number of ways as follows: erythrocyte haemolysis, reduction of blood and liver cholesterol, depression of growth rate, bloat (ruminants), inhibition of smooth muscle activity, enzyme inhibition and reduction in nutrient absorption (Cheeke, 1971). Saponins have also been reported to alter cell-wall permeability and therefore produce some toxic effects when ingested (Belmar et al., 1999). Saponins have been shown to bind to the cells of the small intestine, thereby affecting the absorption of nutrients across the intestinal wall (Johnson et al., 1986). The effect of saponins on chicks have been reported to reduce growth, feed efficiency and interfere with the absorption of dietary lipids, cholesterol, bile acids and vitamins A and E (Jenkins and Atwal, 1994).

Although the toxicological properties of plant saponins have long been recognized, there is renewed interest in these biologically active plant components because recent evidence suggests that saponins possess hypocholesterolemic, anti-carcinogenic and immune-stimulatory properties. There is enormous structural diversity within this chemical class, and only a few are toxic. Most of the saponins occur as insoluble complexes with 3-b-hydroxysteroids; these complexes interact with bile acid and cholesterol, forming large mixed micelles. In addition, they form insoluble saponin–mineral complexes with iron, zinc and calcium, hence saponins lower nutrient availability. Recent evidence suggests that legume saponins may possess anticancer activity and be beneficial for hyperlipidaemia. In addition, they reduce the risk of heart diseases in humans. Epidemiological studies suggest that saponins may play a role in protection from cancer.

10.6   MAJOR ANTI-NUTRIENT COMPOUNDS IN GREEN LEAFY VEGETABLES

Green leafy vegetables play significant role in human nutrition especially as a source of vitamins, minerals, carotenoids and dietary fibre. The varieties of leafy vegetables are diverse, ranging from leaves of annuals and shrubs to tree leaves. However, the main constraint to their nutritional exploitation is the presence of some anti-nutritional and toxic principles such as nitrate, oxalate and saponin (Gupta and Wagle, 1988).

10.6.1   NITRATES AND NITRITES

Nitrogen is the main limiting factor for most field crops, and nitrate is the major form of nitrogen absorbed by crop plants. Farmers often use nitrogen fertilizers to increase crop yields. Consequently, many vegetables and forage crops accumulate high levels of nitrate. In particular, leafy vegetables such as spinach, lettuce and celery contain nitrate at significant levels (Maynardet et. al., 1976). Nitrite content in vegetables is usually very low compared with nitrate (Hunt and Turner, 1994). Vegetables are generally considered the largest source of dietary nitrate. Vegetables can be categorized according to their nitrate content. High nitrate content vegetables (>1000 mg·kg^{-1}) belong to the following families: Brassicaceae (the highest nitrate accumulating vegetables), Chenopodiaceae (beetroot, spinach), Asteraceae (lettuce) and Apiaceae (celery). Most common vegetables are in the medium range for nitrate content (100–1000 mg·kg⁻¹), including peppers, garlic, potatoes and carrots at the lower end, and green beans, cabbage and turnip at the upper end. Vegetables notable for their low nitrate content (<100 mg kg⁻¹) are onions and tomatoes. Nitrate content also varies across the plant: leaf > stem > root. Leaf and stem tissues accumulate the most nitrate, followed by roots (Lorenz, 1978).

The potential health hazards of nitrate and nitrite are well studied. Nitrate can generally be considered to be of relatively lower toxicity than nitrite. However, about 5% of dietary nitrate is converted to nitrite in humans by bacterial and mammalian metabolic pathways. Potentially carcinogenic N-nitroso compounds can then be formed from nitrite.

Nitrate intake may cause methaemoglobinaemia (also known as 'blue baby disease'). This is a condition in which the nitrites in blood interact with haemoglobin and iron. Nitrites convert the haemoglobin-iron compound to methaemoglobin by oxidizing the iron molecule. The methaemoglobin molecule does not carry oxygen. This reduces the amount of oxygen inside blood which is dangerous. Infants are at high risk of complications from nitrates and nitrites because of their young digestive systems. Infants have a higher pH inside their stomach. When infants ingest nitrates, a higher pH leads to a larger conversion of the nitrates to nitrites. More nitrites in the stomach increase their chance of developing methaemoglobinaemia. Though this condition is very rare, certain foods increase the risk. High nitrate vegetables such as carrots, green beans, spinach, squash and beets should not be introduced before 3 months of age.

The discovery that dietary (inorganic) nitrate has important vascular effects came from the relatively recent realization of the 'nitrate-nitrite-nitric oxide (NO) pathway'. Dietary nitrate has been demonstrated to have a range of beneficial vascular effects, including reducing blood pressure, inhibiting platelet aggregation, preserving or improving endothelial dysfunction, enhancing exercise performance in healthy individuals and patients with peripheral arterial disease. Pre-clinical studies with nitrate or nitrite also show the potential to reduce arterial stiffness, inflammation and intimal thickness (Lidder and Webb, 2013). Although these suggest reduction in cardiovascular risk with diets high in nitrate-rich vegetables, others have suggested possible small positive and negative associations with dietary nitrate and cancer, but these remain unproven.

Nitrate content in a plant represents a dynamic balance between rates of absorption, assimilation and translocation (Maynard et al., 1976). Therefore, it is influenced by environmental, agricultural and genetic factors. Of the factors studied, in general, nitrogen fertilization and light intensity have been identified as the major factors which influence nitrate levels in vegetables (Cantliffe, 1973). In particular, light intensity and nitrate content in soils before or at harvest are known to be critical factors in determining nitrate levels in spinach (Maynard et al., 1976).

10.6.2   OXALIC ACID

Another common anti-nutritional factor is oxalic acid (Kumar et al., 2006). Oxalic acid, dicarboxylic acid or its salts (oxalates) are widely distributed in plant food. These oxalates are mostly calcium salts. Many vegetables contain oxalic acid, especially leafy greens such as spinach, kale, chard, parsley, collards and beet greens. Spinach has the highest levels of oxalic acid—750 mg per 100 g serving (Table 10.3). Oxalates are known to interfere with calcium absorption by forming insoluble salts with calcium. Oxalate content of a single food group varies on the basis of the time of year, the type of soil it is grown in and a host of other factors specific to the growing conditions of the plant.

TABLE 10.3   Oxalate Content of Green Leafy Vegetables.

Raw vegetable	Oxalate content (%)
Spinach	0.75
Beet greens	0.61
Okra	0.15
Parsley	0.10
Leeks	0.09
Collard greens	0.07

Oxalates exist in two forms, soluble and insoluble. Oxalic acid forms water-soluble salts with potassium, sodium and ammonium ions, and insoluble salts with calcium, magnesium and iron ions. The soluble form appears to be of most concern in terms of health effects as insoluble oxalate is not absorbed and is simply excreted in the faeces. Our bodies produce oxalic acid, often synthesizing other substances such as vitamin C into oxalic acid, so whether you eat foods that contain it or not, your body maintains a naturally-occurring level of oxalic acid. Oxalic acid has been shown to bind with calcium (and magnesium, iron, sodium and potassium) in the intestine, thus interfering with the absorption of these. Oxalates affects calcium and magnesium metabolism and react with proteins to form complexes which have an inhibitory effect in peptic digestion.

Cooking has a relatively small impact on the oxalate content of foods. Repeated food chemistry studies have shown no statistically significant lowering of oxalate content following the blanching or boiling of green leafy vegetables. A lowering of oxalate content by about 5–15% is the most one should expect when cooking a high-oxalate food.

10.7 PROCESSING REDUCES ANTI-NUTRIENTS

Processing generally improves the nutrient profile of legume seed by increasing in vitro digestibility of proteins and carbohydrates and at the same time there are reductions in some ANCs (Hajos and Osagie, 2004). Most anti-nutritional factors are heat-labile, such as protease inhibitors and lectins, so thermal treatment would remove any potential negative effects from consumption. On the other hand, tannins, saponins and phytic acid are heat stable but can be reduced by dehulling, soaking, germination and/or fermentation. Many traditional methods of food preparation such as fermentation, cooking and malting increase the nutritive quality of plant foods through reducing certain anti-nutrients such as phytic acid, polyphenols and oxalic acid (Hotz and Gibson, 2007). The effect of different domestic cooking and processing methods on anti-nutrient content, oxalate in selected green leafy vegetables and pulses was reported by Mbah et al. (2012), who reported the effect of cooking methods for green leafy vegetables and pulses via blanching, pressure cooking, open pan cooking, drying, boiling and sprouting on oxalate in raw and cooked samples. Blanching for 10 min (green leafy vegetables) and sprouting for 48 h (pulses) were found to be the best methods for lowering the phytic acid and oxalate contents. The average reduction ranges on cooking for oxalate were 40.4% (spinach) and 115.9% (*bathua*) for green leafy vegetables and 24.9% (green pea and Bengal gram) for pulses. Cooking methods showed considerable reduction in oxalate and phytate. The reduction of anti-nutrients on cooking is expected to enhance the nutritional value of these green leafy vegetables and pulses.

All processing steps result in significant decrease in total phenolic content (TPC) and 2,2-diphenyl-1-picrylhydrazyl (DPPH) free radical scavenging activity (DPPH) (Xu and Chang, 2008). The soaking and atmospheric boiling treatments decreased, whereas pressure boiling and steaming increased the oxygen radical absorbing capacity (ORAC). Steaming treatments resulted in a greater retention of TPC, DPPH and ORAC values as compared with boiling treatments. However, TPC and DPPH in cooked vegetables differed significantly between atmospheric and pressure boiling. Pressure processes significantly increased ORAC values in both boiled and steamed vegetables compared with atmospheric processes. Greater TPC, DPPH and ORAC values were detected in boiling water than in soaking and steaming water. Steam processing exhibited several advantages in retaining the integrity of the legume vegetable appearance and texture of the cooked product, shortening process time and greater retention of antioxidant components and activities.

REFERENCES

Abara, A. E. Tannin content Content of *Dioscorea bulbufera*. *J. Chem. Soc. Niger.* **2003,** *28*, 55–56

Almeida, N. G.; Calderon, A. M.; de la Barca; Valencia, M. E. Effect of Different Heat Treatments on the Antinutritional Activity of *Phaseolus vulgaris* (variety; ojo de cabra) lectin. *J. Agri. Food Chem.* **1991,** *39*, 1627–1630.

Anderson, J. W.; Major, A. W. Pulses and Lipaemia, Short and Long Term Effect: Potential in the Prevention of Cardiovascular Disease. *J. Br. Nutr.* **2002,** *88*(3): S263.

Belmar, R.; Nava-Montero, R.; Sandoval-castro, C.; McNab, J. M. Jack Bean (*Canavalia ensiformis* L. DC.) in Poultry Diets: antinutritional Factors and Detoxification Studies— A Review. *World Poult. Sci. J.* **1999,** *55*, 37–59.

Butler, L. G. Effect of Condensed Tannins on Animal Nutrition. In: *Chemistry And Significance of Condensed Tannins*. Hemingway, R.W.; J.J. Karchesy Eds., Plenum Press, New York, 1989, pp: 391–402.

Cantliffe, D. J. Nitrate Accumulation in Table Beets and Spinach as Affected by Nitrogen, Phosphorous, and Potassium Nutrition and Light Intensity. *Agron. J.* **1973,** *65*, 563–565.

Champ, M. M. Non-Nutrient Bioactive Substances of Pulses. *Br. J. Nutr.* **2002,** *88*, S307–S319.

Cheeke, P. R. Nutritional and Physiological Implications of Saponins: A Review. *Can. J. Anim. Sci.* **1971,** *51*, 621–623.

Choudhury, A.; Maeda, K.; Murayama, R.; Dimagno, E. P. Character of a Wheat Amylase Inhibitor, Preparation and Effects on Fasting Human Pancreatic Obiliary Secretions and Hormones. *Gastroenterology* **1996,** *111*, 1313–1320.

Einat, H.; Belmaker, R. H. The Effects of Inositol Treatment In Animal Models Of Psychiatric Disorders. *J. Affective Disorders* **2001,** *62*, 113–121.

Erdman, J. W. Oilseed Phytates: Nutritional Implications. *J. Am. Oil Chemist. Soc.* **1979,** *56*, 736–741.

Friedman, M. Tomato Glycoalkaloids: Role in the Plant and the Diet. *J. Agri. Food Chem.* **2002,** *50*, 5751–5780.

Gupta, K.; Wagle, D. S. Nutritional and Antinutritional Factors of Green Leafy Vegetables. *J. Agri. Food Chem.* **1988,** *36*(3): 472–474

Hajos, G.; Osagie, A. U. In: Muzquiz M., Hill G.D., Cuadrado C.; Pedrosa M. M.; Burbano, C. (Eds.) Proceedings of the fourth international workshop on antinutritional factors in legume seeds and oilseeds, EAAP publication No. 110, Toledo, Spain, 2004, pp. 293–305.

Hotz, C.; Gibson, R. S. Traditional Food-Processing and Preparation Practices To Enhance the Bioavailability of Micronutrients in Plant-Based Diets. *J. Nutr.* **2007,** *137*(4): 1097–100.

Hunt, J.; Turner M. K. A survey of Nitrite Concentrations in Retail Fresh Vegetables. *Food Additives Contaminants* **1994,** *11*(3): 327–332.

Jenkins, K. J.; Atwal, A. S. Effects of Dietary Saponins on Faecal Bile Acids and Neutral Sterols and Availability of Vitamins A And E in the Chick. *J. Nutr. Biochem.* **1994,** *5*, 134–137.

Jilka, C.; Strifler, B.; Fortner, G. W. In Vivo Antitumor Activity of the Bitter Melon (*Momordica charantia*). *Cancer Res.* **1983,** *43*(11): 5151–5155.

Jing, R W.; Li, Y. B.; Chang, F. C.; Hing, J. L.; Shih, J. C.; Chia, Y. W. (2013). Cucurbitane Triterpenoid from *Momordica charantia* Induces Apoptosis And Autophagy In Breast Cancer Cells, In Part, Through Peroxisome Proliferator-Activated Receptor Γ Activation. Evidence-Based Complementary and Alternative Medicine 2013: http://dx.doi.org/10.1155/2013/935675

Johnson, L. T.; Gee, J. M.; Price, K.; Curl, C.; Fenwick, G. R. Influence of saponins in gut permeability and active nutrient transport in vitro. *J. Nutr.* **1986,** *116*, 2270–2272.

Jordinson, M. I., El-Hariry, D.; Calnan, J. Calam; Pignatelli, M. (1999). *Viciafaba* agglutinin, the lectin present in broad beans, stimulates differentiation of undifferentiated colon cancer cells. *Gut* **1999,** *44*, 709–714.

Kumar, R.; Singh, M. Tannins: Their adverse role in ruminant nutrition. *J Agri Food Chem* **1984,** *32*, 447–453.

Kumar, R.; Nataraju, S.; Jayaprehash, C.; Sabapathy, S. N.; Bawa, A. S. (2006). Effect of heat treatment on the stability of oxalic acid in selected plant foods. *Ind J Nutr Dietetics.* **2006,** *43,* 337–340.

Lidder, S.; Webb, A. J. Vascular effects of dietary nitrate (as found in green leafy vegetables and beetroot) via the nitrate-nitrite-nitric oxide pathway. *Br J Clin Pharmacol* **2013,** *75*(3) 677–696.

Liener, I. E. Legume toxins in relation to protein digestibility-a review. *J Food Sci.* **1976,** *41*, 1076–1081.

Liener, I. E. Heat labile antinutritional factors. In: Advances in legume science, Summerfield, R. J.; Bunting A. H. (Eds.) Kew London, Royal Botanic Gardens, 1980; pp 157–170.

Liener, I. E.; Kakade M. L. Protease inhibitors. In: Toxic constituents of plant food stuffs (Editor: I.E. Liener) Academic Press: New York, 1980; pp 7–71.

Lorenz, O. A. Potential nitrate levels in edible plant parts. In: Nitrogen in the environment; Nielsen D. R.; J. G. MacDonald (Eds.), Academic press, 1978; pp 201–219

Mathers, J. C. Pulses and Carcinogenesis: Potential for Prevention of Colon, Breast and Other Cancers. *Br. J. Nutr.* **2002,** *88,* 272–279.

Maynard, D. N.; Barker, A. V.; Minotti, P. L.; Peck, N. H. Nitrate Accumulation in Vegetables. *Adv. Agron.* **1976,** *28,* 71–118.

Mbah, B. O.; Eme, P. E.; Paul, A. E. Effect of drying techniques on the proximate and other nutrient composition of *Moringa oleifera* Leaves from Two Areas in Eastern Nigeria. *Pak. J. Nutr.* **2012,** *11*(11), 1044.

McNaughton, S. A.; Marks, G. C. Development of a Food Composition Database For The Estimation Of Dietary Intakes Of Glucosinolates, The Biologically Active Constituents Of Cruciferous Vegetables. *Br. J. Nutr.* **2003,** *90*(3), 687–697.

Minihane, A. M.; Rimbach, G. Iron Absorption And The Iron Binding And Anti-Oxidant Properties Of Phytic Acid. *Int. J. Food Sci. Technol.* **2002,** *37,* 741–748.

Oliveira, A. C.; Vidal, B. C.; Sgarbieri, V. C. Lesions of intestinal epithelium by ingestion of bean lectins in rats. *J. Nutr. Sci. Vitaminol.* **1989,** *35*(4): 315–322.

Phillippy, B. Q. Inositol Phosphates in Food. *Adv. Food Nutr. Res.* **2003,** *45,* 1–60.

Ray, R. B.; Raychoudhuri, A.; Steele, R.; Nerurkar, P. Bitter Melon (*Momordica charantia*) extract inhibits breast cancer cell proliferation by modulating cell cycle regulatory genes and promotes apoptosis. *Cancer Res.* **2010,** *70*(5): 1925–1931.

Santiago J. G., Levy-Benshimol, A.; Carmona, A. Effect of *Phaseolus vulgaris* lectins on glucose absorption, transport and metabolism in rat everted intestinal sacs. *J. Nutr. Biochem.* **1993,** *4,* 426–430.

Shahidi, F.; Naczk M. Phenolic Compounds In Fruits And Vegetables. In: *Phenolics in Food and Nutrceuticals.* CRC, LLC, 2004, pp 131–156.

Shamsuddin, A. M. Anti-Cancer Function of Phytic Acid. *Int. J. Food Sci. Technol.* **2002,** *37*(7): 769–782.

Shi, J.; Arunasalam, K.; Yeung, D.; Kakuda, Y.; Mittal, G.; Jiang, Y. Saponins from Edible Legumes: Chemistry, Processing, and Health Benefits. *J. Med. Food.* **2004,** *7,* 67–78.

South P. K.; Miller D. D. Iron Binding By Tannic Acid; Effects of Selected Ligands. *Food Chem.* **1998,** *63,*167–172.

Tan, M. J.; Ji, M. Y.; Nigel, T.; Cordula, H. B.; Chang, Q. K.; Chun, P. T.; Tong, C.; Hans, C. W.; Erns, C. W.; Alex, R.; David, E. J.; Yang, Y. Anti-Diabetic Activities of Triterpenoids Isolated from Bitter Melon Associated with Activation of the AMPK Pathway. *Chem. Biol.* **2008,** *15*(3), 263–273.

Xu, B. S.; K. C. Chang. Effect of Soaking, Boiling, and Steaming an Total Phenolic Content And Antioxidant Activities Of Cool Season Food Legumes. *Food Chem.* **2008,** *110,* 1–13.

Yeh, C. T.; Yen G. C. Effects of Phenolic Acids on Human Phenol Sulfo-Transferases in Relation to Their Antioxidant Activity. *J. Agric. Food Chem.* **2003,** *51,* 1474–1479.

Zeiger, E. Solanine and Chaconine. Review of Toxicological Literature. Integrated Laboratory Systems, USA, 1998.

CHAPTER 11

MINIMAL PROCESSING OF VEGETABLES

O. P. CHAUHAN[1] AND A. NATH[2]

[1]Defence Food Research Laboratory, Siddarthanagar, Mysore, India, E-mail: opchauhan@gmail.com

[2]Project Directorate for Farming Systems Research, Modipuram, Meerut, Uttar Pradesh, India, E-mail: amitnath2005@gmail.com

CONTENTS

11.1 INTRODUCTION

In recent years, there is increased demand for fresh-cut fruits and vegetables throughout the world due to their convenience and ready-to-use nature. On account of the consumer's increasing awareness and more interest towards food as prevention and cure of diseases as well as general well-being, there is a rapid growth in the fresh-cut fruit and vegetable industry. Various organizations such as WHO, FAO, USDA, EFSA and so on suggest increased consumption of fruits and vegetables in diet for decreasing the risk of cardiovascular diseases, cancer, hypertension and so on (Ana et al., 2006). In all the stages of food production and distribution chain, modern food processing techniques for maintaining quality and preventing undesirable microbial growth are required for making the food safe for consumption. Minimally processed fruits and vegetables offer great convenience to the users. At the same time, minimally processed fruits and vegetables are highly susceptible to physico-chemical and microbial spoilage. Several factors affect the quality of cut fruits and vegetables including cultivar selection deciding the shelf life and overall quality of minimal processed products (Artés et al., 2009). It is well established that depending on the season, fresh produce may contain a high contamination at the time of harvest and it may range up to 3–7 log units, soil condition, agro-climatic conditions and so on as well as type of the fresh produce. Pathogenic microorganisms associated with fresh vegetables can cause severe food-borne disease outbreaks. It has been reported that fresh produce is the fourth highest cause of all food-borne diseases since 1990 in the United States. Lettuce and sprouts are considered as the most frequent causes of food-borne diseases (CFSAN, 2006). Fresh-cut fruit and vegetable industry is still not fully matured and is still working on to meet the increasing consumer's demand for 'quick' and convenient products with better retention of nutritional value, fresh-like attributes in terms of flavour, colour and so on without or minimal use of chemical additives (Jongen, 2002). Fresh produce are biologically active, whereas processed foods are biologically inactive entities with no respiration and metabolic activities.

For healthy, palatable and easy-to-prepare plant foods, fresh-cut fruits and vegetables emerged to fulfil new consumer's demands. 'Minimal processing' describes non-thermal technologies to process food in a manner to guarantee the food safety and preservation as well as to maintain, as much as possible, the fresh-like characteristics of fruits and vegetables (Manvell, 1997). Towards the end of 20th century, minimally processed (MP) fresh produce had acquired importance. To highlight lightly processed foods including partially processed foods and high moisture refrigerated foods, the term 'minimal processing' was used for the first time by Rice (1987), and this definition was readily accepted by the people working in this area. Minimal processing of vegetables accelerates metabolic activities leading to faster physiological deterioration, biochemical changes and microbial proliferation resulting in degradation of the colour, texture, flavour and overall quality of the cut produce. The storage temperature plays a major role in determining the shelf life of MP produce. There are many other preservation techniques available for shelf life extension of MP vegetables and fruits such as phyto-sanitation measures using chlorine or ozone wash as well as other non-thermal techniques such as ultrasonics, UV light, plasma processing, use of antioxidants, modified atmosphere packaging (MAP) and so on. Nonetheless, new techniques for maintaining quality (Singh et al., 2008; Nath et al., 2010; Singh et al., 2011; Singh et al., 2012; Nath et al., 2011; Chauhan et al., 2011a, 2011b; Nath et al., 2012; Deka et al., 2013; Lamare et al., 2013; Patel et al., 2013b; Nath et al., 2013a, 2013b) and inhibiting undesired microbial growth (Nath et al., 2013b) are required in all the steps of the production and distribution chain as microorganisms have the tendency to adapt to the environment (Ana et al., 2006). Minimal processing operations such as peeling, cutting and grating lead to cell breakdown and release of intracellular materials, for example oxidizing enzymes accelerating the decay of the product. In addition, the cut surfaces of any processed vegetable support better microbial growth. It has been found that in cut produce, the use of UV-C light is beneficial in reducing tissue browning and other metabolic reactions (Lamikanra and Bett-Garber, 2005). Water-jet cutting can be another alternative as a non-contact cutting method which utilizes a concentrated stream of high-pressure water to cut through a wide range of produce. However, phyto-sanitation is the basic step in minimal processing, and therefore, necessary guidelines for MP vegetables specify a phyto-sanitation step to remove foreign materials such as dust, dirt, pesticide residues, microorganisms and so on which are responsible for quality loss and spoilage of the fresh-cut produce (Sapers, 2003).

11.2 KEY REQUIREMENTS FOR MINIMAL PROCESSING OF VEGETABLES

The requirements are as follows:

1. Good quality raw material (appropriate variety, cultivation practices, maturity, harvesting and storage conditions)

2. Strict hygiene, good manufacturing practices (GMP) and HACCP
3. Low temperatures during processing and packaging
4. Careful cleaning and/or washing before and after peeling
5. Water of good quality (sensory, microbiology, pH) used in washing
6. Mild additives in washing for disinfection or browning prevention
7. Gentle spin drying after washing
8. Gentle peeling
9. Gentle cutting/slicing/shredding
10. Correct packaging materials and packaging methods
11. Appropriate temperature and humidity during distribution and retailing

11.3 MINIMAL PROCESSING STEPS

11.3.1 WASHING, PEELING AND CUTTING

Washing is a mandatory step in minimal processing. To remove contaminants such as pesticide residues, parasites, foreign dirt, dust and other extraneous material, vegetables should be washed well before processing and consuming. However, certain precautions need to be taken while washing and cutting to minimize the loss of nutrients. Before cutting, vegetables should be washed thoroughly with potable water. Cutting of vegetables into small pieces causes greater exposure of surface area to the atmosphere, resulting in loss of vitamins due to oxidation as well as microbial spoilage. Therefore, vegetables should be immediately processed after cutting. Cut vegetables should not be soaked in water for long due to leaching of water-soluble minerals and vitamins into soaking media. Minimal processing of vegetables has two purposes as follows:

1. Keeping the produce in fresh condition without losing its nutritional quality
2. Ensuring product shelf life sufficient to make distribution feasible (Huxsoll and Bolin, 1989).

The microbiological, sensory and nutritional shelf life of minimally processed vegetables should be at least 4–7 days, but preferably up to 21 days depending on the market requirement. Ready-to-use vegetables typically involve peeling, slicing, dicing or shredding operations prior to packaging and storage (Barry-Ryan and O-Beirne, 1999). These operations are critical and determine the shelf life of fresh-cut commodities. Wounding stresses result in metabolic activation with increased respiration rate and in some cases ethylene production (Varoquaux and Wiley, 1997). Some vegetables such as potatoes, carrots, onion, garlic, ginger, bottle gourd, cucumber and so on need peeling. There are several peeling methods available, but it is normally done mechanically on an industrial scale (e.g. rotating carborundum drums), chemically or using high-pressure steam peelers (Wiley, 1994). Peeling should be as gentle as possible. The ideal method would be that of hand peeling with a sharp knife. Carborundum-peeled potatoes must be treated with a browning inhibitor, whereas water washing is enough for hand-peeled potatoes. It should resemble knife peeling when mechanical peeling is used. Carborundum, steam peeling or caustic acid disturb the cell walls of a vegetable enhancing the possibility of microbial growth and enzymatic changes (Ahvenainen and Hurme, 1994). With a first stage of rough peeling and then a second stage of finer knife peeling, Carborundum and knife peeling can be combined. Chemical peeling is successful in case of onion. Sharp stainless steel knives or blades must be used for performing the process of cutting and shredding. Mats and blades used in slicing should also be disinfected, using chemical such as 1% hypochlorite solution (Ahvenainen, 1996).

Vegetables should be carefully cleaned and washed before processing. A second wash must be done after peeling and/or cutting (Ahvenainen and Hurme, 1994). For example, Chinese cabbage and white cabbage must be washed after shredding, whereas carrot must be washed before grating. Washing after peeling and cutting removes microbes and tissue fluid, thus, reducing microbial growth and enzymatic oxidation. Instead of dipping into still water, washing in flowing or air-bubbling water is preferable. The microbiological quality of the washing water used must be good and its temperature should be low, preferably below 5°C (Hurme, 1994).

To reduce microbial counts and to retard enzymatic activity, preservatives can also be used in washing water, and thereby improving the shelf life of the cut produce. O'Beirne (1995) reported that 100–200 mg of chlorine or citric acid per litre is effective in washing water before or after peeling and/or cutting to extend shelf life. However, vegetable material should be properly rinsed if chlorine is used. Rinsing reduces the chlorine concentration to the level of that in drinking water and means that sensory quality is not compromised. Alternatives to chlorine include chlorine dioxide, peracetic acid, ozone, tri-sodium phosphate and hydrogen peroxide. Ozone washing is a preferred practice over chemical washing. Washing water should be removed gently from the product using centrifugal methods. The centrifugation time and rate should be chosen carefully so that the process removes free water and should not damage vegetable tissues (Ahvenainen, 1996).

11.3.2 ADDITIVE TREATMENTS

Enzymatic browning of cut tissues is a major problem in MP products. Washing with water is not effective in preventing discoloration and browning. Traditionally, sulphites have been used to prevent browning (Wiley, 1994). However, the use of sulphites or sulphur compounds has disadvantages, in particular, dangerous side effects for asthmatics. Enzymatic browning requires four different components: oxygen, an enzyme, copper and a substrate. Browning reactions have generally been assumed to be a direct consequence of PPO action on polyphenols, although some have attributed at least a partial role to the action of phenol peroxidase (POD) on polyphenols (Innocenti Degl et al., 2005). At least one component must be removed from the system to prevent browning. Theoretically, 2, 5-diphenyloxazole PPO—catalysed browning of vegetables can be prevented by the following:

1. Heat or reaction inactivation of the enzyme
2. Exclusion or removal of one or both of the substrates (oxygen and phenols)
3. Lowering the pH to 2 or more units below the optimum
4. Adding compounds that inhibit PPO or prevent melanin formation.

Probably, ascorbic acid is the most frequently studied alternative to sulphite. This compound is a highly effective inhibitor of enzymatic browning, primarily because of its ability to reduce quinones back to phenolic compounds before they can undergo further reaction to form pigments. The shelf life of fresh-cut eggplants was extended up to 12 days by using 0.5% ascorbic acid solution for 2 min before being packed in polyethylene bags (Li et al., 2014). Citric acid acts as a chelating agent and acidulant, and both these characteristics inhibit PPO. Protease enzymes have been found to be effective browning inhibitors for potatoes. There is no substitute for sulphites in preventing browning; alternatives are usually ascorbic acid-based combinations. Ascorbic acid also acts as an oxygen scavenger, removing molecular oxygen in polyphenol oxidase reactions. Polyphenol oxidase inhibition by ascorbic acid has been attributed to the reduction of enzymatically formed o-quinones to their precursor diphenols (Rico et al., 2007). A typical combination may include the following:

1. A chemical reductant (e.g. ascorbic acid)
2. An acidulant (e.g. citric acid)
3. A chelating agent (e.g. EDTA)

11.3.3 BIOCONTROL AGENTS

Microbial safety is of prime concern in MP products. In the pathogen growth control, use of biocontrol agents is a recently emerging area. The examples include lactic acid bacteria (LAB) which compete with, and thus inhibit, pathogen growth. LAB can produce both metabolites, such as lactic and acetic acids, or bacteriocins (Trias et al., 2008). Bacteriocins such as nisin can contribute in dealing with certain cold-tolerant gram-positive bacteria. Studies of the use of LAB have suggested using them in combination with other preservation techniques such as follows:

1. Reduction of the total microflora in the product by such procedures as washing using sanitizers, heat treatment or irradiation
2. Addition of a bacteriocin-producing bio-control culture to achieve a target initial bacterial count ($CFU \cdot ml^{-1}$)
3. Storage of the product under refrigerated conditions.

Product shelf life would then be determined by the growth of the biocontrol culture. If the product suffered temperature abuse during storage or distribution, for example, the biocontrol culture would grow more rapidly, thus preventing pathogen growth. Such cultures will be a fruitful source of further research.

11.3.4   USE OF EDIBLE COATINGS

An alternative is provided by edible coatings on fresh-cut commodities to modified atmosphere storage by reducing quality changes and quantity losses through modification and control of the internal atmosphere of the individual fruits (Park, 1999) and vegetables (de Jesus Davila-Avina et al., 2011; Patel et al., 2013a). One possible 'packaging' method for extending the post-harvest storage of MP fruit and vegetables are the use of edible coatings (Nath et al., 2013a). These are thin layers of material that can be eaten by the consumer as part of the whole food product. Coatings have the potential to reduce moisture loss, restrict oxygen entrance, lower respiration, retard ethylene production, seal in flavour volatiles as well as they can carry additives (such as antioxidants) for retarding discoloration and microbial growth (Baldwin et al., 1995). During processing, handling and storage, coating films can act as barriers to moisture and oxygen (Xu et al., 2007). Moreover, on account of their natural intrinsic activity or the incorporation of antimicrobial compounds, they can retard food deterioration by inhibiting the growth of microorganisms, (Cha and Chinnan, 2004). Nath et al. (2013b) studied the different plant extracts coating for extending the shelf life of *Khasi* mandarin (*Citrus reticulata* Blanco) fruits during ambient storage, and they reported that *Ocimum sanctum* was highly effective in inhibiting spore germination of *Penicillium brevicompactum* (96.5%) as well as extended the shelf life up to 28 days at ambient condition ($20\pm2°C$ and $65\pm5\%$ RH).

Normally, edible films are made of proteins or polysaccharides that can help to maintain moisture, thereby improving shelf life. However, the hydrophilic nature of these compounds limits their ability to provide desired edible film functions. Chitosan possesses excellent film-forming properties and can be applied as an edible surface coating to fruits and vegetables. It have been described that chitosan coatings limit fungal decay and delay the ripening of several commodities. In addition, chitosan is an ideal preservative coating for fresh fruits and vegetables due to its biochemical properties and its use in food is particularly promising because of its biocompatibility, non-toxicity and antimicrobial action (Ghaouth et al., 1992).

11.3.5 PACKAGING AND STORAGE

Packaging is a key operation in producing minimally processed vegetables. The most studied packaging method for prepared raw vegetables is modified atmosphere packaging (MAP). MAP is a preservation technique already in use by the fresh-cut industry. MAP technology offers the possibility of delaying the respiration rate and extending the shelf life of fresh produce and is used in the fresh and fresh-cut food industry throughout the world (Caleb et al., 2013). It implies altering the atmospheric gases surrounding a commodity to produce a composition different from that of air (Al-Ati and Hotchkiss, 2002). The basic principle in MAP is that a modified atmosphere can be created passively by using suitable selectively-permeable packaging materials, or actively by using a specified gas mixture together with selectively-permeable packaging materials. The aim of both is to create an optimal gas balance inside the package, in which the respiration activity of a product is as low as possible whilst ensuring that oxygen (O_2) concentration and carbon dioxide (CO_2) levels are not detrimental to the product. On the whole, the gas composition is maintained at 2–5% CO_2, 2–5% O_2 levels, the rest being nitrogen. In inhibiting enzymatic browning, preventing anaerobic fermentation reactions and anaerobic microbial growth, high oxygen MAP treatment has been found to be particularly ineffective (Kader et al., 1989). Carbon monoxide (CO) gas atmosphere has also been found effective in inhibiting mushroom PPO (Rico et al., 2007). Antala et al. (2014) reported that sapota fruits shelf life was extended up to 35 days at 11°C when packed in 25 μ LDPE bags. They also reported that the shelf life of sapota fruit could be increased up to 49 days by packaging in 25 μ LDPE bags having gas concentrations of 5% O_2 and 10% CO_2 and stored at 6°C.

Low temperature can be considered an important hurdle to control the microbial growth during storage. Storage at 10°C or above allows most bacterial pathogens to grow rapidly on fresh-cut vegetables. Storage temperature is an important factor in MAP or vacuum packaging of fruits and vegetables (Nath et al., 2010; Singh et al., 2011; Nath et al., 2011; Nath et al., 2012; Deka et al., 2013). At temperatures above 3°C, toxin production by *Clostridium botulinum*, or growth of other pathogens such as *Listeria monocytogenes*, is possible due to increased oxygen consumption in the package. Processing, transport, display and intermediate storage should be at the same low temperature (preferably 2–4°C). Temperature abuses should be avoided. Higher temperatures accelerate spoilage and facilitate pathogen growth. Fluctuating temperatures can cause in-pack condensation leading to spoilage of the commodity. Temperature abuse is a widespread problem in the distribution chain, whether in storage, transportation, retail display and consumer handling. It may be necessary to restrict shelf life of MP products to 5–7 days at a temperature of 5–7°C, to avoid psychrotrophic pathogens to multiply and produce toxin. If the shelf life of vacuum or MAP products is greater than 10 days and there is a risk that the storage temperature will be over 3°C, products should meet one or more of the following controlling factors:

1. A minimum heat treatment such as 90°C for 10 min
2. A pH of 5 or less throughout the food
3. A salt level of 3.5% (aqueous) throughout the food
4. Water activity (aw) value of 0.97 or less throughout the food.
5. Any combination of heat and preservative factors which has been shown to prevent growth of toxin production by *C. botulinum*.

Practically, if the aim is to keep minimally processed produce in fresh-like state, the last mentioned factors, and mainly various preservative factors, are the only possibilities to increase shelf life and assure microbiological safety of MAP or vacuum-packed fresh-cut produce.

11.4   EFFECTS OF MINIMAL PROCESSING ON VEGETABLES

11.4.1   METABOLISM AND PRODUCE PHYSIOLOGY

Minimal processing causes profound effect on produce physiology leading to changes in respiration rate, metabolic and enzymatic reactions. The physical damage caused during minimal processing leads to disruption of the cellular membrane, liberating enzymes and their substrates in direct contact of atmosphere (Chauhan et al., 2007). This stimulates loss of quality and spoilage of cut produce in no time. Among these reactions, increase in respiration rate, ethylene synthesis, phenylalanine ammonia lyase activity, peroxidase, catalases and polyphenol oxidase activities are the most important ones. These physiological changes result in implications for shelf life and quality of the processed products (Soliva Fortuny and Martin-Belloso, 2003; Oms-Oliu et al., 2010). The increases in rate of respiration and ethylene production are physiological and biochemical effects and are inversely correlated to the storage life of the minimally processed products. The increase in ethylene production from mechanical injury was found to enhance senescence in cut plant tissues (Abeles et al., 1992). Ethylene resulting from the physical action of minimal processing was reported to accelerate the loss of chlorophyll from spinach (Abe and Watada, 1991). Significant impact is provided on the quality and shelf life of cut produce by several enzymes such as hydrolases, ethylene biosynthetic enzymes and cell wall degrading enzymes. Minimal processing leads to enhanced enzymatic reactions causing loss of quality (Brecht, 1995). In plant tissues, the cellular compartmentalization provides better contact between the ethylene-generating systems (Watada et al., 1990) and also an increase in the synthesis and activity of 1-aminocyclopropane-1-carboxylic acid (ACC) synthase, which culminates in the accumulation of acid ACC, a precursor of ethylene (Hyodo et al., 1985). Increase in enzymatic reactions cause enhanced enzymatic browning in cut tissues of vegetables being a common feature occurring as a result of decompartmentalization of substrates and oxidative enzymes, due to larger exposure of tissues to oxygen. Basically, browning occurs because of oxidation of products of phenyl propanoid metabolism, such as phenols and possibly other substrates, the reactions being catalysed by phenolases such as polyphenol oxidase and peroxidase (Brecht, 1995). In the case of minimally processed lettuce, increased levels of ethylene production were found to cause higher levels of oxidative browning due to enhanced polyphenol oxidase activity. In certain cases, use of ethylene absorbers has not been found effective in preventing the onset of browning (Howard and Griffin, 1993).

Lipoxidase is another important enzyme which catalyses peroxidation leading to the formation of numerous bad-smelling aldehydes and ketones. Ethylene production can also increase and contributes to the neosynthesis of enzymes involved in fruit maturation leading fruit or vegetable softening. Cutting accelerates the respiration activity of produce by 20 to 70% or more depending on the produce, cutting grade, type of cutting, temperature and so on. It leads to anaerobic respiration causing the formation of ethanol, ketones and aldehydes if the packaging conditions are anaerobic. Compartmentalization of the cell begins to fail once a physical stress or deteriorative process (e.g. wounding response or senescence) sets in (Marangoni et al., 1996) causing mixing of polyphenol substrates (e.g. catechin, polyphenols) with polyphenol oxidase and/or phenol peroxidases (Innocenti Degl et al.,

2005). Peter and David (2008) studied the biochemical basis of appearance and texture changes in fresh-cut fruit and vegetables and reported that the loss of desirable texture in fresh-cut products is a major problem. This is largely due to disruption of cell wall which results in declining cell wall strength and reducing intercellular adhesion. In some species the process is exacerbated by wound-response ethylene. However, wounding, water loss and ripening-related turgor changes are also important contributors to textural deterioration. In fresh-cut vegetables, water loss and damage-induced lignification are common problems (Chauhan et al., 2011a). Nath et al. (2011) studied the changes in post-harvest phytochemical qualities of broccoli florets during ambient and refrigerated storage and reported that the broccoli packed in prolypopylene micro-perforated film showed significantly ($p<0.05$) lower loss in weight, ascorbic acid, chlorophyll, ß-carotene and total antioxidant activity during 144 h refrigerated storage.

The intensity of the wound response is affected by a number of factors. Species and variety of vegetables, O_2 and CO_2 concentrations, water vapour pressure and the presence of inhibitors stand out as the most significant (Brecht, 1995). Wounding of vegetable tissues induces a number of physiological disorders that need to be minimized to get fresh-like quality products. In fresh-cut commodities, due to excessive tissue softening and cut surface browning, the greatest hurdle to the commercial marketing is the limited shelf life.

11.4.2   PATHOGENIC AND SPOILAGE MICROORGANISMS

Cut produce are highly prone to microbial degradation and spoilage. In the processing and marketing of MP products, the presence and activity of pathogens can adversely affect the quality and shelf life of minimally processed fresh produce and is the prime criteria of consideration. The exudates released from the cut tissue are an excellent medium for the growth of microorganisms (Burns, 1995). The occurrence of food-borne diseases (FBDs), due to pathogens in minimally processed products, increases the risk of food poisoning due to the fact that in most cases these products are consumed as such without any thermal treatment (Nguyen and Carlin, 1994). Therefore, the reports of human infections associated with MP foods have heightened the concern of public health agencies as well as consumers (Vanetti, 2004). In recent years, the cases related to

FBDs have increased significantly, both in developed as well as developing countries (Feitosa et al., 2008). The reasons of FBDs are primarily the inadequate process of washing and sanitizing the surface of vegetables as well as post-processing contamination. Therefore, it is mandatory to follow strict phyto-sanitation protocols to reduce the microbiological risk in MP products (Yaun et al., 2004).

During minimal processing operations such as peeling, cutting and shredding, surface of the produce is exposed to the air and to contamination with bacteria, yeasts and moulds. In minimally processed vegetables, most of which fall into the low acid range category (pH 5.8–6.0), high humidity and the large number of cut surfaces can provide ideal conditions for the growth of microorganisms. Pectinolytic strains of *Pseudomonas* are responsible for bacterial soft rot during cold storage of minimally processed leafy vegetables. An increase in storage temperature and carbon dioxide concentration in the package shift the micro-flora towards lactic acid bacteria. The high initial load of microbes makes it difficult to establish the cell number threshold beyond which the product can be considered spoiled. Many studies show that a simple correlation does not exist between spoilage chemical markers such as pH, lactic acid, acetic acid, carbon dioxide, sensory quality and total microbial cell load. In fact, different minimally processed vegetable products seem to possess different spoilage patterns in relation to the characteristics of the raw materials.

As minimally processed fresh vegetables are not thermally processed, they must be handled and stored at refrigerated temperatures, at 5–6°C or below to get a sufficient shelf life and microbiological safety. Some pathogens such as *L. monocytogenes*, *Yersinia enterocolitica*, *Salmonella* spp., *Aeromonas hydrophila* and so on, may still survive or even proliferate at low temperatures. The normal spoilage organisms in refrigerated MP products are also usually psychrotrophic, and therefore, have a competitive advantage over most pathogens. Larson and Johnson (1999) reported presence of *C. botulinum* toxin in fresh-cut cantaloupe and honeydew melons after 9 days storage at 15°C packed under passive modified atmosphere conditions, the botulinal toxin was not detected in samples incubated at 7°C. Psychrotrophic pathogens such as *L. monocytogenes, A. hydrophila* and *Y. enterocolitica* and mesophiles such as *Salmonella* spp., *Staphylococcus* spp. and microaerophilic *Campylobacter jejuni* need further investigation because their emergence in these products represents a highly concerning subject. New techniques for maintaining quality while inhibiting undesired microbial growth are demanded in all the steps of the production and distribution chain of fresh-cut produce, as microorganisms tend to adapt to environment and survive in the presence of previously effective control methods (Rico et al., 2007).

11.5 QUALITY CHANGES

During minimal processing, the produce changes from a relatively stable state having a shelf life of several weeks or months to a very short shelf life, as short as few days at low temperatures. Many cells and tissues are broken during peeling and grating operations and intracellular products, such as oxidizing enzymes, are released. Minimally processed produce deteriorates owing to physiological ageing, biochemical changes and microbial spoilage, resulting in degradation of colour, texture and flavour of the cut produce. Certain additive (e.g. calcium dip) and packaging techniques such as MAP can be used to maintain the quality of cut MP produce during storage (Izumi and Watada, 1995; Artes et al., 1999).

It is self-evident that vegetables intended for pre-peeling and cutting must be easily washable, peelable and their quality must be of first class. The correct and proper storage of vegetables and careful trimming before processing are vital for the production of prepared vegetables of good quality. The study of various cultivar varieties of eight different vegetables showed that not all varieties of the specified vegetable can be used for the manufacture of prepared vegetables. The correct choice of variety is particularly important for carrot, potato and onion. For example, with carrot, the variety which gives the juiciest grated product cannot be used in the production of grated products which should have a shelf life of several days. Potato is another important example, in which poor colour and flavour become problems if the variety is wrong. Furthermore, the results showed that climatic conditions, soil conditions, agricultural practices, for example fertilization and harvesting conditions, can also significantly affect the behaviour of vegetables, particularly that of potatoes, in minimal processing.

Nutrient losses during MP operations are an important aspect. However, washing has not been found to decrease vitamin content (vitamin C and carotenes) of grated carrot, shredded Chinese cabbage or peeled potatoes, significantly. When plant tissues are wounded, nutrient losses may also be accelerated (Klein, 1987). Wounding induces signals that elicit physiological and biochemical responses in both adjacent and distant tissues (Saltveit, 1997). The earliest physiological responses to wounding include a transient increase in ethylene production and an enhanced rate of respiration, which may be interlinked with the wound healing response of the tissue (Brecht, 1995). Ethylene can in turn stimulate other physiological processes, resulting in accelerated membrane deterioration, loss of vitamin C and chlorophyll, abscission, toughening and undesirable flavour changes in a range of horticultural crops (Kader, 1985).

11.6 MICROBIOLOGICAL SPECIFICATIONS

Microbiological specifications of minimally processed commodities largely depend on the specific country (Table 11.1). Strict microbial standards need to be followed at the time of production as well as retail as microbial safety is the major concern in minimally processed commodities.

TABLE 11.1 Microbial Specifications for Minimally Processed Fruits and Vegetables (Germany).

Particulars	Limits
Microbial count at production site	$<5 \times 10^6$ g^{-1}
Microbial counts at retail site	$<5 \times 10^7$ g^{-1}
Escherichia coli	$<10^2$ g^{-1}
Salmonella sp.	Absent in 25 g sample
Listeria monocytogenes	$<10^2$ g^{-1}
Storage life	7 days including days of production at 2–7°C

11.7 CONCLUSION

There is a need for continuing further research to develop minimally processed vegetable products with high sensory quality, microbiological safety and nutritional value. It is possible to reach 7–8 days shelf life at refrigerated temperatures (5°C), but the shelf life need to be further extended using appropriate upcoming technologies so that the products can be marketed for a longer durations. There is a drastic need for more information about the growth of pathogenic bacteria or nutritional changes in minimally processed vegetables with long shelf life. Hurdle technology using natural preservatives, for example inhibitors produced by lactic acid bacteria, and the matching of correct processing methods and ingredients to each other, needs to be developed further in the minimal processing of fresh produce. Use of modern non-thermal technologies for the development of minimally processed products is also a crucial emerging area for achieving microbial safety with extended shelf life.

REFERENCES

Abe, K.; Watada, A. E. Ethylene Absorbent to Maintain Quality of Lightly Processed Fruits and Vegetables. *J. Food Sci.* **1991**, *56,* 1493–1496.

Abeles, F. B.; Morgan, P. W.; Saltveit, Jr., M. E. *Ethylene in Plant Biology;* Academic Press: San Diego, CA, 1992; pp 414.

Ahvenainen, R. New Approaches in Improving the Shelf Life of Minimally Processed Fruit and Vegetables. *Trends Food Sci. Tech.* **1996**, *71,* 179–187.

Ahvenainen, R.; Hurme, E. Minimal Processing of Vegetables in Minima/Processing of Foods (V77 Symposium Series No. 142) (Ahvenainen, R., Mattila-Sandholm, T., Ohlsson, T., Eds), pp 17–35, *Tech. Res. Centre Fin.*, Espoo, Finland, 1994.

Al-Ati, T.; Hotchkiss, J. H. Application of Packaging and Modified Atmosphere to Fresh-cut Fruits and Vegetables. In *Fresh-cut Fruits and Vegetables. Science, Technology and Market;* Lamikanra, O. Ed.; CRC Press: Boca Raton, FL, 2002.

Ana Allende; Tomás-Barberán, F. A.; Gil, M. I. Minimal Processing for Healthy Traditional Foods. *Trends Food Sci. Tech.* **2006**, *17,* 513–519.

Antala, D. K.; Satasiya, R. M.; Akabari, P. D.; Bhuva, J. V.; Gupta, R. A.; Chauhan, P. M. Effect of Modified Atmosphere Packaging on Shelf Life of Sapota Fruit. *Int. J. Agric. Sci. Technol.* **2014**, *2*(1), 32–38.

Artes F.; Conesa, M. A.; Hernandez, S.; Gil, M. I. Keeping Quality of Fresh-cut Tomato. *Postharvest Biol. Technol.* **1999**, *17,* 153–162.

Artés, F.; Gómez, P.; Aguayo, E.; Escalona, V.; Artes-Hernandez, F. Sustainable Sanitation Techniques for Keeping Quality and Safety of Fresh-cut Plant Commodities. *Postharvest Biol. Technol.* **2009**, *51,* 287–296.

Baldwin, E. A.; Nisperos Carriedo, M. O.; Baker, R. A. Use of Edible Coatings to Preserve Quality of Lightly (and Slightly) Processed Products. *Crit. Rev. Food Sci. Nutr.* **1995**, *35,* 509–524.

Barry-Ryan, C.; O'Beirne, D. Ascorbic Acid Retention in Shredded Iceberg Lettuce as Affected by Minimal Processing. *J. Food Sci.* **1999**, *64*(3), 498–500.

Brecht, J. K. Physiology of Lightly Processed Fruits and Vegetables. *Hortic. Sci.* **1995**, *30,* 18–22.

Burns, J. K. Lightly Processed Fruits and Vegetables: Introduction to the Colloquium. *Hortic. Sci.* **1995**, *30,* 14–17.

Caleb, O. J.; Mahajan, P. V.; Al-Said, A.-J. F.; Umezuruike Linus Opara. Modified Atmosphere Packaging Technology of Fresh and Fresh-cut Produce and the Microbial Consequences-A Review. *Food Bioprocess Technol.* **2013**, *6,* 303–329.

CFSAN. Guide to Minimize Microbial Food Safety Hazards of Fresh-cut Fruits and Vegetables. U.S. Department of Health and Human Services Food and Drug Administration Center for Food Safety and Applied Nutrition (CFSAN). Available at http://www.cfsan.fda.gov/wdms/guidance.html. Accessed 22.12.07.

Cha D. S. and M. S. Chinnan 2004. Biopolymer-based Antimicrobial Packaging: *Crit. Rev. Food Sci. Nutr.* **2004**, *44,* 223–237.

Chauhan, O. P.; Raju, P. S.; Bawa, A. S. Pre-cut Fruits and Vegetables: Pre- and Post Harvest Considerations. *Fresh Prod.* **2007**, *1*(2) 82–93.

Chauhan, O. P.; Raju, P. S.; Ravi, N.; Singh, A.; Bawa, A. S. Effectiveness of Ozone in Combination with Controlled Atmosphere on Quality Characteristics Including Lignification of Carrot Sticks. *J. Food Eng.* **2011a**, *102,* 43–48.

Chauhan, O. P.; Raju, P. S.; Singh, A.; Bawa, A. S. Shellac and Aloe Vera Gel Based Surface Coatings for Maintaining keeping Quality of Apple Slices. *Food Chemistry* **2011b**, *126,* 961–966.

de Jesus Davila-Avina, J. E.; Villa-Rodriguez, J.; Cruz-Valenzuela, R.; Rodriguez-Armenta, M.; Espino-Díaz, M.; Ayala-Zavala, J. F.; Olivas-Orozco, G. I.; Heredia, B.; González-Aguilar, G. Effect of Edible Coatings, Storage Time and Maturity Stage on Overall Quality of Tomato Fruits. *Am. J. Agric. Biol. Sci.* **2011**, *6*(1), 162–171.

Deka, B. C.; Nath, A.; Lamare, R. L.; Patel, R. K. Quality and Shelf-life of Sohshang (*Elaegnus latifolia* L.) Fruits in Different Packages During Storage. *J. Hill Farm.* **2013**, *26*(2), 21–25.

Feitosa, T. L.; Bruno, M.; Borges, M.; De, F. Segurança Microbiológica dos Alimentos. In *Ferramentas da Ciência e Tecnologia para a Segurança dos Alimentos;* Bastos, M. S. R. Eds.; Embrapa Agroindústria Tropical: Banco do Nordeste do Brasil: Fortaleza, Brazil, 2008; pp 21–39.

Ghaouth, E. l.; Ponnampalam, R. A.; Castaigne, F.; Arul, J. Chitosan Coating to Extend the Storage Life of Tomatoes. *Hortic. Sci.* **1992**, *27,* 1016–1018.

Howard, L. R.; Griffin, L. E. Lignin Formation and Surface Discoloration of Minimally Processed Carrot Sticks. *J. Food Sci.* **1993**, *58,* 1065–1067.

Hurme, E.; Ahvenainen, R.; Kinnunen, A.; Skytta, E. Factors Affecting the Quality Retention of Minimally Processed Chinese Cabbage in Proceedings of the Sixth International Symposium of the European Concerted Action Program COST 94 'Post-harvest Treatment of Fruit and Vegetables'. Current Status and Future Prospects, Commission of the European Community, Brussels, Belgium, 1994.

Huxsoll, C. C.; Bolin, H. R. Processing and Distribution Alternatives for Minimally Processed Fruits and Vegetables. *Food Technol.* **1989**, *43,* 124–128.

Hyodo, H.; Tanaka, K.; Yashisaka, J. Induction of L-aminocyclopropane-1-carboxylic Acid (ACC) Synthase in Mesocarp Tissue of Winter Squash Fruit and the Effects of Ethylene. *Plant Cell Physiol.* **1985**, *26,* 161–167.

Innocenti Degl, E.; Guidi, L.; Paradossi, A.; Tognoni, F. Biochemical Study of Leaf Browning in Minimally Processed Leaves of Lettuce (Lactuca sativa L. var. Acephala). *J. Agric. Food Chem.* **2005**, *52,* 9980–9984.

Izumi, H.; Watada, A. E. Calcium Treatments Affect Storage Quality of Shredded Carrots. *J. Food Sci.* **1995**, *6*, 789–793.

Jongen, W. Introduction. In *Fruit and Vegetable Processing;* Jongen, W., Ed.; Woodhead Publishing Limited/CRC Press LLC: Cambridge, UK/Boca Raton, FL, 2002.

Kader, A. A. Ethylene-induced Senescence and Physiological Disorders in Harvested Horticultural Crops. *Hortic. Sci.* **1985**, *20*, 54–57.

Kader, A. A.; Zagory, D.; Kerbel, E. L. Modified Atmosphere Packaging of Fruits and Vegetables. *Cm. Rev. Food Sci. Non.* **1989**, *28*(1), 1–30.

Klein, B. P. Nutritional Consequences of Minimal Processing of Fruits and Vegetables. *J. Food Qual.* **1987**, *10*, 179–193.

Lamare, R. L.; Deka, B. C.; Nath, A.; Patel, R. K. Dynamics of Physico-Chemical Values in *Sohshang* (*Elaegnus latifolia* L.) Across Maturity. *Ind. J. Hill Farm.* **2013**, *26*(2), 49–53.

Lamikanra, O.; Bett-Garber, K. Fresh-cut Fruit Moves into the Fast Lane. Agricultural Research 2005. 2005 (http//www.ars.usda. gov/is/AR/archive/aug05/fruit0805.pdf).

Larson, A. E.; Johnson, E. A. Evaluation of Botulinal Toxin Production in Packaged Fresh-cut Cantaloupe and Honeydew Melons. *J. Food Prot.* **1999**, *62*, 948–952.

Li, X.; Jiang, Y.; Li, W.; Tang, Y.; Yun, J. Effects of Ascorbic Acid and High Oxygen Modified Atmosphere Packaging During Storage of Fresh-cut Eggplants. *Food Sci. Technol.* **2014**, *20*(2), 99–108.

Manvell, C. Minimal Processing of Food. *Food Sci. Technol. Today* **1997**, *11:* 107–111.

Marangoni, A. G.; Palma, T.; Stanley, D. W. Membrane Effects in Postharvest Physiology. *Postharvest Biol. Technol.* **1996**, *7*, 193–217.

Nath, A.; Deka, B. C.; Paul, D.; Mishra, L. K. Ambient Storage of Capsicum Under Different Packaging Materials. *Bioinfolet.* **2010**, *7*(3), 266–270.

Nath, A.; Bagchi, B.; Misra, L. K.; Deka, B. C. Changes in Post-Harvest Phytochemical Qualities of Broccoli Florets During Ambient and Refrigerated Storage. *Food Chem.* **2011**, *127*, 1510–1514.

Nath, A.; Deka, B. C.; Singh, A.; Patel, R. K.; Paul, D.; Misra, L. K.; Ohza, H. Extension of Shelf Life of Pear Fruits Using Different Packaging Materials. *J. Food Sci. Technol.* **2012**, *49*(5), 556–563.

Nath, A.; Deka, B. C.; Jha, A. K.; Paul, D.; Misra, L. K. Effect of Slice Thickness and Blanching Time on Different Quality Attributes of Instant Ginger Candy. *J. Food Sci. Technol.* **2013a**, *50*(1), 197–202.

Nath, A.; Barman, K.; Chandra, S.; Baiswar, P. Effect of Plant Extracts on Quality of Khasi Mandarin (*Citrus reticulata* Blanco) Fruits During Ambient Storage. *Food Bioprocess Technol.* **2013b**, *6*(2), 470–474.

Nguyen-The, C.; Carlin, F. The Microbiology of Minimally Processed Fresh Fruits and Vegetables. *Crit. Rev. Food Sci. Nutr.* **1994**, *34*, 371–401.

O'Beirne, D. Influence of Raw Material and Processing on Quality of Minimally Processed Vegetable- in Progress Highlight U95 of ELI Contract A/RI-CT92-0125 'Improvement of the Safety and Quality of Refrigerated Ready-to-eat Foods using Novel Mild Preservation Techniques', Commission of the European Community, Brussels, Belgium, 1995.

Oms-Oliu, G.; Rojas-Graú, M. A.; González, L. A.; Varela, P.; Soliva-Fortuny, R.; Hernando, M. I. H.; Munuera, I. P.; Fiszman, S.; Martín-Belloso, O. Recent Approaches

Using Chemical Treatments to Preserve Quality of Fresh-cut Fruit: A Review. *Postharvest Biol. Technol.* **2010**, *57,* 139–148.

Park, H. J. Development of Advanced Edible Coatings for Fruits. *Trend Food Sci. Technol.* **1999**, *10,* 254–260.

Patel, D. R.; Soni, A. K.; Kabir, J. Agrawal, N. Influence of Wax Coating on Shelf Life of Pointed Gourd (*Trichosanthes dioca* Roxb.). *Karnataka J. Agric. Sci.* **2013a**, *26*(3), 393–398.

Patel, R. K.; Maiti, C. S.; Deka, B. C.; Deshmukh, N. A.; Nath, A. Changes in Sugars, Pectin and Antioxidants of Guava *(Psidium guajava)* Fruits During Fruit Growth and Maturity. *Ind. J. Agric. Sci.* **2013b**, *83*(10), 1017–21.

Peter M.; Toivonen, A.; David, A. B. Biochemical Bases of Appearance and Texture Changes in Fresh-cut Fruit and Vegetables. *Postharvest Biol. Technol.* **2008**, *48,* 1–14.

Rice J. International Trends in Food Packaging. *Food Proc.* **1987**, *48,* 76–81.

Rico D.; Martın-Dianaa, A. B.; Barat, J. M.; Barry-Ryan, C. Extending and Measuring the Quality of Fresh-cut Fruit and Vegetables: A Review. *Trend Food Sci. Technol.* **2007**, *18,* 373–386.

Saltveit, M. E. Physical and Physiological Changes in Minimally Processed Fruits and Vegetables. In *Phytochemistry of Fruits and Vegetables;* Tomas-Barberan, F. A., Robins, R. J., Eds.; Oxford University Press: London, 1997; pp 205–220.

Sapers, G. M. Washing and Sanitizing Raw Materials for Minimally Processed Fruit and Vegetable Products. In *Microbial Safety of Minimally Processed Foods;* Novak, J. S., Sapers, G. M., Juneja, V. K., Eds.; CRC Press LLC: Boca Raton, FL, 2003; pp 221–222.

Singh, A.; Nath, A.; Buragohain, J.; Deka, B. C. Quality and Shelf Life of Strawberry Fruits in Different Packages During Storage. *J. Food Sci. Technol.* **2008**, *45*(5), 439–442.

Singh, A.; Yadav, D. S.; Patel, R. K.; Nath, A.; Bhuyan, M. Wax Coating and Padding Materials Influence Quality and Shelf-life of Purple Passion Fruit During Storage. *Ind. J. Hortic.* **2011**, *68*(2), 46–50.

Singh, A.; Deka, B. C.; Patel, R. K.; Nath, A.; Mulieh, S. R. Effect of Pruning Time, Severity and Tree Aspects on Harvesting Period and Fruit Quality of Low Chilling Peach (*Prunus persica*). *Ind. J. Agric. Sci.* **2012**, *82*(10), 862–866.

Soliva-Fortuny, R. C.; Martín-Bellos O. New Advances in Extending the Shelf-life of Fresh-cut Fruits: A Review. *Trend Food Sci. Technol.* **2003**, *14,* 341–353.

Trias R.; Baneras, L.; Badosa, E.; Montesinos, E. Bioprotection of Golden Delicious Apples and Iceberg Lettuce Against Foodborne Bacterial Pathogens by Lactic Acid Bacteria. *Int. J. Food Microbiol.* **2008**, *123,* 50–60.

Vanetti, M. C. D. Segurança microbiológica em produtos minimamente processados. In *Encontro nacional sobre processamento mínimo de frutas e hortaliças , 3, 2004, Viçosa. Palestras, Resumos e Oficinas;* CEE: Viçosa, Brazil, 2004; pp 30–32.

Varoquaux, P.; Wiley, R. C. Biological and Biochemical Changes in Minimally Processed Refrigerated Fruits and Vegetables. In *Minimally Processed Refrigerated Fruits and Vegetables;* Wiley, R. C., Ed.; Chapman and Hall: New York, 1997; pp 226–268.

Watada, A. E.; Abe, K.; Yamuchi, N. Physiological Activities of Partially Processed Fruits and Vegetables. *Food Technol.* **1990**, *44,* 116–122.

Wiley, R. C. Preservation Methods for Minimally Processed Refrigerated Fruits and Vegetables. In *Minimally Processed Refrigerated Fruits and Vegetables;* Wiley R C., Ed.;
Chapman and Hall: New York, USA, 1994; pp 66–134.

Xu, W. T.; Huang, K. L.; Guo, F.; Qu, W.; Yang, J. J.; Liang, Z. H. Postharvest Grapefruit
Seed Extract and Chitosan Treatments of Table Grapes to Control *Botrytis cinerea. Post-
harvest Biol. Technol.* **2007**, *46,* 86–94.

Yaun, B. R.; Summer, S. S.; Eifert, J. D.; Marcy, J. E. Inhibition of Pathogens on Fresh
Produce by Ultraviolet Energy. *Int. J. Food Microb.* **2004**, *90,* 1–8.

CHAPTER 12

NOVEL NON-THERMAL PROCESSING OF VEGETABLES

O. P. CHAUHAN[1,*], K. CHITRAVATHI[1], L. E. UNNI[1], and I. N. DOREYAPPA GOWDA[2]

[1]Defence Food Research Laboratory, Siddarthanagar, Mysore 570011, India

[2]ICAR-IIHR-Central Horticultural Experiment Station, Kodagu 571248, India, Tel: 91 8212473879, Fax: +91 821 2473468, *E-mail: opchauhan@gmail.com

CONTENTS

12.1   INTRODUCTION

One of the most important constituent of diet is vegetables. They are highly nutritious and are easily susceptible to microbial deterioration resulting in heavy post-harvest losses. Moreover, vegetables which are available as seasonal and regional surpluses are wasted in large quantities due to absence of facilities, know-how for proper handling, distribution, marketing and storage. Although thermal pasteurization and sterilization preserves vegetables in various forms by inactivating microbes and enzymes, thermal energy may trigger unwanted reactions leading to formation of unwanted by-products, loss of essential vitamins, nutrients, characteristic flavour and colour of food. Therefore, preservation of these perishables is a very challenging and innovative field. During wars, the major developments and needs in food processing and preservation started, when extended shelf life of foods became a necessity. Over the years, consumers became more and more learned about adding food preservatives and their adverse effects on long-term health. Increasing consumer demand for minimally processed fruits and vegetables, additive free, shelf stable products had thrown challenge food scientists to explore other physical preservation methods as alternative to traditional treatments such as freezing, canning or drying that rely on heating or cooling operations. However, to ensure a high degree of food safety, these technologies have helped. But heating and cooling methods may contribute to the degradation of various quality attributes. Consequently, research and development and finally the commercialization of several innovative non-thermal food preservation techniques took place. These are called non-thermal processing of foods as the name proclaims that the temperature of food remains under ambient conditions. They are under intense research to evaluate their potential as an alternative or complimentary process to traditional methods of food preservation. By these processing techniques, significant shelf life extension and minimum changes in the physical and chemical properties of foods are demonstrable. Non-thermal technologies for preservation of vegetables encompass high hydrostatic pressure, high pressure carbon dioxide (HPCD), ultrasound, cold plasma, electron beam processing, pulse-electric field, oscillating magnetic fields (OMF), pulse-light, ozonation and irradiation (Table 12.1).

TABLE 12.1 Overview of Novel Non-Thermal Processing of Vegetables.

Technology	Principle	Application in vegetables
High-pressure processing (HPP)	HPP subjects foods, with or without packaging, to pressures between 100 and 1000 MPa or 80,000–130,000 pounds (1–20 min)	Preservation of vegetables in fresh cut, paste, puree and juice form.
		Surface decontamination of vegetables
High-pressure carbon dioxide	The food is in contact with either (pressurized) sub- or supercritical CO_2 for a certain amount of time in a batch, semi-batch or continuous manner. Supercritical CO_2 is at a temperature and pressure above its critical point values and exists as a single phase and has the unique ability to diffuse through solids like a gas and dissolve materials like a liquid. In addition, it can readily change in density upon minor changes in temperature or pressure	Surface decontamination of vegetables
Pulsed electric field (PEF) processing	PEF processing involves the application of pulses of high voltage (20–80 kV/cm) to foods placed between two electrodes	Permeablization of pulp for high juice yield
		Drying rate enhancement by increase in heat-mass transfer
Oscilating magnetic field	Food is subjected to a high intensity, moderate frequency magnetic field in the range of 5–50 T and 5–500 kHz frequencies to destroy or inactivate microorganisms in a mainly non-electrically conductive environment	Surface decontamination of vegetables
Ultrasound processing	Energy generated by high frequency sound waves 0.1–20 MHz	Surface decontamination of vegetables
		Enhanced juice and oil recovery
Pulsed light processing	High intensity flashes of broad spectrum white light pulsed several times from a high power xenon lamp for about 0.1–3 ms per some sources or about 100 μs to 10 ms per other sources inactivates microbes with remarkable rapidity and effectiveness	Surface decontamination of vegetables

TABLE 12.1 *(Continued)*

Technology	Principle	Application in vegetables
Irradiation	It is a physical treatment of food with high-energy ionizing radiation (5–10 MeV)	Prevention of germination and sprouting of potatoes, onion and garlic.
		Disinfestation by killing or sterilizing insects which infests grains, dried fruits, vegetables or nuts.
		Retardation of ripening and ageing of vegetables.
Electron beam processing/ electronic pasteurization	In electron beam pasteurization, electrons are accelerated in a strong electrical field to near the speed of light and the high energy-electrons act on the surface of the object, which may result in breaking chemical bonds when neutralizing microorganisms or creating bonds between molecules	Surface decontamination of vegetables
Cold plasma	Plasma is more or less an electrified gas with a chemically reactive media that consists of a large number of different species such as electrons, positive and negative ions, free radicals, gas atoms and molecules in the ground or any higher state of any form of excited species	Surface decontamination of vegetables
Ozonation	As a potent oxidizing agent, ozone damages unsaturated lipids in microbial cell envelope, the lipopolysaccharide of gram negative bacteria, intracellular enzymes and cellular structures	Surface decontamination of vegetables

12.2   HIGH-PRESSURE PROCESSING

A non-thermal processing technology known as high pressure processing (HPP) is proven to have great potential in food preservation, while keeping the food chemistry essentially intact. HPP employs intense pressure (about 400–600 MPa) to inactivate pathogenic and spoilage microorganisms as well as enzymes. High pressure processing can be conducted at ambient or refrigerated temperatures, thereby eliminating thermally induced cooked off-flavours. The technology is especially beneficial for heat-sensitive products. This unique characteristic is due to its limited effect on non-covalent bonds which has boosted the commercial value of pressure-processed products. Furthermore, the technique is independent of the size, shape or composition of products (Knorr 1995). The concepts of HPP were coined a century before, the relatively lesser incidence of microbial flora at the sea bed could be the motivation behind the framing of antimicrobial effects of high hydrostatic pressures. The basis of high hydrostatic pressure is the Le Chatelier principle, according to which any reaction, conformational change, or phase transition that is accompanied by a decrease in volume will be favoured at high pressures, whereas reactions involving an increase in volume will be inhibited. The high pressure system consists of high pressure vessel and its closure, pressure generator, temperature-controlling system and material-handling system. The high pressure can be generated by following:

1. Direct compression which allows very fast compression and is generated by pressurizing a medium with small diameter end of piston whose large diameter end is driven by a low-pressure pump.
2. Indirect compression uses a high pressure intensifier to pump a pressure medium from a reservoir into a closed high pressure vessel until the desired pressure is reached.
3. Heating of the pressure medium which utilizes expansion of the pressure medium with increasing temperature to generate high pressure. It is used when high pressure is used in combination with high temperature; therefore, it requires very accurate temperature control within the entire internal volume of the pressure vessel.

By membrane damage and cellular effects on the fluid transport of nutrients, mechanism of microbial inactivation has shown that HPP can interrupt cell reproduction and functions (Unni et al. 2011). Moreover, due to structural damage, microbial enzymes can be denatured, and the extent of microbial inactivation depends upon a number of interacting factors including type and number of microorganisms, temperature, duration of treatment and composition of the suspension of the media and food. The baro-sensitivity of microorganisms increased in order of gram positive bacteria, yeasts, gram negative bacteria. High pressure can exert a drastic change in enzymatic reactions. Inactivation of enzymes is mainly influenced by temperature of pressurization, pH and substrate concentration. Pectin esterase can be inactivated when pressurized to 3000–4000 atm. the action is reversible but the reactivation is minimized by storing at 0°C. The activity of polyphenol oxidase increases five times when pressurized at 4000 atm but further increase in pressure does not increase the activity. But the pressurization of apples, bananas, sweet potatoes does not result in activation (Asaka and Hayashi, 1991). HHP also has some serious limitations, such as (i) the occurrence of pressure resistant vegetative bacteria after successive pressure treatments, (ii) the large investment costs (due to the high pressures involved), (iii) at present non-continuous nature of the process and (iv) regulatory and product safety-related issues which need to be further clarified (Devlieghere et al., 2004; Estrada-Girón et al., 2005). These drawbacks are hampering widespread implementation of HHP preservation by the food industry. The study of chemical and microbiological changes of foods processed by using high pressure will determine their safety and quality, but commercial feasibility must include design and construction of equipment.

Popularity has been gained by researches into the application of HPP when Hite (1899) demonstrated that the shelf life of milk and other food products could be extended by pressure treatments. The technology applies high hydrostatic pressure to materials by compressing the surrounding water and transmitting pressure throughout the product uniformly and rapidly (Hayashi et al., 1989). High pressure processing is an emerging technology with potential for optimizing intake of nutrient and non-nutrient phytochemicals in vegetables. Retention of organoleptic attributes and other characteristics of freshness, combined with increased convenience and extended shelf life, will no doubt increase the appeal of vegetable products preserved using HPP. HPP may be of benefit for improving bioavailability of certain vegetable carotenoids, as is suggested by increased in vitro availability of lutein in green beans following HPP (Mc-Lnerney et al. 2007). Unni et al. (2013) studied the effect of HPP on garlic paste and observed retention of antioxidant activity in the order of 600 MPa (18.4%) > 400MPa (18.2%) > 200MPa (16.29%). Some of the effects of high hydrostatic pressures on vegetables are given in Table 12.2.

TABLE 12.2   Effect of High Pressure on Some Quality Attributes of Vegetables.

Product	Conditions	Parameter	Observation	Source
Broccoli	180–210 MPa, −(16–20°C)	Protein, enzymatic activity and sensory evaluation	The protein content decreased after the high-pressure treatments. Peroxidase and polyphenoloxidase enzymes could not be inactivated. After 30 days of frozen storage at −20°C the flavour of broccoli was not acceptable to consumers, but the texture remained quite firm. The vacuole membrane was destroyed, and an internal disorganized cell was observed after pressure shift freezing treatment	Prestamo et al. (2004)
Potato	210–300 MPa	Phase transition	In the pressure range 210–240 MPa, a metastable ice(I) modification area was observed, as the nucleation of ice(I) crystals in the thermodynamically stable region of ice(III) was reached. A significant degree of super cooling was obtained before freezing the tissue water to ice(III). Phase transition and freezing times for the different freezing paths were compared for the processes such as freezing at atmospheric pressure, pressure-assisted freezing and pressure-shift freezing	Schluter et al. (2004)
Eggplants	–	Sensory quality	High-pressure-assisted freezing resulted in lower quality damage in comparison with conventional air-freezing techniques	Otero et al. (1998)

TABLE 12.2 *(Continued)*

Product	Conditions	Parameter	Observation	Source
Tomato puree	100–600 MPa for 12 min at 20°C	Lycopene	High pressure affected the total lycopene content and the percentage of the presumptive 13 *cis*-isomer, both in lycopene solution and tomato puree. At higher storage temperature, the loss of total lycopene and percentage of 13 cis-isomer was greater. The highest stability of lycopene was found when tomato puree was pressurized at 500 MPA and stored at $4 \pm 1°C$	Qiu et al. (2006)
Chinese cabbage (midribs)	100–700 MPa	Texture	Texture of samples frozen at 200, 340 and 400 MPa was comparatively intact in relation to samples frozen at 100 and 700 MPa. Release of pectin and histological damage in midribs frozen at 200 and 340 MPa were less than midribs frozen at 100 and 700 MPa	Fuchigami et al. (1998)
Carrot	100–700 MPa	Texture	Textural properties of carrots pressurized at 200–400 MPa at $-20°C$ were found to be more acceptable, pectin release and histological damage were also lower in samples frozen at 100 and 700 MPa	Fuchigami and Teramoto (2003a) Fuchigami et al. (1997b)
Carrot, tomato and broccoli	500–800 MPa for 25 or 75°C	Quality attributes	No effect on chlorophyll a and b content in broccoli, no effect on lycopene and a-carotene in tomato and no appreciable changes on antioxidants and water soluble nutrients	Butz et al. (2002)

12.3 HIGH PRESSURE CARBON DIOXIDE

The use of HPCD is also proposed as an alternative non-thermal pasteurization technique for foods (Spilimbergo, 2002). In HPCD, for a certain amount of time in a batch, semi-batch or continuous manner, the food is in contact with either (pressurized) sub-or supercritical CO_2. Supercritical CO_2 is at a temperature and pressure above its critical point values ($T_c = 31.1°C$, $P_c = 7.38$ MPa) and exists as a single phase. It has the unique ability to diffuse through solids such as a gas and dissolve materials like a liquid. In addition, it can readily change in density upon minor changes in temperature or pressure. Subcritical (gaseous or liquid) CO_2, on the other hand, is CO_2 at a temperature or pressure below its thermodynamical critical point values. The HPCD technique presents some advantages over HPP related to the milder conditions it employs. Besides the environmentally benign nature of the HPCD process (CO_2 is nontoxic), the CO_2 pressures applied for preservation purposes are much lower (generally < 20 MPa) as compared with the hydrostatic pressures employed in HHP (300–600 MPa). Hence, this makes it easier to control and manage pressure in the HPCD technique (Spilimbergo, 2002). In addition, capital expenditure could considerably be reduced. Different steps of hypothetical inactivation mechanism can be summarized as follows: (i) solubilization of pressurized CO_2 in the external liquid phase, (ii) cell membrane modification, (iii) intracellular pH (pH) decrease, (iv) key enzyme inactivation/cellular metabolism inhibition due to pH lowering, (v) direct (inhibitory) effect of molecular CO_2 and HCO^{3-} on metabolism, (vi) disordering of the intracellular electrolyte balance and (vii) removal of vital constituents from cells and cell membranes. That most of these steps will not occur consecutively, must, however, be noted. They might rather take place simultaneously in a very complex and interrelated manner (Spilimbergo et al., 2003; Damar and Balaban, 2006) (Table 12.3).

TABLE 12.3 Important Opportunities and Drawbacks of the High-Pressure Carbon Dioxide (HPCD) Preservation Technique.

Opportunities	Threats
Natural image	Inactivation mechanism not entirely clear
High fresh-like organoleptic quality	Extraction of/interaction with food ingredients
Spores can be inactivated when combined with, for example, heat, pressure pulsing and acid environment	Occurrence of HPCD-resistant mutants still needs to be investigated

TABLE 12.3   *(Continued)*

Opportunities	Threats
Applicable to acid foods	Inactivation conditions required dependent on type of food and type of microorganism
Up-scaling of equipment has been developed for other applications (e.g., extraction)	Large investment costs
Continuous processing for liquid foods is possible	Processing-related problems for solid foods (discontinuous in nature, limited diffusion of CO_2, packaging after treatment)

12.4   PULSED ELECTRIC FIELD PROCESSING

PEF involves the discharge of high voltage electric pulses (up to 70 KV/cm) into the food product, which is placed between two electrodes for a few microseconds. An external electric field is used to exceed a critical trans-membrane potential of 1 V. This results in a rapid electric breakdown and conformational changes of cell membranes, which leads to the release of intracellular liquid and cell death. The concept of pulsed power, involves dissipation of electrical energy from a capacitor for storage of the charge at low power levels. The stored energy from the capacitor can then be discharged almost instantaneously in a pulsed mode at higher levels of power. Pulsed power supply units and a treatment chamber are required by the generation of pulsed electric fields. Several models are available including bench top, lab scale or pilot-scale capacities. The operating conditions include pulsed electric field (PEF) intensity at 35–70 kV/cm, pulsed width 2–15 µs, pulsed rate 1 Hz, electrode gap 0.95–0.51 cm, chamber volume 20–8 cm^3 and fluid food flow rate at 1200–600 cm^3/min (Jeyamkondan et al., 1999). Microbial cells which are exposed to an external electrical field for a few microseconds respond by an electrical breakdown and local structural changes of the cell membrane. In consequence of the so called *electroporation*, a drastic increase in permeability is observed which in the irreversible case is equivalent to a loss of viability. For several liquid products such as pea soup, the inactivation of microorganisms would be achieved at PEF range of 25–45 kV/cm. In general, the bactericidal effect of PEF is inversely proportional to the ionic strength, and it further increases with electrical resistivity. Apart from this, the factors affecting microbial inactivation include electric field strength, pulse frequency, pulse duration, pulse shape, temperature of the medium and maturity of the bacteria. Inactivation increases greatly beyond a critical value which is specific to the organism and type of food (Knorr et al., 1994). Mass transfer phenomenon is strongly dependable for juice recovery processes from fruits and vegetables. Pulsed electric fields (PEF) can be applied as an alternative method for cell disintegration. Biological tissues exposed to high electric field pulses develop pores in the cell membrane resulting in increased membrane permeability and a facilitated loss of the cell content (Knorr et al., 2001). PEF causes the formation of large, permanent pores in cellular tissues, which can be used to improve juice yield, increase concentrations of functional components and enhance the characteristics of dried produce. It can support or replace conventional processing techniques such as enzymatic maceration and mechanical disintegration. Negative aspects of the PEF treatment include the following facts: spores are not sensitive to PEF treatment, up-scaling of PEF equipment is still under development, the method is rather limited to liquid products and its efficiency depends on the electrical conductivity of food (Devlieghere et al., 2004). Shayanfar et al. (2013) reported that application of low electric field strength is helpful in formation of reversible pores in cellular membranes of the tissues facilitating permeation of texturizing or anti-freezing agents from medium to the tissues and thereby maintaining the cellular integrity after freezing and thawing.

12.4.1 MECHANISM OF MICROBIAL INACTIVATION

Pulsed electric is a variation of pulsed electric field technology. Electric fields and light are both electromagnetic radiation. However, the mechanism of inactivation due to electric fields appears to be distinctly different. Spores do not appear to be inactivated by pulsed electric fields. PEF sterilization requires electric fields of no less than 8 kV/cm. PEF exposure exhibits the characteristics survival tail and conforms to the standard logarithmic decay of microbes subjected to lethal mechanisms such as radiation, biocides and heating. The mechanism by which PEL kills bacteria and spores appears to be due to the effects of UV combined with new disinfection mechanism (Wekhof, 2000). Damage to the nucleic acid and other components of the cell is caused by the exposure PEL. The instantaneous heating of the cell results in rupture of the cell wall or lysis (Gomez-Lopez et al., 2005). As the lethality of light pulses is different at different wavelengths, the full spectrum or selected wavelength may be used to treat the foods. Wavelengths, producing undesirable products in foods, are eliminated through glass or liquid filters. Light pulses produce photochemical or photo thermal reactions in foods. The ultraviolet-rich light causes change in photochemical reactions, whereas visual and infrared light causes photo thermal changes. Ultraviolet light has been shown to inactivate pathogens and indicator organisms (Chang et al., 1985). The antimicrobial effects of these wavelengths are primarily mediated through absorption by highly conjugated carbon-to-carbon double bond systems in proteins and nucleic acids (Jay, 1996).

12.5   OSCILLATING MAGNETIC FIELD

In the non-thermal processing of foods, the application of high intensity magnetic field is one of the latest innovations. Materials have different responses towards magnetic fields, and when the magnetization is equal along the three orthogonal axes, x, y and z, the particle is defined to have isotropic susceptibility. If the magnetization is unequal along the x, y and z axes, the magnetic susceptibility is known as anisotropic susceptibility. Carbon atoms as such show isotropic susceptibility, whereas, two carbon atoms bonded together exhibit anisotropic susceptibility. Most of the organic compounds are diamagnetic in nature during which the magnetic field applied causes lesser intensity of magnetization under normal conditions compared with the induction in vacuum. Usually applied magnetic fields could either be static or oscillating types. In the case of oscillating magnetic field applied in the form of pulses, it causes reversion in the change with each pulse, and intensity of each pulse decreases with time to about 10% of initial intensity (Martin et al., 1997). Generation of high-intensity magnetic field is usually carried out by supplying current to electrical coils. The inactivation of microorganisms usually requires magnetic flux densities from 5–50 T. Oscillating magnetic field of this density can be generated by (i) superconducting coils, (ii) coils that produce DC fields and (iii) coils energized by the discharge of energy stored in the capacitors. Usually, oscillating magnetic field circuits use capacitor banks for storage of energy. A capacitor is charged from a voltage source. An oscillating current is generated between the plates of the capacitor and the same generate OMF. The frequency of the magnetic field is determined by the capacitance of the capacitor and also the resistance and inductance of the coil. The magnetic field changes polarity, as the current changes direction. For generating a magnetic field of 70 T, a capacitor with stored energy of 1.25 MJ is required (Pothakamury et al., 1993).

According to this, very limited studies have been carried out on the application and commercialization of OMF. Retardation of growth and reproduction of microorganisms under OMF might be due to changes in DNA synthesis, a change in the orientation of bio-molecules and to the applied magnetic field or a change in the ionic drift across the plasma membrane to a direction parallel or perpendicular to the applied magnetic field, or a change in the ionic drift across the plasma membrane. The efficiency of inactivation is likely to be affected by the magnetic field factors such as intensity and pulse product properties such as ionic concentration, temperature and pH along with physiological factors such as microbial cell concentration (Mishra et al., 2011). Many foods have significant electric resistivity greater than $10–25$ Ω/cm which is conducive for the application of oscillating magnetic fields. Hofmann (1985) observed the inactivation of microorganisms exposed to a flux density greater than 2 T, a single pulse with flux density between 5–50 T and frequency of 5–500 kHz reduces the number of microorganisms by at least two log cycles. During preservation of food by OMF, for a total exposure time ranging from 25 to 10 ms, the food is sealed in a plastic bag and subjected to 1–100 pulses of OMF with frequency between 5 and 500 kHz at a temperature of $-50°C$. The duration of each pulse include 10 oscillations. Metal packages as such cannot be used in magnetic fields. Food preservation by application of magnetic field is safe to perform. The high-intensity magnetic field exists only within the coil and immediate vicinity; thus, an operator positioned at a reasonable distance from the coil is out of danger.

The OMF process could produce attributes of foods which are fresh by retaining thermo-labile nutrients, reduced energy requirements for processing, inactivation of microorganisms and in some cases, the potential treatment of foods inside flexible packages. Application of OMF technology on commercial scale could not gain commercial testing and acceptance due to cost factor on one end and due to inconsistent results on microbial growth kinetics (Mishra et al., 2011).

12.6 ULTRASOUND PROCESSING

Nowadays, for industrial processing, power ultrasound is considered to be an emerging and promising technology. The use of ultrasound in processing creates novel and interesting methodologies which are often complementary to classical techniques. The uses of ultrasound are broadly classified into two categories. Low energy, diagnostic ultrasound applications involve the use of frequencies higher than 100 kHz at intensities below 1 W/cm². This uses such small power levels that are ultrasonic wave and causes no physical or chemical alterations in the properties of biomaterial through which it passes. It is generally regarded as non-destructive. This technique is successfully used as non-invasive monitoring and non-destructive testing of food processes and product, respectively.

Ultrasound refers to pressure waves with a frequency of 20 kHz or more (Brandum et al., 1998). Generally, ultrasound equipment uses frequency from 20 kHz to 10 MHz high power ultrasound at lower frequencies (20–100 kHz), which is referred to as power ultrasound, has the ability to cause cavitation and has application in food processing to inactivate microbes. In food processing, high-intensity ultrasound at low frequencies, from 20 to 100 kHz, is useful in inactivating microorganisms.

Ultrasound promotes the compression and expansion of the medium particles, resulting in the production of a high amount of energy, when propagated through a biological structure (Butz and Tauscher, 2002). The inactivation of microorganisms is a consequence of cavitation. Cavitation is the formation, growth and collapse of bubbles that generate a localized mechanical and chemical energy (Gogate and Kabadi, 2009; Rastogi, 2011). Mechanical vibrations and acoustic streaming are generated within the liquid, when ultrasound waves pass through a liquid medium.

Ultrasonic technology has a bright future in food industry for bacterial inactivation and also as a processing aid during extractions and filtrations. There are a large number of potential applications of high intensity ultrasound in food processing such as freezing, cutting, drying, tempering, bleaching, sterilization and extraction affected by cavitation phenomena and mass transfer enhancement. The applications can be extended for food physical characterization also to proximate the food composition and measurement of particle sizes. However, the area needs further elaboration with specific emphasis on combination processing such as mano-thermo-sonication. Ultrasound offers several advantages in terms of productivity, yield and selectivity, with better processing time, enhanced quality, reduced chemical and physical hazards and is environmental friendly (Chemat et al., 2011). Ultrasounds can cause quite a number of physiological effects in the microbial cells leading towards cell disintegration facilitating release of matrix components. The cavitation effects as such increases the surface area in contact between the solid and liquid phases causing improved mass transfer and also permeability of cell membrane to ions. By mechanical effects of ultrasound providing an easier penetration of solvent into cellular materials and improved mass transfer, the solvent extraction of organic compounds from vegetable sources is significantly improved. In some cases, sonication increased the efficiency of extraction at lower temperatures producing a purer product in a shorter time (Mason, 1998). A number of other benefits in food processing include improved protein recovery and enzyme extractions. As such, the sonication and the cavitation effect can lead towards better liberation of phenolic compounds and anthocyanins in vegetables.

12.7   PULSED LIGHT PROCESSING

Pulsed light is a non-thermal method of food preservation that involves the use of intense and short duration pulses of broad-spectrum white light. The food material to be treated is exposed to a least one pulse of light having energy density in the range of about 0.01–50 J/cm² at the surface. Pulsed light kills microorganisms using short-time high frequency pulses of an intense broad spectrum, rich UV-C light. The wavelength distribution is adjusted such that at least 70% of the electromagnetic energy within the range from 170 to 2600 nm is used. The food material to be sterilized is exposed to at least one pulse of light (typically 1–20 flashes per s) with a duration range from 1 μs to 0.1 s (Dunn et al., 1991). The antimicrobial effects of pulsed light are well known to be primarily mediated by its UV light component. The latter is absorbed by highly conjugated double-bond systems in bio-molecules, leading to their chemical modification. It is considered that ultraviolet plays a crucial role in the microbial cell inactivation as a major portion of the pulsed light spectrum covers UV light. It was also found that that there is no killing effect if a filter is used to remove ultraviolet (UV) wavelength region lower than 320 nm (Takeshita et al., 2003). The ultraviolet spectrum comprises of three wave ranges:

Long-wave ultraviolet-A (320–400 nm), Medium-wave ultraviolet-B (280–320 nm) and short-wave ultraviolet-C (200–280 nm). Mechanisms that have been proposed to explain the lethality of pulsed light treatment are related to ultraviolet (UV) part of the spectrum which include photochemical and photo thermal effect (Anderson et al., 2000). Mitchell et al. (1992) suggests that the germicidal effect of UV light has been attributed primarily to a photo chemical transformation of pyrimidine bases in the DNA of bacteria, viruses and other pathogens to form dimmers. The formation of such bonds prevents DNA unzipping for replication, and the organism becomes incapable of reproduction. Without sufficient repair mechanisms, such damage results in mutation, impaired replication and gene transcription and ultimately the death of organism (Mc Donald et al., 2002). As in the case of proteins and nucleic acids, UV light is highly effective against conjugated carbon–carbon double bond systems (–C=C–). The biological effects of pulsed light could be either photo-thermal or photochemical effects. In the photo-thermal effects, due to rise in temperature, cell disruption is caused. In the photo-chemical effects, the modification of DNA by generation of dimmers prevents mitosis and protein synthesis ultimately resulting in reduction in microbial load. During the generation of high intensity pulsed light, high energy density electrical storage capacitors are used to generate intense pulses of light. For the generation of light pulses, gas-filled flash lamps are used. These flashes of light are of broad spectrum containing wave lengths from UV to infrared range. The flash lamps are usually filled with xenon or krypton inert gases due to their high efficiency of electrical to optical energy conversion (Dunn et al., 1991). The decontamination effect of pulsed light on several minimally processed vegetables has been studied by several researchers. Log reductions between 0.56 and 2.04 can be achieved with mesophilic, aerobic organisms after treating spinach, celeria, radish, iceberg lettuce, white cabbage, carrots, green bell pepper and soy bean sprouts with 2700 pulses per side at both sides (Gomez-Lopez et al., 2005). The differences in log reductions between samples may be related to differences in resistance of the natural microbial populations affecting each vegetable, the site of microorganisms on and in the samples (shadow effects) and protective substances in vegetables. The disinfected containers are filled with processed commercially sterile vegetables which are subsequently sealed at the top with a sterile lid. The lids may also be disinfected using light pulses. Certain pumpable foods such as juices have also been subjected to processing by light pulses. The food product flows through the chamber surrounding the pulsed light source. The treatment chamber may be designed to include a reflector assembly as an outer wall or external reflection to reflect back the illumination traversing the food product. The most crucial challenge of PL processing is to optimize treatment conditions, while assuring appropriate quality, to extend shelf-life of food products.

12.8 IRRADIATION

Irradiation processing of food involves controlled application of energy from ionizing radiations such as gamma rays, X-rays and electron beam for food preservation. It is a much older technology and is also called as 'cold pasteurization'. The forms of ionizing energy which may be used in food processing include gamma rays (cobalt-60 or cesium-137), X-rays and accelerated electrons (electron beams).

While any food is subjected to irradiation processing, the following aspects are considered:

1. Wholesomeness of irradiated food product for consumption
2. Possibility of included radioactivity
3. Microbiological safety
4. Safety to chemical changes
5. Nutritional adequacy

Food irradiation was first used in United States of America to inactivate human *Trichinella spiralis*, which contaminate pork muscle. Today, more than 40 countries have approved irradiated food for consumption. Although the use of irradiation is spreading worldwide undoubtedly, the negative reactions including lengthy regulatory procedures for granting approval for the irradiated food makes its commercial use marginal (Ehlermann, 2009). Ionizing radiation is used as these radiations are capable of converting atoms and molecules to ions by removing electrons. These are energetic charged particles such as electron, high-energy photons such as X-rays or gamma rays. Absorbed dose is measured in units of kilo greys (kGy) where one grey is the energy absorption of 1 J/kg. Radiation is the treatment with relatively low doses (0.1–8 kGy) to eliminate pathogenic organisms such as tape worm and trichina and microorganisms other than viruses. Radicidation refers to treatment with doses of less than 0.4 kGy. Radurization is the treatment with doses of about 0.4–10 kGy which is sufficient to enhance the keeping quality and shelf life of the food. Radappertization is an irradiation treatment with doses of about 10–50 kGy to bring about virtually complete sterilization. It is also termed as irradiation sterilization or commercial sterility with the resulting product being shelf stable under normal conditions (Table 12.4).

TABLE 12.4 Vegetables and Spices Approved for Radiation Preservation by Ministry of Health and Family Welfare under Food Safety Standards Authority of India. (*Source:* Training manual for food safety regulators, 2010.)

S. no.	Name of foods	Dose of irradiation (kGy)		
		Minimum	Maximum	Overall average
1.	Onions	0.030	0.09	0.06
2.	Potatoes	0.06	0.15	0.10
3.	Ginger, garlic and small onions	0.03	0.15	0.09
4.	Spices	6	14	10

Irradiation finds its wide applications in shelf life extension, sprouting inhibition in bulbs and tubers, disinfestations of stored food products, reduction of the largely unavoidable pathogens that contaminate the food in raw state without any production of toxic compounds as well as no vital loss in nutritional aspects. The treated tubers and bulbs with ionizing energy resulted in remarkable morphological and histological changes in dormant buds which are subsequently modified as deformed buds and subsequently necrosis occurs at growing points during subsequent storage. The extent and area of radiation-induced necrosis in growing points and their adjacent cells varies with irradiation dose (Matsuyama and Umeda, 1983). To inhibit sprouting of potato, yam, artichoke, sweet potato, ginger, sugar beet, table beet, turnip, carrot, onion and garlic treatment of ionizing energy at 0.05–0.15 kGy has been shown. The irradiation doses below 0.15 kGy have minor effects on quality attributes of these vegetables. However, doses above 0.15 kGy may induce undesirable effects such as decreased wound-healing ability, tissue darkening, increased sugar content in potatoes, decreased vitamin content and increased susceptibility to post-harvest pathogens (Matsuyama and Umeda, 1983). Lettuce (*Lactuca sativa*), cabbage (*Brassica oleracea*) and celery (*Apium graveolens*) were artificially contaminated with *Vibrio cholerae* El Tor 01 Inaba and irradiated at 0.50, 0.75, 1.00 kGy and doses of less than 0.75 kGy were sufficient to eliminate an initial contamination of 10^5 cells/g of *V. cholerae* during 7-days storage under refrigeration (5–10°C) (Rubio et al., 2001). Hydration rate of dehydrated vegetables, increasing the yield of juice from grapes without affecting wine making, increasing the rate of drying fruits such as prunes, reducing in cooking time of dried beans, increasing the loaf size of bread and reducing the amount of barely needed in beer production by increasing the yield of malted grain, reducing the flatulence producing propensity of beans are other potential applications. Food irradiation is increasingly being recognized as an effective method for ensuring the hygienic quality of food. Regulatory and safety aspects of the irradiation process, equipment and cost factors, and necessary consumer acceptance contribute to hesitant commercialization of food irradiation. Food irradiation processing requires the same infrastructure as other physical process such as canning, freezing, and drying and irradiation requires less energy. The food and drug administration has issued a final regulation, effective from 18 April 1986, permitting the irradiation of fresh foods, including fruits and vegetables at doses up to 1 kGy (100 krad) for the inhibition of growth and maturation and for insect disinfection (Kader, 1986) (Table 12.5).

TABLE 12.5   Effects of Ionizing Radiation on Fresh Fruits and Vegetables.

Dose (kGy)	Observed effects
0.05–0.15	Sprout inhibition in tuber, bulb and root vegetables, inhibition of growth in asparagus and mushrooms
0.15–0.75	Insect disinfestation
0.25–0.50	Delayed ripening of some tropical fruits such as banana, mango and papaya
> 1.75	Control of post-harvest disease
1.0–3.0	Accelerated softening, development of off flavours in some fruits and vegetables
> 3.0	Excessive softening, abnormal ripening, incidence of some physiological disorders, impaired flavour

The induced activity decays rapidly by a factor of 10 during first 24 h following irradiation. All ionizing radiation produces same chemical changes but differences related to their penetration powers, and hence the dimensions and densities of the food being irradiated (Murray, 1990). The difference in effect on chemical reactions in foods between gamma and electron type of irradiation is smaller than difference observed in biological effects (Haayashi, 1991). By law, most irradiated foods (spices are exempt) must carry the phrase 'treated with radiation' or some comparable wording on their label, with a visual symbol called the 'radura'. ICGFI (International Consultative Group on Food Irradiation), an inter-governmental body with a membership of 46 governments in 1999, has as one of its mandates the function to provide information to Member States of the FAO, WHO, and IAEA and to the three organizations themselves on the safe and proper use of food irradiation technology. Ionizing radiation interacts with the food by transferring energy to electrons which sometimes can leave molecule and forms a positive ion if the radiation is high enough. To induce chemical and biological changes, this ionization is about one thousand times likely. The chemical complexity of foods makes a precise prediction of the changes produced by the irradiation almost as impossible as the prediction of changes produced by heat (Table 12.6).

TABLE 12.6   Radiation. (*Sources:* Foodstuffs are Generally Irradiated with Gamma Radiation, Electronic Beams and X-rays.)

Type of radiation	Source of radiation	Nature of particles	Penetrating power	Safety	Maximum limit to use
γ rays	^{60}Co, ^{137}Ce	Photons	Penetrates to depth of several feet	Source is stored down in a pool of water when not in use	1.25 MeV
Electrons	Electron accelerators	Electrons	Low penetrating power	Generator can be simply switch on or off.	10 MeV
X-rays	Electron beam in a heavy metal target	X-rays	High penetrating power	Requires heavy shielding for safety	5 MeV

12.9   ELECTRON BEAM PROCESSING/ELECTRONIC PASTEURIZATION

Food irradiation can be accomplished using either radioactive isotopes or using accelerated electrons (electron beam/electronic pasteurization).

Electron beam technology is a versatile technology that works on the principle of bombarding objects and surfaces with low energy (<300 keV) electrons to inactivate microorganisms. In other words, electrons are accelerated in a strong electrical field to near the speed of light, and the high energy-electrons act on the surface of the object, which may result in breaking chemical bonds when neutralizing microorganisms or creating bonds between molecules. In order to generate accelerated electrons, linear accelerators employ commercial electricity. These electrons are propelled to just below the speed of light and are used to bombard the food item. The accelerated electrons during E-beam irradiation (electronic pasteurization) damage the nucleic acids by direct 'hits'. In addition, damage to the nucleic acids can also occur when the radiation ionizes an adjacent molecule, which in turn reacts with the genetic material. Water is very often the adjacent molecule that ends up producing a lethal product (Grecz et al., 1983). Electron beam–based pasteurization has a higher dose rate (103–106 Gy/s), which translates to a higher energy deposition. This technology may substitute chemical processes or gamma irradiation with an environmentally sound, sustainable and compact solution. The penetration capacity of the E-beam source does depend on its energy. However, compared with X-rays and gamma rays from radioisotope sources, E-beam has a limited (approximately 3 in.) penetration capacity. To overcome penetration depth issues, most E-beam pasteurization facilities have linear accelerators positioned on the top and bottom of a conveyor belt to maximize penetration. The advantage of this solution is that it can be integrated into any production process, and as it works with low-energy electrons, it requires only simple, local radiation shielding, and production personnel no longer have to work with hazardous materials and technologies.

12.10 COLD PLASMA

Plasma is fourth state of matter that is energetically distinguishable from solids, liquids and gases and can be either thermal or non-thermal. Plasma is a mixture of heavy molecules, atoms, free radicals and ions and generated light from electrons and photons generated by excitation of gas by electric discharges. It is a novel non-thermal food-processing technology that uses energetic, reactive gases to inactivate contaminating microbes on fruits and vegetables. This flexible sanitizing method uses electricity and a carrier gas, such as air, oxygen, nitrogen or helium; antimicrobial chemical agents are not required. The primary modes of action are due to UV light and reactive chemical products of the cold plasma ionization process. A wide array of cold plasma systems that operate at atmospheric pressures or in low-pressure treatment chambers are under development (Niemira, 2011). The inactivation efficiency of plasma on inert surfaces will depend greatly on the equipment design and process conditions such gas type, flow rate and pressure and so forth. For use in sanitizing foods, cold plasma has been evaluated to a limited extent (Mishra et al., 2011).

Cold plasma has found applications in sanitization of the surface of fresh produce, liquid products as well as equipment surface. Cold plasma is thought to be an effective alternative to thermally based microbial inactivation methods having comparable microbial inactivation properties against major pathogens while having little or no effect on product's nutritional or other quality. However, cell density has been shown to affect the efficacy of cold plasma processing with upper cell layers acting as physical barriers to underlying cells in complex bacterial multilayer structure. Montenegro et al. (2002) employed direct current corona discharges for reduction of *E. coli* 0157:H7 in apple juice. The treated apple juice for 40 s at a frequency less than 100 Hz with 4000 pulses of 9000 V peal voltage, the number of cell reduction was more than five log CGU per g. Among food processors, the application of cold plasma is increasingly finding acceptance for surface sterilization and combating the problems of biofilms. Its wider trials on sensitive constituents of foods, namely lipids and vitamins, need to be addressed (Table 12.7).

Reductions of greater than five logs can be obtained for pathogens such as *Salmonella, Escherichia coli* O157:H7, *Listeria monocytogenes* and *Staphylococcus aureus*. Depending on the food treated and the processing conditions, effective treatment times can range from 120 s to as little as 3 s. Key limitations for cold plasma are the relatively early state of technology development, the variety and complexity of the necessary equipment and the largely unexplored impacts of cold plasma treatment on the sensory and nutritional qualities of treated foods. Moreover, depending on the type of cold plasma generated, the antimicrobial modes of action for various cold plasma systems vary. Optimization and scale up to commercial treatment levels require a more complete understanding of these chemical processes. Nevertheless, this area of technology shows promise and is the subject of active research to enhance efficacy.

TABLE 12.7 Response of Microorganisms to Exposure to One Atmosphere Uniform Glow Discharge Plasma in the Parallel Plate Reactor Configuration.

Micro-organisms (concentration)	Surface[a]	Exposure time	Log_{10} reduction	D Values[b]
Bacteria				
Escherichia coli	Polypropylene	30 s	$\geq 10^5$	$D_1 = 6$ s
(7×10^6)				$D_2 = 2$
E. coli	Glass	70 s	$\geq 10^5$	$D_1 = 33$
(1×10^6)				$D_2 = 10$
E. coli	Agar	5 min	$\geq 10^6$	$D_1 = 70$
(8×10^7)				$D_2 = 17$
Staphylococcus aureus	Polypropylene	60 s	$\geq 10^6$	$D_1 = 7$
(1×10^7)				$D_2 = 2$
S. aureus	Filter paper	30 s	$\geq 10^5$	ND[c]
(1×10^6)				
Pseudomonas aeruginosa	Polypropylene	30 s	$\geq 10^6$	ND[c]
(2×10^7)				
Endospores				
Bacillus pumillus spores	Paper strips	3 min	$\geq 10^5$	$D_1 = 1.8$ min
(1×10^6)				$D_2 = 12$ s
B. niger spores	Paper strips	55 min	$\geq 10^5$	$D_1 = 5.5$ min
(1×10^6)				$D_2 = 12$ s
Yeast				
Saccharomyces cerevisiae	Glass slides	5 min	$\geq 10^5$	$D_1 = 3$ min
(1.5×10^6)				$D_2 = 30$ s
Candida albicans	Glass slides	3.5 min	$\geq 10^5$	$D_1 = 2.1$ min
(1.5×10^6)				$D_2 = 30$ s
Viruses				
Bacteriophage Phi X 174	Glass slides	15 min	$\geq 10^6$	$D_1 = 6.8$ min
(2.5×10^7)				$D_2 = 1.8$ min

[a]Polypropylene was used to stimulate microorganism deposited on porous surfaces, and glass slides were representative of nonporous surfaces.
[b]D value is described as the time necessary to reduce the population of cells 1log_{10} or 90%. These values were determined from plots of the number of survivors versus time. Kill curves were biphasic in nature; D values are presented for the first (D_1) and subsequent (D_2) portions of the curves.
[c]Not determined.

12.11 OZONATION

Ozonation emerged as a potential anti-senescence measure having a vital role in lowering the post-harvest physiology of fresh-cut fruits and vegetables (Zhang et al., 2005). Ozone, an approved sanitizing agent by USFDA (2001), owes its origin as a phytosanitary measure. Within the post-harvest handling industry, ozone-based treatments of fresh vegetables and fruits had been used. Owing to its rapid decomposition to oxygen that alleviates concerns about toxic residues, it has attracted much attention for food safety uses. Ozone is one of the most potent sanitizers known and is effective against a wide spectrum of microorganisms at relatively low concentration (Khadre et al., 2001). Interest in ozone has expanded in recent years among consumer as ozone is considered as 'greener' food additives and an environmental friendly technology. Ozone does not form residues upon treatment. Ozone as such is a strong oxidant and a potent disinfecting agent (Khadre et al., 2001). Ultraviolet radiation and corona discharge methods can be used to generate ozone. In the corona discharge method, two electrodes are used which are separated by a ceramic dielectric medium with a narrow discharge gap. A certain fraction of collisions occur when the electrons have sufficient kinetic energy to dissociate the oxygen molecule, and a molecule of ozone can be formed from each oxygen atom. If air is passed through generator as feed gas, 1–3% ozone gas can be obtained which can be enhanced up to 6% yield by use of oxygen as flow gas (Guzel-Seydim et al., 2004a).

The mechanism involved in microbial inactivation by ozone is attributed to its oxidation reactions with cellular components of microorganisms (Guzel-Seydim et al., 2004b). When a cell becomes stressed by viral, bacterial or fungal attack, its energy level is reduced by the outflow of electrons and becomes electropositive. The third atom of oxygen is possessed by ozone, which is electrophlic, that is, ozone has a small free radical electrical charge in the third atoms of oxygen which seeks to balance itself electrically with other material with a corresponding unbalanced charge. Diseased cells, viruses, harmful bacteria and other pathogens carry such a charge and so attract ozone and its by-products. As the normal healthy cells possess a balanced electrical charge and strong enzyme system, they cannot react with ozone or its by-products. Due to its very high oxidation reduction potential, ozone acts as an oxidant of the constituent elements of cell walls before penetrating inside microorganisms and oxidizing certain essential components, for example, unsaturated lipids, enzymes, proteins, nucleic acids and so on. When a large part of the membrane barrier is destroyed causing a leakage of cell contents, the bacterial or protozoan cells lyse (unbind) results in gradual or immediate destruction of the cell. Most of the pathogenic and food-borne microbes are susceptible to this oxidizing effect (Mahapatra et al., 2005).

Relatively few produce handlers and processors have used ozone for water disinfection, surface sanitation, cold room air treatment and other post-harvest applications such as final rinses of whole, trimmed-in-the-field, peeled or minimally processed produce. The use of ozone in the processing of foods has recently come to the forefront as a result of the recent approval by the U.S. Food and Drug Administration approving the use of ozone as an anti-microbial agent for food treatment, storage and processing. The fact that ozone is 52% stronger than chlorine and has been shown to be effective over a much wider spectrum of microorganisms than chlorine and other disinfectants is the potential utility of ozone to the food industry. Complementing the effectiveness, is the fact that ozone, unlike other disinfectants, leaves no chemical residual and degrades to molecular oxygen upon reaction or natural degradation. It is both an asset and a liability that the fact that ozone has a relatively short half-life. This is particularly true in treatment of drinking water where ozonation is employed to enhance filtration and provide primary disinfection but requires the addition of chlorine as the terminal disinfectant to maintain a residual in the distribution system. Skog and Chu (2001) reported that ozone ($0.4 \ \mu l \cdot l^{-1}$) improved the quality of broccoli and cucumber stored at 3°C, prolonging the storage life.

12.12 CONCLUSION

The latest application in food preservation is the non-thermal processing. In view of growing demand of heat-sensitive bioactive compounds, there is an urgent need of popularizing many non-thermal technologies for commercial usage. Furthermore, consumers demand the foods that are lightly processed, additive free, natural and fresh. This has prompted the food-processing industries to select the milder preservation techniques including high hydrostatic processing, high pressure carbondioxide, pulsed electric fields, irradiation, cold plasma, oscillating magnetic fields, pulsed lights, ultra sounds, ozonation and so on, which ensures the fresh-like flavour and texture of foods with high retention of vitamins and nutrients. Food scientists are fascinated by the never ending fight against food spoilage organisms and food pathogens and are often encouraged by promising results of in vitro experiments. These newly emerging technologies aids in not only preserving foods but also improves the rheological and functional properties of foods. But still the commercialization of these techniques is limited due to high-cost capitalization, difficulties in fabricating vessels to withstand high pressures, high frequencies and so on. Lack of consumer awareness about irradiation makes the irradiated foods confined to research works. The risk of dielectric breakdown limits pulsed electric field processing only to liquid foods. However, in order to overcome these difficulties, further researches and efforts are being made. Food engineering has a challenging task ahead in making these technologies and equipments available for introduction in the developing countries. However, to meet this high expectation, consumers and stakeholders must be convinced about the improvements of these new technologies. This will require convincing data and provision of clear, objective and unbiased information including the potentially negative aspects of the technology and their limitations. The equipment's required for non-thermal processing must be cost effective so that treated processed products would be cheaper.

REFERENCES

Anderson, J. G.; Rowan, N. J.; MacGregor, S. J.; Fouracre, R. A.; Farish, O. Inactivation of Food Borne Enteropathogenic Bacteria and Spoilage Fungi Using Pulsed-Light. *IEEE Trans. Plasma Sci.* **2000,** *28*(1), 83–88.

Asaka, M.; Hayashi, R. Activation of Polyphenol Oxidase in Pear Fruits by High Pressure Treatment. *Agric. Biol. Chem.* **1991,** *55,* 2439–2440.

Brandum, J.; Egebo, M.; Agerskov, G.; Busk, H. Online Pork Carcass Grading with Auto form Ultrasound System. *J. Anim. Sci.* **1998,** *76,* 1859–1868.

Butz, P.; Edenharder, R.; Fernandez, G. A.; Fister, H.; Merkel, C.; Tauscher, B. Changes in Functional Properties of Vegetables Induced by High Pressure Treatment. *Food Res. Int.* **2002,** *35,* 295–300.

Chang, J. C. H.; Ossoff, S. F.; Lobe, D. C.; Dorfman, M. H.; Dumais, C. M.; Qualis, R.G.; Johnson, J. D. UV Inactivation of Pathogenic and Indicator Microorganisms. *Appl. Environ. Microbiol.* **1985,** *49,* 1361–1365.

Chemat, F.; Khan, M. K. Applications of Ultrasound in Food Technology: Processing, Preservation and Extraction. *Ultrason. Sonochem.* **2011,** *18*(4), 813–835.

Damar, S.; Balaban, M. O. Review of Dense Phase CO_2 Technology: Microbial and Enzyme Inactivation, and Effects on Food Quality. *J. Food Sci.* **2006,** *71,* 1–11.

Devlieghere, F.; Vermeiren, L.; Debevere, J. New Preservation Technologies: Possibilities and Limitations. *Int. Dairy J.* **2004,** *14,* 273–285.

Dunn, J.; Clark, R. W.; Asmus, J. F.; Pearlman, J. S.; Boyer, K.; Pairchaud, F.; Hofmann, G. Methods for Preservation of Foodstuffs. Maxwell Laboratories, Inc., U.S. Patent **1991,** *5,* 235–240.

Ehlermann, D. A. The RADURA-Terminology and Food Irradiation. *Food Control* **2009,** *20*(5), 526–528.

Estrada-Girón, Y.; Swanson, B. G.; Barbosa-Canovas, G. V. Advances in the Use of High Hydrostatic Pressure for Processing Cereal Grains and Legumes. *Trends Food Sci. Technol.* **2005,** *16,* 194–203.

Fuchigami, M.; Teramoto, A. Texture and Structure of High Pressure Frozen Gellan Gum Gel. *Food Hydrocolloids* **2003a,** *17,* 895–899.

Fuchigami, M.; Kato, N.; Teramoto, A. High Pressure Freezing Effects on Textural Quality of Carrots. *J. Food Sci.* **1997b,** *62,* 804–808.

Fuchigami, M.; Kato, N.; Teramoto, A. High Pressure Freezing Effects on Textural Quality of Chinese Cabbage. *J. Food Sci.* **1998,** *63,* 122–25.

Gomez-Lopez, V. M.; Devlieghere, F.; Bonduelle, V.; Debevere, J. Factors Affecting the Inactivation of Microorganisms by Intense Light Pulses. *J. Appl. Microbiol.* **2005,** *99,* 460–470.

Grecz, N.; Rowley, D. B.; Matsuyama, A. The Action of Radiation on Bacteria and Viruses. In *Preservation of Foods by Ionizing Radiation;* CRC Press: Boca Raton, FL, 1983; Vol. 2.

Guzel-Seydim, Z.; Bever, P.; Greene, A. K. Efficacy of Ozone to Reduce Bacterial Populations in the Presence of Food Components. *J. Food Microbiol.* **2004a,** *21,* 475–479.

Guzel-Seydim, Z. B.; Greene, A. K.; Seydim, A. C. Use of Ozone in Food Industry. *LWT-Food Sci. Technol.* **2004b,** *37,* 453–460.

Haayashi, T. Comparative Effectiveness of Gamma Rays and Electron Beams in Food Irradiation. In *Food Irradiation;* Throne, S., Ed.; Elsevier Applied Science: New York, 1991; pp 169–206.

Hayashi, R.; Kawamura, Y.; Nakasa, T.; Okinaka, O. Application of High Pressure to Food Processing: Pressurization of Egg White and Yolk and Properties of Gel Formed. *Agric. Biol. Chem.* **1989,** *53,* 2935–2939.

Hite, B. H. The Effect of Pressure in the Preservation of Milk. *W. Va. Agric. Exp. Stn. Bull.* **1899,** *58,* 15–35.

Hofmann, G. A. Deactivation of Microorganisms by an Oscillating Magnetic Field. U.S. Patent 4,524,079, 1985.

Jay, J. M. *Modern Food Microbiology;* Chapman Hall: New York, 1996.

Jeyamkondan, S.; Jayas, D. S.; Holley, R. A. Pulsed Electric Field Processing of Foods: A Review. *J. Food Prot.* **1999,** *62,* 1088–1096.

Kader, A. A. Potential Applications of Ionizing Radiation in Postharvest Handling of Fresh Fruits and Vegetables. *Food Technol.* **1986,** *40,* 117–121.

Khadre, M.; Yousef, A.; Kim, J. G. Microbiological Aspects of Ozone Applications in Food: A Review. *J. Food Sci.* **2001,** *66,* 1242–1252.

Knorr, D. Hydrostatic Pressure Treatment of Food: Microbiology. In: *New Methods of Food Preservation*; Gould, G.W., Ed.; Blackie Academic & Professional Chapman & Hall: 1995; pp 159–175. Bishopbriggs.

Knorr, D.; Geulen, M.; Grahl, T.; Sitzmann, W. Food Application of High Electric Field Pulses. *Trends Food Sci. Technol.* **1994,** *5,* 71–75.

Knorr, D.; Angersbach, A.; Eshtiaghi, M. N.; Heinzn, V.; Lee, D. U. Processing Concepts Based on High Intensity Electric Field Pulses. *Trends Food Sci. Technol.* **2001,** *12,* 129–135.

Mahapatra, A. K.; Muthukumarappan, K.; Julson, J. L. Applications of Ozone, Bacteriocins and Irradiation in Food Processing: A Review. *Crit. Rev. Food Sci. Nutr.* **2005,** *45*(6), 447–461.

Martin, O.; Vega-Mercada, H.; Qin, B. L.; Chang, F. J.; Barbosa-Canovas, G. V.; Swanson, B. G. Inactivation of *E. coli* Suspended in Liquid Eggs Using Pulsed Electric Fields. *J. Food Process. Preserv.* **1997,** *21,* 193.

Mason, T. J. Power Ultrasound in Food Processing—The Way Forward. In *Ultrasound in Food Processing;* Povey, M. J. W., Mason, T. J., Eds.; Blackie Academic and Professional: London, 1998; pp 103−126.

Matsuyama, A.; Umeda, K. Sprout Inhibition in Tubers and Bulbs. In *Preservation of Food by Ionizing Radiation;* Josephson, E. S., Peterson, M. S., Eds.; CRC Press: Boca Raton, FL. 1983; Vol. 3, p 159.

Mc-Donald, K. F.; Curry, R. D.; Hancock, P. J. Comparison of Pulsed and CW Ultraviolet Light Sources to Inactivate Bacterial Spores on Surfaces. *IEEE Trans. Plasma Sci.* **2002,** *30,* 1986–1989.

McLnerney, J. K.; Seccafien, C. A.; Stewart, C. M.; Bird, A. R. Effects of High Pressure Processing on Antioxidant Activity, and Total Carotenoid Content and Availability, in Vegetables. *Innovative Food Sci. Emerging Technol.* **2007,** *8,* 543–548.

Mishra, N. N.; Kadam, S. U.; Pankaj, S. K. An Overview of Non-Thermal Technologies in Food Processing. *Indian Food Ind.* **2011,** *30,* 44–51.

Mitchell, D. L.; Jen, J.; Cleaver, J. E. Sequence Specificity of Cyclobutane Pyrimidine Dimers in DNA Treated with Solar (Ultraviolet B) Radiation. *Nucleic Acids Res.* **1992,** *20;* 225–229.

Montenegro, J.; Ruan, R.; Ma, H.; Chen, P. Inactivation of E. coli 0157: H7 Using a Pulsed Non-Thermal Plasma System. *J. Food Sci.* **2002,** *67,* 646–648.

Niemira, B. A. Cold Plasma Decontamination of Foods. *Annu. Rev. Food Sci. Tech.* **2011,** *3,* 125–142.

Otero, L.; Solas, M. T.; Sanz, P. D.; Elvira, C.; Carrasco, J. A. Contrasting Effects of High Pressure Assisted Freezing and Conventional Air Freezing on Egg Plant Tissue Microstructure. *Food Res. Technol.* **1998,** *206*(5), 338–342.

Pothakamury, U. R.; Barbosa Canovas, G. V.; Swanson, B. G. Magnetic Field Inactivation of Microorganisms and Generation of Biological Changes. *Food Technol.* **1993,** *47*(12), 85–93.

Prestamo, G.; Palomares, L.; Sanz, P. Broccoli (*Brasica oleracea*) Treated Under Pressure-Shift Freezing Process. *Eur. Food Res. Technol.* **2004,** *219*(6), 598–604.

Qiu, W.; Jiang, H.; Wang, H.; Gao, Y. Effect of High Hydrostatic Pressure on Lycopene Stability. *Food Chem.* **2006,** *97*(3), 516–523.

Rubio, T.; Araya, E.; Avendaño, S.; Lopez, L.; Espinoza, J.; Vargas, M. Effect of Ionizing Radiation on Fresh Vegetables Artificially Contaminated with Vibrio Cholerae. In *Irradiation to Control Vibrio Infection from Consumption of Raw Seafood and Fresh Produce*; International Atomic Energy Agency, Vienna, Austria. 2001, 23–30.

Schluter, O.; Benet, G. U.; Heinz, V.; Knorr, D. Metastable States of Water and Ice During Pressure-Supported Freezing of Potato Tissue. *Biotechnol Prog.* **2004,** *20*(3), 799–810.

Shayanfar, S.; Chauhan, O. P.; Toepfl, S.; Heinz, V. The Interaction of Pulsed Electric Fields and Texturizing Anti-Freezing Agents in Quality Retention of Defrosted Potato Strips. *Int. J. Food Sci. Technol.* **2013,** *48,* 1289–1295.

Skog, C. L.; Chu, L. J. Effect of Ozone on Qualities of Fruits and Vegetables in Cold Storage. *Can. J. Plant Sci.* **2001,** *81*(4), 773–778.

Spilimbergo, S. A Study about the Effect of Dense CO_2 on Microorganisms. Ph.D. Thesis, University of Padova, Italy, 2002

Spilimbergo, S.; Bertucco, A.; Lauro, F. M.; Bertoloni, G. Inactivation of Bacillus Subtilis Spores by Supercritical CO_2 Treatment. *Innovative Food Sci. Emerging Technol.* **2003,** *4,* 161–165.

Takeshita, K.; Shibato, J.; Sameshima, T.; Fukunaga, S.; Isobe, S.; Arihara, K.; Itoh, M. Damage of Yeast Cells Induced by Pulsed Light Irradiation. *Int. J. Food Microbiol.* **2003,** *85*(1), 151–158.

Training Manual for Food Safety Regulators. In *Emerging Issues in Food Processing. Food Safety and Standards Authority of India*; 2010; pp 108–129.

Unni, L. E.; Chauhan, O. P.; Raju, P. S.; Bawa, A. S. High Pressure Processing of Foods: Present Status and Future Strategies. *Int. J. Food Ferment. Technol.* **2011,** *1,* 49–62.

Unni, L. E.; Chauhan, O. P.; Raju, P. S. High Pressure Processing of Garlic Paste: Effect on the Quality Attributes. *Int. J. Food Sci. Technol.* **2013,** *49*(6), 1579–1585.

USFDA. *Analysis and Evaluation of Preventative Control Measures for the Control and Reduction/Elimination of Microbial Hazards on Fresh and Fresh-Cut Produce.* US Food and Drug Administration Center for Food Safety and Applied Nutrition, 2001.

Wekhof, A. Pharma-Food, January, Huttig Verlag Heidelberg Germany. *Escherichia coli. Water Res.* **2000,** *39,* 2921–2925.

Zhang, L.; Lu, Z.; Yu, Z.; Gao, X. Preservation of Fresh-Cut Celery by Treatment of Ozonated Water. *Food Control* **2005,** *16*(3), 279–283.

CHAPTER 13

ENCAPSULATION OF BIOACTIVE COMPOUNDS FROM VEGETABLES FOR BETTER NUTRACEUTICAL DELIVERY

K. NARSAIAH, REKHA RAWAT, and SHEETAL BHADWAL

AS & EC Division, Central Institute of Post-Harvest Engineering and Technology (CIPHET), Ludhiana141004, Punjab, India

CONTENTS

13.1   INTRODUCTION

The positive correlation between diet and health has increased consumer demand for more information related to healthy diets, including fruits and vegetables, with functional characteristics that help to delay the aging processes and reduce the risk of various diseases, mainly cardiovascular diseases and cancer, as well as other disorders. Vegetables provide a major source of healthy diet because of their highly bioactive secondary metabo-lites that could represent useful leads in the development of new func-tional ingredient (Singh et al., 2007). These bioactive compounds include phenolic acids, flavones, flavonols, organosulphurs and others. The types and levels of bioactive component vary distinctly between different species of vegetables and even cultivar. In addition, climatic, agronomic and harvest conditions, in case of vegetables, also significantly influence the levels of these phytochemicals (Tiwari and Cummins, 2011).

Bioactive compounds are generally highly susceptible to environ-mental processing and gastrointestinal conditions, and therefore subject to rapid inactivation or degradation. Protected coating or encapsulation would be an effective means of delivering stable forms of bioactive compounds as it slows down the degradation processes or prevents degra-dation until the product is delivered at the required sites (McClements and Lesmes, 2009). To protect the core material from adverse environmental conditions, such as undesirable effects of light, moisture and oxygen, encapsulation is a process of confining active compounds within a matrix or membrane in particulate form, thereby contributing to an increase in the shelf life of the product. In addition, encapsulation also promotes a controlled liberation of the active agent and improves final product quali-ties (Chan et al., 2009). Controlled-delivery could enhance bioavailability of an active compound by customizing the release mechanism or rate in gastro-intestinal tract. Therefore, encapsulation is an effective tool for improving the delivery of bioactive compounds and to develop the value-added products.

Various methods are employed for the encapsulation of bioactive compounds such as spray-cooling/chilling, spray-drying, liposome entrap-ment, fluidized bed coating, coacervation and so on. As every bioactive compound has its own characteristic molecular structure, however, none of them could be considered as a universally applicable method. They show great variation in terms of molecular weight, polarity, solubility and others which implies that different encapsulation procedures should be applied to meet the particular physicochemical and molecular require-ments for a specific bioactive compound (Augustin and Hemar, 2009). Therefore, the present review focuses on various approaches used for the encapsulation of bioactive molecules from vegetables for better nutraceu-tical delivery.

13.2 TYPES OF BIOACTIVE COMPOUNDS

The most commonly applied bioactive compounds of vegetables that are already encapsulated in industrial applications are polyphenols, carotenoids and organosulphur. These bioactive compounds are found in various vegetables such as sweet potato, red-leafed lettuce, cruciferous vegetables, red capsicum, carrot, cauliflower, tomato, onion, garlic and others.

13.2.1 POLYPHENOLS

Polyphenols comprise one of the most abundant and ubiquitous groups of plant metabolites and are vital part of both human and animal diets. They represent a wide group of compounds, which are classified according to the features of their chemical structures (the number of phenol rings and the structural elements that connect these rings). Polyphenols include phenolic acids (hydroxybenzoic acids and hydroxycinnamic acids), flavonoids (flavonols, flavones, isoflavones, flavanones, anthocyanidins and flavanols), stilbenes and lignans (Manach et al., 2004). They have a high spectrum of biological activities such as antioxidant, anti-inflammatory, antibacterial and antiviral functions (Bennick, 2002). Reports suggest that plant polyphenols can slow the progression of certain cancers, decrease the risks of neurodegenerative disease, cardiovascular diseases, diabetes or osteoporosis, suggesting that plant polyphenols might act as potential chemopreventive and anti-cancer agents in humans (Arts and Hollman, 2005). These compounds are widely used in food, pharmaceutical, cosmetics and active packaging applications. However, most plant-derived phenolic compounds are highly volatile, unstable and sensitive to oxygen, light and heat during processing, utilization and storage. Consequently, the delivery of these compounds needs product formulators and manufacturers to provide protective mechanisms that can maintain the active molecular form until the time of consumption and delivers this form to the physiological target within the organism (Chen et al., 2006). The unpleasant taste of polyphenols is another drawback, and it needs to be masked before incorporation into food products. The exploitation of encapsulated polyphenols instead of free compounds can overcome the drawbacks of their instability, alleviate unpleasant tastes or flavours as well as improve the bioavailability and half-life of the compound (Fang and Bhandari, 2010).

13.2.2 CAROTENOIDS

Carotenoids, mainly found in fruits and vegetables that typically have 40-carbon molecules and multiple conjugated double bonds, are a class of natural pigments (Qian et al., 2012).These are the most common pigments found in nature and play an important role in human diet as precursors of vitamin A and as antioxidants. Among different types of carotenoids, β-carotene, lycopene, lutein and zeaxanthin are the most abundant pigments. As carotenoids are well known active ingredients, they are widely used in the food, cosmetic and pharmaceutical industries as natural colorants (Martin et al., 2007). However, they are highly susceptible to degradation due to chemical, mechanical and thermal stresses encountered during food processing and storage. In addition, they are insoluble in water due to their hydrophobic nature which makes it difficult to disperse them in water. It is thus necessary to develop effective delivery systems to improve the bioavailability and stability of carotenoids in foods (Silva et al., 2011).

13.2.3 ORGANOSULPHURS

Organosulphurs are another class of bioactive components found in vegetables. Organosulphurs include allicin, glucosinolates, allyl methyl-methylallyl and trans-1 propenyl-thiosulfinate, S-allyl-L-cysteine, diallyldisulfide, diallylsulfide, diallyltrisulfide and so on. In the intact vegetable, allicin is not present as such but is produced in the presence of the enzyme, alliinase during cutting or crushing (Williams et al., 2013). Allicin is chemically unstable and rapidly breaks down to give the unsaturated disulphide, ajoene. Glucosinolates are sulphur-containing natural plant products found in Brassica vegetables. They occur in the plant in conjunction with the hydrolytic enzyme, myrosinase which breakdown glucosinolates into various by-products such as isothiocyanate and sulphoraphane. Organosulphurs have received much attention for its perceived anti-carcinogenic activity, anti-obesogenic, anticoagulant and fibrinolytic activity (Santhosha et al., 2013).

13.3 TYPES OF WALL MATERIALS INVOLVED IN ENCAPSULATION

Wall material is used to entrap the active component and leads to the formation of microcapsule of small size with diameter of about 5–300 μm (Gibbs et al., 1999). Structurally, microcapsule is of different types depending upon the core material and distribution of shell molecules. It may contain continuous core region having bioactive ingredients with a continuous shell, or it may have irregular geometry with small particles of core material/bioactive ingredient dispersed throughout the matrix of shell/wall material (Vilstru, 2001). It is desirable that the wall material should not influence the sensory properties, colour, texture or flavour of food products, and it must remain stable over the range of environmental conditions. It affects the size, shape and stability of microcapsules as well as encapsulation efficiency. From polymers, such as sugars, gums, proteins, natural and modified polysaccharides, lipids and synthetic polymers, shell materials for encapsulation of bioactive components are often selected (Table 13.1). A new alternative employed to increase the protection of active ingredient is the use of blends of polymers to form shell walls. The modifiers which are used in combination with wall material to achieve the desired encapsulation are surfactants, plasticizers, anti-oxidants, oxygen scavengers, dyes, pigments and ultraviolet (UV) absorbers.

TABLE 13.1 Food Grade Wall Material Applied for Microencapsulation.

Wall materials	Examples
Polysaccharides	Gum arabic, modified starches, hydrolysed starches, alginates, pectin, carrageenan, chitosan, cellulosic materials, maltodextrin, xanthan gum, β-cyclodextrin, inulin
Fats and waxes	Hydrogenated vegetable oil, bees wax, phosphotidyl choline, carnauba wax, candelilla wax, glycerides
Proteins	Gelatin, whey proteins, sodium caseinate, gluten, vegetable protein, animal protein
Other materials	Polyvinylpyrrolidone, paraffin, shellac and inorganic materials

The functional performance of the microcapsulation depends on the morphology, the chemical nature and the surface characteristics of the polymeric shell influenced by the process parameters (Yadav et al., 2007).

Therefore, the choice of coating material is determined on the basis of objective of microencapsulation process (taste masking, enteric protection, time-release) and the desired barrier properties. In general, a hydrophobic core is usually protected by a hydrophilic shell, and vice versa. By selection of the appropriate coating material, it is possible to design effective microencapsulation systems which can deliver the active component in the appropriate place within the human body after consumption, for example, the oral cavity for flavour molecules or the gastro-intestinal tract for vitamins and nutraceuticals.

Each type of carrier has advantages and disadvantages in terms of properties, cost and encapsulation efficiency. Due to their low toxicity, low immunogenicity and chemical stability, alginate gels formed in the presence of Ca^{2+} are commonly utilized as a physical barrier for microencapsulation, but its major drawback is its high cost. Pectins are health-promoting soluble fibres and frequently used to enhance viscosity and gel strength of food products (Thakur et al., 1997). Polysaccharides such as starch, modified starches have good ability to absorb volatiles from the environment which makes it suitable candidate for encapsulation of flavour. Gum arabic is another generally used wall material because of its viscosity, solubility and emulsification characteristics. For entrapping flavours, oils at ambient temperatures, they can then be used. The cheapness of inulin with many health benefits makes it a good matrix molecule for capsules (de Vos et al., 2010). Protein based wall materials have been used with success in the microencapsulation of hydrolysed casein, essential oil and vitamins. However, their solubilities in cold water, reactivity with carbonyls and high cost limit their application in food industry. Lipids are also gaining interest because of their suitability for several administration routes. In addition, they are easily digested in the intestine by lipases, which cause the active ingredients to be released in the vicinities of their site of action (Favaro-Trindade et al., 2011).

At present, the use of novel wall materials such as milk protein, fruit and vegetable extract, having nutraceutical/functional properties, is in trend for the encapsulation of bioactive components. Milk proteins are natural vehicles of bioactive compounds and demonstrate the excellent surface and self-assembly properties, gelation and pH responsive gel swelling and so on. These properties are helpful in controlled release, biocompatibility biodegradability and bioavailbility (Livney, 2010). Proteins have functional properties suitable for microencapsulation, such as solubility, water and fat absorption, emulsion stabilization, gelation, foaming, plus good film-forming and organoleptic properties. The information of protein physical–chemistry properties would facilitate realization of the potential role of food proteins as substrate for the development of nutraceutical delivery systems in the form of hydrogel, micro- or nano-particles (Chen et al., 2006).Whey protein microbeads, manufactured using a cold-set gelation process, have been used to encapsulate riboflavin and are well suited to act as sorbents for encapsulation (O'Neill et al., 2014).

13.4   TYPES OF ENCAPSULATION PROCEDURES

There are various encapsulation procedures that can be used for the encapsulation of bioactive compounds. However, the most commonly applied technologies are emulsification, coacervation, spray-drying, spray-cooling, freeze-drying, nanoparticles, inclusion technologies and so on.

13.4.1   SPRAY-DRYING

Spray-drying, a well-known technology in the food industry, is at the present the most commonly used microencapsulation method for bioactive compounds (Vos et al., 2010). , When adequately performed, it is a fast and relatively cheap method that is highly reproducible. The principle of spray-drying is dissolving the core in a dispersion of a chosen matrix material which is subsequently atomized in heated air. This promotes fast removal of the solvent, and the powdered particles are then separated from the drying air at the outlet at a lower temperature. The relative ease and also the low cost are the main reasons for the broad application of spray-drying in industrial settings. Ahmed et al. (2010) encapsulated flours from purple-fleshed sweet potato by spray-drying using combinations of various levels of ascorbic acid and maltodextrin and evaluated their effects on bioactive components. They observed higher total phenolic content, antioxidant capacity and water solubility index in encapsulated flours than non-encapsulated flour.

Guadarrama-Lezama et al. (2012) encapsulated capsicum extract containing carotenoid and polyphenolic compound content. Spray-drying with a gum arabic and maltodextrin biopolymers encapsulated the extracts. Similarly, Xue et al. (2013) used spray-drying method to encapsulate the tomato lycopene with zein as coating material. The results of controlled release showed that zein particles could protect most of the lycopene from being released in the stomach. 5-methyltetrahydrofolic acid (5-MTHF) is an alternative to folic acid but fortification of foods with 5-MTHF is considered problematic, because it is highly sensitive to normal food-processing operations. Microencapsulation of 5-MTHF was performed by Shrestha et al. (2012) using pectin and sodium alginate by spray-drying. They studied thermal degradation of 5-MTHF at different temperatures and found better stability of encapsulated 5-MTHF as compared with the free form. Encapsulation of polyphenols by spray-drying has also been reported in the following cases: anthocyanins from black carrots with maltodextrins (Ersus and Yurdagel, 2007), bioactive compounds from red pepper by-products with gum arabic as wall material (Romo-Hualde et al., 2012) and eugenol rich clove extract in solid lipid carriers (Cortés-Rojas et al., 2014).

13.4.2 SPRAY-CHILLING

Spray-chilling (or spray-congealing) is the process of atomizing a molten liquid and cooling the resulting droplets to form prills or powders that are solid at room temperature. Commonly spray-chilled carriers for encapsulation include fats, waxes, polyethylene glycols, fatty acids and fatty alcohols (Kjaergaard, 2001). Thus far, it is only applied for dry products to conserve enzymes, flavours, minerals and proteins.

13.4.3 FREEZE-DRYING

Freeze-drying in combination with matrix molecules has been proposed as an alternative for spray-drying of heat sensitive bacteria (Augustin and Hemar, 2009). However, freeze-drying technique is less attractive than other techniques because the operating costs of freeze-drying are 50 times higher than spray-drying, and the storage and transport of particles produced is extremely expensive; the commercial applicability is also severely restricted by the long processing time (Madene et al., 2006). Lim et al. (2014) encapsulated lutein and β-carotene in freeze-dried emulsions with layered interface and trehalose as glass former and observed better protection towards carotenoid losses during storage.

13.4.4   EMULSIFICATION

The process of dispersing one liquid in a second immiscible liquid is defined as emulsification. By incorporating the core material in the first liquid, bioactive component could be encapsulated. Addition of a surfactant induces encapsulation by forming vesicles, micelles, reverse micelles and bilayers around the bioactive molecules. To enhance packing efficacy, bulk emulsification approaches are also applied in some cases. It is entrapping the bioactive molecules in fat-droplets or water–oil–water emulsions. Simple as well as very complex emulsions could be produced depending on the type of molecules used for forming the emulsion (Augustin and Hemar 2009). In a recent study carried out by Paz et al., 2014, water-soluble β-carotene formulations was developed by high temperature, high-pressure emulsification method to protect the active compound from degradation and overcome the low bioavailability due to a low solubility in aqueous media. Chitosan microbeads were prepared by emulsion technique and loaded with thyme polyphenols and observed that release of encapsulated polyphenols in simulated gastrointestinal fluids was prolonged to 3 h (Trifkovic et al., 2014).

13.4.5   COACERVATION

Coacervation is a phase-separation process for the production of insoluble complexes, based on the simultaneous dissolution of polyelectrolytes with opposite charges, induced by changes in the medium. Normally, the two biopolymers used include a protein molecule and a polysaccharide molecule (Jun-xia et al., 2011). For compounds which are sensitive to high temperatures and to certain organic solvents, this process is an alternative to microencapsulation. It is normally used to encapsulate solid or liquid ingredients that are insoluble in water (Gouin, 2004). Coacervation encapsulation can be achieved simply with only one colloidal solute, such as gelatin, or through a more complex process, such as with gelatine and

gum acacia, soy protein, polyvinyl alcohol, gelatin/carboxymethylcellu-
lose, β-lactoglobulin/gum acacia and guar gum/dextran. Xiao-Ying et al.
(2011) encapsulated lutein using the complex coacervation method and
found an improvement in stability against light, humidity and temperature.

13.4.6 COAXIAL AIR FLOW NOZZLES

Coaxial-air-flow bead generator is used to produce small alginate microcapsules in a controllable manner. The basic principle of the instrument is
the use of a coaxial pressurized air stream which facilitates the pulling of
droplets from a needle tip into the gelling bath. With the help of this technique, it is possible to produce small-sized capsules with diameter of around
50 μm or more. In pursuit of simple and cost-effective microencapsulation system, microencapsulator with multiple air jet droplet generatorfor
microencapsulation of sensitive functional ingredients for incorporation in
food products was developed at CIPHET, Ludhiana (Narsaiah and Oberoi,
2011). In addition, for production of microcapsules, they also developed
an autoclavable microencapsulation system with multistage break up
two-fluid nozzle (Narsaiah et al., 2011).Mandge et al., (2012) prepared
microcparticles of bioactive components through two-fluid glass nozzle.
Similarly, to encapsulate nisin with sodium alginate (Narsaiah et al., 2012)
and to form hybrid hydrogel matrix, this two-fluid nozzle was also used
(Wilson et al., 2012).

13.4.7 LIPOSOMES

Liposomes are artificial microscopic vesicles with an aqueous core enclosed
in one or more phospholipid bilayers (Sharma and Sharma, 1997). Its
structure is similar to cell membrane with hydrophilicity and lipophilicity
and suitable for use as the carrier for one or more active pharmaceutical
ingredients. Liposomes can be utilized in the entrapment, delivery and
release of water soluble, lipid-soluble and amphiphilic materials owing to
the possession of both lipid and aqueous phases. They can transport drugs
for site-specific delivery to enhance bioavailability and efficacy, sustains
low-release, reduce adverse reactions and improve patient compliance
(Date et al., 2007). In addition, liposomes have been extensively used

as an adjuvant for vaccine antigens against bacterial and viral infections and tumours (Sinha and Khullar, 1997). A number of reports suggested that liposomes enhance the bioactivity and bioavailability of bioactive compounds. Curcumin is the principle curcuminoid of the turmeric, which exhibits anti-HIV, antitumour, antioxidant, and anti-inflammatory activities. However, on account of the low water solubility of curcumin and low stability against gastrointestinal fluids and alkaline pH conditions, it is poorly absorbed from the gastrointestinal tract after oral administration. To enhance the bioavailability and food functionality of curcumin, liposome-encapsulated curcumin can be prepared (Takahashi et al., 2009). Narsaiah et al. (2013) encapsulated pediocin in hybrid capsules of alginate plus guar gum incorporated with nanoliposomes and found it more effective in inhibiting bacterial growth as compared with free pediocin.

13.4.8   NANOPARTICLES

Nanoparticles have attracted increasing attention as an efficient and non-toxic alternative lipophilic colloidal drug carrier prepared either with physiological lipids or lipid molecules used as common pharmaceutical excipients. Recently, Luo et al. (2013) prepared zein and zein/carboxymethyl chitosan nanoparticles for encapsulation of two bioactive compounds, namely indole-3-carbinol (I3C) and 3,30-diindolylmethane (DIM) from cruciferous vegetables. Controlled release of I3C and DIM nanoparticle formulations provided and demonstrated protection against ultraviolet light. Woranuch and Yoksan (2013) reported an increase in thermal stability of eugenol by encapsulating active ingredient into chitosan nanoparticles through an emulsion–ionic gelation cross-linking method. Eight-fold higher thermal stability and 2.7-fold greater radical scavenging activity than that containing naked eugenol was shown by encapsulated eugenol. Similarly, Choi et al. (2009) also revealed that encapsulation of eugenol into polycaprolactone nanoparticles could enhance its stability against light oxidation. Coradini et al. (2014) encapsulated curcumin in lipid-core nanocapsules and reported that the in vitro antioxidant activity of polyphenols against hydroxyl radicals was enhanced by nanoencapsulation. To inhibit lipid peroxidation and DNA damage as well as to improve endogenous antioxidants defences, curcumin has the ability to scavenge free radicals.

13.4.9   INCLUSION ENCAPSULATION

Molecular inclusion is generally obtained by using cyclodextrins (CDs) as the encapsulating agents. The exterior part of the cyclodextrin molecules is hydrophilic, whereas the interior part is hydrophobic. Through a hydrophobic interaction, this structure characteristic makes CDs a satisfactory medium for encapsulation of less polar molecules into the apolar internal cavity (Bhandari et al., 1999). One excellent advantage of the inclusion of polyphenols in CDs is the improvement of their water solubility, especially for the less water-soluble phytochemicals. Yao et al. (2014) developed a myricetin/hydroxypropyl-β-CD inclusion complex and reported that the oral bioavailability of the myricetin/HP-β-CD inclusion complex in rats was effectively increased 9.4-fold over free myricetin, and its antioxidant activity was also improved. The improved antioxidant efficacy of the inclusion complex may come from the protection of the compound against rapid oxidation by free radicals (Mercader-Ros et al., 2010), which may in part be explained by an increase in their solubility in the biological moiety (Ding et al., 2003). Myricetin is a naturally occurring flavonoid commonly found in vegetables and reported to exhibit various effects such as anti-oxidative, anti-carcinogenic, prevention of platelet aggregation and cytoprotective effects.

13.4.10   OTHER METHODS

In a recent study carried out by Chen et al. (2014), recombinant human H-chain ferritin (rHuHF) was prepared and used to encapsulate β-carotene, by taking advantage of the reversible dissociation and reassembly characteristic of apoferritin in different pH environments. Results indicated that β-carotene molecules were successfully encapsulated within protein cages with a β-carotene/protein molar ratio of 12.4:1. These β-carotene-containing apoferritin nanocomposites were water-soluble, and the thermal stability of the β-carotene encapsulated within apoferritin nanocages was markedly improved as compared with free β-carotene, upon such encapsulation.

13.5   CONCLUSION

In order to protect the core material from degradation, encapsulation technology, in which tiny particles or droplets are surrounded by a functional coating, has been developed as an effective approach by reducing its reactivity to its outside environment, controlling transfer rate of the core material to the outside environment, tailoring the release of the core material slowly over time, or at a particular time and to mask an unwanted flavour or taste of the core material. Therefore, this technology is a good alternative that provides high productivity and, at the same time, satisfies an adequate quality of the bioactive compound in the final food products.

13.6 ACKNOWLEDGEMENTS

Financial support by the National Fund for Basic, Strategic and, Frontier Application Research in Agriculture, Indian Council of Agricultural Research through project entitled 'Microencapsulation methods for bacteriocins for their controlled release' is acknowledged.

REFERENCES

Ahmed, M.; Akter, M. S.; Lee, J. C.; Eun, J. B. Encapsulation by Spray Drying of Bioactive Components, Physicochemical and Morphological Properties from Purple Sweet Potato. *LWT-Food Sci. Technol.* **2010**, *43*, 1307–1312.

Arts, I. C. W.; Hollman, P. C. H. Polyphenols and Disease Risk in Epidemiologic Studies. *Am. J. Clin. Nutr.* **2005**, *81*, 317–325.

Augustin, M. A.; Hemar, Y. Nano- and Micro-Structured Assemblies for Encapsulation of Food Ingredients. *Chem. Soc. Rev.* **2009**, *38*, 902–912.

Bennick, A. Interaction of Plant Polyphenols with Salivary Proteins. *Crit. Rev. Oral Biol. Med.* **2002**, *13*, 184–196.

Bhandari, B. R.; D'Arcy, B. D.; Padukka, I. Encapsulation of Lemon Oil by Paste Method Using β-cyclodextrin: Encapsulation Efficiency and Profile of Oil Volatiles. *J. Agric. Food Chem.* **1999**, *47*, 5194–5197.

Chan, E. S.; Lee, B. B.; Ravindra, P.; Poncelet, D. Prediction Models for Shape and Size of Ca-Alginate Macrobeads Produced Through Extrusion-Dripping Method. *J. Colloid Interface Sci.* **2009**, *338*, 63–72.

Chen, L.; Remondetto, G. E.; Subirade, M. Food Protein Based Materials as Nutraceutical Delivery Systems. *Trends Food Sci. Technol.* **2006**, *17*, 272–283.

Chen, L.; Bai, G.; Yang, R.; Zang, J.; Zhou, T.; Zhao, G. Encapsulation of β-carotene within Ferritin Nanocages Greatly Increases its Water-Solubility and Thermal Stability. *Food Chem.* **2014**, *149*, 307–31.

Choi, M. J.; Soottitantawat, A.; Nuchuchua, O.; Min, S. G.; Ruktanonchai, U. Physical and Light Oxidative Properties of Eugenol Encapsulated by Molecular Inclusion and Emulsion–Diffusion Method. *Food Res. Int.* **2009**, *42*, 148–156.

Coradini, K.; Lima, F. O.; Oliveira, C. M.; Chaves, P. S.; Athayde, M. L.; Carvalho, L. M.; Beck, R. C. R. Co-encapsulation of Resveratrol and Curcumin in Lipid-Core

Nanocapsules Improves Their in vitro Antioxidant Effects. *Eur. J. Pharm. Biopharm.* **2014,** *88*(1), 178–185.

Date, A. A.; Joshi, M. D.; Patravale, V. B. Parasitic Diseases: Liposomes and Polymeric Nanoparticles Versus Lipid Nanoparticles. *Adv. Drug. Delivery Rev.* **2007,** *59,* 505–521.

de Vos, P.; Faas, M. M.; Spasojevic, M.; Sikkema, J. Encapsulation for Preservation of Functionality and Targeted Delivery of Bioactive Food Components. *Int. Dairy J.* **2010,** *20,* 292–302.

Ding, H.; Chao, J.; Zhang, G.; Shuang, S.; Pan, J. Preparation and Spectral Investigation on Inclusion Complex of β-Cyclodextrin with Rutin. *Spectrochim. Acta Part A.* **2003,** *59,* 3421–3429.

Ersus, S.; Yurdagel, U. Microencapsulation of Anthocyanin Pigments of Black Carrot (Daucus carota L.) by Spray Drier. *J. Food Eng.* **2007,** *80,* 805–812.

Fang, Z.; Bhandari, B. Encapsulation of Polyphenols a Review. *Trends Food Sci. Technol.* **2010,** *21,* 510–523.

Favaro-Trindade, C. S.; Heinemann, R. J. B.; Pedroso, D. L. Developments in Probiotic Encapsulation. CAB Reviews: Perspectives in Agriculture, Veterinary Science. *Nutr. Nat. Resour.* **2011,** *6,* 1–8.

Gibbs, B. F.; Kermash, S.; Alli, I.; Mulligan, C. N. Encapsulation in the Food Industry: A Review. *Int. J. Food Sci. Nutr.* **1999,** *50,* 213–224.

Gouin, S. Microencapsulation: Industrial Appraisal of Existing Technologies and Trends. *Trends Food Sci. Technol.* **2004,** *15,* 330–334.

Guadarrama-Lezama, A.Y.; Dorantes-Alvarez, L.; Jaramillo-Flores, M. E.; Perez-Alonso, C.; Niranjan, K.; Gutiérrez-López, G. F.; Alamilla-Beltran, L. Preparation and Characterization of Non-Aqueous Extracts from Chilli (Capsicum annuum L.) and Their Micro-encapsulates Obtained by Spray-Drying. *J. Food Eng.* **2012,** *112,* 29–37.

Jun-xia, X.; Hai-yan, Y.; Jian, Y. Microencapsulation of Sweet Orange Oil by Complex Coacervation with Soybean Protein Isolate/Gum Arabic. *Food chem.* **2011,** *125,* 1267–1272.

Kjaergaard, O. G. Prilling. In *Microencapsulation of Food Ingredients;* Vilstrup, P., Ed.; Leatherhead Publishing: Surrey, England, 2001; pp 197e214.

Lim, A. S. L.; Griffin, C.; Roos, Y. H. Stability and Loss Kinetics of Lutein and β-carotene Encapsulated in Freeze-Dried Emulsions with Layered Interface and Trehalose as Glass Former. *Food Res. Int.* **2014,** *62,* 403–409.

Livney, Y. D. Milk Proteins as Vehicles for Bioactives. *Curr. Opin. Colloid Interface Sci.* **2010,** *15*(1–2), 73–83.

Luo, Y.; Wang, T. T. Y.; Teng, Z.; Chen, P.; Sun, J.; Wang, Q. Encapsulation of indole-3-car-binol and 3,30-diindolylmethane in Zein/Carboxymethyl Chitosan Nanoparticles with Controlled Release Property and Improved Stability. *Food Chem.* **2013,** 139, 224–230.

Madene, A.; Jacquot, M.; Scher, J.; Desobry, S. Flavour Encapsulation and Controlled Release—A Review. *Int. J. Food Sci. Technol.* **2006,** *41,* 1–21.

Manach, C.; Scalbert, A.; Morand, C.; Remesy, C.; Jimenez, L. Polyphenols: Food Sources and Bioavailability. *Am. J. Clin. Nutr.* **2004,** *79,* 727–747.

Mandge, H. M.; Narsaiah, K.; Wilson, R. A.; Jha, S. N.; Manikantan, M. R. Effect of Process Parameters and Nozzle Diameter on Size of Microcparticles Prepared with Two Fluid Glass Nozzle. *Crop Improv.* **2012,** (Special Issue), 1183–1184.

Martin, A.; Mattea, F.; Gutierrez, L.; Miguel, F.; Cocero, M. J. Co-Precipitation of Carotenoids and Bio-Polymers with the Supercritical Anti-Solvent Process. *J. Supercrit. Fluids* **2007**, *41*, 138–147.

McClements, D.; Lesmes, U. Structure-Function Relationships to Guide Rational Design and Fabrication of Particulate Food Delivery Systems. *Trends Food Sci. Technol.* **2009**, *20*, 448–457.

Mercader-Ros, M. T.; Lucas-Abellan, C.; Fortea, M. I.; Gabaldon, J. A.; Nunez-Delicado, E. Effect of HP-b-Cyclodextrins Complexation on the Antioxidant Activity of Flavonols. *Food Chem.* **2010**, *118*, 769–773.

Narsaiah, K.; Oberoi, H. S. Microencapsulator with Multiple Air Jet Droplet Generator for Production of Microcapsules (Indian patent application no. 10/DEL/2011 dated 04 January 2011) 2011.

Narsaiah, K.; Jha, S. N.; Manikantan, M. R. An Autoclavable Microencapsulation System with Multistage Break Up Two Fluid Nozzle (Indian patent application no. 3014/ DEL/2011 dated 21 October 2011) 2011.

Narsaiah, K.; Jha, S. N.; Wilson, R. A.; Mandge, H. M.; Manikantan, M. R. Optimizing Microencapsulation of Nisin with Sodium Alginate and Guar Gum. *J. Food Sci. Technol.* **2012**, *51*(12), 4054–4059.

Narsaiah, K.; Jha, S. N.; Wilson, R. A.; Mandge, H. M.; Manikantan, M. R.; Malik, R. K.; Vij, S. Pediocin Loaded Nanoliposomes and Hybrid Alginate-Nanoliposome Delivery Systems for Slow Release of Pediocin. *Bionanoscience* **2013**, *3*, 37–42.

O'Neill, G. J.; Egan, T.; Jacquier, J. C.; O'Sullivan, M.; O'Riordan, E. D. Whey Microbeads as a Matrix for the Encapsulation and Immobilization of Riboflavin and Peptides. *Food Chem.* **2014**, *160*, 46–52.

Paz, E. D.; Martín, A.; Bartolome, A.; Largo, M.; Cocero, M. J. Development of Water-Soluble β-carotene Formulations by High Temperature, High-Pressure Emulsification and Antisolvent Precipitation. *Food Hydrocolloids* **2014**, *37*, 14–24.

Qian, C.; Decker, E. C.; Xiao, H.; McClements, D. J. Physical and Chemical Stability of β-carotene-enriched Nanoemulsions: Influence of pH, Ionic Strength, Temperature, and Emulsifier Type. *Food Chem.* **2012**, *132*, 1221–1229.

Romo-Hualde, A.; Yetano-Cunchillos, A. I.; Gonzalez-Ferrero, C.; Saiz-Abajo, M. J.; Gonzalez-Navarro, C. J. Supercritical Fluid Extraction and Microencapsulation of Bioactive Compounds from Red Pepper (Capsicum annum L.) by-products. *Food Chem.* **2012**, *133*, 1045–1049.

Santhosha, S. G.; Prakasha, J.; Prabhavathi, S. N. Bioactive Components of Garlic and Their Physiological Role in Health Maintenance: A Review. *Food Biosci.* **2013**, *3*, 59–74.

Sharma, A.; Sharma, U. S. Liposomes in Drug Delivery: Progress and limitations. *Int. J. Pharm.* **1997**, *154*, 123–140.

Shrestha, A. K.; Arcot, J.; Yuliani, S. Susceptibility of 5-methyltetrahydrofolic Acid to Heat and Microencapsulation to Enhance its Stability During Extrusion Processing. *Food Chem.* **2012**, *130*, 291–298.

Silva, H. D.; Cerqueira, M. A.; Souza, B. W. S.; Ribeiro, C.; Avides, M. C.; Quintas, M. A. C. Nanoemulsions of β-carotene Using a High-Energy Emulsification Evaporation Technique. *J. Food Eng.* **2011**, *102*, 130–135.

Singh, J.; Upadhyay, A. K.; Prasad, K.; Bahadur, A.; Rai, M. Variability of Carotenes, Vitamin C, E and Phenolics in Brassica Vegetables. *J. Food Compos. Anal.* **2007**, *20*, 106–112.

Sinha, R. K.; Khuller, G. K. The Protective Efficacy of a Liposomal Encap-Sulated 30 kDa Secretary Protein of Mycobacterium Tuberculosis H37Ra Againsttuberculosis in Mice. *Immunol. Cell Biol.* **1997**, *75*, 461–466.

Takahashi, M.; Uechi, S.; Takara, K.; Asikin, Y.; Wada, K. Evaluation of an Oral Carrier System in Rats: Bioavailability and Antioxidant Properties of Liposome-Capsulated Curcumin. *J. Agric. Food Chem.* **2009**, *57*, 9141–9146.

Thakur, B. R.; Singh, R. K.; Handa, A. K. Chemistry and Uses of Pectin—A Review. *Crit. Rev. Food Sci. Nutr.* **1997**, *37*(1), 47–73

Tiwari, U.; Cummins, E. Factors Influencing Levels of Phytochemicals in Selected Fruit and Vegetables During Pre- and Post-Harvest Food Processing Operations. *Food Res.* **2011**, *50*(2), 497–506.

Trifkovic, K. T.; Milasinovic, N. Z.; Djordjevic, V. B.; Krusic, M. T. K.; Knezevic-Jugovic, Z. D.; Nedovic, V. A.; Bugarski, B. M. Chitosan Microbeads for Encapsulation of Thyme (Thymus serpyllum L.) Polyphenols. *Carbohydr. Polym.* **2014**, *111*, 901–907.

Vilstru, P. *Microencapsulation of Food Ingredients;* Leatherhead Food RA Publishing: Surrey, UK, 2001.

Vos, P. D.; Faas, M. M.; Spasojevic, M.; Sikkema, J. Encapsulation for Preservation of Functionality and Targeted Delivery of Bioactive Food Components. *Int. Dairy J.* **2010**, *20*, 292–302.

Williams, D. J.; Edwards, D.; Hamernig, I.; Jian, L.; James, A. P.; Johnson, S. K.; Tapsell, L. C. Vegetables Containing Phytochemicals with Potential Anti-Obesity Properties: A Review. *Food Rev. Int.* **2013**, *52*, 323–333.

Wilson, R. A.; Narsaiah, K.; Mandge, H. M.; Jha, S. N.; Manikantan, M. R. Controlled Release of Bactericin form Hybrid Hydrogel Matrix. *Crop Improv.* **2012**, (Special Issue), 1293–1294.

Woranuch, S.; Yoksan, R. Eugenol-Loaded Chitosan Nanoparticles: I. Thermal Stability Improvement of Eugenol Through Encapsulation. *Carbohydr. Polym.* **2013**, *96*, 578–585.

Xiao-Ying, Q.; Zhi-Ping, Z.; Jian-Guo, J. Preparation of Lutein Microencapsulation by Complex Coacervation Method and its Physicochemical Properties and Stability. *Food Hydrocolloids* **2011**, *25*, 1596–1603.

Xue, F.; Li, C.; Liu, Y.; Zhu, X.; Pan, S.; Wang, L. Encapsulation of Tomato Oleoresin with Zein Prepared from Corn Gluten Meal. *J. Food Eng.* **2013**, *119*, 439–445.

Yadav, M. P.; Igartuburu, J. M.; Yan, Y.; Nothnagel, E. A. Chemical Investigation of the Structural Basis of the Emulsifying Activity of Gum Arabic. *Food Hydrocolloids* **2007**, *21*(2), 297–308.

Yao, Y.; Xie, Y.; Hong, C.; Li, G.; Shen, H.; Ji, G. Development of a myricetin/hydroxypropyl-β-cyclodextrin Inclusion Complex: Preparation, Characterization, and Evaluation. *Carbohydr. Polym.* **2014**, *110*, 329–337.

CHAPTER 14

FERMENTATION TECHNOLOGY IN VEGETABLES

SUNITA SINGH[1] AND SHRUTI SETHI[2]

[1]*Division of Food Science and Postharvest Technology, ICAR-Indian Agricultural Research Institute, New Delhi 110012, India*

CONTENTS

14.1   INTRODUCTION

Fermented foods and beverages are defined as products that have been subjected to the action of microorganisms or enzymes to undergo desirable biochemical changes. Due to their unique flavour, aroma and textural attributes, fermented foods apart from having a good preserving effect are preferred. Fermented foods are nutritionally rich as compared with unfermented counterparts. Food fermentations may enhance the protein content or improve the balance of essential amino acids or their availability, increase the contents or availability of vitamins such as thiamine, riboflavin, niacin or folic acid (Steinkraus, 1997). It may also reduce the contents of indigestible materials in plant foods, such as cellulose, hemicellulose and polygalacturonic and glucuronic acids thus improving the bioavailability of mineral and trace elements (Kalantzopoulos, 1997).

Traditionally, several kinds of vegetables are used for fermentation. But the industry has preferred in some kind of vegetables with the availability of starter culture and standard process of protocols. Some of the processes are still being researched after increased understanding of their importance to health. The reasons for slow increase in popularity of fermented vegetables may be many. One of the foremost is essential requirement of the starter culture(s) for a process itself. Well-identified cultures for such fermentation processes are still under the scientific domain and research and need to be shown a way to industry for commercializations. Olives, sauerkraut, dill pickles, tomato-based drink, cabbage juice and carrot juice, *gundruk, sinki, kanji* and other such traditional fermented products still need a way out to establishing the product for commercialization.

14.2 MICROBIAL STARTERS FOR FERMENTATION

There are about 21 different commercial vegetable fermented products available in Europe; the most economically available among all these are the fermented olives, cucumbers (pickles), sauerkraut and kimchi (Caplice and Fitzerald, 1999). There are a range of fermented vegetables which are made worldwide with a major vegetable as ingredient using specific microorganisms for fermentation (Table 14.1). Most vegetable fermentations occur as a consequence of providing growth conditions that favour the lactic acid bacteria. These bacteria are present on fresh vegetables in very low numbers, accounting for only 0.15–1.5% of the total population (Buckenhüskes, 1997), but spoilage may set in due to favourable pH for the spoilage microflora (Fig. 14.1).

TABLE 14.1 Examples of Acid-Fermented Vegetables Produced in Different Regions of the World.

Product name	Country	Major ingredients	Microorganisms	Usage	Reference
Almagro eggplants	Spain, Mediterranean countries	Eggplant	*Lactobacillus fermentum, L. brevis, L. plantarum*	Pickle	Sánchez et al. (2000)
Gundruk	India, Nepal	Leafy green vegetable (mustard)	*Pediococcus* and *Lactobacilluscellobiosus, L. plantarum*	Side dish	Battcock and Azam-Ali (1998), Tamang et al. (2005)
Burong mustasa	Philippines	Mustard	*Lb. brevis, Pediococcuscerevisiae*	Salad, side dish	Karovičová and Kohajdová (2003)
Camuoi	Vietnam	Eggplant, salt, sugar	*Lb. fermentum, Lb. pentosus, L. brevis*	Side dish	Nguyen et al. (2013); Sesena et al. (2001)
Duamuoi	Vietnam	Mustard, beet, salt	*Lb. fermentum, Lb. pentosus, Lb. plantarum*	Side dish	Lee (1997)
Dakguadong	Thailand	Mustard leaf, salt	*Lb. plantarum*	Salad, side dish	Lee (1997)
Kanji	India	Black carrot	*Lactobacillus plantarum, Lactobacillus para-casei subsp. paracasei, Leuconostocmesenteroides subsp. mesenteroides, Pediococcuspentosaceaceus, and Lactobacillus delbrueckii subsp. delbrueckii*	Beverage	Sura et al. (2001); Sethi (1990a)

TABLE 14.1 *(Continued)*

Product name	Country	Major ingredients	Microorganisms	Usage	Reference
Khalpi	Nepal, India	Cucumber	*L. fallax, P. pentosaceus, Lactobacillus brevis, Lactobacillus plantarum*	Pickle	Tamang et al. (2005)
Kimchi	Korea	Korean cabbage, radish, various vegetables, salt	*L. mesenteroides, Lb. brevis, Lb. plantarum*	Salad, side dish	Kim and Chun (2005)
Sauerkraut	Europe, USA	Cabbage, salt	*Leuconostocmesenteroides, Lactobacillus brevis, Lactobacillus plantarum*	Salad, side dish	Caplice and Fitzgerald (1999)
Sinki	Nepal, India	Radish tap roots	*Lactobacillus brevis, Lactobacillus plantarum, Lbcasei, L. fallax*	Soup, pickle	Tamang et al. (2005)
Shalgam	Turkey	Turnip, carrot, salt	*Lb. brevis, Lb. delbrueckii subsp. delbrueckii and Lb. fermentum*	Beverage	Tanguler and Erten (2012)

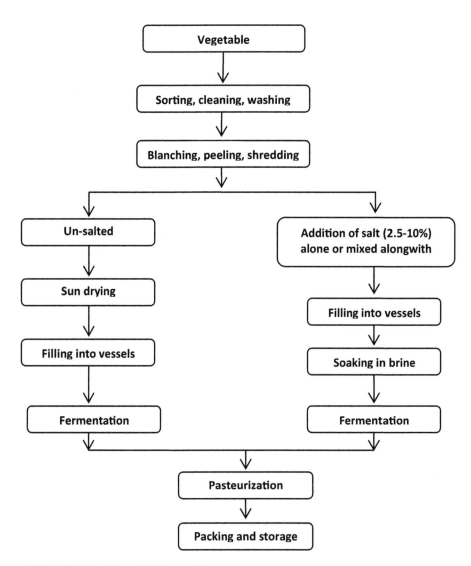

FIGURE 14.1   Vegetable fermentations.

The lacto-fermentation of various vegetables has been performed traditionally, and this process began long back to give shelf stable acidified and acceptable fermented foods and drinks. Some of these have been refined and are in use by the industry for their health benefits, as discussed below.

The lactic acid bacteria are mainly involved in these vegetable fermentations. They are non-sporulating, gram positive, catalase negative cocci or rods. They are generally non-motile and divide in only one plane. These organisms are obligate carbohydrate fermenters with main product of this fermentation being lactic acid along with production of volatile acids and carbon dioxide (Carr et al., 1973; ICMSF, 1980; Krieg and Holt, 1984). The lactic acid bacterial group comprises of the genera, *Lactococcus, Leuconostoc, Pediococcus, Lactobacillus,* as well as the lactic *Streptococcus* species (Stamer, 1979; Stiles, 1994; Vanderzant and Splittstoesser, 1992). The group is usually divided into two physiological subgroups based on their products of glucose metabolism; homofermentative and heterofermentative (Baxter et al., 1983; Jacquet et al., 1995; Settanni and Moschetti, 2010). From hexose, the homofermentative species ferment available sugars to produce lactic acid primarily by using the Embden-Meyerhof pathway (also known as glycolysis). The heterofermentative LAB are those that produce CO_2, lactic acid, acetic acid, ethanol and mannitol from hexoses (Kandler, 1983; Martinez et al., 1963; Peterson and Fred, 1920; Wood, 1961).

Overall, lactic acid bacteria are a small part (2–4 log CFU g^{-1}) of the autochthonous microbiota of raw vegetables, and by the species of vegetables, temperature and harvesting conditions, their cell density is mainly influenced by the species of vegetables, temperature and harvesting conditions (Buckenhüskes, 1997). McFeeters (2004) highlights the importance of studies on the use of lactic acid bacteria starters to get reliable and controlled fermentation processes for the development of these fermented products. Only a few cultures have been used for vegetable fermentations out of which *L. plantarum* is used most frequently. Owing to its ability to survive in low pH and high-salt conditions prevailing in these products, it is one of the most studied microorganisms in fermented products (Kuratsu et al., 2010).

The fermentation of vegetables is frequently characterized by a succession of hetero- and homo-fermentative lactic acid bacteria when occurring spontaneously, together with or without yeasts, which are responsible for multi-step fermentation processes. *Leuconostoc mesenteroides, L. brevis* and *L. plantarum* are the major organisms isolated from fermenting cabbages, black carrots and beet roots (Karovičová and Kohajdová, 2003; Sura et al., 2001; Kingston et al., 2010); *Pediococcus* spp., *Leuconostoc* spp., *Lactococcus* spp., *Lactobacillus delbrueckii* and *L. plantarum* are isolated from table olives (Harris, 1998), and *Leuc. mesenteroides, Leuconostoc pseudomesenteroides, Lactococcus lactis, L. brevis and L. plantarum* from Korean kimchi (Kim and Chun, 2005). In some cases, the spontaneous lactic acid fermentation may occur at a very slow pace, and this may lead to undesirable variations of the sensory properties of fresh vegetables, owing to spoilage and pathogenic microorganisms (Buckenhüskes, 1997).

The main criteria for selecting starters for vegetable fermentation are the (i) rate of growth; (ii) rate and total production of acids that in turn, affect the pH changes; and (iii) environmental adaptation/tolerance (Karovicova et al., 1999; Gardner et al., 2001). The main environmental factors which affect growth and acidification by lactic acid bacteria are temperature and degree of exposure to air and concentration of fermentable carbohydrates, buffering capacity, pH and presence of naturally inhibitory compounds in raw vegetables (Demiÿr et al., 2006).

In the initial stages of the fermentation process, Gram-negative bacteria and yeasts dominate the microbiota of vegetables as displayed in Table 14.2. They are inhibited by the high concentration of NaCl (5–7%, w/v), other chemical compounds or because of the rapid growth of autochthonous lactic acid bacteria (Fleming et al., 1975). Spontaneous lactic acid fermentation of vegetables may take 4–6 days for considerably reducing the product pH or it may not always lead to desirable nutritional, texture and sensorial changes depending on the prevailing strains (Aukrust et al., 1994).

TABLE 14.2 Sequence of Predominant Microorganisms During Spontaneous Fermentation of Brined Vegetables. (*Source:* Fleming (1982).)

Stage	Predominant microorganisms
Initial fermentation	Gram positive and negative bacteria
Propagative fermentation	Lactic acid bacteria and yeasts
Post-fermentation	Oxidative yeasts, fungi and bacteria

14.3   POTENTIAL OF LACTIC ACID BACTERIAs AS PROBIOTICS

The lactic acid bacteria (*Leuconostoc mesenteroides, Lactobacillus brevis, Lactobacillus plantarum, Pediococcus cerevisiae* etc.) involved in vegetable fermentations serve as probiotics and possess several potential health or nutritional benefits such as improvement in nutritional value of food (Kalantzopoulos, 1997), control of intestinal infections (Nout and Ngoddy, 1997), control of certain cancers (Klaenhammer, 1995) and the control of serum cholesterol levels (Kaur et al., 2002). The metabolites produced by LAB that give them antimicrobial properties are listed in Table 14.3. Probiotics are those microorganisms that, when used in large amounts in the preparation of foods, are able to survive the passage through the upper digestive tract and adhere to intestinal cells, helping in the intestinal balance (Campos Perez and Mena, 2012). To avoid the dissemination of resistant microorganisms, bacterial strains intended to be used as probiotics in food systems need to be carefully checked for their susceptibility to antibiotics, The possibility of interchange of plasmids containing genes for antibiotic resistance among microorganisms is considered an important issue especially in food and drug industries (Danielsen, 2002; Delgado et al., 2007).

TABLE 14.3   Antimicrobial Compounds Produced by Lactic Acid Bacteria. (Source: Holzapfel et al., 1995.)

Antimicrobial compound	Target organisms
Lactic acid	Putrefactive and Gram-negative bacteria, certain fungi
Acetic acid	Putrefactive bacteria, clostridia, certain yeasts and fungi
Hydrogen peroxide	Pathogens and spoilage organisms
Reuterin	Wide spectrum of bacteria, moulds and yeasts
Diacetyl	Gram-negative bacteria
Fatty acids	Wide spectrum of bacteria
Nisin	Certain lactic acid and gram-positive bacteria

Many researchers reported about the main health benefits associated with the ingestion of probiotic microorganisms and these can be summarized as alleviation of lactose intolerance, prevention and reduction of diarrhoea symptoms, treatment and prevention of allergy, reduction of the risk associated with mutagenicity and carcinogenicity, hypocholesterolemic effect, inhibition of *Helicobacter pylori* and intestinal pathogens, prevention of inflammatory bowel disease and modulation of the immune system (Akin et al., 2007; Goldin, 1998; Holzapfel et al., 1998; Mattila et al., 1992; Ouwehand et al., 2003).

With the positive aspects of fermentation, we still need to go through the actual safety concerns in the fermented foods/beverages made from vegetables. The safety of food fermentation is related to various aspects. The fermenting food substrates having a high population of desirable and edible microorganisms may become resistant to the invasion by spoilage-causing or food-poisoning microorganisms. The fermentations involving the production of lactic acid should also be safe. Thus, the use of a starter culture(s) not only involves exploiting the metabolite tailored advantages into the product but also provides the right kind of metabolites produced during its growth and metabolism. Time and again, these concerns have been emphasized and plenty of literature is available to show the negative and positive sides of fermentations occurring in vegetables. Way ahead to improving these processes can be made possible by identifying/improving starter cultures for vegetable fermentations towards safer products for a consumer.

14.4 TECHNOLOGY OF FERMENTED VEGETABLE PRODUCTS

14.4.1 OLIVES

Table olives can be considered one of the main fermented vegetables in the world. The world production of table olives for the 2013–2014 was about 15.51 million tonnes (FAO stat, 2015) with Spain being the largest producer followed by Italy. Green olives are processed either with fermentation or without fermentation. Elimination of oleuropein, a bitter compound in olives, is achieved by treating olives with an alkaline treatment in diluted lye solution (sodium hydroxide). The lye concentrations vary from 2 to 3.5%, depending on the ripeness of the olives, the temperature, the variety and the quality of the water. This is followed by repeated water wash to remove any remaining lye residue. Fermentation is carried out by covering the olives with brine. Brine concentrations are 9–10% initially that decline to 5% owing to the exosmosis of olive juice (Table 14.4). The initial phase of fermentation is characterized by predominance of enterobacteriaceae. The population of *Pediococcus* and

Leuconostoc increases with simultaneous elimination of enterobacteria-ceae in the brine as the pH of the medium reaches 6. In the final phase of fermentation, there is a rapid growth of *Lactobacillus plantarum* and reduction of pH to 4. Reduction in the level of fermentable substrates leads to a decrease in the LAB populations. The number of yeasts also varies during the process (10^4–10^6 CFU ml^{-1}). Fermentative yeasts do not cause deterioration, but oxidant yeasts consume lactic acid and raise the pH levels and may therefore hamper the fermentation. Ideal conditions for the fermentation of olives are 5% salt and 4.5 pH.

TABLE 14.4　Fermentation of Vegetables with Salt. (*Source:* Fleming (1982).)

Fermented vegetable(s)	Salt concentration used during	
	Fermentation	Storage
Cabbage	2–3% (dry salt)	2–3% (dry salt)
Cucumber, olive, chilli peppers	5–8% (salt solution)	8–16% (salt solution)
Bell pepper, onion, cauliflower	16–26% (salt solution)	16–26% (salt solution)

Microorganisms such as *Leuconostoc mesenteroides, Pediococcus acidilactici* and *Lactobacillus brevis* in addition to *Lactobacillus plantarum* are considered starters in olive fermentation (Hutkins, 2008). *L. plantarum* as a starter has evolved with physiological and genetic mechanisms that make it salt and osmotolerant in olive brines that contain more than 1 M salt, or osmolalities above 2 osm (Hutkins, 2008). The selection of starters for olive fermentation is based on diverse criteria including homo- and hetero-fermentative metabolism, acid production, salt tolerance, flavour development, temperature, oleuropein-splitting capability and bacteriocin production (Panagou et al., 2008). Furthermore, the ability of the starter to grow at low temperatures must be considered essential in cold regions, as maintaining brine temperature is complex and expensive (Durán et al., 1999).

Olive fruits are a rich source of phenolic compounds, the most abundant being oleuropein, that decreases during ripening and processing. The oleuropein gets degraded to non-bitter compounds during alkaline treatment and fermentation. Moreover, the lye treatment causes loss of oil from olives. Several studies (Keys et al., 1986; Willett et al., 1995; Solfrizzi et al., 1999) highlight the importance of consumption of olives towards cardioprotective and anti-inflammatory effects. It is also believed that olives provide phenols for pharmacological properties (D'Angelo et al., 2001) and protect human cells against oxidative stress and injury. During spontaneous and controlled fermentation, the phenols are found to reduce by 55 and 46%, respectively (Othman et al., 2009).

The main commercial preparations, (i) treated green olives (Spanish style), (ii) naturally black olives (Greek style) and (iii) ripe olives processed by alkaline oxidation (Californian style), are among the three processing methods defined by the Standard (IOOC, 2004). The Greek style olives differ from California style black olives due to the fact that the black colour is due to oxidation and are not lye treated (Nychas et al., 2002). They are fermented by LABs, yeasts and non-lactic bacteria (*Pseudomonas* and *Enterobacteriaciae*). The fermented product is less acidic (<0.6% acidity) and also contains acetates, citrates, malate, CO_2 and ethanol. Presence of various other organisms such as aerobic bacteria, fungi and spore-forming bacteria cause spoilage in fermented olives. A major disadvantage in making Spanish style olives as highlighted by Ruiz-Barba and Jimenez-Diaz (1995) using *L. plantarum* as a starter is that it utilizes B complex vitamins: pantothenic acid, biotin and vitamin B_6. Once these vitamins get depleted the organism stops to ferment, leading to incomplete fermentations and setting in of spoilage. Yeasts have also been reported to contribute to such vitamins during the fermentative process.

14.4.2 SAUERKRAUT

Sauerkraut is a German word that simply means sour white cabbage. For sauerkraut preparation, shredded cabbage (*Brassica oleracea*) is fermented by various lactic acid bacteria, including *Leuconostoc, Lactobacillus* and *Pediococcus*. Cabbage used for sauerkraut preparation must be sound, ripe and well-leafed with a total sugar level of about 24%. For continuing the fermentation, the shredded cabbage is mixed with 2.25–2.5% salt along with 1.5% mustard (optional). Within a few hours, brine is formed by exuding of the water from the cabbage by exosmosis and fermentation starts. Optimum temperature for fermentation is about 18°C with a final acidity of 1.0–1.8% with pH value 4.1 or lower. The maximum acidity of about 1.5% lactic acid maybe obtained in 4–6 weeks. Fermented cabbage has a high antioxidant potential regarding health-promoting properties

(Kusznierewicz et al., 2010) as it is rich in antioxidants such as vitamin C and ascorbigen (Wagner et al., 2008; Harmaum et al., 2008; Peñas et al., 2010b). The major concern in raw cabbage is the presence of glucosinolates, which are antinutritive for human consumption. Apart from sinigrin, the major glucosinolate in cabbage, a number of other glucosinolates, namely progoitrin, glucoiberin, glucobrassicin, glucoraphanin, gluconapin, 4-OH-glucobrassicin and 4-MeO-glucobrassicin are also present (Tolonen et al., 2004). The breakdown of glucosinolates is catalysed by myrosinase, an enzyme released from damaged plant cells (Verhoeven et al., 1997), which spontaneously cyclizes them to goitrin. Goitrin reduces the production of thyroid hormones such as thyroxine (McMillan et al., 1986) and body functions are also affected by it. However, an intake of up to 20 mg of goitrin (as from milk obtained from cow fed on cruciferous forage) is not anti-nutritive (Van Etten, 2012). Therefore, lowering of glucosinolates to non-harmful levels is one of the major concerns in substrates used for fermentation. Fermentation of the cabbage to sauerkraut helps to reduce the levels of these glucosinolates besides producing various other beneficial biomolecules. The fermentative process is of about 35 days to sauerkraut and leads to lowering of the pH and breakdown of glucosinolates. At pH below 4, the glucosinolates get decomposed to nitriles instead of isothiocyanates. But the quantity of nitriles is so low (< 10 ion ratio/100 gfw fermented product) that it does not affect human health (Tolonen et al., 2002).

Sauerkraut is also reported to contain high levels of glucosinolate (GLS) breakdown products (Ciska and Pathak, 2004), such as isothiocyanates (ITC), indole-3-carbinol (I3C), indole-3-acetonitrile (I3ACN) and ascorbigen (ABG) (Ciska and Pathak, 2004; Martinez-Villaluenga et al.,

2009; Peñas et al., 2010b). These breakdown products have been linked to reduction in cancer (Das et al., 2000; Mithen, 2001) by inhibition of tumour cell growth and stimulation of apoptosis (Johnson, 2002). Earlier reports also claim that the level of vitamin B increases during sauerkraut fermentations, whereas vitamins C and A are preserved (Fleming et al., 1995).

Starter culture used for sauerkraut preparation also influences the production of glucosinolate breakdown products as they produce myrosinase (Tolonen et al., 2004). Amongst the starter cultures studied (*Lactobacillus plantarum, Lactobacillus sakei, Leuconostoc mesenteroides, Pediococcus pentosaceus, Lactococcus lactis* N8 and *L. lactis* LAC67), it was observed that when *L. sakei* was used, the amount of total breakdown products in sauerkraut was 2–3 times that of other strains. Goitrin was found in relatively high concentrations (480–1090 mg 100 g^{-1} FW) in all end products. Observations of the antimicrobial effects of cabbage juice containing nisin and glucosinolate breakdown products showed a difference in type of inhibitions on indicator strains in the fermented end products. The juice fermented with *L. sakei* strain effectively inhibited the growth of *Candida lambica*, whereas *Listeria monocytogenes* was not affected. *Escherichia coli* was found to be strongly inhibited in all cabbage juices. *L. lactis* N8 produced up to 1400 IU·ml^{-1} of nisin in cabbage brine within 24 h which was detectable even after 13 days at a level of 250 IU·ml^{-1}.

Biogenic amine production is also influenced by LAB starters having amino acid decarboxylation activity (*Lactobacillus* spp., *L. mesentroides, Pediococcus damnosus*) (Halasz et al., 1994; Morreno-Arribas et al., 2003). The recommended level of histamine, tyramine, putrescine and cadaverine in sauerkraut can be 10, 20, 50 and 25 mg/kg (Kunsch et al., 1989). A significant increase in the total and individual biogenic amine concentration was found to be higher in spontaneous fermentations as compared with the fermentation carried out by starter cultures during sauerkraut fermentation (Halasz et al., 1994; Spicka et al., 2002). Thus, selection of appropriate starters for sauerkraut fermentation still remains a challenge in aspects of safety of the fermentation process. Here, level of salt addition to the fermenting product also influences the biogenic amine production. Lower salt concentration (0.5%) has been shown to result in reduced amine production as compared with higher salt (1.5%) content. *L. mesenteroides* starter and 0.5% NaCl were the optimal fermentation conditions for producing sauerkrauts with the lowest biogenic amine contents (Penas et al., 2010b; Kalac et al., 2000). For preparation of sauer-kraut with lowered biogenic amine content, it is recommended to have the initial temperatures of 15–20°C during fermentation and pasteurization of the product as soon as total acidity reaches the value 9–10 g (as lactic acid) per kg or at pH value 4.0–3.8 (Künsch et al., 1989).

14.4.3   KIMCHI

Kimchi refers to various traditional fermented vegetables of Korean origin. It mainly constitutes Chinese cabbage (*Brassica pekinensis*) and radish, with the addition of other seasonings such as garlic, onion, ginger, red pepper, mustard, parsley and salt (Kim et al., 2012). Cabbage is cut and placed in 5–7% salt for 12 h, which favours an increase of the concen-tration of NaCl up to 2.0–4.0% of the total weight. Cabbage is then rinsed several times with fresh water and drained followed by addition of season-ings. The fermentation is carried out by autochthonous lactic acid bacteria, which vary depending on the main ingredients, temperature (5–30°C) and concentration of salt. In general, *Leuconostoc mesenteroides* starts the fermentation which is later inhibited by the increasing concentration of lactic acid. Acid-tolerant species such as *L. brevis* dominate in the initial fermentation stage, being replaced by *L. plantarum* during later stage of fermentation. Nevertheless, the best tasting kimchi is obtained before the overgrowth of *L. plantarum* and *L. brevis*, at an optimal pH of 4.5 and lactic acid content of 0.4–0.8%. After fermentation, kimchi is left to ripen for several weeks under refrigeration conditions (Lee et al., 2011).

14.4.4   CUCUMBERS

Fully ripe, regular-shaped cucumbers are washed, drained and, eventually, sliced for making fermented pickles. They are dipped in brine (5–7% of NaCl) in suitable containers with addition of calcium chloride (0.1–0.4%) to get a crisp texture (Fleming et al., 1995). Most of the times, sponta-neous lactic acid fermentation starts owing to the native microflora and lasts 2–3 weeks depending on the temperature (20–27°C). The final pH ranges between 3.1 and 3.5 with 1.5% lactic acid and little or no residual sugar. LAB synthesize several bacteriocins and antimicrobial peptides during fermentation which prove inhibitory to spoilage bacteria. Carbon dioxide may be produced due to malo-lactic fermentation by *L. plantarum*. Cucumbers may be purged with air to remove CO_2 to avoid gaseous/ bloater spoilage. Potassium sorbate (~0.04%) or 0.16% acetic acid can be used to restrict the growth of aerobic microorganisms in air-purged cucumber fermentations, particularly moulds and yeasts. A commercial starter culture of *L. plantarum* that does not decarboxylate malic acid (and hence does not contribute to the formation of bloaters) has been developed (Daeschel et al., 1984). Khalpi is the most popular cucumber pickle in the Himalayan region. Here the cucumbers are cut into pieces, sun-dried for 2 days, put into a bamboo vessel and left to ferment at room temperature for 3–7 days (Tamang et al., 2005).

14.4.5 PROBIOTIC DRINKS

Microorganisms used as probiotics need to survive at high cell numbers and should not impart unsuitable modifications of the sensory properties of the fermenting juice. To develop organoleptically acceptable and safe probiotic non-dairy beverages, the careful selection of the matrix, probiotic strain and the addition of other ingredients are of immense importance (Mousavi et al., 2010). Desirable properties of fermented vegetable juices can be achieved by choosing *Lactobacillus* strains suitable for the lactic acid fermentation of individual raw materials (Demir et al., 2004). To assess the suitability of a particular strain to be used as a starter, its ability for acid production, change of pH, decrease of nitrate concentration and production of biogenic amines are important (Karovicova et al., 1999). Probiotic beverages so developed can help to increase the intake of vegetables in the human diet as healthier alternatives.

Tomato, carrot, cabbage, artichokes and reed beet juices have been found to be suitable for development of probiotic drinks with a viable cell population above ca. 8 log CFU ml[-1] (Valerio et al., 2006; Rivera-Espinoza and Gallardo-Navarro, 2010). Nutrients such as yeast extract that are essential for growth and survival of *L. plantarum* and *L. acidophilus* strains in beetroot juice have been added to get optimal growth and acidification (Rakin et al., 2007).

Yoon et al. (2006) have developed a probiotic cabbage juice by use of lactic acid bacteria (*Lactobacillus plantarum* C3 and *Lactobacillus delbrueckii* D7). They inoculated the cabbage juice with a 24-h-old lactic culture for fermentation at 30°C. After 4 weeks of cold storage at 4°C, the viable cell counts of *L. plantarum* and *L. delbrueckii* were still 4.1×10^7 CFU·ml^{-1} and 4.5×10^5 CFU·ml^{-1}, respectively. They concluded that these vegetable juices could serve as functional beverages for vegetarians and lactose-intolerant individuals.

Kefir-like beverages (KLBs) were produced by back-slopping technique using vegetables such as carrot, fennel, onion and tomato juices by Corona et al. (2016). They inoculated the vegetable juices with a mixed culture inoculum consisting of approximately 10^9 CFU·g^{-1} of *Lactobacillus*, *Lactococcus*, *Leuconostoc* and *Saccharomyces*. Onion and tomato juices retained a high antioxidant activity after lactic fermentation. Carrot kefir-like beverage (KLB) was the product mostly appreciated by the panellists.

Carrot juice (*Daucus carota L.*) is one of the most popular vegetable juices and represents a rich source of natural β-carotene if prepared from orange carrots. It is difficult to preserve vegetable juices such as carrot because of low acidity and high concentration of spoilage and spore-forming bacteria. They are often fermented as the best means for their preservation. Fermentation may be spontaneous or may be controlled using a starter culture (Liepe and Junker, 1984). Lactic acid bacteria such as *Lactobacillus plantarum*, *Lactobacillus brevis*, *Lactobacillus xylosus* and so on are mainly used as starter cultures (Schobinger, 1987). During the lacto-fermentation, high and low esterified pectins of vegetable mash are depolymerized because of the action of pectolytic enzymes such as polygalacturonase, pectinlyase and pectinesterase, thus increasing the juice yield and β-carotene content in the juice (Demiyr, 2000; Demiÿr et al., 2006; Sakellaris et al., 1988; Karam and Belarbi, 1995; Wong, 1995). The thermal preservation of the product is easy and shelf life can be increased (Liepe and Junker, 1984; Acar, 1998). Demir et al. (2004) and Demiÿr et al. (2006) demonstrated that an initial bacteria concentration of 3×10^5 CFU/g mash resulted in the most acceptable fermented carrot juice. By altering the population of the starter, the product acidity could be adjusted. In certain cases, the fermentation process with native LAB is completed in 6–7 days (Sethi, 1990a, 1990b) or 7–10 days (Sura et al., 2001) and results in a beverage high in nutritional value and possessing cooling and soothing properties (Sura et al., 2001).

Sliced black carrot   **Kanji**

Beside orange carrot, black carrot is also used in fermentation. As of late, it received wide attention due to high antioxidant activity and other functional benefits. Traditionally, black carrot is used for preparation of probiotic beverage kanji (Koley et al., 2014) which is considered to possess the potential of preventing infectious and malignant diseases (Sura et al., 2001) with the predominance of *L. plantarum* (Kingston et al., 2010). Optimized growth conditions in batch modes can add value to lactic drinks (Sethi, 1990a; Karovicova et al., 1999). The drink can be prepared under optimized conditions by using *L. lactis*, a nisin producer, to prepare kanji (Singh et al., 2013, 2016) along with a starter from sourdough in from purple carrots, having an acceptable flavour and a final pH of 3.3 (Singh et al., 2015) (Fig. 14.2).

Shalgam is a traditional lactic acid beverage of Turkey, nearly similar to *kanji* produced from the lactic acid fermentation of black carrot, turnip, rock-salt, sourdough, bulgur flour and potable water (Tanguler and Erten, 2012). There are two main processing methods for commercial production: the traditional method and the direct method (Erten et al., 2008). In the traditional method the mixture of bulgur flour, salt, sourdough and water is left for fermentation by LAB and yeasts at room temperature for 3–5 days. Three to five times, the fermented mixture is extracted with water. The extracts obtained from this fermentation are mixed with chopped black carrot, salt and sliced turnip for second fermentation to proceed for 3–10 days at ambient temperature. After the completion of fermentation, the juice is ready for marketing. In direct production method, fermentation of dough fermentation is not carried out. The chopped black carrots, salt and sliced turnip are mixed with bakers' yeast (*Saccharomyces cerevisiae*) or sourdough and allowed to ferment at ambient temperature (Erten et al., 2008).

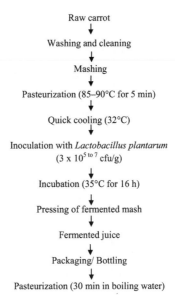

Raw carrot
↓
Washing and cleaning
↓
Mashing
↓
Pasteurization (85–90°C for 5 min)
↓
Quick cooling (32°C)
↓
Inoculation with *Lactobacillus plantarum*
$(3 \times 10^{5 \text{ to } 7} \text{ cfu/g})$
↓
Incubation (35°C for 16 h)
↓
Pressing of fermented mash
↓
Fermented juice
↓
Packaging/ Bottling
↓
Pasteurization (30 min in boiling water)

FIGURE 14.2   Flow diagram of fermented carrot juice.
Source: Demir et al., 2004; Demiÿr et al., 2006 international, vol.

14.5   INNOVATIVE VEGETABLE-BASED FERMENTED PRODUCTS

In recent times, a novel protocol for the manufacture of fermented smoothies was standardized by Di Cagno et al. (2011). White grape juice and aloe vera extract were mixed with red (cherries, blackberries, prunes and tomatoes) or green (fennels, spinach, papaya and kiwi) fruits and vegetables, and were subjected to fermentation with mixed autochthonous starters, consisting of *L. plantarum*, *W. cibaria* and *L. pentosus* strains. Lactic acid fermentation by selected starters led to an increase of antioxidant compounds and enhanced sensory attributes of the product.

Joshi et al. (2008) developed appetizers from lactic acid fermented vegetables. The vegetables carrot, radish and cucumber were fermented using sequential culture of lactic acid bacteria, namely *Lactobacillus plantarum* (NCDC020), *Pediococcus cerevisiae* (NCDC038) and *Streptococcus lactis* var *diacetylactis* (NCDC 061) with a mixture of fruit pulps. Out of the three vegetables and pulps, radish-based appetizer having 20% radish + 10% apricot had the highest overall acceptability and was rated the best.

14.6 PROSPECTS

The beneficial role of fermented vegetable products in preventing certain diseases has recently received considerable attention. Although a few products have been commercialized worldwide, but still to prove the effectiveness, clinical research is needed on some of the traditional fermented foods and beverages specific to certain regions. Lot of attention has been directed towards bacteria of the genus *Lactobacillus* and *Bifidobacterium* (Saavedra et al., 2010) on their proven application as probiotics. Growth of these *Lactobacillus* and related species involved in fermenting vegetables has also been modelled across various studies (Singh et al., 2015, 2016; Zwiettering et al., 1990). These studies have not exploited them further as starter cultures per se after studying their biomass. The future outlook could be modelling of these specific strains for use in fermentations by upscaling their growth and harnessing their full potential towards various health benefits and development of safe fermented products. Further improvements in the area of fermented vegetable product development may be led by in vitro or in vivo studies and these studies also can recommend such foods in the diet of people.

REFERENCES

Acar, J. Aktueller Stand Der Karottensaftherstellung. *Flüssigesobst* **1998,** *4,* 196–198.

Aukrust, T. W.; Blom, A.; Sandtorv, B. F.; Slinde, E. Interactions Between Starterculture and Raw Material in Lactic Acid Fermentation of Sliced Carrot. *Lebensmittel Wissenschaftund Technol.* **1994,** *27,* 337–341.

Akin, M. B.; Akin, M. S.; Kirmaci, Z. Effects of Inulin and Sugar Levels on the Viability of Yoghurt and Probiotic Bacteria and the Physical and Sensory Characteristics in Probiotic Ice-cream. *Food Chem.* **2007,** *104,* 93–99.

Battcock, M.; Azam-Ali, S. Fermented Fruits and Vegetables. A Global Perspective. FAO Agricultural Services Bulletin No. 134. Rome, Food and Agriculture Organization of the United Nations. 1998.

Baxter, J.; Gibbs, P. A.; Blood, R. M. Lactic Acid Bacteria: Their Role in Food Preservation—A Literature Study. *Br. Food Manuf. Ind. Res. Assoc.* **1983,** ISSN: 0144–2074.

Buckenhüskes, H. J. Fermented Vegetables. In *Food Microbiology: Fundamentals and Frontiers,* 2nd ed.; Doyle, P. D., Beuchat, L. R., Montville, T. J., Eds.; ASM Press: Washington DC, 1997; pp. 595–609

Campos Perez, A. I.; Mena, A. L. *Lactobacillus*; Nova Biomedical: New York, US, 2012.

Caplice, E.; Fitzgerald, G. F. Food Fermentations: Role of Microorganisms in Food Production and Preservation. *Int. J. Food Microbiol.* **1999,** *50,* 131–149.

Carr, J. G.; Curring, C. V.; Whiting, G. C. *Lactic Acid Bacteria in Beverages and Food;* Academic Press: New York, 1973.

Ciska, E.; Pathak, D. Glucosinolate Derivatives in Stored Fermented Cabbage. *J. Agric. Food Chem.* **2004,** *52,* 7938–7943.

Corona, O.; Randazzo, W.; Miceli, A.; Guarcello, R.; Francesca, N.; Erten, H.; Moschetti, G.; Settanni, L. Characterization of Kefir-Like Beverages Produced from Vegetable Juices. *LWT—Food Sci. Technol.* **2016,** *66,* 572–581.

Daeschel, M. A.; McFeeters, R. F.; Fleming, H. P.; Klaenhammer, T. R.; Sanozky, R. B. Mutation and Selection of Lactobacillus Plantarum Strains that do not Produce Carbon Dioxide from Malate. *Appl. Environ. Microbiol.* **1984,** *47,* 419–420.

D'Angelo, S.; Manna, C.; Migliardi, V.; Mazzoni, O.; Morrica, P.; Capasso, G.; Pontoni, G.; Galletti, P.; Zappia, V. Pharmacokinetics and Metabolism of Hydroxytyrosol, a Natural Antioxidant from Olive Oil. *Drug Metab. Dispos.* **2001,** *29,* 1492–1498.

Danielsen, M. Characterization of the Tetracycline Resistance Plasmid pMD5057 from Lactobacillus Plantarum 5057 Reveals a Composite Structure. *Plasmid* **2002,** *48,* 98–103.

Das, S.; Tuagi, A. K.; Haur, H. Cancer Modulation by Glucosinolates: A Review. *Curr. Sci.* **2000,** *79,* 1665–1671.

Delgado, S.; Flórez, A. B.; Mayo, B. Antibiotic Susceptibility of Lactobacillus and Bifidobacterium Species from the Human Gastrointestinal Tract. *Curr. Microbiol.* **2007,** *50,* 202–207.

Demir, N.; Acar, J.; Bahceci Savas, K. Effects of Storage on Quality of Carrot Juices Produced with Lactofermentation and Acidification. *Eur. Food Res. Technol.* **2004,** *218,* 465–468.

Demiyr, N. *Researches on the Application of Total Liquefaction and Other Related Techniques for the Production of Carrot Juices;* PhD diss, Hacettepe University: Ankara, Turkey, 2000.

Demiÿr, N.; Bachçeci, K. S.; Acar, J. The Effects of Different Initial Lactobacillus Plantarum Concentrations on Some Properties of Fermented Carrot Juice. *J. Food Process. Preserv.* **2006,** *30,* 352–363.

Di Cagno, R.; Minervini, G.; Rizzello, C. G.; De Angelis, M.; Gobbetti, M. Effect of Lactic Acid Fermentation on Antioxidant, Texture, Color and Sensory Properties of Red and Green Smoothies. *Food Microbiol.* **2011,** *28,* 1062–1071.

Durán Quintana, M. C.; García García, P.; Garrido Fernández, A. Establishment of Conditions for Green Table Olive Fermentation at Low Temperature. *Int. J. Food Microbiol.* **1999,** *51,* 133–143.

Erten, H.; Tanguler, H.; Canbas, A. A Traditional Turkish Lactic Acid Fermented Beverage: Shalgam (Salgam). *Food Rev. Int.* **2008,** *24,* 352–359.

FAO stat. 2015, http://faostat3.fao.org/browse/Q/QC/E/olives (accessed April 28, 2016).

Fleming, H. P. Fermented Vegetables. In *Economic Microbiology;* Rose, A. H., Ed.; Academic Press: London, England, 1982; Vol. 7, 228–258.

Fleming, H. P.; Etchells, J. L.; Thompson, R. L.; Bell, T. A. Purging of CO_2 from Cucumber Brines to Reduce Bloater Damage. *J. Food Sci.* **1975,** *40,* 1304–1310.

Fleming, H. P.; Kyung, K.-H.; Breidt, F. Jr Vegetable Fermentations. In *Biotechnology;* Rehm, H.-J., Reed, G., Eds.; VCH Publishers: New York, NY, 1995; pp 29–661.

Gardner, N. J.; Savard, T.; Obermeier, P.; Caldwell, G.; Champagne, C. P. Selection and Characterization of Mixed Starter Cultures for Lactic acid Fermentation of Carrot, Cabbage, Beet and Onion Vegetable Mixtures. *Int. J. Food Microbiol.* **2001,** *64,* 261–275.

Goldin, B. R. Health Benefits of Probiotics. *Br. J. Nutr.* **1998**, *80,* S203–207.

Halasz, A.; Barath, A.; Sarkadi-Simon, L.; Holzapfel, W. Biogenic Amines and Their Production by Microorganisms in Food. *Trends Food Sci. Technol.* **1994**, *51,* 42–49.

Harmaum, B.; Hubbermann, E. M.; Zhu, Z.; Schwarz, K. Impact of Fermentation on Phenolic Compounds in Leaves of Pakchoi Brassica L. ssp. chinesis var. communis) and Chinese Leaf Mustard (Brassica junceacoss*). J. Agric. Food Chem.* **2008**, *56,* 148–157.

Harris, L. The Microbiology of Vegetable Fermentations. In *Microbiology of Fermented Foods,* 2nd ed.; Wood, B. J. B., Ed.; Blackie Academic and Professional: London, 1998; pp 46–72.

Holzapfel, W. H.; Geisen, R.; Schillinger, U. Biological Preservation of Foods with Reference to Protective Cultures, Bacteriocins, and Food-Grade Enzymes. *Int. J. Food Microbiol.* **1995**, *24,* 343–362.

Holzapfel, W. H.; Haberer, P., Snel, J.; Schillinger, U.; Huis In't Veld, J. H. Overview of Gut Flora and Probiotics. *Int. J. Food Microbiol.* **1998**, *41,* 85–101.

Hutkins, R. W. *Institute of Food Technologists Series: Microbiology and Technology of Fermented Foods (1);* Wiley-Blackwell: Hoboken, US, 2008.

ICMSF. International Commission on Microbiological Specifications for Foods (1980). *Microbial Ecology of Foods;* Academic Press, Inc.: San Diego, Californiam, 1980; 1, pp 205–225.

IOOC. International Olive Oil Council Trade Standard Applying to Table Olives. 2004.

Jacquet, C.; Catimel, B.; Brosch, R.; Buchrieser, C.; Dehaumont, P.; Goulet, V.; Lepoutre, A.; Veit, P.; Rocourt, J. Investigations Related to the Epidemic Strain Involved in the French Listeriosis Outbreak in 1992. *Appl. Environ. Microbiol.* **1995**, *61,* 2242–2246.

Johnson, I. T. Glucosinolates in the Human Diet. Bioavailability and Implications for Health. *Phytochem. Rev.* **2002**, *1,* 183–188.

Joshi, V. K.; Sharma, S.; Rana, R. Preparation and Evaluation of Appetizer's from Lactic Acid Fermented Vegetables. *J. Biotechnol.* **2008**, *136S,* S743–750.

Kalac, P.; Spicka, J.; Knzek, M.; Pelikanova, T. The Effects of Lactic Acid Bacteria Inoculants on Biogenic Amines Formation in Sauerkraut. *Food Chem.* **2000**, *70,* 355–359.

Kalantzopoulos, G. Fermented Products with Probiotic Qualities. *Anaerobe* **1997**, *3,* 185–190.

Kandler, O. Carbohydrate Metabolism of Lactic Acid Bacteria. *Antonie van Leeuwenhoek* **1983**, *49,* 2099–2224.

Karam, N. E.; Belarbi, A. Detection of Polygalacturonases and Pectinesterases in Lactic Acid Bacteria. *World J. Microbiol Biotechnol.* **1995**, *11,* 559–563.

Karovičová, J.; Kohajdová, Z. Lactic Acid Fermented Vegetable Juices. *Hortic. Sci.* **2003**, *30,* 152–158.

Karovicova, J.; Drdak, M.; Greif, G.; Hybenova, E. The Choice of Strains of Lactobacillus Species for the Lactic Acid Fermentation of Vegetable Juices. *Eur. Food Res. Technol.* **1999**, *210,* 53–56.

Kaur, I. P.; Chopra, K.; Saini, A. Probiotics: Potential Pharmaceutical Applications. *Eur. J. Pharm. Sci.* **2002**, *15,* 1–9.

Keys, A.; Menotti, A.; Karvonen, M.; Aravanis, C.; Blackburn, H.; Buzina, R.; Djordjevic, B. S.; Dontas, A. S.; Fidanza, F.; Keys, M. H. The Diet and 15-year Death Rate in the Seven Countries Study. *Am. J. Epidemiol.* **1986**, *124,* 903–915.

Kim, M.; Chun, J. Bacterial Community Structure in Kimchi, a Korean Fermented Vegetable Food, as Revealed by 16S rRNA Gene Analysis. *Int. J. Food Microbiol.* **2005,** *103,* 91–96.

Kim, J.; Bang, J.; Beuchat, L. R.; Kim, H.; Ryu, J.-H. Controlled Fermentation of Kimchi Using Naturally Occurring Antimicrobial Agents. *Food Microbiol.* **2012,** *32,* 20–31.

Kingston, J. J.; Radhika, M.; Roshini, P. T.; Raksha, M. A.; Murali, H. S.; Batra, H. V. Molecular Characterization of Lactic Acid Bacteria Recovered from Natural Fermentation of Beet Root and Carrot Kanji. *Ind. J. Microbiol.* **2010,** *50,* 292–298.

Klaenhammer, T. R. Genetics of Intestinal Lactobacilli. *Int. Dairy J.* **1995,** *5,* 1019–1058.

Koley, T. K.; Singh, S.; Khemariya, P.; Sarkar, A.; Kaur, C.; Chaurasia, S. N. S.; Naik, P. S. Evaluation of Bioactive Properties of Indian Carrot (Daucus carota L.): A Chemometric Approach. *Food Res. Int.* **2014,** *60,* 76–85.

Krieg, N. R.; Holt, J. G. *Bergey's Manual of Systematic Bacteriology;* Williams and Wilkins: Baltimore, Maryland, 1984.

Kunsch, U.; Schärer, H.; Temperli, A. Biogene Amine als Qualitätsindikatoren von Sauerkraut. In XXIV. Vortragstagung der Deutschen Gesellschaftfür Qualitätsforschung. Qualitätsaspekte von Obst und Gemüse. Ahrensburg, Germany, 1989.

Kuratsu, M.; Hamano, Y.; Dairi, T. Analysis of the Lactobacillus Metabolic Pathway. *Appl. Environ. Microbiol.* **2010,** *76,* 7299–7301.

Kusznierewicz, B.; Lewandowska, J.; Kruszyna, A.; Piasek, A.; Smiechowska, A.; Namiesnik, J.; Bartoszek, A. The Antioxidative Properties of White Cabbage (Brassica oleracea var. capitata F. Alba) Fresh and Submitted to Culinary Processing. *J. Food Biochem.* **2010,** *34,* 262–285.

Lee, C. H. Lactic Acid Fermented Foods and Their Benefits in Asia. *Food Control* **1997,** *8,* 259–269.

Lee, H.; Yoon, H.; Ji, Y.; Kim, H.; Park, H.; Lee, J. Functional Properties of Lactobacillus Strains from Kimchi. *Int. J. Food Microbiol.* **2011,** *145,* 155–161.

Liepe, H. U.; Junker, M. *Gemüsesaefte Flüssigesobst* **1984,** *50,* 304–307.

Martinez, G.; Barker, H. A.; Horecker, B. L. A Specific Mannitol Dehydrogenase from Lactobacillus Brevis. *J. Biol. Chem.* **1963,** *238,* 1598–1603.

Martinez-Villaluenga, C.; Peñas, E.; Frias, J.; Ciska, E.; Honke, J.; Piskula, M. K.; Kozlowska, H.; Vidal-Valverde, C. Influence of Fermentation Conditions on Glucosinolates, Ascorbigen, and Ascorbic Acid Content in White Cabbage (Brassica oleracea var. capitata cv. Taler) Cultivated in Different Seasons. *J. Food Sci.* **2009,** *74,* C62–C67.

Mattila, S. T.; Matto, J.; Saarela, M. Lactic Acid Bacteria with Health Claims-Interations and Interference with Gastrointestinal Flora. *Int. Dairy J.* **1992,** *9,* 25–35.

McFeeters, R. F. Fermentation Microorganisms and Flavor Changes in Fermented Food. *J. Food Sci.* **2004,** *69,* 35–37.

McMillan, M.; Spinks, E. A.; Fenwick, G. R. Preliminary Observations on the Effect of Dietary Brussels Sprouts on Thyroid Function. *Hum. Toxicol.* **1986,** *5,* 15–19.

Mithen, R. F. Glucosinolates and Their Degradation Products. *Adv. Bot. Res.* **2001,** *35,* 213–262.

Morreno-Arribas, M. V.; Polo, M. C.; Jorganes, F.; Munoz, R. Screening of Biogenic Amine Production by Acid Lactic Bacteria Isolated from Grape must and Wine. *Int. J. Food Microbiol.* **2003,** *84,* 117–123.

Mousavi, Z. E.; Mousavi, S. M.; Razavi, S. H.; Emam-Djomeh, Z.; Kiani, H. Fermentation of Pomegranate Juice by Probiotic Lactic Acid Bacteria. *World J. Microbiol. Biotechnol.* **2010,** *27,* 123–128.

Nguyen, D. T. L.; Van Hoorde, K.; Cnockaert, M. A Description of the Lactic Acid Bacteria Microbiota Associated with the Production of Traditional Fermented Vegetables in Vietnam. *Int. J. Food Microbiol.* **2013,** *163,* 19–27.

Nout, M. J. R.; Ngoddy, P. O. Technological Aspects of Preparing Affordable Fermented Complementary Foods. *Food Control* **1997,** *8,* 279–287.

Nychas, G. J. E.; Panagou, E. Z.; Parker, M. L.; Waldron, K. W.; Tassou, C. C. Microbial Colonization of Naturally Black Olives During Fermentation and Associated Biochemical Activities in the Cover Brine. *Lett. Appl. Microbiol.* **2002,** *34,* 173–177.

Othman, N. D.; Roblain, N. C.; Thonart, P.; Hamdi, M. Antioxidant Phenolic Compounds Loss During the Fermentation of Chtoui Olives. *Food Chem.* **2009,** *116,* 662–669.

Ouwehand, A. C.; Salvadori, B. B.; Fondén, R.; Mogensen, G.; Salminen, S.; Sellars, R. Health Effects of Probiotics and Cultures-Containing Dairy Products in Humans. *IDF Bull.* **2003,** *380,* 4–19.

Panagou, E. Z.; Schillinger, U.; Franz, C. M. A. P.; Nychas, G. J. E. Microbiological and Biochemical Profile of cv. Conservolea Naturally Black Olives During Controlled Fermentation with Selected Strains of Lactic Acid Bacteria. *Food Microbiol.* **2008,** *25,* 348–58.

Peñas, E.; Frias, J.; Sidroand, B.; Vidal-Valverde, C. Impact of Fermentation Conditions and Refrigerated Storage on Microbial Quality and Biogenic Amine Content of Sauerkraut. *Food Chem.* **2010b,** *123,* 143–150.

Peterson, W. H.; Fred, E. B. The Fermentation of Glucose, Galactose, and Mannose by Lactobacillus Pentoaceticus. sp. *J. Biol. Chem.* **1920,** *41,* 431–450.

Rakin, M.; Vukasinovic, M.; Siler-Marinkovic, S.; Maksimovic, M. Contribution of Lactic Acid Fermentation to Improved Nutritive Quality Vegetable Juices Enriched with Brewer's Yeast Autolysate. *Food Chem.* **2007,** *100,* 599–602.

Rivera-Espinoza, Y.; Gallardo-Navarro, Y. Non-Dairy Prebiotic Products. *Food Microbiol.* **2010,** *27,* 1–11.

Ruiz-Barba, J. L.; Jimenez-Diaz, R. Availability of Essential B-Group Vitamins to Lactobacillus Plantarum in Green Olive Fermentation. *Appl. Environ. Microbiol.* **1995,** *61,* 1294–1297.

Saavedra, J. M.; Faap, M. D.; Dattilo, A. M. Use of Probiotics and Prebiotics in Children. In *Probiotics: A Clinical Guide;* Floch, M. H., Kim, A. S., Eds.; Pub. Slack Incorporated: Thorofare, 2010; pp 141–180.

Sakellaris, G.; Niklaropoulos, S.; Evangelopoulos, A. E. Polygalacturonase Biosynthesis by Lactobacillus Plantarum: Effect of Cultural Conditions on Enzyme Production. *J. Appl. Bacteriol.* **1988,** *65,* 397–404.

Sánchez, I.; Palop, L.; Ballesteros, C. Biochemical Characterization of Lactic Acid Bacteria Isolated from Spontaneous Fermentation of 'Almagro' Eggplants. *Int. J. Food Microbiol.* **2000,** *59,* 9–17.

Schobinger, U. *Frucht-Und Gemuesesaefte, 2. Auflage. 637;* Verlageugen Ulmer: Stuttgart, Germany, 1987.

Sesena, S.; Sanchez-Hurtado, I.; Gonzalez Vinas, M. A.; Palop, L. Contribution of Starter Culture to the Sensory Characteristics of Fermented Almagro Eggplants. *Int. J. Food Microbiol.* **2001,** *67,* 197–205.

Sethi, V. Lactic Fermentation of Black Carrot Juice for Spiced Beverage. *Indian Food Packer* **1990a,** *3,* 7–12.

Sethi, V. Evaluation of Different Vegetables for Lactic Fermentation. *Indian J. Agric. Sci.* **1990b,** *60,* 638–640.

Settanni, L.; Moschetti, G. Non-Starter Lactic Acid Bacteria Used to Improve Cheese Quality and Provide Health Benefits. *Food Microbiol.* **2010,** *27,* 691–697.

Singh, S.; Gupta, S.; Pal, R. K.; Kaur, C. In *Production of a Biopreservative 'Nisin': Assay and Quantification from a Cell-Free Extract.* Paper Presented and in Book of Abstracts : Lactic Acid Bacteria, 7th Asian Conference, Sept 6–8, India Habitat Centre: New Delhi, 2013.

Singh, S.; Sethi, S.; Gupta, S.; Kaur, C. Official Report-In house-Project 'Development of Starter Cultures for Fermented Functional and/or Healthy Vegetable and Fruit Drinks'. IARI.: New Delhi, 2015.

Singh, S.; Dash, S.; Kooliyottil, R.; Saha, S.; Gupta, S.; Upadhyay, D.; Holmes, L. Nisin Production in a Two Liters Bioreactor Using Lactococcus Lactis NCIM 2114. *J. Med. Biol. Sci. Res.* **2016,** *2,* 21–26.

Solfrizzi, V.; Panza, F.; Torres, F.; Mastroianni, F.; Del Parigi, A.; Venezia, A.; Capurso, A. High Monounsaturated Fatty Acids Intake Protects Against Age-Related Cognitive Decline. *Neurology* **1999,** *52,* 1563–1569.

Spicka, J.; Kalac, P.; Bover-Cid, S.; Krizek, M. Application of Lactic Acid Bacteria Starter Cultures for Decreasing the Biogenic Amine Levels in Sauerkraut. *Eur. Food Res. Technol.* **2002,** *215,* 509–514.

Stamer, J. R. The Lactic Acid Bacteria: Microbes of Diversity. *Food Technol.* **1979,** *33,* 60–65.

Steinkraus, K. H. Classification of Fermented Foods: Worldwide Review of Household Fermentation Techniques. *Food Control* **1997,** *8,* 311–317.

Stiles, M. E. Potential for Biological Control of Agents of Foodborne Disease. *Food Res. Int.* **1994,** *27,* 245–250.

Sura, K.; Garg, S.; Garg, F. C. Microbiological and Biochemical Changes During Fermentation of Kanji. *J. Food Sci. Technol.* **2001,** *38,* 165–167.

Tamang, J. P.; Tamang, B.; Schillinger, U.; Franz, C. M. A. P.; Gores, M.; Holzapfel, W. H. Identification of Predominant Lactic Acid Bacteria Isolated from Traditionally Fermented Vegetable Products of the Eastern Himalayas. *Int. J. Food Microbiol.* **2005,** *105,* 347–356.

Tanguler, H.; Erten, H. Occurrence and Growth of Lactic Acid Bacteria Species During the Fermentation of Shalgam (salgam), A Traditional Turkish Fermented Beverage. *LWT—Food Sci. Technol.* **2012,** *46,* 36–41.

Tolonen, M.; Taipale, M.; Viander, B.; Pihlava, J. M.; Korhonen, H.; Ryhanen, E. L. Plant Derived Biomolecules in Fermented Cabbage. *J. Agric. Food Chem.* **2002,** *50,* 6798–6803.

Tolonen, M.; Rajaniemia, S.; Pihlavaa, J. M.; Johanssonc, T.; Sarisb, P. E. J.; Ryhanen, E. L. Formation of Nisin, Plant-Derived Biomolecules and Antimicrobial Activity in Starter Culture Fermentations of Sauerkraut. *Food Microbiol.* **2004,** *21,* 167–179.

Valerio, F.; De Bellis, P.; Lonigro, S. T.; Morelli, L.; Visconti, A.; Lavermicocca, P. In vitro and in vivo Survival and Transit Tolerance of Potentially Probiotic Strains Carried by Artichokes in the Gastrointestinal Tract. *Appl. Environ. Microbiol.* **2006,** *72,* 3042–3045.

Van Etten, C. H. Goitrogens. In *Toxic Constituents of Plant Foodstuffs;* Liener, I., Ed.; Elsevier Pub.: Washington, DC, 2012; p 121.

Vanderzant, C.; Splittstoesser, D. *Compendium of Methods for the Microbiological Examination of Foods;* American Public Health Association: USA, 1992.

Verhoeven, D. T. H.; Verhagen, H.; Goldbohm, R. A.; van den Brandt, P. A.; van Poppel, G. A Review of Mechanisms Underlying Anti-Carcinogenicity by Brassica Vegetables. *Chem.-Biol. Interact.* **1997,** *103,* 79–129.

Wagner, P. H.; Konishi, T.; Rahman, M. M.; Nakahara, M.; Matsugo, S. Free Radical Scavenging and Antioxidant Activity of Ascorbigen Versus Ascorbic Acid: Studies in vitro and in Cultured Human Keratinocytes. *J. Agric. Food Chem.* **2008,** *56,* 11694–11699.

Willett, W. C.; Sacks, F.; Trichopoulou, A.; Drescher, G.; Ferro-Luzzi, A.; Helsing, E.; Trichopoulos, D. Mediterranean Diet Pyramid: A Cultural Model for Healthy Eating. *Am. J. Clin. Nutr.* **1995,** *61* (Suppl. 6), 1402–1406S.

Wong, W. S. D. *Food Enzymes;* Chapman and Hall: New York, 1995; pp 212–236.

Wood, W. A. Fermentation of Carbohydrates. In *The bacteria, Metabolism;* Gunsalus, I. C., Stanier, R. Y., Eds.; Academic Press Inc.: New York, 1961; Vol. 2, pp 59–149.

Yoon K. Y.; Woodams E. E.; Hang Y. D. Production of probiotic cabbage juice by lactic acid bacteria. *Bioresour Technol.* **2006,** 97(12), 1427-1430.

Zwiettering, M. H.; Jongenburger, I.; Rombouts, F. M.; van'T Riet, K. Modeling of the Bacterial Growth Curve. *Appl. Environ. Microbiol.* **1990,** *56,* 1875–1881.

CHAPTER 15

RECENT DEVELOPMENT IN DEHYDRATION OF VEGETABLE CROPS

V. R. SAGAR[1], SUDHIR SINGH[2], AND SURESH KUMAR P.[3]

[1]*Division of Postharvest Technology, Indian Agricultural Research Institute, New Delhi, India*

[2]*Indian Institute of Vegetable Research, Varanasi, India*

[3]*ICAR—National Research Centre for Banana, Tiruchirapalli, Tamil Nadu, India*

CONTENTS

15.1   INTRODUCTION

India ranks the second in fruits and vegetables production in the world, after China. In India, more than 40 kinds of vegetables belonging to different groups, such as Solanaceous, Cucurbitaceous, Leguminous, Cruciferous, root crops and leafy vegetables are grown in tropical, subtropical and temperate regions. Vegetables are important protective foods and highly beneficial for the maintenance of fresh and prevention of disease. Dietary fibre concentrates from vegetables showed a high total dietary fibre content and better insoluble and soluble dietary fibre ratios than cereal bran's (Grigelmo-Miguel and Martin- Belloso 1999). Vegetables are important sources of essential dietary nutrients such as vitamins, minerals and fibre. As the moisture content of fresh vegetables is more than 80%, they are classified as highly perishable commodities (Orsat et al., 2006). Keeping the product fresh is the best way to maintain its nutritional value, but most storage techniques require low temperatures, which are difficult to maintain throughout the distribution chain (Sagar and Kumar, 2010).

India is the largest producer of vegetables of which 15–20% of fruits and vegetables are lost at various points of supply chain, thus leading to huge post-harvest losses. On the other hand, drying is a suitable alternative for post-harvest management especially in developing countries such as India where cold chain management is poorly established during whole supply chain of vegetables at the time of handling and distribution.

Drying is one of the simplest low-cost technologies to enhance the shelf life of vegetables and to increase the aesthetic quality of vegetables. Drying simultaneously combines heat and mass transfer. The fundamental aspect of drying is to reduce the availability of water in food to such an extent which is not favourable for undesirable microorganisms and favourable for minimized rates of chemical reactions. Drying not only facilitates reduction in bulk of fresh material, easy transport because of reduced weight and volume but also increases the availability of vegetables during off season.

The drying operation in convective heating involves moisture transfer from wet material to heated air, which may be reflected as a transport of moisture from the material core to its surface, followed by evaporation from the surface of the material and dissipation of water vapour into the bulk of the drying air. Drying as such without suitable additive treatments causes irreversible structural damage to the cellular structure of vegetables. The damage to the cellular structure further prevents the proper drying, and rehydration of dried products is affected.

The terms *dried* and *dehydrated* are used interchangeably, USDA lists dehydrated foods as those which contain no more than 2.5% water (dry basis), whereas dried foods apply to any food product with more than

2.5% water (dry basis). It is estimated that over 20% of the world perishable crops are dried to increase the shelf life and to promote food security (Grabowski et al., 2003; Sagar and Kumar, 2010). Vegetables and their products are dried to enhance the storage stability to minimize the packaging requirements and to reduce the transportation cost.

The market for dehydrated vegetables and fruits is important for most countries worldwide (Funebo and Ohlsson, 1998). For example, US $ 7.6 billion worth of dehydrated vegetables, instant dried soup, and seaweed were consumed annually in Japan, excluding uses in restaurants and institutions. In China, the production of dehydrated vegetables is worth about US $ 800 million, including US $ 420 million for dehydrated red pepper, about 60–70% (about 230,000 t) for export (Liu, 2003). In Europe, the market for dehydrated vegetables was estimated to be worth US $ 260 million in the early 1990s (Tuley, 1996). In the United States of America, a large market exists for dehydrated grape (raisin), garlic, onion and tomato (Liu, 2003). The growth and popularity of convenience foods in many Asian countries has created increasing market demand for high-quality dehydrated fruits and vegetables. The demand is expected to increase over the next decades in all emerging economies of the world.

Nonetheless, in India, vegetables are dried on a very small scale. Drying process has not gained popularity in India as consumers have biggest concern on the perception of quality, and majority of Indian consumers are satisfied with homemade vegetable dishes. Drying processes can be channelized to improve the sensory perception of vegetables in terms of flavour, colour and body and texture of rehydrated vegetables along with handling of large quantities of vegetables during peak season of production. This would prevent the enormous post-harvest losses of perishable vegetables and would help in maintaining nutritional security as well as prevent the high prices of offseason vegetables (Sagar and Kumar, 2010).

Consumers today are highly pertinent to quality and safety of processed food products. Therefore, alternate technologies which offer convenience of traditional drying without compromising on quality are of much demand. The technologies include low-temperature drying, freeze drying and vacuum drying. The most applicable method of drying includes freeze, vacuum, osmotic, cabinet or tray, fluidized bed, spouted bed, ohmic, microwave and combination thereof (George et al., 2004). All the drying methods except freeze drying apply heat during drying through conduction, convection and radiation. These are the basic techniques to force water to vaporize, whereas forced air is applied to encourage the removal off the vapour. A large number of food and biomaterials are dehydrated in a variety of units with diverse processing conditions. The choice of drying method depends on various factors such as the type of product, availability of dryer, cost of dehydration and final quality of desiccated product. Energy consumption and quality of the dried produces are other critical areas in the selection of a drying process. To reduce the cost of fossil fuel, electrical energy is an alternate source of energy for drying applications, especially in which electricity is generated by a renewable energy source such as hydro power or wind power (Raghavan et al., 2005 and Raghavan and Orsat, 1998).

15.2   DEHYDRATION

The drying of vegetables has been principally accomplished by convective drying (Nijhuis et al., 1998). The selection of dryer should be based on the entire manufacturing process. Raw materials, intermediate product, final product specifications and characteristics need to be clearly defined. The selection of dryer should include the production capacity, initial moisture content of the product, particle-size distribution, drying characteristics of the product, maximum allowable product temperature, explosion characteristics, moisture isotherms and physical nature of the product (Vega-Mercado et al., 2001). The selection of dryer depending on the unit operation is mentioned in Figure 15.1.

Dehydration in general for practical reasons could be grouped in to two, namely traditional and modern methods. The traditional methods basically involves drying using heat, such as solar drying, sun drying, cabinet drying, drum drying, tunnel drying, fluidized bed driers and spray driers, whereas the modern methods involve low heating along with combination of other process variables such as vacuum, microwave, pneumatic pressure, pulsed electric field, ohmic heating and so forth (Sagar and Kumar, 2010). In other words, the dehydration industry could be divided into four Gs—first generation includes cabinet and bed-type dryers (kiln, tray, truck tray, rotary flow conveyor, tunnel); second generation (spray and drum dries); third generation (freeze and osmotic dehydration); and the fourth generation includes hybrid and alternate drying methods (high vacuum, microwaves, refractance window, hurdle approach, pulsed electric field).

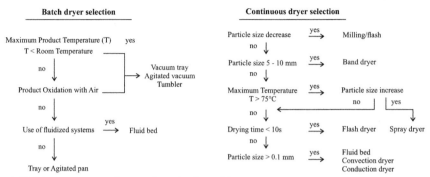

FIGURE 15.1   Flow diagram for selection of dryer.

15.2.1   TRADITIONAL METHODS OF DRYING

15.2.1.1   SOLAR DRYING

Solar drying is probably the oldest industrial process in which plenty of sunshine is available, even currently being followed. Its uses dates back to centuries and has been used with many different products including fruit, meat, fish and plants. However, earlier solar dryers have limitations of requirement of larger space area, high recurring costs, no control on drying and no control of insect infestation. Nowadays, solar dryers are designed to trap maximum solar energy, provision for the circulation of hot air across the trays as well as escape of humid air from dryers. The trays can be put together at definite distances for effective drying of vegetables.

15.2.1.2   CONVECTIVE TRAY DRYING

Similar to solar and sun drying, convective dryers are simplest and most convenient method of dehydration of vegetables. The drying process takes place in an enclosed, heated chamber, and hot air is allowed to pass through the product which is normally kept in trays (Kumar et al., 2013). This can be either batch or continuous type and could be used for relatively low-value crops. The air may be heated by direct or indirect methods. Direct methods of drying involve contact of dried air with vegetables to be dried, whereas indirect methods did not allow the air to be contaminated as it totally oxidized. Kiln-type driers, cabinet/pan driers belong to the direct heating type, whereas continuous belt tunnel driers, belt trough driers, air lift drier and fluidized bed driers belong to the indirect heating category (Sagar and Kumar, 2010).

A number of studies that have addressed the problems are associated with conventional convective drying. Some important physical properties of the products are changed, such as loss of colour (Chua et al., 2000), change of texture, chemical changes affecting flavour and nutrients and shrinkage (Mayor and Sereno, 2004; Kumar and Sagar, 2014). Rehydration time takes longer time, and the quality of rehydrated vegetables is also unacceptable due to improper rehydration and existence of leathery body (Khraisheh et al., 2004). The high temperature of the drying process has also detrimental factors affecting the quality of dried vegetables. The drying process time can be regulated depending upon the lowering the process temperature. It has great potential for improving the quality of dried products (Nindo et al., 2003 and Beaudry et al., 2004). However, the total operating time and processing cost is increased.

15.2.1.2.1 *Tray Drying of Vegetables*

Tray driers have insulated cabin made up of iron, provision for circulating hot air through fan and source for heating the air. There has been provision for handling of fresh vegetables already cut into slices and put into thin layers in trays, Fresh air is sucked by the fan through the heater coils and is then blown across the trays for drying. Air is heated by indirect methods such as electrical coils. Heated air is passed across the trays and in between the trays. There has been the provision for removal of humid air from the dryer. These dryers are normally preferred for small-scale operation. Normally, vegetables are dried at 50–60°C, and drying should be completed in 6–8 h. Dried vegetables should have final moisture content of 3–4% moisture. Dried vegetables are cooled to room temperature and are suitable packed in plastic pouches and subsequently heat sealed.

Singh and Singh (2015) standardized the drying process of vegetables using blanching treatment with permitted additives for retention of green, ascorbic acid with acceptable sensory properties along with good rehydration ratio and aesthetic quality in rehydrated vegetables.

Drying of Okra

Freshly harvested okra deteriorates vary fast, and shelf life under ambient storage is not more than 2 days. However, large quantities of okra can be dried during peak production and can be stored in sealed polypropylene pouches at ambient temperature for 6–8 months without adversely affecting the quality.

Flow diagram of drying of Okra

Okra
↓
Sorting, Grading and Washing
↓
Blanching in boiling water containing 0.1% magnesium oxide and 0.5% sodium bicarbonate for 2-3 min
↓
Cutting into 0.25-0.3 cm slices
↓
Drying in cabinet dryer at 55-60°C for 6-8 hr
Cooling at room temperature
↓
Packaging in polypropylene pouches and storage at ambient temperature (20-25°C) for 6-8 months

Drying of Indian Beans

Indian beans are store house of nutrients such as carbohydrates, proteins, minerals, cholesterol-lowering fibre and free from fat. Indian beans can be effectively dried in a tray drier at 55–60°C for 10–12 h to reduce the final moisture to less than 1% with good rehydration property.

Flow diagram for drying of Indian beans

Indian beans
↓
Washing in running water
↓
Cutting into 2.0-2.5 cm size with sharp stainless steel knife
↓
Blanching in boiling water containing 0.1% magnesium oxide for 1-1.5 min followed by dipping into 1% sodium sulphite solution for 10 min
↓
Drying in cabinet dryer
↓
I stage of drying at 60°C for 3-4 hrs
↓
II stage of drying at 50-55°C for 8-10 hrs
↓
Cooling at room temperature
↓
Packaging in polypropylene pouches and storage at room temperature at 20-25°C for 5-6 months

Drying of Tomato powder

Tomato is a rich source of lycopene, an antioxidant with immunostimulatory properties, a good source of ascorbic acid, b-carotene, phenolics such as ferulic, chlorogenic and caffeic acid and less loss of red lycopene under starch gelatinization process due to fewer *cis*-isomers. Tomato-drying process involves blanching of tied tomatoes in muslin cloth in boiling water for 2 min followed by immediate cooling in chilled water. Blanched tomatoes are cut into thin slices of 0.25–0.3 cm and dipped into 2% gelatinized starch solution for 10 min and subsequently dried in cabinet dryer at 45–55°C for 8–10 h. Dried tomato slices are ground in a grinder and passed through a 25–30 mesh sieve to get fine tomato powder.

Flow diagram for drying of Tomato powder

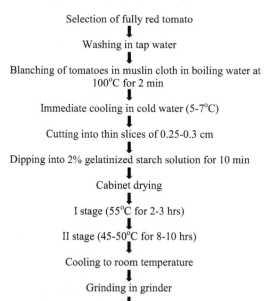

Selection of fully red tomato
↓
Washing in tap water
↓
Blanching of tomatoes in muslin cloth in boiling water at
100°C for 2 min
↓
Immediate cooling in cold water (5-7°C)
↓
Cutting into thin slices of 0.25-0.3 cm
↓
Dipping into 2% gelatinized starch solution for 10 min
↓
Cabinet drying
↓
I stage (55°C for 2-3 hrs)
↓
II stage (45-50°C for 8-10 hrs)
↓
Cooling to room temperature
↓
Grinding in grinder
↓
Packaging in poly propylene pouches and storage at 15-20°C for 4-5 months

Drying of Cabbage leaves

In this process, cabbage leaves cut into 1–1.5 cm shreds are blanched in boiling water containing 0.1% magnesium oxide followed by dipping into 1% sodium sulphite solution for 10 min. Cabbage-shredded leaves are dried in a cabinet dryer at 50–55°C for 5–7 h.

Flow diagram of drying of cabbage leaves

Cabbage heads
↓
Washing and cutting into 1-1.5 cm in
length with sharp edged stainless steel knife
↓
Blanching of cabbage shreds in 0.1% Magnesium oxide at 100 °C for
30 sec followed by dipping into 1% sodium sulphite solution
for 10 min
↓
Drying in cabinet dryer at 50-55°C for 5-7 hrs
↓
Cooling and packaging in polypropylene pouches for 5-6 months at 20-25°C

Drying of Fenugreek leaves

Fenugreek leaves are rich sources of minerals such as calcium, magnesium, potassium and iron along with dietary fibre. Immediately after harvest and handled for 2–3 h, fenugreek leaves start wilting; as a result, market price gets substantially reduced. Large quantities of fenugreek leaves after removal of stalk are blanched and dried at 50–55°C for 4–5 h to reduce the final moisture content less than 1%.

Flow diagram of Fenugreek leaves

Fenugreek leaves
↓
Sorting, grading and removal of field weeds
↓
Washing thoroughly in water 3-4 times to remove dirt
particles
↓
Blanching in 0.1% magnesium oxide,
0.1% sodium bicarbonate and 0.5% potassium metabisulphite
solution at 80°C for 30-60 sec
↓
Drying in cabinet dryer at 55-60°C for 4-5 hrs
↓
Cooling and packaging in polypropylene pouches and
storage for 6-7 months at 20-25oC

Drying of Bathua leaves

Bathua leaves are store houses of many nutrients such as protein, dietary fibre, along with good for functioning of liver, spleen, gall bladder and act as appetizer and help in controlling hypertension. Bathua leaves free from stalks are cleaned and thoroughly washed 3–4 times in running water and thereafter blanched and dried in cabinet drier at 50–60°C for 4–5 h.

Flow diagram of drying of *Bathua* leaves

Bathua leaves
↓
Sorting, grading, cleaning and washing in
running water 3-4 times
↓
Blanching in 0.1% magnesium oxide,
0.1% sodium bicarbonate and 0.5% potassium
metabisulphite solution at 80°C for 30-60 sec
↓
Drying in cabinet dryer at 55-60°C for 4-5 hrs
↓
Cooling and packaging in polypropylene pouches and
storage for 6-7 months at 20-25°C

Drying of Spinach leaves

Spinach leaves contain many health promotional and many disease prevention properties. It is also rich sources of several polyphenolic antioxidants such as lutein, zeaxanthin, b-carotene and vitamins such as vitamin A and vitamin C, many vitamin B complex such as thiamine, folate, riboflavin, niacin and rich sources of minerals such as iron, potassium, magnesium, copper and zinc.

Flow diagram of drying of Spinach leaves

Spinach leaves
↓
Sorting, grading, cleaning of field weed and thoroughly washing
3-4 times in water
↓
Blanching with 0.1% magnesium oxide, 0.1% sodium
bicarbonate and 0.5% potassium metabisulphite solution at 100°C
for 30 sec
↓
Drying in cabinet dryer at 50-55°C for 5-6 hrs
↓
Cooling and packaging in polypropylene pouches and storage
for 8-10 months at 20-25°C

Drying of Amaranth leaves

Amaranth leaves are rich sources of many vitamins such as vitamins A, C, D, E, K, B complex and folate along with fairly good sources of minerals such as calcium, magnesium, phosphorus, potassium, zinc, copper and selenium. Amaranth leaves free from stalk are cleaned and washed 3–4 times in running water and afterwards blanched and dried in a cabinet dryer at 50–55°C for 4–5 h.

Flow diagram of drying of Amaranth leaves

Amaranth leaves
↓
Sorting, grading, cleaning of field weed and thoroughly
washing 3-4 times in running water
↓
Blanching with 0.1% magnesium oxide,
0.1% sodium bicarbonate and 0.5% potassium
metabisulphite solution at 100°C for 15-20 sec
↓
Drying in cabinet dryer at 50-55°C for 4-5 hrs
↓
Cooling and packaging in polypropylene pouches and
storage for 8-10 months at 20-25°C

Drying of Moringa leaves

Moringa leaves contain significant quantities of proteins, vitamins such as vitamins A–C and minerals such as calcium, iron and phosphorus and fairly high amounts of essential amino acids with right balance of minerals and vitamins. The regular consumption of moringa leaves controls diabetes and hypertension in adults. Moringa leaves are blanched in water containing 0.1% magnesium oxide followed by dipping into 1% sodium sulphite solution. Moringa leaves are dried in a cabinet dryer at 50–55°C for 5–7 h. As a result, the final moisture is reduced to less than 1%.

Flow diagram for drying of Moringa leaf

Moringa leaves
↓
Washing
↓
Blanching with 0.1% magnesium oxide at 100°C for 30 sec
100°C for 30 sec followed by dipping into 1% sodium
sulphite solution for 10 min
↓
Cabinet drying at 50-55°C for 5-7 hrs till the final moisture of 1%
↓
Blending in grinder and passing to 30-35 mesh sieve
↓
Packaging in polypropyline pouches and storage at
20-25°C for 6-7 months

Drying of Moringa pod

Similar to moringa leaves, moringa pods are rich sources of proteins, vitamins, minerals and bioactive compounds. Tender moringa pods are cut into small pieces are blanched in boiling water containing 0.1% magnesium oxide for 2–3 min followed by dipping into sodium sulphite solution for 10 min. Moringa pods are dried in a cabinet dryer at 55–60°C for 5–6 h.

Flow diagram of drying of Moringa pod

Moringa pod

↓

Washing and cutting into 5-6 cm size

↓

Blanching in 0.1% magnesium oxide at 100°C for 2-3 min
followed by dipping into 1% sodium sulphite solution

↓

Drying in cabinet dryer at 55-60°C for 5-6 hrs

↓

Cooling and packaging in polypropylene pouches and
storage at 20-25°C for 6-8 months

15.2.1.3 ROLLER DRYING

Roller drying process involves placing of vegetable slurry, paste or concentrated fluid as thin film upon the smooth surface internally steam heated and rotating metal drum. The film of dried product is continuously scrapped off by a stationery knife located opposite the point of application of the feed. The drums used for roller driers are 24–48 in. in diameter and 12 ft in length. Drums are carefully machined both from inside and outside so that the thickness of the drum throughout its length is the same. This assists uniform heat transfer during drying. The speeds of drums are adjustable and usually range between 1 and 36 rpm depending upon the concentration of vegetables slurry and degree of dryness required. The dried vegetable slurry or concentrated dried mass is removed after 3/4 to 7/8 of a revolution. The concentrated vegetable slurry remains with the contact of roller for about 3 s or less. One of the double drums remains mounted on a stationery bearing and the other is movable. The spacing between the drums is about 0.02–0.04 in. Steam is fed into the centre of drum at one end of the shaft through hub at about 90 psi of dry saturated steam (150°C). The condensate that moves to the bottom of hub of drum is removed by pump or siphon. The efficiency of a drier is affected by the temperature of feed rate, distance between the rollers, speed of rollers and steam pressure. Typically, the products such as mashed potato, purees and tomato concentrates are dried by the drum driers. This technique has been used extensively in the past, but its use for food processing is decreasing as other methods of drying that cause less heat damage have gained popularity.

15.2.1.4   SPRAY DRYING

Spray drying is a process of instantaneous removal of moisture from a liquid being taken place. Vegetable juices are clarified first followed by converting into fog-like mist by atomisation to produce large surface area. The atomized liquid is then exposed to a flow of hot air in a drying chamber. The hot air has the function of supplying heat for the evaporation and also acts as carrier of the vapour and the powder. The atomized mist comes into the contact of hot air which immediately evaporates moisture and the solids are recovered as fine hollow spherical powder with some occluded air. Air inlet temperature is generally maintained at 200–215°C for drying of concentrated vegetable juices (Hall, 1996). Uniformity of drop size and homogeneity of the spray jet are important consideration in designing nozzle. Pneumatic two-fluid nozzle, pressure nozzle and cone nozzle are most commonly used. Drying through spray drying may be either single stage, two-stage or three stage. The dried product is often agglomerated to facilitate its rapid dispersion in water. Pneumatic nozzle type driers mostly worked with single-stage drying. Two-stage system comprises the spray dryer followed by a vibro-fluidized bed system. Three-stage processes improve the properties of dried powder by instant reconstitution because the fluid bed works as a dryer–agglomerator, controlling particle agglomeration (Master, 2004). Some examples of products that have been spray dried are instant coffee, tea and powdered milk. Spray drying can be combined with a fluidized-bed dryer, which will convert the liquid directly into an agglomerated product. A large disadvantage with this process is the size of the equipment required to achieve drying. Furthermore, not all materials can be dried in this way; for instance, very oily materials might require special preparation to remove excessive levels of fat before atomization.

15.2.1.5   OSMOTIC DEHYDRATION

Osmosis is known as movement of water from higher concentration to lower concentration through semipermeable membrane. The blanched cut/cored/sliced vegetable pieces are osmotically diffused in 5–6% sodium chloride solution at 50–60°C for 120–150 min. Around 2–5% moisture is removed during osmotic diffusion process which resulted in uniform drying and subsequent optimum rehydration. Although it does not remove enough moisture to be considered as a dried product, the process has the advantage of requiring little energy (Kumar et al., 2013). It works well as a pretreatment prior to applying other drying methods. The application of osmotic dehydration in vegetables has achieved greater attention in recent past years as a technique for production of intermediate moisture foods or as a pre-treatment prior to drying in order to reduce energy consumption or heat damage (Jayaraman and Gupta, 1992).

In osmotic dehydration process, vegetables with high moisture content are immersed into concentrated solution of sugar and salt solution, which initiates three types of counter current mass transfer.

1. Water outflow from vegetables to the surroundings solution through a semi-permeable membrane.
2. A solute transfer from solution to the product.
3. A leaching out of the water-soluble component along with water from the product to solution,

Nsonzi and Ramaswamy (1998) modelled the mass-transfer process with respect to moisture loss and solids gain and reported that even though the moisture loss and solids gain occurred at the same time, the rate of moisture loss was much higher than the rate of solids gain. The advantage of osmotic dehydration is its lower energy use and lower product thermal damage as lower temperatures are used allowing the retention of nutrients (Shi et al., 1997). Lenart (1996) described the main advantages of using osmotic dehydration, namely as the reduction of process temperature, reduction of 20–30% energy consumption and shorter drying time following osmotic dehydration.

The driving force for the diffusion of water from the tissue into the solution is provided by the higher osmotic pressure of the hypertonic solution. The rate of mass transfer during osmotic diffusion is generally low.

Singh and Singh (2015) standardized the osmo-air drying of vegetables with good retention of green colour, rehydration ratio and good aesthetic quality in rehydrated vegetables.

15.2.1.5.1 Process for Osmo-Air Drying of Bitter Gourd

Bitter gourd is rich sources of many vitamins such as A, B1, B2 and C and also presence of minerals such as calcium, phosphorus, copper and potassium. The regular consumption of bitter gourd provides many health benefits such as antidotal, antilypolytic, lypogenic, antipyretic tonic, appetizing, stomachic, antibilious, purgative, anti-inflammatory, antiflatulent and higher healing capacity. Washed bitter gourd is cut into thin slices of 0.25–0.3 cm by slicing/grating machine and slices are blanched in boiling water containing 0.1% magnesium oxide for 1 min followed by dipping into 1% sodium sulphite solution for 10 min. Blanched bitter gourd slices are osmotically diffused in 4–5% sodium chloride solution at 50–60°C for 60–90 min. The slices are dried in a cabinet dryer at 50–60°C for 10–12 h to reduce the final moisture to less than 1%.

Flow diagram of osmo-air drying of Bitter gourd

Bitter gourd
↓
Washing and cutting into slices (0.25-0.3 cm)
↓
Blanching of bitter gourd slices in boiling water containing 0.1% magnesium oxide for 1 min followed by dipping into 1% sodium sulphite solution for 10 min
↓
Osmotic diffusion treatment in 4-5% sodium chloride solution at 50-60°C for 60-90 min
↓
Drying in cabinet dryer
↓
I stage of drying at 60 C for 3-4 hrs
↓
II stage of drying at 50-55°C for 8-10 hrs
↓
Cooling at room temperature
↓
Packaging in polypropylene pouches and storage at room temperature at 20-25°C for 6 months

15.2.1.5.2 Osmo-Air Drying of Cauliflower Florets

Cauliflower florets are rich sources of carbohydrates, proteins, dietary fibres, vitamins such as Vitamin A (750 IU) and vitamin C (55–60 mg/100 g) and rich sources of minerals such as sodium, potassium, magnesium, iron, calcium and phosphorus. It also contained fairly large amounts of major carotenoids such as lutein, neoxanthin, violaxanthin and β-carotene. Washed cauliflower florets are blanched in boiling water for 30–40 s followed by dipping into 0.2% potassium metabisulphite solution for 10 min. Blanched florets are after osmotic diffusion treatment in 4–5% sodium chloride solution at 50–60°C for 45–60 min are dried in a cabinet dryer at 55–60°C for 8–10 h to reduce the final moisture content to less than 1%.

Flow diagram of osmo-air drying of Cauliflower florets

Cauliflower florets
↓
Washing
↓
Blanching in boiling water for 30-45 sec followed by dipping into
0.2% potassium metabisulphite solution for 10 min
↓
Osmotic diffusion treatment in 4-5% sodium chloride
solution at 50-60°C for 45-60 min
↓
Drying in cabinet dryer
↓
I stage of drying at 60°C for 3-4 hrs
↓
II stage of drying at 50-55°C for 8-10 hrs
↓
Cooling at room temperature
↓
Packaging in polypropylene pouches and storage
at room temperature at 20-25°C for 6-7 months

15.2.1.5.3   Osmo-Air Drying of Carrot Slices

Carrots are rich sources of vitamins, minerals and dietary fibre and fairly good sources of both hydrophilic as well as lipophilic antioxidants such as b-carotene, a-carotene, lutein and lycopene. Blanched carrot slices are osmotically diffused in 50–55% sugar at 50–60°C for 90 min and subsequently dried in a cabinet dryer at 50–55°C for 8–10 h to reduce the final moisture to less than 2%.1

Flow diagram of osmo-air drying of carrot slices

Carrot
↓
Grading, washing and cutting into slices (0.2-0.25 cm)
↓
Blanching in 20% sugar solution at 80-85°C for
4-5 min followed by dipping into 0.2% potassium
metabisulphite solution for 10 min
↓
Osmotic diffusion of 50–55% sugar solution
at 50-55°C for 90 min
↓
Cabinet drying of carrot slices
↓
I stage of drying at 60°C for 3-4 hrs
↓
II stage of drying at 50-55°C for 8-10 hrs
↓
Cooling at room temperature
↓
Storage in air tight plastic containers at 10-15°C for 4-6 months

15.2.1.5.4   Osmo-Air Drying of Ivy Gourd

Ivy gourd fruits are rich sources of β-carotene, a major vitamin A precursor from plant sources along with good sources of minerals such as calcium, phosphorus, iron, copper and potassium. Ivy gourds are cut into 0.3–0.35 cm slices and blanched in 0.1% magnesium oxide followed by dipping into 1% sodium sulphite solution for 10 min. Blanched ivy gourd slices are osmotically diffused in 3–4% sodium chloride solution at 50–55°C for 1 h and thereafter dried in a cabinet dryer at 50–60°C for 7–8 h.

Flow diagram of osmo-air drying of Ivy gourd

Ivy gourd
↓
Washing and cutting into 0.3-0.35 cm slices
↓
Blanching treatment at 100°C for 30 sec in 0.1% magnesium oxide followed by dipping in 1% sodium sulphite solution for 10 min
↓
Osmotic diffusion treatment in 3-4% sodium chloride solution at 50-55°C for 1 hr
↓
Drying in cabinet dryer at 50-60°C for 7-8 hrs
↓
Cooling and packaging in polypropylene pouches and storage at 20-25°C for 6-8 months

15.2.1.5.5   Osmo-Air Drying of Pointed Gourd

Pointed gourd fruits are rich sources of vitamins A, B1, B2 and C and minerals such as calcium, phosphorus, iron, copper and potassium. The consumption of pointed gourd helps in purification of blood, enhancement in digestion process and thus stimulating the normal function of liver. The process for osmo-air dried pointed gourd involves cutting into 0.2–0.25 cm thin slices followed by blanching in boiling water containing 0.1% magnesium oxide for 30 s and dipping into 1% sodium sulphite solution for 10 min. The blanched pointed gourd slices are osmotically diffused in 4–5% sodium chloride solution at 55–60°C for 1 h and dried in a cabinet dryer at 50–60°C for 8–10 h to reduce the final moisture to less than 1%.

Flow diagram of osmo-air drying of Pointed gourd

Pointed gourd
↓
Washing and cutting into 0.2-0.25 cm slices
↓
Blanching treatment at 100°C for 30 sec in 0.1% magnesium
oxide followed by dipping in 1% sodium sulphite
solution for 10 min
↓
Osmotic diffusion treatment in 4-5% sodium chloride
solution at 55-60°C for 1 hr
↓
Drying in cabinet dryer at 50-60°C for 8-10 hrs
↓
Cooling and packaging in polypropylene pouches and
storage at 20-25oC for 6-8 months

15.2.1.5.6 Process for the Manufacture of Green Chilli Powder

Green chillies are acceptable due to presence of pungent capsinoids present in pericarp and placenta of fruits. Green chillies are an important source of vitamin C. Chillies also help in digestion by stimulating saliva and gastric juice flow. The process for green chilli powder involves cutting the chillies uniformly into 1–1.5 cm size and blanching in 0.5% magnesium carbonate solution at 100°C for 5 min followed by dipping into 0.75% potassium metabisulphite solution for 10 min. The blanched chilli pieces are osmotically diffused in 4–5% sodium chloride solution at 55–60°C for 90–120 min and dried in a cabinet dryer at 50–55°C for 8–10 h. Dried chilli flakes are blended and sieved through 25–30 mesh sieve size green chilli powder.

Process for the manufacture of green Chilli powder

Green chillies
↓
Washing and cutting uniformly 1.0-1.5 cm size
↓
Blanching in 0.5% magnesium carbonate solution at
100°C for 5 min followed by dipping in
0.75% potassium metabisulphite solution for 10 min
↓
Osmotic diffusion treatment in 4-5% sodium chloride
solution at 55-60°C for 90-120 min
↓
Drying in cabinet dryer at 55°C for 8-10 hrs
↓
Blending of chilli flakes and sieving to 25-30 mesh size
↓
Packaging in polypropylene pouches and storage
at 20-25°C for 6 month

15.2.2.1   FLUIDIZED-BED DRYING

Fluidized-bed dryers have the important features of producing particles that are of uniform size and being able to maintain constant temperatures. By setting the operating conditions within narrow limits, scale-up from laboratory to commercial-sized units can be readily accomplished. The technique involves augment particulate solids in an upward-flowing gas stream, usually of hot air. Fluidization mobilizes the solid particulates, thus creating intimate contact between the dry, hot carrier gas and the solids. Drying occurs by convection. Fluidization is dependent on the characteristics of the particles: size distribution, density, shape and viscosity. Fluidized-bed drying is usually carried out as a batch process and requires relatively small, uniform and discrete particles that can be readily fluidized. Thus, small vegetable pieces, such as whole peas, are well suited for this process, whereas powders would be inappropriate as they would clog up.

15.2.2.2   VACUUM DRYING

Vacuum drying is an important process for heat-sensitive materials. The process of vacuum drying can be considered according to physical condition used to add heat and to remove water vapour. Low temperature can be used under vacuum for certain methods that might discolour or decompose at high temperature. The reduction in pressure causes the expansion and escape of gas occluded into pores (Kumar and Sagar, 2014). When the pressure is restored, the pores can be occupied by the osmotic solution, increasing the available mass-transfer surface area. Vacuum pressure (50–100 mbar) is applied to the system for the shorter time to achieve the desirable result. Vacuum-drying technology is not suited on larger scale operation due to high cost.

15.2.2.3   FREEZE DRYING

Vegetables contain natural volatile flavour which is lost during processing. Vegetable-based products which are too sensitive to withstand any heat are often freeze dried. In the freeze-drying process, the vegetable as such or cut into small pieces after blanching treatment are frozen. The frozen water is removed by sublimation and desorption under vacuum. Frozen water is normally sublimated at -92 to $-96°C$ and at pressure of 0.042–0.062 mbar. Different vegetables such as pea, carrot after blanching in 15–20% sugar syrup for 15–20 min are surface dried to remove surface moisture and freeze dried to produce easily rehydrated carrot shreds and green pea with sweet taste. Frozen vegetables become porous freeze-dried vegetables which are easily rehydrated. There has been little or no loss of flavour and aroma during freeze drying. The overall sensory quality of freeze-dried vegetables is reported to be very good because of retention of volatile flavour compounds due to less degradative processes, absolutely free from enzymatic browning, protein denaturation. Four potential rate-limiting steps have been identified: the external transfer of heat to the outer surface of the material from the heat source, the internal transfer of heat within the material, the external mass transfer of water vapour from the surface, and the internal mass transfer within the material. During the drying cycle, the thickness of the dried layer increases, thus slowing down the sublimation rate (Cohen and Yang, 1995). The long processing time requires additional energy to run the compressor and refrigeration units, which makes the process very expensive for commercial use. Therefore, freeze drying is most often used for vegetables suitable for instant vegetable soup mix for good sensory properties.

Singh and Singh (2015) standardized the process for osmo-freeze drying of vegetables with good retention of green colour, rehydration ratio and good aesthetic quality for development of instant soup mix and instant vegetable dessert mix with longer shelf life.

15.2.2.3.1   Osmo-Freeze Drying of Green Pea

Green peas are one of the most nutritious leguminous vegetables rich in health benefitting phytonutrients, minerals, vitamins and antioxidants. Fresh tender peas are relatively low in calories in comparison with cowpea and beans. Green peas are good sources of proteins, vitamins and soluble and insoluble fibres and excellent sources of folic acid and ascorbic acid. The sweetness in green peas is lost very rapidly during storage. Freeze drying can retain sweetness and texture of green pea for longer time. The process for osmo-freeze drying of green pea involves blanching in sugar syrup at 100°C for 2–3 min followed by immediate cooling in chilled water for 20–30 min. Blanched green peas are freeze dried at 0.042–0.062 mbar and at the temperature of −92 to −98°C. Freeze-dried green peas are acceptable during ambient storage for 6–8 months.

Flow diagram of osmo-freeze drying of green pea

Green pea

↓

Blanching in 20% sugar syrup at 100°C for 2-3 min

↓

Immediate cooling in chilled water for 20-30 min

↓

Surface moisture drying at room temperature

↓

Freeze drying at 0.042-0.062 mbar and -92 to -98°C

↓

Cooling and packaging in polypropylene pouches and storage at 20-25°C for 6-8 months

15.2.2.3.2   Osmo-Freeze Drying of Carrot Flakes

Carrot is rich sources of vitamins such as vitamin A, C, K, thiamine, riboflavin, pyridoxine and folates along with lipophilic antioxidants such as lycopene and lutein. The regular consumption of carrot significantly controls diabetes due to low glycaemic index as well as helpful in lowering cholesterol level due to increased dietary fibre. Freeze-dried carrot flakes are more acceptable in instant vegetable soup mix. The process for the manufacture of osmo-freeze-dried carrot flakes involves washing, grating, blanching in 20% sugar syrup at 100°C for 2–3 min followed by immediate cooling in chilled water for 20–30 min. Carrot flakes are freeze dried at 0.042–0.062 mbar and at temperature of −92 to −98°C.

Flow diagram for manufacture of osmo-freeze drying of Carrot flakes

Carrot

↓

Washing and grating to form flakes

↓

Blanching in 20% sugar syrup at 100°C for 2-3 min

↓

Immediate cooling in chilled water for 20-30 min

↓

Surface moisture drying at room temperature

↓

Freeze drying at 0.042-0.062 mbar and -92° to -98°C

↓

Cooling and packaging in polypropylene pouches and storage at 20-25°C for 6-7 months

15.2.2.3.3   Freeze Drying of Bottle Gourd

Bottle gourd fruits are good source of nutrition with minimum calories and fat. Furthermore, the consumption of bottle gourd controls urinary disorder due to flushing of excess water from body through urine. It also prevents premature greying of hair and reduced ageing effect along with helpful in early curing of jaundice and inflammation of kidneys. The conventional drying of bottle gourd in a cabinet dryer results in darkening of dried bottle gourd pieces to dark brown and black colour and adversely affects the taste of dried bottle gourd pieces. However, the quality of freeze dried bottle gourd pieces is much superior due to good colour along with a good taste. The process of freeze drying of bottle gourd pieces involves peeling and cutting into small pieces and subsequent cooking at 100°C for 15–20 min. The cooked bottle gourd pieces are immediately cooled in chilled water and freeze dried at 0.042–0.062 mbar and at − 92 to − 98°C for 6–7 months.

Flow diagram for freeze drying of Bottle gourd

Bottle gourd fruits
↓
Peeling of bottle gourd fruits
↓
Cutting into small pieces
↓
Cooking of bottle gourd pieces at 100°C for 15-20 min
↓
Immediate cooling of bottle gourd pieces in chilled water to room temperature
↓
Freeze drying at 0.042-0.062 mbar and -92 to -98°C
↓
Cooling and packaging in polypropylene pouches and storage at 20-25°C for 4-5 months

15.2.2.3.4   Freeze Drying of Curry Leaves

Curry leaves are rich sources of carbohydrates, energy, fibre along with many vitamins and minerals. Curry leaves are natural flavouring agents with a number of important health benefits which make our foods more healthy and tasty along with pleasing aroma. Curry leaves also contain antioxidant property and have the ability to control diarrhoea, gastro-intestinal tract problems such as indigestion, excessive acid secretion, peptic ulcers, dysentery and controlling diabetes along with unhealthy cholesterol imbalance. Curry leaves are believed to have cancer-fighting properties and are known to protect liver. Freeze-dried curry leaves are dark in green colour with pleasing natural taste. The process for freeze-dried curry leaves involves blanching treatment followed by freeze drying at 0.042–0.062 mbar and −92 to −98°C. Freeze-dried curry leaves are acceptable for 4–5 months at ambient storage.

Flow diagram for freeze drying of Curry leaves

Curry leaves
↓
Sorting and washing with running water
↓
Blanching in boiling water with 0.1% magnesium oxide for 15 sec followed by dipping into 1% sodium sulphite solution for 10 min
↓
Freeze drying at 0.042-0.062 mbar and -92 to -98°C
↓
Cooling and packaging in polypropylene pouches and storage at 20-25°C for 4 - 5 months

15.2.2.3.5 Freeze Drying of Onion Rings

Onions are a very significant source of antioxidants. The presence of most abundant flavonoid, quercetin is linked to preventing cancer. It also helps in reducing urinary blood infections promoting prostate health and lowering blood pressure. The presence of phytochemicals such as disulphide, trisulphide, cepaene and vinyldithins are responsible for good health, anti-cancer and antimicrobial properties. Onions are very low in sodium and free from fat or cholesterol. The presence of fibre, folic acid and vitamin B complex help the body to make more healthy new cells. The dried onions are preferred for aroma in instant soups and curries. Freeze-dried onion flakes are produced by cutting into 0.3–0.5 cm thin rings and freeze drying at 0.042–0.062 mbar and −92 to −98°C. Freeze-dried onion rings are suitable for 5–6 months during low-temperature storage (10–15°C).

Flow diagram of freeze drying Onion rings

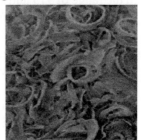

Onion
↓
Peeling of onion
↓
Washing and cutting into thin rings of 0.3 - 0.5 cm
↓
Freeze drying at 0.042-0.062 mbar and -92 to 98°C
↓
Cooling and packaging in polypropylene pouches and storage at 10-15°C for 5-6 months

15.2.2.3.6   Freeze Drying of Garlic Pieces

Garlic is effective for treatment of bronchitis, hypertension, tuberculosis, liver disorders, dysentery, flatulence, colic, removal of intestinal worms, rheumatism, diabetes and fevers. The regular consumption of garlic assists in prevention of cancers of different parts such as lung, prostate, breast, stomach, rectal cancer and colon. It is also effective in controlling atherosclerosis, coronary heart diseases, cholesterol and osteoarthritis. Garlic gives distinct taste in soups and curries. Freeze-dried garlic pieces give most acceptable garlic flavour in instant formulations. The process for freeze drying of garlic pieces involves peeling, washing, cutting into 0.2–0.3 cm length and freeze drying at 0.042–0.062 mbar and at −92 to −98°C. Freeze-dried garlic pieces are most acceptable during storage at 10–15°C for 5–6 months.

Flow diagram of freeze drying of Garlic pieces

Garlic flakes
↓
Peeling
↓
Washing
↓
Cutting into thin pieces of 0.2 − 0.3 cm
↓
Freeze drying at 0.042-0.062 mbar and -92°C to -98°C
↓
Cooling and packaging in polypropylene pouches
and storage at 10-15°C for 5-6 months

15.2.2.4   MICROWAVE DRYING

Microwaves are part of electro-magnetic spectrum in the frequency range falling between radio and infrared region. Microwave heating takes place in the range of 915 and 2450 MHz. Convention heating processes generate a temperature gradient from the outside surface to the centre of the product whereas, in microwave heating, the surface temperatures are often lower due to evaporative cooling than those at the centre of the product. The predominant development has been focused recently on cooking of vegetables at home as well as preheating the vegetables for catering establishments. The dimensions and shape of vegetables govern the heating of vegetables. Appropriate size selection in relation to product dielectric properties reduces non-uniformity of heating. Unevenly shaped pieces of food may experience uneven temperature when exposed to microwaves. The thermal properties such as conductivity and specific heat as determined by product composition also play a vital role in achieving uniformity of microwave heating (Thompkinson, 2000).

Microwaves are generated by special oscillator tubes called magnetrons and klystron. These are devices that convert low-frequency electrical energy into hundreds and thousands of megacycles. The electro-magnetic energy at microwave frequency is conducted through a coaxial tube or wave guide at appoint of usage. Both magnetrons and klystron are electron tubes which generate microwaves.

Magnetron is a cylindrical diode with a ring of resonant cavities that acts as an anode structure. The cavity is the space in the tube which becomes excited in a way that makes as a source for the oscillation of microwave energy. The magnetron is a vacuum valve in which the electron, emitted by the cathode turn around under the action of a continuous electric field. The movement produces electromagnetic radiation.

Klystron is a vacuum tube in which oscillations are generated by alternative action of electron beam. This results in periodic action of electron beams. Klystron uses the transit time between two designated points to produce the modulated electron stream which then delivers pulsating energy to a cavity resonator and sustain oscillation within the cavity.

The heating effect of microwave is achieved by the transfer of energy to a dipole within the product. Water that exhibits dipolar behaviour is present in large quantities in vegetables in the range of 90–98%. The microwaves upon entering the product interact with regions of positive and negative charges of water molecules. The dipoles rotate to align with the alternating field, but as this is rapidly changing from oscillating, the dipoles are constantly on the move and generate heat due to molecular friction. The elevated temperature produced in the area of high water concentration transfers heat by conduction/convection to other parts of the product. The dissolved salts of positive and negative ions also interact with an electric field by migrating towards positively charged regions and disrupt hydrogen bonds with water to generate additional heat (Zhang et al., 2005).

Microwave-drying process consists of three drying periods. (1) A heating-up period in which MW energy is converted into thermal energy within the moist materials, and the temperature of the product increases with time. Once the moisture-vapour pressure in food becomes more than that of the environment, the material starts to lose moisture, but at relatively smaller rates. (2) Rapid drying period, during which a stable temperature profile is established, and thermal energy converted from microwave energy is used for the vaporization of moisture. In porous food structures, rates of moisture vaporization at different locations in foods depend, to a large extent, upon the local rates of thermal energy conversion from microwave. (3) Reduced drying rate period, during which the local moisture is reduced to a point when the energy needed for moisture vaporization is less than thermal energy converted from microwave. Local temperature then may rise above the boiling temperature of water. Even though loss factors of the food materials decrease with moisture reduction and the conversion of microwave energy into heat is reduced at lower moisture content, product temperature may still continue to rise, resulting in overheating or charring. During microwave-drying processes, the heating period is relatively short and moisture loss is small. Dielectric heating with microwave energy has found industrial applications in drying of fruits and vegetables. There is a renewed interest in exploring the unique characteristics of microwave heating for drying heat-sensitive materials (Funebo and Ohlsson, 1998).

The use of microwave energy in drying offers additional advantages in reduced drying times and assists conventional drying in later stages by specifically targeting the internal residual moisture (Osepchuk, 2002). In dried carrots, great quality is obtained in terms of colour, shrinkage and rehydration property (Wang and Xi, 2005). In general, microwave drying can meet the four major requirements in drying of foods: speed of operation, energy efficiency, cost of operation and quality of dried products (Gunasekaran, 1989).

Process	Experiment type	Functions
Tempering	Batch	Delivering energy into top and bottom cavities
	Continuous	Open-ended cavities with microwave suppression tunnels at both ends
Cooking	Continuous/steam	Long cavity with magnetron in upper compartment and steam injection manifold at back of cavity
	Ladder-line applicator	Having a Teflon conveyor belt and slow energy distribution across belt
Dehydration	Continuous	Power unit installed at top of cavity carrying belt suitable for drying of potato chips and thin sliced vegetables
Baking	Continuous	Combination of microwave energy and hot air using conveyor belt having 896-MHz power units

15.2.2.5 BALL DRYING

The material to be dried is added to the top of the drying chamber through a screw conveyor. The conveyor assures a constant rate of product addition, but it can be bypassed and the material can be added directly to the drying chamber. Heated air is also added continuously to the chamber. The material within the drying chamber comes into direct contact with heated balls made from ceramic or other heat-conductive material. Drying occurs primarily by conduction. The large screw within the chamber rotates during the entire drying process, and the speed of rotation governs the dwell time of the product within the chamber (Cohen and Yang, 1995). When the product arrives at the bottom of the chamber, it is separated from the balls and collected. Except for temperature, the most important variable to control is that of rotational speed. Relatively small particles such as vegetable pieces must be used. If the material has excessive sugar content, as is the case with fruit in syrup, the material tends to stick to the drying balls and cannot be separated. The ball-drying process can be run at somewhat lower temperatures (70°C) than all of the other techniques described except for freeze drying (Cohen and Yang, 1995). However, sanitation can be a concern at low temperatures because of the extended length of the drying process (Table 15.1).

TABLE 15.1 Properties of Various Drying Methods.

Dehydration method	Produces suited	Remarks
Solar drying	Vegetables, meat and Fish	Simple and low cost to operate. But it requires larger space, slow and labour intensive
Convection drying	Low-value products	It is continuous and commonly preferred for drying of vegetables and difficult to control drying
Drum drying	Liquids especially milk	It requires more skill to operate
Spray drying	Puree, liquids, vegetable-based soup, tomato, tea and coffee	It is preferably suitable for liquid and spherical products
Fluidized bed drying	Pea, mushroom and small particles	Suitable for batch operation and provides uniform drying
Freeze drying	High-value products	It is slow and expensive and suitable for high valued processed products

TABLE 15.1 *(Continued)*

Dehydration method	Produces suited	Remarks
Explosive puffing	Suitable for small particles	It imparts honey-comb structure but generates more heat and thus loss of product integrity
Osmotic drying	Fruits and vegetables	It is basically a two-step process but gives superior dried products
Microwave and dielectric heating	High-value products	It is slow and expensive
Microwave-augmented freeze drying	High-value products	It is rapid and good quality end product. Very expensive method
Ball drying	Vegetables and small pieces	It is operated relatively at lower temperature but is rapid and good-quality end products

15.3 RELATIONSHIP AMONG MATRIX MOBILITY AND QUALITY PARAMETERS

Fresh foods can be regarded as a matrix consisting of carbohydrates, proteins, fats, water and components dissolved in water. In fresh foods, the molecular mobility of compounds within the water phase is high, and therefore they are sensitive to chemical, enzymatic, microbial and physical deterioration. Drying of fresh foods leads to a reduced water content and therefore dissolved components get concentrated to crystalized materials such as sugars due to sublimation or evaporation of water, depending on the drying method. During the concentrating process, the amorphous components of the concentrated solution will turn into a glass when the maximum concentration is reached. If the matrix is cooled sufficiently fast, and/or crystallization inhibitors are used, components that would naturally crystallize may also remain in the amorphous, glassy state. A glass can be characterized by its phase transition, which occurs in a range of temperatures, between the so-called onset and the endset. To use a single value, it is common to use the average between the onset and endset, as measured in a differential scanning calorimetry (DSC) scan, as the glass transition temperature (T_g). This second-order transition is accompanied by an increased molecular mobility and a drastic drop of the elastic modulus and can be measured by means of DSC (as a change in heat capacity), dynamic mechanical spectroscopy (as a change in elastic modulus) or magnetic resonance spectroscopy (as a change in molecular mobility). The glass transition temperature decreases with increasing water content: Water acts as a plasticizer/softener for sugars. Drying of foods is a way to increase the glass transition temperature. Sugars in their turn can act as a plasticizer for higher molecular weight carbohydrates. The glass transition temperatures of anhydrous carbohydrates increase with increasing molecular weight, which provides a tool for manipulating the glass transition temperature of food products, that is, increasing it by adding high-molecular weight carbohydrates.

15.4 EFFECT OF DRYING ON QUALITY OF VEGETABLES

In the past, research and development in drying focused mainly on the process and technology. Food drying was performed principally to extend the shelf-life, without much considerations for retaining quality attributes. Due to an increase in consumer demand for improved quality of processed products, efforts have been directed at developing high-quality dried foods. This can be achieved by using novel drying technologies, improving and optimizing existing drying methods, and maximizing quality attributes such as structure, colour, flavour, nutrition, bulk density, shearing and puncture strength and rehydration ratio and so forth (Beaudry et al., 2003; Okos et al., 1992, Sablani, 2006a; Rahman, 2005).

An air velocity and size is essential to limit the resistance to air drying to the interior of the vegetables (Mulet et al., 1989; Marinos-Kouris and Maroulis, 1995). Thus, the diffusion of water prevails to the resistance, and the resistance at the exterior of the product is not very important. The effect of air humidity on the acceleration of the drying progress is essential, in general, as lower than that of air temperature. There used to be an acceleration of the drying process due to the decrease of the air humidity of the drying air from 40 to 20%.

The effects of drying parameters, such as air temperature, relative humidity of drying air, air velocity and particle size on the progress of the drying process of various vegetables, were investigated (Krokida et al., 1998). A drying model was developed, incorporating the effect of the above parameters on the drying process. The drying model predicted successfully the drying of several vegetables. Temperature of drying is the most important factor of drying rate for all the examined materials, whereas the effect of air velocity and air humidity is considered lower than that of air temperature.

Conventional air drying is the most frequently used method of dehydration in the food industry. Dried products are characterised by low porosity or high apparent density. Hence, the rehydration property is poor. Hot-air drying also results in significant colour changes with low sorption capacity. Vacuum-dried materials have relatively higher porosity and are better in colour and aroma. Microwave-assisted vacuum drying combines the advantages of both vacuum and microwave drying in terms of improved energy consumption and product quality (Sagar and Kumar, 2010). Osmotic dehydration has also proven to be very effective for partial dehydration of many heat-sensitive materials. Freeze drying offers superior taste, aroma retention, no or little shrinkage and better rehydration quality of dried products. However, freeze drying is recognized as the most expensive drying process for manufacturing (Ratti, 2005). Heat pump and low-pressure-superheated steam drying have also been shown to be suitable methods for high-quality heat sensitive food and bioproducts (Okos et al., 1992, Sablani, 2006a).

15.4.1   QUALITY ATTRIBUTES

There are several changes in quality parameters in dried products during drying and storage. The extent of the changes depends on the care taken in preparing the material before dehydration and on the process used. Major-quality parameters associated with dried food products include colour, visual appeal, shape of product, flavour, microbial load, retention of nutrients, porosity-bulk density, texture, rehydration properties, water activity, freedom from pests, insects and other contaminants, preservatives, and freedom from taints and off-odours (Ratti, 2005). The state of the product, such as glassy, crystalline or rubbery, is also important. These quality parameters can be classified into four major groups: (i) physical, (ii) chemical, (iii) microbial and (iv) nutritional quality. Greater stability and quality can be achieved by maintaining the fresh or optimum conditions of the raw materials (Perera, 2005).

15.4.2   INFLUENCE OF DRYING ON PHYSICAL QUALITY

Physical changes, such as structure, case hardening, collapse, pore formation, cracking, rehydration, caking and stickiness, can also influence the
quality of final dried products. Shrinkage of biological materials during
drying takes place simultaneously with moisture diffusion and thus may
affect the moisture removal rate (Wang and Brennan, 1995). The density
and porosity are important physical properties characterizing the texture
and the quality of dry and intermediate moisture foods (Schubert, 1987).
Porosity is an important parameter in the prediction of diffusional properties of cellular foods (e.g. fruits and vegetables) during drying.

Hot-air drying usually destroys the cell structure and thereby took
more time for dehydration whereas solid state of water during freeze
drying protects primary structure and shape of products, keeping the cells
almost intact, with high porosity end products (Saguy et al., 2004). At a
low drying rate, the amount of shrinkage bears a simple relationship to the
amount of moisture removed. The moisture content at the centre of a piece
is never very much higher than that at the surface; internal stresses are
minimized and the material shrinks down fully onto a solid core. Towards
the end of drying, shrinkage is reduced so that the final size and shape of
the material is fixed before drying is completed. At a high drying rate, the
outer layers of the material become rigid, and their final volume is fixed
early in the drying. As drying proceeds, the tissues split and rupture internally forming an open structure (Van Arsdel et al., 1973).

Experiment on green beans, potato and peas revealed that the shape of
the produce such as parallelpiped, cylinder and sphere has definite impact
on the drying kinetics of produces under a fluidized dryer (Senadeera
et al., 2003). Pre-treatments given to foods before drying or optimal drying
conditions are used to create a more porous structure so as to facilitate
better mass transfer rates. Maintaining level moisture gradients in the
solid, which is a function of drying rate, can reduce the extent of crust
formation; the faster the drying rate, the thinner the crust (Achanta and
Okos, 1996). Depending on the end use, hard crust and pore formation
may be desirable or undesirable. Rahman (2001) has outlined the current
knowledge on the mechanism of pore formation in foods during drying
and related processes. The glass transition theory is one of the concepts
proposed to explain the process of shrinkage and collapse during drying
and other related processes. According to this concept, there is negligible

collapse (more pores) in a material when it is processed below the glass transition. The higher the process temperature above the glass transition temperature (T_g), the higher the structural collapse. The methods of freeze and hot-air drying can be compared on the basis of this theory. In freeze drying, as the temperature is below T_g' (maximally freeze-concentrated T_g), the material is in the glassy state. Hence shrinkage is negligible. As a result, the final product is very porous. In hot-air drying, on the other hand, as the temperature of drying is above T_g' or T_g, the material remains in a rubbery state, and substantial shrinkage occurs. Hence, the food produced from hot-air drying is dense and shrivelled (Peleg, 1996; Sablani and Rahman, 2002).

Recent experimental results show that the concept of glass transition may not be valid for freeze-drying of all types of biological materials, indicating the need for incorporation of other concepts such as surface tension, structure, surrounding pressure and mechanisms of moisture transport (Meda and Ratti, 2005; Sablani and Rahman, 2002). Rahman (2001) hypothesized that as capillary force is the main factor responsible for collapse, then counterbalancing this force will cause formation of pores and lower shrinkage. This counterbalancing takes place because of generation of internal pressure, variation in moisture transport mechanism and surrounding pressure. Another factor could be the strength of the solid matrix (i.e. ice formation, case hardening and matrix reinforcement). Quality parameters such as volume, shrinkage, apparent density, colour and rehydration behaviour of carrots dried in SS under vacuum were superior to air vacuum–dried carrots (Devahastin et al., 2006). Starch gelatinization in potato slices occurred more rapidly in SS drying than in hot-air drying, leading to glossy appearance on the surface. Basil leaves, dried in SS under vacuum, rendered a product with a higher retention of the original volatile compounds than conventional air drying (Iyota et al., 2001).

15.4.3   INFLUENCE OF DRYING ON CHEMICAL COMPOUNDS

Browning, lipid oxidation, colour loss and change in flavour in foods can occur during drying and storage. Browning reactions modify the colour, reduce the nutritional content and solubility, produce off-flavours and bring about textural changes. Browning reactions can be categorized as enzymatic or non-enzymatic (Salunkhe et al., 1991). Enzymatic browning of foods is undesirable because it develops undesirable colour and produces off-flavour. The application of heat, sodium chloride solution, sulphur dioxide or sulphites and acids can help in controlling the changes in enzymatic and non-enzymatic browning. The major disadvantage of using this treatment for food products is its adverse destructive effect on vitamin B or thiamine. Organic acids such as citric, malic, phosphoric and ascorbic are also employed to lower pH, thus reducing the rate of enzymatic browning. Dipping in osmotic solution can inhibit enzymatic browning in fruits. This treatment can also reduce the moisture content with osmotic pre-concentration. There are three major types of non-enzymatic reaction: (i) Maillard reaction, (ii) caramelization, and (iii) ascorbic acid oxidation (Salunkhe et al., 1991). Factors that can influence non-enzymatic browning are water activity, temperature, pH and the chemical composition of foods. Browning tends to occur primarily at the mid-point of the drying period. This may be because of migration of soluble constituents towards the centre region. Browning is also more severe near the end of the drying period when the moisture level of the sample is low and less evaporative cooling is taking place.

Rapid drying through the 15–20% moisture range can minimize the time for Maillard browning. In carbohydrate foods, browning can be controlled by removing or avoiding amines and conversely, in protein foods, by eliminating the reducing sugars. Sulphur treatment can prevent the initial condensation reaction by forming non-reactive hydroxy-sulphonate sugar derivatives. In caramelization, heating of sugars produces hydroxyl-methyl furfural, which polymerizes easily. This reaction may be slowed by sulphite, which reacts with sugars to decrease the concentration of the aldehydic form. Discolouration of ascorbic acid containing vegetables can occur because of formation of dehydroascorbic acid and diketogluconic acids from ascorbic acid during the final stages of drying. Sulphur treatment can prevent this browning due to reactivity of bisulphite towards carbonyl groups present in the breakdown products (Salunkhe et al., 1991).

Fatty foods are prone to develop rancidity at very low moisture content (i.e. less than monolayer moisture). Lipid oxidation is responsible for rancidity, development of off-flavours and the loss of fat-soluble vitamins and pigments in many foods, especially in dehydrated foods (Sablani, 2006a). Factors that influence the oxidation rate include moisture content, type of substrate (fatty acid), extent of reaction, oxygen content, temperature, presence of metals or natural antioxidants, enzyme activity, UV light, protein content and free amino acid content. Moisture content plays a big part in the rate of oxidation. Air-dried foods are less susceptible to lipid oxidation than freeze-dried products due to lower porosity (Sablani, 2006b).

15.4.4   EFFECT OF DRYING ON MICROBIAL SAFETY

Dried food products are considered safe with respect to microbial hazard. There is a critical water activity (a_w) below which no microorganisms can grow. Pathogenic bacteria cannot grow below a_w of 0.85–0.86, whereas yeast and moulds are more tolerant to a reduced water activity of 0.80. Usually, no growth occurs below a_w of about 0.62 (Sablani, 2006a). Reducing the water activity inhibits microbial growth but does not result in a sterile product. The heat of the drying process does reduce total microbial count, but the survival of food spoilage organisms may give rise to problems in the rehydrated product. The type of microflora present in dried products depends on the characteristics of the products, such as pH, composition, pre-treatments, types of endogenous and contaminated microflora and method of drying. Brining (addition of salts) in combination with drying decreases the microbial load. The dried products should be stored in polypropylene pouches under low temperature and lower relative humidity to protect them from microbial and fungal spoilage (Rahman et al., 2000).

15.4.5   NUTRITIONAL QUALITY

Fruits, vegetables and their products in the dried form are good sources of energy, minerals and vitamins. However, during the process of dehydration, there are changes in nutritional quality (Sablani, 2006a). Large quantities of vitamins such as A, C and thiamine are heat sensitive and sensitive to oxidative degradation. Sulphite treatment can significantly destroy thiamine and riboflavin which are light sensitive. Pre-treatments such as blanching and dipping in sulphite and other solutions are used to reduce the loss of vitamins during drying. As much as an 80% decrease in the carotene content of some vegetables may occur if vegetables are processed without enzyme inactivation. However, if the product is adequately blanched then carotene loss can be reduced to 5%, depending on the product. Steam or hot gas blanching retains higher amounts of vitamin C in spinach compared with hot-water blanching (Ramesh et al., 2001). Blanching in sulphite solution can retain more ascorbic acid in okra (Inyang and Ike, 1998). The salts of metabisulphite treatment were also able to reduce oxidation of carotenoid in carrots, and L-cysteine–HCl retained the larger quantities of ascorbic acid (Mohamad and Hussein, 1994). Bed processing that is superheated steam fluidized is used for both drying and inactivation of antinutritional factors in soybean such as trypsin inhibitor and urease.

The retention of vitamin C in freeze-dried products is significantly higher than that of oven and sun-dried products. Microwave and vacuum drying methods can also reduce the loss of ascorbic acid due to low levels of oxygen. Shade drying, in the absence of light, can also be effective for the retention of nutrients (Sablani, 2006a).

15.4.6 INFLUENCE OF STORAGE ON QUALITY

A significant loss of nutrients occurs in dried fruits and vegetables during storage. This loss can be attributed to the storage temperature, packaging, pH, exposure to oxygen, porosity, light and presence of organic acids (Kumar and Sagar 2015, 2009; Sagar and Kumar, 2009). The extent of losses depends on the type of vitamins and storage conditions such as the exposure to oxygen and light (Sablani, 2006b). During storage of spaghetti, for example, no loss of thiamine and niacin was observed, but riboflavin was susceptible to storage time and light (Watanabe and Ciacco, 1990).

In some situations, the method of dehydration can also influence the loss of nutrients. Kaminski et al. (1986) reported a rapid degradation of carotenoids in freeze-dried carrots. They observed that air-drying was more efficient for carotene preservation when stored at room temperature. Freeze-dried products are generally more porous. This facilitates oxygen transfer and promotes rapid oxidation of carotene. Cinar (2004) reported that the highest pigment loss was in carrot stored at 40°C (98.1 g/100 g) whereas the lowest loss was in sweet potato kept at 4°C (11.3 g/100 g) during 45 days of storage.

15.4.7 EFFECTS OF DEHYDRATION ON SENSORY CHARACTERISTICS

Sensory properties of dried foods are also important in determining quality. These include colour, aroma, texture and taste. Aroma and flavour change because of loss of volatile organic compounds, the most common quality deterioration for dried products. Low-temperature drying is used for foods that have high economic value such as flavouring agents, herbs and spices (Salunkhe et al., 1991, Singh et al., 2006). Low-temperature drying is important for heat-sensitive produces. Singh et al. (2006) found that a low-temperature dryer caused minimal damages to leafy vegetables when drying and thereby retain more nutrients than other dryers. Processing in the absence of oxygen preserves many components which are sensitive to oxidation (Feng et al., 1999). Microwave drying can reduce the process time and improve the quality (Beaudry et al., 2003).

The properties of dried vegetables are influenced by chemical and physical changes. Chemical changes mainly affect sensory properties such as colour, taste and aroma, whereas physical changes mainly influence the handling properties such as swelling capacity and cooking time (Kumar and Sagar, 2009). Heat treatment of fruits and vegetables often reduce the number of original volatile flavour compounds, while introducing additional volatile flavour compounds through the auto-oxidation of unsaturated fatty acids and thermal decomposition, and/or initiation of Maillard reactions. Several factors such as the initial content of solids, type of solids, sample thickness, sample matrix, initial volatile concentration and drying rate are considered to control the retention of volatiles.

The loss of volatile flavour compounds after dehydration may occur because of the inactivation of volatile forming enzymes as well as loss of the precursors. The alliinase activity of fresh onions is decreased up to 90% after hot-air drying, whereas freeze-dried onions still showed 45% activity (Freeman and Whenman, 1975). The flavour characteristics of food products containing unsaturated fatty acids can be drastically affected when these lipids are in direct contact with oxygen. Initiators such as light, metallic ions and heat can initiate auto oxidation reactions. Karel and Yong (1981) summarized the effects of the water content on lipid oxidation. They suggested that the water content is the major factor controlling lipid oxidation in dehydrated foods, but its effects are rather complex. The Maillard reaction is also an important source of volatile flavour compounds, which can have a considerable effect on the flavour of dried vegetables and fruits. Nursten (1980) has classified these volatiles into three groups based on their origin: sugar dehydration/fragmentation products, amino acid degradation products (Strecker degradation products) and compounds produced by further interactions (Loch-Bonazzi et al., 1992).

The flavour and textural characteristics of hot-air-dried and freeze-dried French beans, peas and spinach were compared by means of flavour and texture profiling by Sinesio et al. (1995). These authors found that the sensory panel clearly detected benefits of the freeze-drying process over traditional air drying of the vegetables. The samples were rehydrated before presenting to the trained judges. The freeze-dried products, which received higher scores for juiciness and fruity aroma, were natural in colour and texture. The air-dried products were characterized by an unnatural colour, were wrinkled, cracked, tough with a less-juicy texture, and had a burnt and bitter flavour. Sinesio et al., (1995) reported that the air-dried products were less acceptable than the freeze-dried ones. Freeze-dried products obtained higher sensory scores from the judges for fresh green odour than did the air-dried samples. Onion-like odour was lower in the air-dried products.

15.4.8   REHYDRATION KINETICS OF DEHYDRATED PRODUCT

Rehydration is the process of moistening a dry material. In most cases, dried vegetables are soaked in water and subsequently heated to boiling temperature to 4–5 min depending upon the rehydration before cooking or consumption, thus rehydration is one of the important quality attributes of dried products. In practice, most of the changes during drying are irreversible, and rehydration cannot be considered simply as a process reversible to dehydration (Lewicki, 1998). In general, absorption of water is rapid at the beginning and thereafter slows down. A rapid moisture uptake is due to surface and capillary suction. Rehydration can be considered a measure of the injury to the material caused by drying and treatment preceding dehydration (McMinn and Magee, 1997a; Okos et al., 1992). Rehydration of dried plant tissues is composed of three simultaneous processes: the imbibition of water into dried material, the swelling and the leaching of soluble (Lewicki, 1998; McMinn and Magee, 1997b). It has been shown (Steffe and Singh, 1980) that the volume changes (swelling) of biological materials are often proportional to the amount of absorbed water. The gain in volume due to water sorption equalled the volume of imbibed water. It is generally accepted that the degree of rehydration is dependent on the degree of cellular and structural disruption. Drying leads to destruction of cellular structure, a dense structure of collapsed, greatly shrunken capillaries with reduced hydrophilic properties, as reflected by the inability to imbibe sufficient water to rehydrate fully. The water temperature was found to influence the rehydration rates and the equilibrium moisture content in a positive way. The rehydration ratio ranges between 1 and 4 for most of the products. The rehydration ability appeared to show a hysteresis during rehydration due to cellular and structural disruption that take place during dehydration.

During rehydration, absorption of water into the tissue results in an increase in the mass. Simultaneously, leaching out of solute (sugars, acids, minerals and vitamins) occurs, and both phenomena are influenced by the nature of the product and conditions employed for dehydration (Lewicki, 1998; McMinn and Magee, 1997a). Rehydration is influenced by several factors, grouped as intrinsic factors (product chemical composition, pre-drying treatment, product formulation, drying techniques and conditions, post-drying procedure, etc.) and extrinsic factors such as composition of immersion media, temperature, hydrodynamic conditions (Oliveira and Lincanu, 1999). Some of these factors induce changes in the structure and composition of the plant tissue, which results in the impairment of the reconstitution properties (Taiwo et al., 2002). There is no consistency in the literature on rehydration characteristics with regard to the effect of food-to-water ratio, temperature of rehydration, level of agitation and procedure for the determination of moisture content (Lewicki, 1998). Rahman and Perera (1999) and Lewicki (1998) reviewed the factors affecting the rehydration process. These factors are porosity, capillaries and cavities near the surface, temperature, trapped air bubbles, amorphous-crystalline state, soluble solids, dryness, anions and pH of the soaking water. Porosity, capillaries and cavities near the surface enhance the rehydration process, whereas the presence of trapped air bubbles gives a major obstacle to the invasion of fluid. When the cavities are filled with air, water penetrates to the material through its solid phase. In general, temperature strongly increases the early stages of water rehydration. Resistance of crystalline structures to solvation causes development of swelling stresses in the material, whereas amorphous regions hydrate fast. The presence of anions in water affects the volume increase during water absorption (Sablani, 2006b).

Texture of dried products can be influenced by their moisture content, composition, pH, and product maturity. The chemical changes associated with textural changes in fruits and vegetables include crystallization of cellulose, degradation of pectin and starch gelatinization. In meat products, the changes such as aggregation and denaturation of proteins and a loss of water-holding capacity leads to toughening of muscle tissue. The methods of drying and process conditions also influence the texture of dried products. High air temperature causes complex chemical and physical changes to the surface leading to 'case hardening'. Krokida et al. (2001) studied the quality of potato and carrot with different drying methods, namely convective, vacuum, microwave, freeze and osmotic drying. It was found that air, vacuum and microwave-dried materials caused extensive browning in the vegetables, whereas freeze drying seemed to preserve colour changes, resulting the produce with improved colour characteristics.

Earlier attempts to minimize shrinkage and improve rehydration of dried vegetables include flash-drying techniques such as explosive puffing, vacuum puffing, centrifugal fluidized bed drying, high-temperature short-time pneumatic drying (Jayaraman et al., 1990) and other pre-drying treatments such as osmotic pre-treatment (with sucrose and sodium chloride solution), as well as with other additives, namely glycerol, starch, and so forth. Lewicki (1998) reported that soaking in water as well as dipping in starch solution improved the rehydration characteristics of dehydrated onion. Osmotic dewatering of onion before drying decreased its ability to absorb water but increased the loss of soluble solids during rehydration and improved the retention of constitutive dry mater. Neumann (1972) reported better rehydration characteristics of celery with incorporation of polyhydroxy compounds such as sucrose and glycerol. Pre-treatment with a low concentration mixed solution of salt (3%) and sucrose (6%) caused considerable increase in rehydration and reduced shrinkage due to the fact that solutes in this concentration range provide sufficient structural and mechanical strength to withstand the shock during hot-air drying and prevent cell rupture due to shrinkage and improve water uptake during rehydration (Jayaraman et al., 1990). It may be attributed to the fact that the solute (salt and sugar) concentration was such that at this level the osmotic pressure difference between the plant tissue and rehydrating solution was not significant, which in turn offered protection for cell damage (Lewicki, 1998). Osmotic pre-treatment improves nutritional, sensorial and functional properties of the dehydrated food without changing its integrity. It also improves the texture as well as stability of the pigment during dehydration and the storage of dehydrated product (Raoult-Wack, 1994; Rastogi et al., 2002).

15.5 WAY TO CONSERVE ENERGY DURING DRYING

In most cases, some form of through- or cross-flow convective dryer is used for drying purpose. A common feature of these dryers is their efficient use of high energy. There is considerable concern worldwide over global warming which is attributed to greenhouse gases produced by the combustion of fossil fuels. As a result, there is increasing pressure to reduce the energy consumption, particularly in those countries which are signatories to Kyoto protocol. Raghavan et al. (2005) indicated that about 34% of the world produces required artificial drying for at least part of the crop. Routine care in operation should always form an integral part of crop-dryer operation (Sagar and Kumar, 2010). The use of microcomputer-based control system improved energy efficiency by 18% and reduced over drying by 54% (Gullickson, 2001). The combination of drying is best way to reduce the energy consumption, increased the throughput and improving quality (Raghavan et al., 2005). Optimization of energy through mathematical modelling is another important way to reduce energy consumption (Achariyaviriya et al., 2002).

Intermittent-drying (Chua et al., 2003) and electro-drying technologies (Raghavan et al., 2005) are also used to reduce energy consumption. The application of microwave was found to have a major impact on both drying time and the energy consumption. The specific energy consumption for the drying of grapes reduced from 81.15 MJ/kg in the case of convective drying to 7.11–24.32 MJ/kg by combined microwave convective drying (Tulasidas et al., 1995). Wang et al. (2002) described mathematical model of the drying of 4-mm carrot particles in a batch-fluidized bed dryer with microwave heating. Infrared-convective dryers reduce the energy consumption of osmotically pre-treated samples of potato and pineapple (Tan et al., 2001). Heat-pump dryers (Regalado and Madamba 2000) and high-electric field dryers have a great potential for industrial application, particularly for high-value crops because of superior quality of its products, simplicity of design and low energy use.

15.6   CONCLUSION

The dehydrated vegetables are used to manufacture instant vegetable noodles, soups, snacks and fast food. Dehydrated onion is used as condiment and flavouring agent in manufacturing of tomato ketchups, sauces, salad, pickles, chutneys, meat sausages, masala bread and buns, breakfast foods and so forth. Dehydrated garlic is used for aids in digestion and for absorption of food having athelemetic and antiseptic properties and in some medicinal formulations. Many new dimensions came up in drying technology to reduce the energy utilization and operational cost. Among the technologies, osmotic dehydration, vacuum drying, freeze drying, SS drying, HPD drying, MW drying and spray drying are offering great scope for the production of best quality dried produces and powders. Due to their selective and volumetric heating effects, microwaves bring new characteristics to various bioprocessing techniques such as increased rate of drying, enhanced final product quality and improved energy consumption. The quality of microwave-dried commodities often lies between air-dried and freeze-dried products. The rapidity of the process yields better colour and retention of aroma. Quality is further improved when vacuum is used as the thermal and oxidative stress is reduced. Despite having great potential, the high cost for using single-unit operation to dry the produce is hectic and cost ineffective. Therefore, cost-effective drying with suitable additives should be promoted to sophisticated drying system at the same time with minimum cost of drying. Combination of drying with an initial conventional drying process followed by a finish microwave or microwave vacuum process has proven to reduce drying time, while improving product quality and minimising energy requirements. An optimal drying system for the preservation of quality is cost effective, shortens the drying time and causes minimum damage to the product. There is often a decrease in the quality of the dried products because most conventional techniques require high temperatures during the drying process. Processing may also introduce undesirable changes in appearance and will cause modification of the natural 'balanced' flavour and colour. Therefore, dehydration technologies should be focusing on the production of dried products with little or no loss in their sensory characteristics together with the advantages of added convenience along with low processing costs.

REFERENCES

Achanta, S.; Okos, M. R. Predicting the Quality of Dehydrated Foods and Biopolymers–Research Needs and Opportunities. *Drying Technol.* **1996,** *14,* 1329–1368.

Achariyaviriya, A.; Tiansuwan, J.; Soponronnarit, S. Energy Optimization of Whole Longan Drying. Simulation Results. *Int. J. Ambient energy.* **2002,** *23,* 212–220.

Beaudry, C.; Raghvan, G. S. V.; Rennie, T. J. Micro Wave Finish Drying of Osmotically Dehydrated Cranberries. *Drying Technol.* **2003,** *21*(9), 1797–1810.

Beaudry, C.; Raghavan, G. S. V.; Ratti, C.; Rennie, T. J. Effect of Four Drying Methods on the Quality of Osmotically Dehydrated Cranberries. *Drying Technol.* **2004,** *22*(3), 521–539.

Chua, K. J.; Majumdar, A. S.; Chou, S. K.; Hawlader, M. N. A.; Ho, J. C. Convective Drying of Banana, Guava and Potato Pieces: Effect of Cyclical Variations of Air Temperature on Drying Kinetics and Color Change. *Drying Technol.* **2000,** *18*(4/5), 907–936.

Chua, K. J.; Majumdar, A. S.; Chou, S. K. Intermittent Drying of Bioproducts—an Overview. *Bioresour. Technol.* **2003,** *90*(3), 285–295.

Cinar, I. Carotenoid Pigment Loss of Freeze-dried Plant Samples Under Different Storage Conditions. *Food Sci. Technol.* **2004,** *37,* 363–367.

Cohen, S. J.; Yang, C. S. T. Progress in Food Dehydration. *Trends Food Sci. Technol.* **1995,** *61,* 20–25.

Devahastin, S.; Tapaneyasin, R.; Tansakul, A. Hydrodynamic Behavior of a Jet Spouted Bed of Shrimp. *J. Food Eng.* **2006,** *74*(3), 345–351.

Feng, H.; Tang, J.; Mattinson, D. S.; Fellman, J. K. Microwave and Spouted Bed Drying of Frozen Blue Berries. *J. Food Process. Preserv.* **1999,** *23,* 463–479.

Freeman, G. G.; Whenman, R. J. The use of Synthetic -S-1-Propyl-l-Cysteine Sulphoxide and of Aliinase Preparation in Studies of Flavour Changes Resulting from Processing of Onion (Allium cepa L.). *J. Sci. Food Agri.* **1975,** *26,* 1333–1346.

Funebo, T.; Ohlsson, T. Microwave-assisted Air Dehydration of Apple and Mushroom. *J. Food Eng.* **1998,** *38,* 353–367.

George, S. D.; Cenkowski, S.; Muir, W. E. *A Review of Drying Technologies for the Preservation of Nutritional Compounds in waxy skinned fruit,* North Central ASAE/CSAE Conference, Winnipeg, Manitoba, Canada, Sep. MB 04-104, 2004.

Grabowski, S.; Marcotte, M.; Ramaswamy, H. S. Drying of Fruits, Vegetables, and Spices. In *Handbook of Postharvest Technology;* Chakraverty A., Mujumdar A. S., Raghavan G. S. V., Rawaswamy H. S., Ed.; Marcel Dekker: New York, 2003.

Grigelmo, M. N.; Martin-Belloso, O. Comparison of Dietary Fibre from By-products of Processing Fruits and Greens and from Cereals. *LWT—Food Sci. Technol.* **1999,** *32,* 503–508.

Gullickson, G. Discount Grain Drying. *Farm Industry News.* Sep. 23–28, 2001.

Gunasekaran, S. Pulsed Microwave-vacuum Drying of Food Materials. *Drying Technol.* **1989,** *17*(3), 395–412.

Hall, C. W. Expanding Opportunities in Drying Research and Development. *Drying Technol.* **1996,** *14*(6), 1419–1427.

Inyang, U. E.; Ike, C. I. Effect of Blanching, Dehydration Method, Temperature on the Ascorbic Acid, Color, Sliminess and Other Constituents of Okra Fruit. *Int. J. Food Sci. Nutr.* **1998,** *49,* 125–130.

Iyota, H.; Nishimura, N.; Yoshida, M.; Nomura, T. Simulation of Superheated Steam Drying Considering Initial Steam Condensation. *Drying Technol.* **2001,** *19,* 1425–1440.

Jayaraman, K. S.; Das Gupta, D. K. Dehydration of Fruit and Vegetables-recent Developments in Principles and Techniques. *Drying Technol.* **1992,** *10,* 1–50.

Jayaraman, K. S.; Das Gupta, D. K.; Babu Rao, N. Effect of Pretreatment with Salt and Sucrose on the Quality and Stability of Dehydrated Cauliflower. *Int. J. Food Sci. Technol.* **1990,** *25,* 47–60.

Kaminski, E.; Wasowicz, E.; Zawirska, R.; Wower, M. The Effect of Drying and Storage of Dried Carrot on Sensory Characteristics and Volatile Constituents. *Nahrung.* **1986,** *30,* 819–828.

Karel, M.; Yong, S. Autoxidation-Initiated Reactions in Food.' In *Water Activity: Influences on Food Quality;* Rockland, L. B., Stewart, G. F., Eds.; Academic: London, 1981; pp 511–529.

Khraisheh, M. A. M.; McMinn, W. A. M.; Magee, T. R. A. Quality and Structural Changes in Starchy Foods During Microwave and Convective Drying. *Food Res. Int.* **2004,** *37*(5), 497–503.

Krokida, M. K.; Tsami, E.; Maroulis, Z. B. Kinetics on Color Changes During Drying of Some Fruits and Vegetables. *Drying Technol.* **1998,** *16*(3–5), 667–685.

Krokida, M. K.; Maroulis, Z. B.; Saravacos, G. D. The Effect of Method of Drying on Colour of Dehydrated Product. *Intl. J. Food Sci. Technol.* **2001,** *36,* 53–59.

Kumar, P. S.; Sagar, V. R. Influence of Packaging Materials and Storage Temperature on Quality of Osmo-vac Dehydrated Aonla Segments. *J. Food Sci. Technol.* **2009,** *46*(3), 259–262.

Kumar, P. S.; Sagar, V. R. Drying Kinetics and Physico-chemical Characteristics of Osmo-dehydrated Mango, Guava and *aonla* Under Different Drying Conditions. *J. Food Sci. Technol.* **2014,** *51*(8), 1540–1546.

Kumar, P. S.; Sagar, V. R. Effect of Packaging Materials and Storage Temperature on Quality of Osmo-vac Dehydrated Guava Slices During Storage. *Proc. Natl. Acad. Sci., India, Sect. B Biol. Sci.* 2015. DOI: 10.1007/s40011-015-0545-6.

Kumar, P. S.; Choudhary, V. K.; Kanwat, M.; Sankaran, K. M.; Sangeetha, A. Enhancement of Food Drying and Preservation Through Pulsed Electric Fields: A Review. *Beverage and Food World.* **2013,** *40*(3), 17–19.

Lenart, A. Osmo-Convective Drying of Fruits and Vegetables: Technology and Application. *Drying Technol.* **1996,** *14*(2), 391–413.

Lewicki, P. P. Effect of Pre-drying Treatment, Drying and Rehydration on Plant Tissue Properties: A Review. *Int. J. Food Prop.* **1998,** *1,* 1–22.

Liu, L. Entry into Supermarket of Agricultural Products After Entering WTO. *Agric. Prod. Process.* **2003,** *6*(5), 4–5.

Loch-Bonazzi, C.; Wolff, E.; Gilbert, H. Quality of Dehydrated Cultivated Mushrooms (Agaricus bisporus): A Comparison Between Different Drying and Freeze-drying Processes. *Lebensm. Wiss. u-Technol.* **1992,** *25,* 334–339.

Marinos-Kouris, D.; Maroulis, Z. B. Thermophysical Properties of the Drying of Solids. In *Handbook of Industrial Drying;* Mujumdar, A., Ed.; Marcel Dekker: NY, 1995.

Master, K. Current Market Driven Spray Drying Activities. *Drying Technol.* **2004,** *22*(6), 1351–1370.

Mayor, L.; Sereno, A. M. Modelling Shrinkage During Convective Drying of Food Materials. *J. Food Eng.* **2004,** *61*(3), 373–386.

McMinn, W. A. M.; Magee, T. R. A. Physical Characteristics of Dehydrated Potatoes-Part II. *J. Food Eng.* **1997a,** *33*, 49–55.

McMinn, W. A. M.; Magee, T. R. A. Quality and Physical Structure of Dehydrated Starch Based System. *Drying Technol.* **1997b,** *15*(6–8), 49–55.

Meda, L.; Ratti, C. Rehydration of Freeze Dried Strawberries at Varying Temperature. *J. Food Process. Eng.* **2005,** *28*, 233–246.

Mohamad S.; Hussein, R. Effect of Low Temperature Blanching, Cysteine-HCl, N-acetyl-L-cysteine, Na Metabisulphite and Drying Temperatures on the Firmness and Nutrient Content of Dried Carrots. *J. Food Process. Preserv.* **1994,** *18*, 343–348.

Mulet, A.; Berna, A.; Rossello, C. Drying of Carrots. I. Drying Models. *Drying Technol.* **1989,** *7*(3), 537–557.

Neumann, H. J. Dehydrated Celery: Effects of Pre-drying Treatments and Rehydration Procedures on Reconstitution. *J. Food Sci.* **1972,** *37*, 437–441.

Nijhuis, H. H.; Torringa, H. M.; Muresan, S.; Yuksel, D.; Leguijt, C.; Kloek, W. Approaches to Improving the Quality of Dried Fruits and Vegetables. *Trends Food Sci. Technol.* **1998,** *9*, 13–20.

Nindo, C. I.; Sun, T.; Wang, S. W.; Tang, J.; Powers, J. R. Evaluation of Drying Technologies for Retention of Physical Quality and Antioxidants in Asparagus. *Lebensmittel-Wissenschaft und-Technologie.* **2003,** *36*(5), 507–516.

Nsonzi, F.; Ramaswamy, H. S. Osmotic Dehydration Kinetics of Blueberries. *Drying Technol.* **1998,** *16*(3–5), 725–741.

Nursten, H. E. Recent Developments in Studies of the Maillard Reaction. *Food Chem.* **1980,** *6*, 263–277.

Okos, M. R.; Narishman, G.; Singh, R. K.; Weitnauer, A. C. Food Dehydration. In *Handbook of Food Engineering;* Heldman, D. R., Lund, D. B., Eds.; Marcel Dekker: New York, 1992; pp 437–562.

Oliveira, A. R. F.; Lincanu, L. Rehydration of Dried Plant Tissue: Basic Concepts and Mathematical Modeling. In *Processing Foods, Quality, Optimization and Process Assessment;* Oliveira, A. R. F., Oliveira, J. C., Eds.; CRC Press: London, 1999; pp 201–227.

Orsat, V.; Changrue, V.; Raghavan, G. S. V. Microwave Drying of Fruits and Vegetables. *Stewart Post harvest Review.* **2006,** *6*, 4.

Osepchuk, J. M. Microwave Power Applications. *IEEE Trans. Microwave Theory Tech.* **2002,** *50*(3), 975–985.

Peleg, M. On Modelling Changes in Food and Biosolids at and Around Their Tg Temperature Range. *Crit. Rev. Food Sci. Nutr.* **1996,** *36*, 49–67.

Perera, C. O. Selected Quality Attributes of Dried Foods. *Drying Technol.* **2005,** *23*, 717–730.

Raghavan, G. S. V.; Orsat, V. Electro-technology in Drying and Processing of Biological Materials. Keynote presentation at IDS'98, Halkididi, Greece, 1998.

Raghavan, G. S. V.; Rennie, T. J.; Sunjka, P. S.; Orsat, V.; Phaphuangwittayakul, W.; Terdtoon, P. Overview of New Techniques for Drying Biological Materials with Emphasis on Energy Aspects. *Braz. J. Chem. Eng.* **2005,** *22*(2), 195–201.

Rahman, M. S. Toward Prediction of Porosity in Foods During Drying: A Brief Review. *Drying Technol.* **2001,** *19,* 1–13.

Rahman, M. S. Dried Food Properties: Challenges Ahead. *Drying Technol.* **2005,** *23,* 695–716.

Rahman, M. S.; Perera, C. O. Drying and Food Preservation. In *Handbook of Food Preservation;* Rahman, M. S., Ed.; Marcel Dekker: New York, 1999; pp 173–216.

Rahman, M. S.; Guizani, N.; Al-Ruzeiki, M. H.; Al-Khalasi, S. Microflora Changes in Tunas During Convection Air Drying. *Drying Technol.* **2000,** *18,* 2369–2379.

Ramesh, M. N.; Wolf, W.; Tevini, D.; Jung, G. Influence of Processing Parameters on the Drying of Spice Paprika. *J. Food Eng.* **2001,** *49,* 63–72.

Raoult-Wack, A. L. Advances in Osmotic Dehydration. *Trends Food Sci. Technol.* **1994,** *5*(8), 255–260

Rastogi, N. K.; Raghavarao, K. S. M. S.; Niranjan, K.; Knorr, D. Recent Developments in Osmotic Dehydrations: Methods to Enhance Mass Transfer. *Trends Food Sci. Technol.* **2002,** *13*(2), 58–69.

Ratti, C. Freeze Drying of Plant Products: Where we are and Where we are Heading to. *Stewart Postharvest Review:* 4:5: 2005. http://www.stewartpostharvest.com/ December_2005/Ratti.pdf ISSN 1745–9656.

Regalado, M. J. C.; Madamba, P. S. Design and Testing of a Combined Conduction-convection Rotary Paddy Dryer. *Drying Technol.* **2000,** *18*(9), 1987–2008.

Sablani, S. S. Drying of Fruits and Vegetables: Retention of Nutritional/Functional Quality. *Drying Technol.* **2006a,** *24,* 428–432.

Sablani, S. S. Food Quality Attributes in Drying. *Stewart Postharvest Rev.* **2006b,** *2,* 81–85.

Sablani, S. S.; Rahman, M. S. Pore Formation in Selected Foods as a Function of Shelf Temperature During Freeze Drying. *Drying Technol.* **2002,** *20,* 1379–1391.

Sagar, V. R.; Kumar, P. S. Involvement of Some Process Variables in Mass Transfer Kinetics of Osmotic Dehydration of Mango Slices and Storage Stability. *J. Sci. Ind. Res.* **2009,** *68,* 1431–1436.

Sagar, V. R.; Kumar, P. S. Recent Advances in Drying and Dehydration of Fruits and Vegetables: A Review. *J. Food Sci. Technol.* **2010,** *47*(1), 15–26.

Saguy, I. S.; Marabi, A.; Wallach, R. Water Imbibition in Dry Porous Foods. *Proceedings of 9th Intl conference on Engg Food.* Montpellier, France, 2004.

Salunkhe, D. K.; Bolin, H. R.; Reddy, N. R. Dehydration. In *Storage, processing, and nutritional quality of fruits and vegetables (2nd ed) – Vol. II Processed fruits and vegetables,* CRC Press, Inc.: Boca Raton, Florida, 1991; pp 49–98.

Schubert, H. Food Particle Technology. Part I: Properties of Particles and Particulate Food Systems. *J. Food Eng.* **1987,** *6,* 1–32.

Senadeera, W.; Bhandari, B. R.; Young, G.; Wijesinghe, B. Influence of Shapes of Selected Vegetable Materials on Drying Kinetics During Fluidized Bed Drying. *J. Food Eng.* **2003,** *58,* 277–283.

Shi, J. X., Maguer, M. L.; Wang, S. L.; Liptay, A. Application of Osmotic Treatment in Tomato Processing-effect of Skin Treatments on Mass Transfer in Osmotic Dehydration of Tomatoes. *Food Res. Int.* **1997,** *30*(9), 669–674.

Sinesio, F.; Moneta, E.; Spataro, P.; Quaglia, G. B. Determination of Sensory Quality of Dehydrated Vegetables with Profiling. *Ital. J. Food Sci.* **1995**, *1*, 11–19.

Singh, S.; Singh, B. Value Addition of Vegetable Crops. IIVR Technical Bull No. 65, 2015, pp 1–61.

Singh, U.; Sagar, V. R.; Behera, T. K.; Kumar, P. S. Effect of Drying Conditions on the Quality of Dehydrated Selected Vegetables. *J. Food Sci. Technol.* **2006**, *43*(6), 579–582.

Steffe, J. R.; Singh, R. P.; Note of Volumetric Reduction of Short Grain Rice During Drying. *Cereal Chem.* **1980**, *57*, 148–150.

Taiwo, K. A.; Angersbach, A.; Knorr, D. Influence of High Intensity Electric Field Pulses and Osmotic Dehydration on the Rehydration Characteristics of Apple Slices at Different Temperatures. *J. Food Eng.* **2002**, *52*(2), 185–192.

Tan, M.; Chua, K. J.; Majumder, A. S.; Chou, S. K. Effect of Osmotic Pre Treatment and Infra Red Radiation on Drying and Colour Changes During Drying of Potato and Pineapple. *Drying Technol.* **2001**, *19*, 2193–2207.

Thompkinson, D. K. Microwave Heating of Foods. In New concepts in Dairy Technology. Centre of Advanced Studies in Dairy Technology. 2000, pp 38–43.

Tulasidas, T. N.; Raghavan, G. S. V.; Mujumdar, A. S. Microwave Drying of Grapes in a Single Mode Cavity at 2450 MHz –II: Quality and Energy Aspects. *Drying Technol.* **1995**, *13*(8/9), 1973–1992.

Tuley, L. Swell Time for Dehydrated Vegetables. *Int. Food Ingredients* **1996**, *4*, 23–27.

Van Arsdel, W. B.; Copley, M. J.; Morgan, A. I. Food Dehydration. *Drying phenomena* **1973**, *2*, 27–33.

Vega-Mercado, H.; Gongora-Nieto, M. M.; Barbosa-Canovas, G. V. Advances in Dehydration of Foods. *J. Food Eng.* **2001**, *49*, 271–289.

Wang, N.; Brennan, J. G. Changes in Structure, Density and Porosity of Potato During Dehydration. *J. Food Eng.* **1995**, *24*, 61–76.

Wang, J.; Xi, Y. S. Drying Characteristics and Drying Quality of Carrot Using a Two-stage Microwave Process. *J. Food Eng.* **2005**, *68*, 505–511.

Wang, W.; Thorat, B. N.; Chen, G.; Majumdar, A. S. Fluidized Bed Drying of Heat Sensitive Porous Material with Microwave Heating. In Drying-Proceedings of the 13th Intl. Drying Symposium. Beijing, China, 2730, Aug. Vol B: 2002, pp 901–908.

Watanabe, E.; Ciacco, C. F. Influence of Processing and Cooking on the Retention of Thiamine, Riboflavin and Niacin in Spaghetti. *Food Chem.* **1990**, *36*, 223–231.

Zhang, M.; Li, C. L.; Ding, X. L. Effects of Heating Conditions on the Thermal Denaturation of White Mushroom Suitable for Dehydration. *Drying Technol.* **2005**, *23*(5), 1119–1125.

CHAPTER 16

MICROBIAL SAFETY AND QUALITY ASSURANCE IN VEGETABLES

M. MANJUNATH, A. B. RAI AND B. SINGH

ICAR—Indian Institute of Vegetable Research, Jakhini, Shahanshapur, Varanasi 221305, UP, India

CONTENTS

16.1 INTRODUCTION

The basic human need is access to adequate and safe food. It is vital for making a world free of hunger and poverty (World Health Organization, 2013). The food which we are eating is prone to contamination either by harmful microorganisms, chemicals or both. Millions of people are becoming sick and many are succumbing to death worldwide due to the consumption of unsafe food. Food-borne disease is the occurrence of sickness after consumption of food or water contaminated with pathogenic microorganisms or their toxins (Zhao et al., 2014). Food safety is not adequately addressed in spite of rising international awareness of food-borne diseases as a major risk to health and socio-economic development. This is largely due to limited availability of epidemiological data on food-borne diseases, particularly in the developing countries. Wherein, most of the visible forborne outbreaks often go unrecognized, unreported or uninvestigated. In the year 2015, WHO has estimated global burden of food-borne diseases and reports that one in every 10 people become sick from eating contaminated food and thus 0.42 million die every year (WHO, 2015). Fruits and vegetables are essential components of the healthy human diet (Abadias et al., 2008). In recent years, various measures are taken by several countries to enhance the consumption of fruits and vegetables, as they provide essential nutrients, such as vitamins, minerals and fibres, and reduces the risk of certain diseases, namely cardiovascular diseases, obesity, cancer and so on (Buyukunal et al., 2015; Maffei et al., 2013; Velusamy et al., 2010; Taban and Halkman, 2011; Ignarro et al., 2007; Liu, 2003). An enhancement is observed in the vegetable production and consumption in Asia and the Pacific region (Konuma, 2011). The liberalization of global trade also led to the increase in international trade of fruits and vegetables (FAO/WHO, 2008b). Regardless of their nutritional and health benefits, food-borne diseases (from fruits and vegetables) have increased due to the consumption of contaminated fruits and vegetables (Beuchat, 2002; Altekruse and Swerdlow, 1996; Warriner et al., 2009; FAO/WHO, 2008a). Microbial contamination can occur during any time of the production-to-consumption (Aycicek et al., 2006). This may arise from environmental, animal, human sources (FDA, 2001; Zhao, 2005; FAO/FAO 2008b; Sela and Fallik, 2009), soil (e.g. manure, faeces, soil micro-organisms), dust, water and handling during pre- or post-harvest stages (Eni et al., 2010). Fresh produce is inclined to contamination by entero-toxigenic and entero-haemorrhagic *Escherichia coli* (Viswanathan and K, 2001; Oliveira et al., 2010; Gemmell and Schmidt, 2012), *Salmonella* spp. (Badosa et al., 2008), *Listeria monocytogenes* (Johnston et al., 2006), thermo-tolerant *Campylobacter* spp. (Verhoeff-Bakkenes et al., 2011), parasitic and viral pathogens (Abadias et al., 2012). Salad and leafy green vegetables such as lettuce, cucumber, tomato, radish, carrot, cabbage and spinach were identified as high risk group from the point of view of microbiological contamination. Fresh produce is more likely to cause outbreaks of food-borne illness in future (Stine et al., 2011; Mritunjay and Kumar, 2015).

16.2 VEGETABLE-BORNE DISEASE OUT BREAKS

Vegetable-borne disease outbreak that occurred in Japan during the year 1996 is the biggest ever reported disaster in the history of food-borne illness. Three school children died and more than 11,000 people were affected by the disease caused by the contamination of radish sprouts with *E. coli* O157:H7 (Ministry of Health and Welfare of Japan, 1997; Buck et al., 2003). During the year 1993–1997, fruits and vegetables were associated with 1.4–3% of disease outbreaks in the United States of America (Olsen et al., 2000). Nine outbreaks of forborne illness caused by *Salmonella* or *E. coli* O157:H7 due to consumption of fresh vegetable sprouts occurred between 1995 and 1998 (Buck et al., 2003). Consumption of fruits, vegetables and salad items resulted in 60 intestinal disease outbreaks in England and Wales between the years 1992 and 1999. Fruit- and vegetable-associated outbreaks represent 4.3% (60/1408) of the total number of outbreaks of food-borne disease that were reported during that period. In Sweden, between 1992 and 1997, salad containing one or more cooked ingredients caused 4.3% of the 252 reported incidents of food-borne illness (Lindqvist et al., 2000). In India, no systematic work has been done to know about the different kind of foods and causal agents responsible for food-borne diseases (Sudershan et al., 2014). In 2011, an outbreak of food poisoning largely due to Salmonella sp. contaminated potato-bitter gourd vegetable served at a military establishment took place, in which 43 cases were reported sick within a few hours (Kunwar et al., 2013).

16.3 VEGETABLE MICROFLORA

Microflora of vegetables is very diverse (Beuchat, 2002; Oliveira et al., 2011). Common food-borne pathogenic microorganisms include *Listeria monocytogenes, Staphylococcus aureus, Salmonella enterica, E. coli* (Kim et al., 2013; Zhao et al., 2014). Salad vegetables are mostly contaminated with *E. coli, Salmonella* sp., *S. aureaus, Enterobacter* sp., *Klebsiella* sp., *S. typhi, Serratia* sp. and *Pseudomonas aeruginosa, Cryptosporidium oocystis. E. coli* and *Salmonella* sp. are responsible for majority of food-borne disease outbreaks worldwide, causing symptoms of gastroenteritis and chronic infections. These organisms grow on cucumber, carrot, lettuce, cabbage, tomato and other salad vegetables (Lin et al., 1996).

Viswanathan and Kaur (2001) reported the presence of *Salmonella, Serratia, Enterobacter, S. aureus*, faecal *E. coli* and *P. aeruginosa* in vegetables and fruits in India. A very little information is available on microbial ecology of raw fruits and vegetables surfaces. The pH of many vegetables, melons and soft fruits is 4.6 or higher, which is suitable for the growth of pathogenic bacteria. Some vegetables such as fully ripe tomatoes have pH in the range of 3.9–4.5 that prevents or retards growth of enteric pathogens such as *Shigella* and *E. coli* O157:H7 (Buck et al., 2003).

16.3.1   SALMONELLA

Salmonella spp. are gram negative, rod-shaped and non-spore-forming bacteria, most commonly found pathogens in the fresh produce. Several vegetables such as lettuce, melons, tomatoes, cauliflower, sprouts and spinach are prone to its contamination (Thunberg et al., 2002). In 2006, two major tomato-related *Salmonella* outbreaks have occurred in the United States of America. During 1973–1997, *Salmonella* was isolated in 48% of cases in the United States (Sivaplasingham et al., 2004) and in 41% of cases in the UK between 1992–2000 (Health Protection Agency, 2005). Throughout the world, about 93.8 million cases of gastroenteritis caused by *Salmonella* are reported annually (Majowicz et al., 2010). In India, presence of *Salmonella* isolates was reported in tomato and cilantro leaves (Nair et al., 2015).

16.3.2   ESCHERICHIA COLI O157:H7

Escherichia coli are gram negative, rod-shaped and facultatively anaerobic bacteria. Majority of *E. coli* strains are non-pathogenic and are normally found in the intestines of all animals, including humans. From faeces of animal manures, wild birds, gulls and so on, *E. coli* O157:H7 contamination may occur (Kudva et al., 1998; Moller Nielsen et al., 2004; Wallace et al., 1997). This strain produces large amounts of potent toxins that can cause severe damage to the lining of the intestines. It causes haemorrhagic colitis, haemolytic uremic syndrome (HUS), or thrombotic thrombocytopenic purpura (TTP) (www.produceafetyproject.org). Leafy vegetables (spinach and leafy greens) are most common source of *E. coli* O157:H7 infection (Stopforth et al., 2004; www.produceafetyproject.org). Musale et al. (2014) reported the potentially pathogenic *E. coli* in raw vegetables and sprouts which are sold in the local market of Goa.

16.3.3 LISTERIA MONOCYTOGENES

Listeria spp. are psychrotolerant that is they grow at refrigeration temperatures (Wonderling et al., 2004; Beuchat and Brackett, 1990), are ubiquitous organisms and are found in the faeces of livestock, soil, water and vegetation. The genus *Listeria* is Gram-positive, rod-shaped, non-spore-forming, facultative anaerobic bacteria (Liu 2006). These are potential sources of enteric infection (Blakeman, 1985; Mritunjay and Kumar, 2015). Serogroup one of this organism is most predominant one (Harvey and Gilmour, 1993; Heisick et al., 1989). Its occurrence has been reported on pepper, radish, cucumber, potato, leafy vegetables, beansprout, broccoli, tomato, cabbage, lettuce, celery, carrots and spinach (Beuchat, 1998; Arumugasamy et al., 1994; Szabo et al., 2000; Cre'pet et al., 2007; Thunberg et al., 2002). This organism causes listeriosis, a serious disease in pregnant women, aged people and persons with weak immune systems. Listeriosis is characterized by flu-like symptoms and includes persistent fever, nausea, vomiting and diarrhoea. Contaminated raw vegetables are the main cause of this disease outbreak (www.produceafetyproject.org). In India, Dhanashree et al. (2003) reported the presence of this pathogen in cabbage, coriander leaves and spinach leaves.

16.3.4 CYCLOSPORA

Cyclospora is a protozoan pathogen transmitted by faeces-contaminated fresh produce and water. This pathogen has been reported from basil, fresh raspberries and mesclun lettuce (www.produceafetyproject.org). The likely cause of contamination is poor quality of irrigation water and worker hygiene. This pathogen causes watery diarrhoea, weight loss, abdominal bloating, fever, fatigue, nausea and vomiting.

16.3.5 NOROVIRUSES (NORWALK-LIKE VIRUSES)

Norovirus contaminate water and salads. These viruses cause gastroenteritis in humans. Norovirus contamination is high in fresh fruits and vegetables especially when they get exposed to contaminated water that is used for diluting pesticides. Lettuces and soft berries are more prone to contamination by this virus. This virus survives even at refrigeration temperatures (Verhaelen, 2013).

16.3.6 HEPATITIS A

Hepatitis A virus outbreaks occur through contaminated vegetables, fruits and infected workers in food-processing industries (www.produceafety-project.org). Irrigating the crop with sewage-contaminated water led to outbreak of hepatitis A through lettuce and spring onions (Seymour and Appleton, 2001; Josefson, 2003). This virus may cause nausea, fever, abdominal discomfort and jaundice.

16.4 SOURCES OF CONTAMINATION

Microbiological contamination of vegetables can occur directly or indirectly through contact with soil, dust, water during cultivation, harvesting, packaging, storage, transporting and marketing (De Giusti et al., 2010; FAO/WHO, 2008b).

16.4.1 LOCATION

The demand for land is increasing with rise in population, and its area is diminishing. Vegetables are often grown in areas near to the garbage dump, urban sewage run-off, which are potential sources of contamination of the produce with pathogenic microorganisms.

16.4.2 NUTRIENT SOURCE

This is the use of improperly composted poultry and animal manure as a source of nutrients. In addition to this, human and animal faeces, including bird droppings in the growing fields serve as a potential source of microbial contamination. From the use of animal wastes as fertilizer, more than 20% of infections by *Shigella*, *Salmonella*, *Vibrio cholerae* and amoebas may develop (ICMSF, 1998).

16.4.3 WORKERS HYGIENE

Most of the farmers do not know about hygienic production and good agricultural practices. Lack of availability of washrooms on the farms, presence of sick workers in the field and at packaging may contribute to the contamination of the fresh produce.

16.4.4 CONTAMINATED WATER SOURCES

Using polluted or contaminated water or sewage water for irrigation purpose leads to contamination of the produce with pathogenic microbes (FAO/WHO, 2008a). According to the Ministry of environment, Government of British Columbia (2001), upper limit for *E. coli, Enterococci, P. aeruginosa* and faecal coliforms in irrigated water is ≤77, 20, nil and ≤200/100 ml geometric mean, respectively.

16.4.5 USE OF DIRTY HARVEST EQUIPMENT, CONTAINERS AND STORAGE FACILITIES

The usage of unclean equipment will also serve as a source of contamination of fresh produce (WHO, 2008).

16.4.6 CONSUMPTION PATTERNS

Nowadays, both in developed and developing countries, consumption of fresh fruits and vegetables is increasing. The foods which are traditionally consumed after cooking are being eaten raw nowadays (FAO/WHO, 2008a).

16.5 DETECTION OF FOOD-BORNE PATHOGENS

16.5.1 CONVENTIONAL METHODS

It involves growing of microorganisms on agar media. This method includes the following.

- Blending of food product with a selective enrichment medium to increase the population of the target organism
- Plating onto selective or differential agar plates to isolate pure cultures
- *Salmonella* sp.—XA medium (Park et al., 2012)
- *Staphylococcus. Aureus*—Baird-Parker (El-Hadedy and El-Nour, 2012)
- *E. coli*—MacConkey agar (El-Hadedy and El-Nour, 2012)
- Examining the cultures by phenotypic and biochemical analysis

These methods are simple and inexpensive but laborious and time-consuming. It depends on their ability to grow in pre-enrichment media, selective enrichment media and selective plating media. For preliminary identification, 2–3 days are required and more than 7 days for confirmation of the pathogen at species level (Zhao et al., 2014). Sometimes false negative results may also occur due to viable but non-culturable pathogens.

16.5.2   RAPID METHODS

16.5.3   IMMUNOLOGICAL-BASED METHODS

These methods are based on antigen and antibody reaction.

16.5.3.1   ENZYME-LINKED IMMUNOSORBENT ASSAY

ELISA is widely used method for the detection of food-borne pathogens. The most accurate one is the sandwich ELISA involving two antibodies (Zhao et al., 2014). The primary antibody is immobilized onto the walls of microtitre plate wells. Then, the antigens such as bacterial cells, toxins and so on are added and they will bind to the already immobilized primary antibody. Later, an enzyme conjugated secondary antibody is added, and it will bind to the antigen by forming antibody–antigen–antibody sandwich. It can be detected by conversion of colourless substance into coloured product in the presence of enzyme (Zhang, 2013). The most frequently used enzymes are alkaline phosphatase, horse radish peroxidise and beta-galactosidase (Yeni et al., 2014). ELISA is used for the detection of toxins such as Staphylococcal entero-toxins A, B, C and E, botulinum toxins and *E. coli* entero-toxins (Aschfalk and Mülller, 2002; Zhao et al., 2014). Commercial ELISA test kits are also available for the detection of *Salmonella* and other food-borne pathogens (Bolton et al., 2000).

16.5.4 POLYMERASE CHAIN REACTION

This is the most widely used molecular method for the detection of food-borne pathogens (Woan-FeiLaw e al. 2015). It has been used for the detection of *S. aureus, Salmonella* spp, *E. coli* O157:H7, *L. monocytogenes, Shigella* spp. *and Campylobacter jejuni* (Cheah et al., 2008; Lee et al., 2008; Alves et al., 2012; Chiang et al., 2012; Zhou et al., 2013).

16.5.5 REAL-TIME OR QUANTITATIVE POLYMERASE CHAIN REACTION

Unlike normal PCR, it does not require gel electrophoresis for the visualization of PCR products. In this method, PCR products formation could be continuously monitored throughout the reaction by measuring the fluorescent signals produced by the specific labelled probes or intercalating dyes. The intensity of signal is proportional to the amount of PCR products or amplicons (Omiccioli et al., 2009; Zhao et al., 2014). Liming and Bhagwat (2004) detected *Salmonella* in fresh-cut fruits and vegetables by molecular beacon qPCR targeting the invasion associated gene (*iagA*).

16.5.6 DNA MICROARRAY

Microarray is the multi-gene detection technology (Call et al., 2001). Microarrays are widely used for the detection of food-borne pathogens. These are made of glass slides or chips, coated with hundreds of different oligonucleotide probes of length 25–80 bp (Severgnini et al., 2011). The sample nucleic acid fragments such as DNA, mRNA or cDNA are labelled with fluorescent dye and single stranded fragments are obtained through denaturation process. These fragments are allowed to hybridize their corresponding oligonucleotide probes. The fluorescent signals produced by the probe-sample complex are visualized and intensity of signals is proportional to the concentration of nucleic acid fragment (Lauri and Mariani, 2009). Li et al. (2006) detected pathogenic *Shigella* and *E. coli* serotypes by oligonucleotide DNA microarray.

16.5.7 BIOSENSORS

These are analytical devices that convert a biological response to a measurable electrical signal proportional to the concentration of the analytes. A biosensor consists of a bioreceptor and transducer. A receptor could be a tissue, microorganism, enzyme, antibody, organelle and so on, and the transducer may be optical, thermometric, electrochemical and so on (Su et al., 2010). In the rapid detection of food-borne pathogens or their toxins even in very minute quantities such as nanograms, biosensors are very helpful (Rasooly and Herold, 2006).

16.6   PREVENTION OF MICROBIAL CONTAMINATION

Vegetable quality and safety depend on many factors, such as soil, irrigation water, fertilizer, presence of animals on the field and good practices during growing, processing and marketing (Beuchat, 2002; Oliveira et al., 2011). By following hygiene practices during farming, transport and merchandizing of vegetables, as well as making the population aware of vegetable sanitization before consumption would lead to minimizing the risk of contamination and food-borne diseases (Maffei et al., 2013). Washing and rinsing of vegetables prolong the shelf life by reducing the number of microorganisms (Kumar, 2012).

16.6.1   MEASURES TO AVOID CONTAMINATION OF VEGETABLES WITH PATHOGENIC MICROORGANISMS

4.   Protection of farm fields from animal faecal contamination
5.   Use of properly treated manure
6.   Use of clean and contamination free irrigation water
7.   Always keeping harvest and storage equipment clean and dry
8.   Practice of good personal hygiene

16.7 CONCLUSION

Food safety is an important factor for food security, nutritional security and sustainable development. Provision of safe foods to people reduces the food-borne diseases and improves the quality of life. In the betterment of global health status and assuring sustainable development, food safety is a crucial issue (WHO, 2013). So, it should be adequately addressed at international levels. In India, vast majority of the people prefer to buy fresh vegetables from the local markets, which are lacking good infrastructure and hygienic conditions. To avoid the chances of food-borne disease outbreaks, there is an urgent need to make people aware of hygienic practices during handling, transport, storage and marketing.

16.8 ACKNOWLEDGEMENTS

The authors are grateful to the authorities of the Indian Institute of Vegetable Research (ICAR), Varanasi, for providing the facilities towards undertaking this work.

REFERENCES

Abadias, M.; Usall, J.; Anguera, M.; Solsona, C.; Vinas, I. Microbiological Quality of Fresh, Minimally-Processed Fruit and Vegetables, and Sprouts from Retail Establishments. *Int. J. Food Microbiol.* **2008,** *123,* 121–129.

Abadias, M.; Alegre, I.; Oliveira, M.; Altisent, R.; Vinas, I. Growth Potential of *Escherichia coli* O157:H7 on Fresh-Cut Fruits (Melon and Pineapple) and Vegetables (Carrot and Escarole) Stored Under Different Conditions. *Food Control* **2012,** *27,* 37–44.

Altekruse, S. F.; Swerdlow, D. L. The Changing Epidemiology of Foodborne Diseases. *Am. J. Med. Sci.* **1996,** *311,* 23–29.

Alves, J.; Marques, V. V.; Pereira, L. F. P.; Hirooka, E. Y.; Moreirade Oliveira, T. C. R. Multiplex PCR for the Detection of *Campylobacter* spp.and *Salmonella* spp.in Chicken Meat. *J. Food Saf.* **2012,** *32*(3), 345–350.

Arumugasamy, R. K.; Rahamat Ali, G. R.; Hamid, S. N. B. A. Prevalence of *Listeria Monocytogenes* in Foods in Malaysia. *Int. J. Food Microbiol.* **1994,** *23,* 117–121.

Aschfalk, A.; Mülller, W. *Clostridium Perfringens* Toxin Types from Wild Caught Atlantic Cod (*Gadus Morhua* L.) Determined by PCR and ELISA. *Can. J. Microbiol.* **2002,** *48,* 365–368. DOI:10.1139/w02-015.

Aycicek, H.; Oguz, U.; Karci, K. Determination of Total Aerobic and Indicator Bacteria on Some Raw Eaten Vegetables from Wholesalers in Ankara, Turkey. *Int. J. Hyg. Environ. Health* **2006,** *209,* 197–201.

Badosa, E.; Trias, R.; Pares, D.; Pla, M.; Montesinos, E. Microbiological Quality of Fresh Fruit and Vegetable Products in Catalonia (Spain) Using Normalised Plate-Counting Methods and Real Time Polymerase Chain Reaction (QPCR). *J. Sci. Food Agric.* **2008,** *88,* 605–611.

Beuchat, L. R. *Surface Decontamination of Fruits and Vegetables Eaten Raw: A Review World Health Organisation (WHO/FSE/FOS/98.2)*; 1998.

Beuchat, L. R. Ecological Factors Influencing Survival and Growth of Human Pathogens on Raw Fruits and Vegetables. *Microbes Infect.* **2002,** *4,* 413–23.

Beuchat, L. R.; Brackett, R. E. Inhibitory Effects of Raw Carrots on Listeria Monocytogenes. *Appl. Environ. Microbiol.* **1990,** *56,* 1734–1742.

Blakeman, J. P. Ecological Succession of Leaf Surface Microorganisms in relation to biological Control. In Biological control on the phylloplane. Windels, C. E., Lindow, S. E., Eds.; APS Press: Minnesota, 1985; pp 6–30.

Bolton, F. J.; Fritz, E.; Poynton, S. Rapid Enzyme-Linked Immunoassay for the Detection of *Salmonella* in Food and Feed Products: Performance Testing Program. *J. AOAC Int.* **2000,** *83,* 299–304.

Buck, J. W.; Walcott, R. R.; Beuchat, L. R. Recent Trends in Microbiological Safety of Fruits and Vegetables. *Plant health progress,* **2003,** *10,* 1094.

Buyukunal, S. K.; Ghassan, I.; Filiz, A.; Aydin, V. Microbiological Quality of Fresh Vegetables and Fruits Collected from Supermarkets in Istanbul, Turkey. *J. Food Nutr. Sci.* **2015,** *3*(4), 152–159. DOI: 10.11648/j.jfns.20150304.13.

Call, D. R.; Brockman, F. J.; Chandler, D. P. Detecting and Genotyping *Escherichia coli* O157:H7 Using Multiplexed PCR and Nucleic Acid Microarrays. *Int. J. Food Microbiol.* **2001,** *67*(1), 71–80.

Cheah, Y. K.; Noorzaleha, A. S.; Lee, L. H.; Radu, S.; Sukardi, S.; Sim, J. H. Comparison of PCR Finger Printing Techniques for the Discrimination of *Salmonella Enterica* subsp. *Enterica* Serovar Weltevreden Isolated from Indigenous Vegetables in Malaysia. *World J. Microbiol. Biotechnol.* **2008,** *24,* 327–335.

Chiang, Y. C.; Tsen, H. Y.; Chen, H. Y.; Chang, Y. H.; Lin, C. K.; Chen, C. Y. Multiplex PCR and a Chromogenic DNA Macroarray for the Detection of *Listeria monocytogenes, Staphylococcus aureus, Streptococcus agalactiae, Enterobacter sakazakii, Escherichia coli* O157:H7, *Vibrio parahaemolyticus, Salmonella* spp. and *Pseudomonas fluorescens* in Milk and Meat Samples. *J. Microbiol. Methods* **2012,** *88,* 110–116.

Cre′pet, A.; Albert, I.; Dervin, C.; Carlin, F. Estimation of Microbialcontamination of Food from Prevalence and Concentration Data: Application to Listeria Monocytogenes in Fresh Vegetables. *Appl. Environ. Microbiol.* **2007,** *73,* 250–258.

De Giusti, M.; Aurigemma, C.; Marinelli, L.; Tufi, D.; Demedici, D. The Evaluation of the Microbial Safety of Fresh Ready to Eat Vegetables Produced by Different Technologies in Italy. *J. Appl. Microbiol* **2010,** *109,* 996–1006.

Dhanashreea, B.; Ottab, S. K.; Karunasagarb, I.; Goebelc, W.; Karunasagar, I. Incidence of *Listeria* spp. in Clinical and Food Samples in Mangalore, India. *Food Microbiol.* **2003,** *20,* 447–453.

El-Hadedy, D.; El-Nour, S. A. Identification of *Staphylococcus aureus* and *Escherichia coli* Isolated from Egyptian Food by Conventional and Molecular Methods. *J. Genet. Eng. Biotechnol.* **2012,** *10*(1), 129–135.

Eni, A. O.; Oluwawemitan, I. A.; Oranusi, S. Microbial Quality of Fruits and Vegetables Sold in Sango Ota Nigeria. *Afr. J. Food Sci.* **2010,** *4,* 291–296.

Food and Agriculture Organization/World Health Organisation (FAO/WHO). Microbiological Risk Assessment Series: Microbiological Hazards in Fresh Fruit and Vegetables. 2008a. http://www.who.int/foodsafety/publications/micro/MRA_FruitVeges.pdf.

Food and Agriculture Organization/World Health Organisation (WHO/FAO). Microbiological Risk Assessment Series: Microbiological Hazards in Fresh Leafy Vegetables and Herbs. 2008b. http://www.fao.org/ag/agn/agns/jemra/Jemra_Report%20on%20 fresh%20leafy%20vegetables%20and%20herbs.pdf.

Food and Drug Administration (FDA). Analysis and Evaluation of Prevention Control Measures for the Control and Reduction/Elimination of Microbial Hazardson Fresh and Freshcut Produce. Center for Food Safety and Applied Nutrition. 2001. http://www.fda. gov/Food/FoodScienceResearch/SafePracticesforFoodProcesses/ucm091016.htm.

Gemmell, M. E.; Schmidt, S. Microbiological Assessment of River Water Used for the Irrigation of Fresh Produce in a Sub-Urban Community in Sobantu, South Africa. *Food Res. Int.* **2012,** *47,* 300–305.

Harvey, J.; Gilmour, A. Occurrence and Characteristics of Listeria in Foods Produced in Northern Ireland. *Int. J. Food Microbiol.* **1993,** *19,* 193–205.

Health Protection Agency. Health Protection Agency Advisory Committee on the Microbiological Safety of Food Information Paper 'Microbiological Status of RTE Fruit and Vegetables' ACM/745.Food Standards Agency: London, UK, 2005. http://food.gov.uk/ multimedia/pdfs/acm745amended.pdf.

Heisick, J. E.; Wagner, D. E.; Nierman, M. L.; Peeler, J. T. *Listeria* spp. Found on Fresh Market Produce. *Appl. Environ. Microbiol.* **1989,** *55,* 1925–1927.

ICMSF. *Microbial Ecology of Food Commodities. Microorganisms in Foods*; Blackie Ackademic & Professional: London, UK, 1998.

Ignarro, L. J.; Balestrieri, M. L.; Napoli, C. Nutrition, Physical Activity, and Cardiovascular Disease: An Update. *Cardiovasc. Res. 2007, 73*(2), 326e340.

Johnston, L M.; Jaykus Moll, L.A. D.; Martinez, M. C.; Anciso, J.; Mora, B.; Moe, C. L. A Field Study of the Microbiological Quality of Fresh Produce of Domestic and Mexican Origin. *Int. J. Food Microbiol.* **2006,** *112,* 83–95.

Josefson, D. Three Die in US Outbreak of Hepatitis. *BMJ* **2003,** *327*(7425), 1188.

Kim, S. J.; Cho, A. R. Han, J. Antioxidant and Antimicrobial Activities of Leafy Green Vegetable Extracts and Their Applications to Meat Product Preservation. *Food Control* **2013,** *29,* 112–120.

Konuma, 2011. http://www.fao.org/docrep/014/i1909e/i1909e00.pdf.

Kudva, I. T.; Blanch, K.; Hovde, C. J. Analysis of *Escherichia coli* O157:H7 Survival in Ovine or Bovine Manure and Manure Slurry. *Appl. Environ. Microbiol.* **1998,** *64,* 3166–3174.

Kumar, V. Incidence of *Salmonella* sp. and *Listeria monocytogenes* in Some Salad Vegetables Which are Eaten Raw: A Study of Dhanbad City, India. *Int. J. Eng. Sci. Res.* **2012,** *2,* 2277–2685.

Kunwar, C. R.; Singh, M. H.; Mangla, M. V.; Hiremath, M. R. Outbreak Investigation: *Salmonella* Food Poisoning. *Med. J. Armed Forces Ind.* **2013,** *69,* 388–391.

Lauri, A.; Mariani, P. O. Potentials and Limitations of Molecular Diagnostic Methods in Food Safety. *Genes Nutr.* **2009,** *4,* 1–12. DOI:10.1007/s12263- 008-0106-1.

Lee, L. H.; Cheah, Y. K.; Noorzaleha, A. S.; Sabrina, S.; Sim, J. H.; Khoo, C. H. Analysis of *Salmonella* Agona and *Salmonella* Weltevreden in Malaysia by PCR by PCR Fingerprinting and Antibiotic Resistance Profiling. *Antonie Van Leeuwenhoek* **2008,** *94,* 377–387.

Li, Y.; Liu, D.; Cao, B.; Han, W.; Liu, Y.; Liu, F. Development of a Serotype-Specific DNA Microarray for Identification of Some *Shigella* and Pathogenic *Escherichia coli* Strains. *J. Clin. Microbiol.* **2006,** *44,* 4376–4383.

Liming, S. H.; Bhagwat, A. A. Application of a Molecular Beacon Real Time PCR Technology to Detect Salmonella Species Contaminating Fruits and Vegetables. *Int. J. Food Microbiol.* **2004,** *95*(2), 177–187.

Lin, C. M.; Fernando, S. Y.; Wei, C. I. Occurrence of *Listeria monocytogenes, Salmonella* spp. *Escherichia coli* and *E.coli* O157:H7 in Vegetable Salads. *Food Control* **1996,** *7,* 135–140.

Lindqvist, R.; Andersson, Y.; de Jong, B.; Norberg, P. A Summary of Reported Foodborne Disease Incidents in Sweden, 1992–1997. *J. Food Prot.* **2000,** *63,* 1315–1320.

Liu, R. H. Health Benefits of Fruit and Vegetables are from Additive and Synergistic Combinations of Phytochemicals. *Am. J. Clin. Nutr.* **2003,** *78*(3), 517–520.

Liu, D. Identification, Subtyping and Virulence Determination of *Listeria* Monocytogenes, an Important Foodborne Pathogen. *J. Med. Microbiol.* **2006,** *55,* 645–659

Maffei, D. F.; de Arruda Silveira, N. F.; and da Catanozi, M. P. L. M. Microbiological Quality of Organic and Conventional Vegetables Sold in Brazil. *Food Control* **2013,** *29,* 226–230.

Majowicz, S. E.; Musto, J.; Scallan, E.; Angulo, F. J.; Kirk, M.; O'Brien, S. J.; Hoekstra, R. M. The Global Burden of Non-Typhoidal *Salmonella* Gastroenteritis. *Clin. Infect. Dis.* **2010,** *50*(6), 882–889.

Ministry of Health and Welfare of Japan. National Institute of Infectious Diseases and Infectious Disease Control Division. Verocytotoxin producing *Escherichia coli* (enterohaemorrhagic *E. coli*) Infection, Japan, 1996–June 1997. *Infectious Agents Surveillance Report* **1997,** *18,* 153–154.

Moller Nielsen, E.; Skov, M. N.; Madsen, J. J.; Lodal, J.; Brochner Jespersen, J.; Baggesen, D. L. Verocytotoxin-Producing *Escherichia coli* in Wild Birds and Rodents in Close Proximity to Farms. *Appl. Environ. Microbiol.* **2004,** *70,* 6944–6947.

Mritunjay, S. K.; Kumar, V. Potential Hazards of Microbial Contamination Associated with Raw Eaten Salad Vegetables and Fresh Produces Middle-East. *J. Sci. Res.* **2015,** *23*(4), 741–749.

Musale, V.; Kale, S.; Raorane, A.; Doijad, S.; Barbuddhe, S. Incidences of ETEC *E. coli* in Raw Vegetables and Sprouts Sold in Local Market of Goa. *Indian J. Microbiol. Res.* **2014,** *1*(1), 46–51.

Nair, A.; Balasaravanan, T.; Malik, S. V.; Mohan, V.; Kumar, M.; Vergis, J.; Rawool, D. B. Isolation and Identification of *Salmonella* from Diarrhoeagenic Infants and Young Animals, Sewage Waste and Fresh Vegetables. *Vet. World* **2015,** *8*(5), 669–673.

Oliveira, M.; Usall, J.; Vinas, I.; Anguera, M.; Gatius, F.; Abadias, M. Microbiological Quality of Fresh Fresh Lettuce from Organic and Conventional Production. *Food Microbiol.* **2010,** *27,* 679–684.

Oliveira, M. A.; Souza, V. M.; Bergamini, A. M. M.; Martinis, E. C. P. Microbiological Quality of Ready-to-Eat Minimally Processed Vegetables Consumed in Brazil. *Food Control* **2011,** *22*(8), 1400–1403.

Olsen, S. J.; MacKinon, L. C.; Goulding, J. S. Bean, N. H. L. Slutsker (2000). Surveillance for Foodborne Disease Outbreaks-United States, 1993-1997.MMWR March 17 49(SS01):1-51.

Omiccioli, E.; Amagliani, G.; Brandi, G.; Magnani, M. A New Platform for Real-Time PCR Detection of *Salmonella* spp., *Listeria monocytogenes* and *Escherichia coli* O157 in Milk. *Food Microbiol.* **2009,** *26,* 615–622.

Park, S. H.; Ryu, S.; Kang, D. H. Development of an Improved Selective and Differential Medium for Isolation of *Salmonella* spp. *J. Clin. Microbiol.* **2012,** *50*(10), 3222–3226.

Rasooly, A.; Herold, K. E. Biosensors for the Analysis of Food- and Waterborne Pathogens and Their Toxins. *J. AOAC Int.* **2006,** *89*(3), 873–883.

Sela, S.; Fallik, E. Microbiological Quality and Safety of Fresh Produce. In *Postharvest Handling;* Florkowski, W. J., Shewfelt, R. L., Brueckner, B., Prussia, S. E., Eds.; Academic Press Inc.: Philadelphia, PA, 2009; pp 351–398.

Severgnini, M.; Cremonesi, P.; Consolandi, C.; DeBellis, G.; Castiglioni, B. Advances in DNA Microarray Technology for the Detection of Foodborne Pathogens. *Food Bioprocess. Technol.* **2011,** *4,* 936–953.

Seymour, I. J.; Appleton, H. A Review, Foodborne Viruses and Fresh Produce. *J. Appl. Microbiol.* **2001,** *91,* 759–773.

Sivaplasingham, S.; Friedman, C. R.; Cohen, L.; Tauxe, R. V. Fresh Produce: A Growing Cause of Outbreaks of Foodborne Illness in the United States, 1973 Through 1997. *J. Food Protect.* **2004,** *67,* 2342–2353.

Stine, S. W.; Song, I.; Choi, C. I.; Gerba, C. P. Application of Pesticide Sprays to Fresh Produce: A Risk Assessment for Hepatitis A and *Salmonella. Food Environ. Virol.* **2011,** *3,* 86–91.

Stopforth, J. D.; Ikeda, J. S.; Kendall, P. A.; Stofos, J. N. Survival of Acid-Adapted or Nonadapted *Escherichia coli* O157:H7 in Apple Wounds and Surrounding Tissue Following Chemical Treatments and Storage. *Int. J. Food Microbiol.* **2004,** *90,* 51–61.

Su, L.; Jia, W; Hou, C.; Lei, Y. Microbial biosensors: A review. Biosensors and Bioelectronics, In Press, Corrected Proof:, ISSN 0956-5663. 2010.

Sudershan, R. V.; Naveen Kumar, R.; Kashinath, L.; Bhaskar, V.; Polasa, K. Foodborne Infections and Intoxications in Hyderabad, India. *Epidemiol Res. Int.* **2014,** *2,* Article ID 942961, 5.

Szabo, E. A.; Scurrahand, K. J.; Burrows, J. M. Survey for Psychrotrophic Bacterial Pathogens in Minimally Processed Lettuce. *Lett. Appl. Microbiol.* **2000,** *30*(6), 456–460.

Taban, B. M.; Halkman, A. K. Do Leafy Green Vegetables and Their Ready-to-Eat [RTE] Salads Carry a Risk of Foodborne Pathogens? *Anaerobe* **2011,** *17,* 286–287.

Thunberg, R. L.; Tran, T. T.; Bennett, R. W.; Matthews, R. N. Microbial Evaluation of Selected Fresh Produce Obtained at Retail Markets. *J. Food Prot.* **2002,** *65*(4), 677–682, ISSN 0362-028X.

Velusamy, V.; Arshak, K.; Korostynska, O.; Oliwa, K.; Adley, C. An Overview of Food-borne Pathogen Detection: In the Perspective of Biosensors. *Biotechnol. Adv.* **2010,** *28,* 232–254.

Verhaelen, K. Contamination With Norovirus High in Fresh Fruits and Vegetables. 2013. http://www.medindia.net/news/contamination-with-norovirus-high-in-fresh-fruits-and-vegetables-115725-1.htm#ixzz3vjfixgPj.

Verhoeff-Bakkenes, L.; Jansen, H. A. P. M.; in't Veld, P. H.; Beumer, R. R.; Zwietering, M. H.; Van Leusden, F. M. Consumption of Raw Vegetables and Fruits: A Risk Factor for *Campylobacter* Infections. *Int. J. Food Microbiol.* 2011, *144,* 406–412.

Viswanathan, P.; Kaur, R. Prevalence and Growth of Pathogens on Salad Vegetables, Fruits, and Sprouts. *Int. J. Hyg. Environ. Health* **2001,** *203,* 205–213.

Wallace, J. S.; Cheasty, T.; Jones, K. Isolation of Verocytotoxigenic Producing *Escherichia coli* O157 from Wild Birds. *J. Appl. Microbiol.* **1997,** *82,* 399–404.

Warriner, K.; Huber, A.; Namvar, A.; Fan, W.; Dunfield, K. Recent Advances in the Micro-bial Safety of Fresh Fruits and Vegetables. *Adv. Food Nutr. Res.* **2009,** *57,* 155–208.

WHO.2015.http://www.who.int/mediacentre/news/releases/2015/foodborne-disease-estimates/en/.

WHO/FAO. *Microbiological Hazards in Fresh Leafy Vegetables and Herbs: Meeting Report;* World Health Organization: Geneva, 2008; p 138 (Microbiological risk assess-ment series no. 14).

Woan-FeiLaw, J.; AbMutalib, N. S.; Chan, K. G.; Lee, L. H. Rapid Methods for the Detec-tion of Foodborne Bacterial Pathogens: Principles, Applications, Advantages and Limi-tations. *Front. Microbiol.* **2015,** *5,* 1–18.

Wonderling, L. D.; Wilkinson, B. J.; Bayles, D. O. The htrA (degP) Gene of *Listeria monocytogenes* 10403S is Essential for Optimal Growth Under Stress Conditions. *Appl. Environ. Microbiol.* **2004,** *70,* 1935–1943.

World Health Organisation/Food and Agriculture Organization (WHO/FAO). Microbio-logical Hazards in Fresh Leafy Vegetables and Herbs, Microbiological Risk Assessment Series, Meeting Report. 20, Avenue Appia CH-1211 Geneva 27, Switzerland. 2008. http://www.fao.org/ag/agn/agns/jemra/Jemra_Report %20on%20fresh%20leafy%20 vegetables%20an d%20herbs.pdf.

World Health Organization. Advancing Food Safety Initiatives: Strategic Plan for Food Safety Including Foodborne Zoonoses 2013–2022. 2013, ISBN 978 92 4 150628 1. www.produceafetyproject.org.

Yeni, F.; Acar, S.; Polat, O. G.; Soyer, Y.; Alpas, H. Rapid and Standardized Methods for Detection of Foodborne Pathogens and Mycotoxins on Fresh Produce. Food Cont. **2014,** *40,* 359–367 DOI: 10.1016/j.foodcont.2013.12.020

Zhang, G. Food Borne Pathogenic Bacteria Detectionan Evaluation of Current and Devel-oping Methods. *Meducator* **2013,** *1,* 15.

Zhao, Y. Pathogens in Fruit. In *Improving the Safety of Fresh Fruit and Vegetables;* Jonger, W., Ed.; CRC Press: Washington, DC, 2005; pp 44–88.

Zhao, X.; Lin, C. W.; Wang, J.; Oh, D. H. Advances in Rapid Detection Methods for Food-borne Pathogens. *J. Microbiol. Biotechnol.* **2014,** *24,* 297–312.

Zhou, B.; Liu Xiao, J. S.; Yang, J.; Wang, Y.; Nie, F. Simultaneous Detection of Six Food-Borne Pathogens by Multiplex PCR with GEXP Analyzer. *Food Control* **2013,** *32,* 198–204.

CHAPTER 17

PRE- AND POST-HARVEST PESTICIDE CONTAMINATION MANAGEMENT FOR PRODUCTION OF QUALITY VEGETABLES

M. H. KODANDARAM[1,*], Y. BIJEN KUMAR[1], KAUSHIK BANERJEE[2], A. B. RAI[1] AND B. SINGH[1]

[1]Division of Crop Protection, ICAR-Indian Institute of Vegetable Research, Varanasi 221305, UP, India, *E-mail: kodandaram75@gmail.com

[2]National Referral Laboratory, ICAR-National Research Centre for Grapes, Pune 412307, Maharashtra, India

CONTENTS

17.1 INTRODUCTION

In the horticulture sector, vegetables are important constituents and are mostly low gestation and high income generating crops. Diverse agro-climatic conditions of the country permit growing of several vegetables throughout the year. India is the second largest producer of vegetables after China, and a leader in production of okra and peas. In the recent times, India produces about 162.89 million tonnes of vegetables from an area of 9.39 million ha with an average productivity of 17.3 t/ha (NHB, 2014). Though there has been a phenomenal increase in area, production and productivity of vegetables in India in the last one decade, still there is a huge gap between the present production and future requirements. This necessitates enhancing vegetable production for meeting the current and future needs.

One of the major constraints in vegetable production is pest problem. In India, the yield losses due to major pests and diseases in vegetable crops are to the extent of 32–40% (Kodandaram et al., 2013), which needs to be averted. Pesticides have substantially contributed in controlling pests and increasing crop yields. Vegetable farmers use a wide range of pesticides at different levels to reduce losses from pests and diseases. In the country, the average pesticide consumption in vegetables is around 0.56 kg a.i./ha. Around 13–14% of pesticides used in the country are applied to vegetables (Kodandaram et al., 2014b). However, their non-judicious, unregulated and indiscriminate use has increased both the hazards to human health and the pollution to the environment. The presence of high pesticide residue levels in certain vegetables has been a point of serious concern which warrants sincere attention from all sectors—the growers, the consumers and the various government agencies dealing with advisory, regulatory and law enforcement services (Rai et al., 2014d). The pre- and post-harvest management strategies help to minimize the use of chemical pesticides and their toxicity to humans and beneficial organisms, and their effects on non-target species are being developed and implemented at various levels. Many pesticide users do not observe the necessary precautions. With special reference to India, this article is aimed to provide effective means for pesticide use in vegetables to avoid pre- and post-harvest pesticide contamination during vegetable production.

17.2 CHANGING PEST SCENARIO IN VEGETABLES

During the last one decade there is a considerable shift in pest status on account of the changes in the cropping pattern, ecosystems, habitat, climate and introduction of the input of intensive high-yielding varieties/ hybrids and other technological changes. An increasing trend have been shown by the incidence of several sucking insect pests such as whitefly (*Bemisia tabaci*) and leaf hopper (*Amrasca bigutella bigutella*) on okra and brinjal; mite (*Polyphagotarsonemus latus*) and thrips (*Scirothrips dorsalis*) on chilli; fruit fly (*Bactrocera cucurbitae*) and red spider mite (*Tetranychus urticae*) on cucurbits; mealy bug (*Phenacoccus solenopsis*) on brinjal, tomato, okra and cucurbits and aphids (*Aphis gossypii*) under protected conditions; Hadda beetle (*Henosepilachna vigitioctopunctata* and *Epilachna dodecastigma*) on cowpea and bitter gourd; plume moth (*Sphenaeches caffer*) in bottle gourd; gall midge (*Asphondylia capparis*) on chilli and brinjal in southern India (Rai et al., 2014a, 2014b, 2014c). The diamondback moth, *Plutella xylostella,* and fruit and shoot borer, *Leucinodes orbonalis,* have consistently remained the most destructive pest of cruciferous and brinjal, respectively. However, there has been an overall decline in the severity of borer insect pest *Helicoverpa armigera* in vegetable crops. Insect pest and diseases of much significance to vegetable crops and yield loss are given in Table 17.1.

TABLE 17.1 Yield Losses Due to Major Insect Pest and Diseases in Vegetables in India.

Crop/Pest	Yield loss (%)	Crop/Pest	Yield loss (%)
Tomato		*Cabbage*	
Fruit borer	24–65	Diamondback moth	17–99
Late blight	63–85	Cabbage caterpillar	69
Early blight	78–80	Cabbage leaf webber	28–51
Fusarium wilt	10–80	Cabbage borer	30–58
Tomato leaf curl	70	Black rot	5–70
Root knot nematode	27.21	Alternaria blight	70–80
Brinjal		*Cucurbits*	
Fruit and shoot borer	11–93	Fruit fly	
Root knot nematode	16.67	Bitter gourd	60–80

TABLE 17.1 *(Continue)*

Crop/Pest	Yield loss (%)	Crop/Pest	Yield loss (%)
Phomopsis blight	30–50	Cucumber	20–39
Bacterial wilt	90–91	Ivy gourd	63
Chillies		Musk melon	76–100
Thrips	12–90	Snake gourd	63
Mites	34	Sponge gourd	50
Gall midge	40	Cucurbits	
Anthracnose	30–80	Downy mildew	75–80
Chilli leaf curl	60	Powdery mildew	55–60
Root knot nematode	12.85	Bitter gourd mosaic	65
Okra		Bottle gourd mosaic	25–50
Fruit borer	22	Ridged gourd	50
Leafhopper	54–66	Root knot nematode	18.20
Whitefly	54	*Legume vegetables*	
Shoot and fruit borer	23–54	Cow pea pod borer	10–60
Root knot nematode	14.10	Root knot and reniform nematode	27.30
Okra yellow vein mosaic	50	Anthracnose of bean	10–65

17.3 PESTICIDE CONSUMPTION AND USAGE IN INDIA AND WORLD

In India, annual consumption of pesticides has a rising trend from 1955–1956 to 1990–1991 and thereafter, it started declining. During 2011–2012, the consumption of technical grade pesticides in the country was around 50,583 t (Anonymous, 2013). India's consumption of pesticides is only 2% of the total world consumption. Several reasons could be ascribed to this trend, including the use of new eco-friendly novel molecules, in which the quantities of newer pesticide molecules that required per unit area are almost 1/8–1/100 times less than the conventional molecules (Bambawale, 2007). Consumption of pesticides in India is around 560 g a.i./ha which is significantly less as compared with other developed countries such as the United States of America, Europe, Japan, China, Taiwan and so on. Lower consumption of pesticides in India can be attributed to the fragmented land holdings, low purchasing power of farmers, dependence on monsoons, inadequate awareness among farmers and low investment capabilities of small and marginal farmers.

Only 25–30% of the total cultivated area in the country is under pesticide cover. Among the different classes of pesticides used in India, the share of insecticides is the highest (60%) followed by fungicides (19%), herbicides (16%), biopesticides (3%) and others (3%). It is estimated that around 13–14% of the total pesticides used in the country are applied on vegetables, of which insecticides account for two-thirds share. Among different vegetable crops, the maximum pesticide usage is in chilli (5.13 a.i kg/ha) followed by brinjal (4.60 a.i kg/ha), cole crops (3.73 a.i kg/ha) and okra (2–3 a.i kg/ha). On the other hand, the global agrochemicalconsumption is dominated by fruits and vegetables, which account for 25% of the total market (Kodandaram et al., 2013).

17.4   STATUS OF PESTICIDE RESIDUES IN VEGETABLES

Pesticide residue is any substance or mixture of substances remaining in food meant for consumption by human being or animals and might include any specified degradation and conversion products, metabolites, reaction products and impurities which are considered to be of toxicological significance. The term 'pesticide residue' includes residues from unknown sources (i.e. background residues) as well as those from known uses or the chemical in question. Food contamination with pesticide residues always attracts public attention, apprehensions of poisoning and health hazards. These are supported by many studies conducted across the country at different points of time and location. Residue contamination in vegetable samples is reported mostly for organochlorine (OC) pesticides in India (Bhanti et al., 2000; Kumari et al., 2004; Bhanti and Taneja 2007), organophosphorus (OP) and synthetic pyrethroids (SPs).

In the harvested crops, pesticide residues may get degraded up to some extent when transported from the point of production to the final consumers. But there is a possibility of detection of pesticide residues either below or above the levels of toxicological significance in the farm gate samples, in which transportation is minimal. Study results involving sample collection from farmers' fields for pesticide residue screening in vegetables have been reported by many researchers (Swarnam and Velmurugan, 2012; Gowda and Somashekar, 2012; Kumari et al., 2004).

Vegetables of different types have been monitored for the presence of organochlorine, organophosphorus and synthetic pyrethroid group of pesticides. The studies reported the contamination of vegetable samples with pesticides ranging from 34% (Swarnam and Velmurugan, 2012) to 100% (Gowda and Somashekar, 2012) of the samples tested. The major pesticides contaminating the vegetables included α-Endosulfan, β-Endosulfan and Endosulfan sulfate from the organochlorine group; α-Cypermethrin, Fenvalerate-I and λ-Cyhalothrin from the pyrethroid group and Chlorpyrifos, Profenophos, Monocrotophos and Acephate from the organophosphorus group. The contamination of OCs in vegetables ranged between 14.5% (Swarnam and Velmurugan, 2012) and 97% (Gowda and Somashekar, 2012), whereas those of Ops and SPs were of 54–83% and 32–60%, respectively. However, various researchers also reported the detection of pesticides above MRL values, which is of great concern. Swarnam and Velmurugan (2012) reported that 15.3% of the tested vegetable samples exceed the prescribed MRLs.

Monitoring studies at different market places by different researchers (Kumari et al., 2002, 2003; Charan et al., 2010; Srivastava et al., 2011) reported significant contaminations of vegetable samples with pesticide residues at different levels. A report from the central Aravalli region reflects about 40.11% of the vegetable samples tested being contaminated with different pesticide residues, with 35.62% samples having residues above the MRLs (Charan et al., 2010). Kumari et al. (2002) reported that with about 23% of the samples having contaminations of organophosphorous compounds above their respective MRL values, contaminations in 100% of the samples were tested either at low or high concentrations. However, many of these detected pesticides either appeared due to their unauthorized applications or through various indirect sources, for example contaminated agro-inputs, drift from adjoining agricultural fields and so on. In another study in Lucknow, 58.33% of the samples tested were without residues, 28.33% of samples contained pesticide residues at or below the MRLs and 13.33% of samples contained pesticide residues above MRL. Brinjal was the most positive followed by cabbage, tomato and okra (Bankar et al., 2012).

In view of their potential toxic and persistent nature, there is a pressing need for the control and monitoring of pesticide residues in environment. The vegetable (potato, tomato, cabbage, cauliflower, spinach and okra) samples collected from the local markets of Jaipur city, Rajasthan, India, were also found to be contaminated with residues of DDT and its metabolites (DDD, DDE), isomers of HCH (α-,β-, and γ-), Heptachlor, Heptachlor epoxide and Aldrin. Some of the detected insecticides exceeded the limit of tolerance prescribed by WHO/FAO (Bakore et al., 2002). During 1997–1998 for pesticide contaminations, 100% of the samples were contaminated with low but measurable amounts of pesticide residues in one of the monitoring studies of 80 winter vegetable samples. Residue levels of organophosphorous insecticides were the highest, followed by carbamates, synthetic pyrethroids and organochlorines. Moreover, 32% of the samples showed contamination with organophosphorous and carbamate insecticides above their respective MRL values (Kumari et al., 2003).

Every possible and feasible way to monitor the contamination of vegetables by pesticides has been tried and reported (Kapoor et al., 2013). In Kothapally and Enkepally villages of Ranga Reddy district, Andhra Pradesh, out of 75 tomato samples, 26 (35%) were contaminated and 4% had residues above the MRLs. In another studies, Sharada (1988) reported from Mysore, Karnataka, that all the tomato samples (514 over the four seasons) tested were positive for pesticide residues with 72% being contaminated with HCH, and 14% with DDT and DDT+HCH residues. Similarly, in case of chillies, the corresponding values were 57, 28 and 15%, respectively. Kapoor et al. (2013) screened around 250 samples, including 70 samples of vegetables from Lucknow, India, and analysed for the presence of Imidacloprid residues. Although Imidacloprid was detected in 24% of the vegetable samples, only 5.71% showed the presence of Imidacloprid above its MRL.

17.5 REGULATION OF PESTICIDE RESIDUES IN INDIA

The regulation of pesticide production and uses in India is governed by two separate bodies, namely the Central Insecticides Board and Registration Committee (CIB and RC) and the Food Safety and Standards Authority of India (FSSAI). The former was established in 1968 under the Department of Agriculture and Co-operation of Ministry of Agriculture and Farmers Welfare, Govt. of India, which is responsible for advising central and state governments on technical issues related to manufacture, use and safety issues related to pesticides. Its responsibilities also include recommending uses of various types of the pesticides depending on their toxicity and suitability, determining the shelf life of pesticides and recommending a minimum gap between the pesticide applications and harvesting of the crops, that is waiting period (http://cibrc.nic.in/cibrc.htm).

The other part of the CIB and RC, the registration committee, is responsible for registering pesticides after verifying the claims of the manufacturer or importer related to the efficacy and safety of the pesticides (http:// cibrc.nic.in/registration_committee.htm). The CIB and RC register pesticides in India and recommend them for various crops. The Registration Committee has granted registration to 260 active ingredients and 503 pesticide formulations as of May 2015. This also includes 16 microbial and 2 plant-based bio-pesticides. A total of 82 combination products have been registered for use in agriculture, which includes 36 insecticides, 27 fungicides, 18 herbicides and one insecticide and fungicide, combination products. In India the use of several old and highly toxic pesticides and their formulations have been banned and restricted. Among this 28 pesticides are banned for manufacture, import and use, four pesticide formulations are banned for import, manufacture and use, two pesticides are banned for use with approval for manufacture exclusively for export, seven pesticides are withdrawn from registration, 18 have been refused registration and 13 are restricted for use in India (http://www.cibrc.nic.in).

The Food Safety and Standards Authority of India (FSSAI) holds the responsibility for setting maximum residue limits (MRL) of various pesticides in food commodities that have been registered by CIB and RC. The MRLs for all the registered pesticides should be set for all the crops they have been registered for. The exceptions for which MRLs are not required include neem-based products, biopesticides and fewer chemical pesticides such as Sulphur.

17.5.1 WAITING PERIOD/PRE-HARVEST INTERVAL

On the basis of the MRL, the pre-harvest intervals (PHI) of pesticides are calculated. PHI is the safe waiting period, which is the minimum time in days that must be provided between the last application of a pesticide and harvesting of the produce so that its residue level at harvest reaches below the MRL. Estimating PHI of a pesticide ideally involves multi-location field trials wherein the pesticides are applied following the guidelines of Good Agricultural Practices (GAPs) in terms of time, dose and frequency of applications. Representative samples are collected from the treated plants and analysed for the residues. The sampling is initiated on the day of the final application and continued at regular time interval till harvest. After a precise estimation of the residues in each sample, the residue data are statistically processed to correlate the dissipation with progress of time. Linear or non-linear kinetics models are applied for the estimation of PHIs depending on the nature of dissipation pattern.

The PHI must therefore be long enough to allow for the pesticide residues in the harvested crop to degrade to a level that is acceptable. So that the MRL for a given crop is not exceeded, it is important to respect the PHI. Residues found in excess of the MRL on food would constitute a violation of the FSSAI regulations and could also pose a risk to consumers' health. In such situations, the harvested crop could be seized, destroyed or forbidden for consumption as well as export.

17.5.2 MAXIMUM RESIDUE LIMITS

Maximum residue limits are the upper legal levels of a concentration for pesticide residues in or on food or feed based upon good agricultural practices and to ensure the lowest possible consumer exposure. The MRLs are established to ensure that the total amount of pesticide residues absorbed through food consumption will not exceed the acceptable daily intake (quantity of a pesticide humans can ingest in one day without any harmful effects) for a pesticide, whichever it may be. At the international level, the Codex Alimentarius Commission of the FAO/WHO decides the MRL. With a primary mandate to establish MRL for pesticides in food, the Codex Committee on Pesticide Residues (CCPR) was formed by the United Nations. In India, the MRL of different pesticide-commodity combinations are prescribed under the Food Safety and Standards Regulations, 2010 (Table 17.2). The new Act authorizes the Food Safety and Standards Authority of India (FSSAI) to specify the limits for use of food additives, crop contaminants, pesticide residues, residues of veterinary drugs, heavy metals, processing aids, mycotoxins, antibiotics and pharmacologically active substances and irradiation of food.

TABLE 17.2   The Indian Maximum Residue Limits (MRL) for Pesticides Recommended for Insect Control in Vegetables.

Sl. no.	Insecticides	MRL or tolerance limits in mg/kg (ppm)						
		Tomato	Brinjal	Chilli	Okra	Cabbage	Cauliflower	Cucurbits
1.	Acetamiprid 20 SP	–	–	0.1	–	0.1	–	–
2.	Buprofezin	–	–	0.01	–	–	–	–
3.	Carbaryl 5 DP	–	–	–	10.0	–	–	–
4.	Carbaryl 50 WP	–	–	5.0	–	–	–	–
5.	Carbosulfan	–	–	0.2	–	–	–	–
6.	Chlorantraniliprole	–	–	–	–	0.03	–	–
7.	Chlorfenpyre	–	–	0.05	–	0.05	–	–
8.	Cypermethrin 0.25 DP	–	0.20	–	–	–	–	–
9.	Cypermethrin 10 EC	–	2.0	–	–	–	–	–
10.	Cypermethrin 25 EC	–	0.20	–	–	–	–	–
11.	Deltamethrin 2.8 EC	0.05	–	–	0.05	–	–	–
12.	Dicofol	–	–	1.0	–	–	–	–
13.	Difenthiuron 50 WP	–	1.0	0.05	–	1.0	–	–
14.	Emamectin benzoate	–	–	–	0.05	–	–	–
15.	Ethion	–	–	–	–	–	–	0.5
16.	Etoxazole	–	–	–	–	–	–	–
17.	Fenazaquin	–	–	0.5	–	–	–	–
18.	Fenpropathrin	–	0.2	0.2	0.5	–	–	–
19.	Fenpyroximate	–	–	1.0	–	–	–	–
20.	Fenvalrate	–	2.0	–	2.0	–	2.0	–
21.	Fipronil	–	–	0.001	–	0.001	–	–
22.	Flumite/Flufenzine	–	0.5	–	–	–	–	–
23.	Hexythiazox	–	–	0.01	–	–	–	–
24.	Indoxacarb 14.5% SC	0.05	–	0.01	–	0.1	–	–
25.	λ-Cyhalothrin 4.9 CS	0.1	0.2	0.5	2.0		–	–
26.	Lufenuron	–	–	–	–	0.3	0.1	–
27.	Methomyl	0.05	–	0.05	–	–	–	–
28.	Milibectin	–	–	0.01	–	–	–	–
29.	Novaluron	0.01	–	0.01	–	0.01	–	–

TABLE 17.2   *(Continue)*

Sl. no.	Insecticides	MRL or tolerance limits in mg/kg (ppm)						
		Tomato	Brinjal	Chilli	Okra	Cabbage	Cauliflower	Cucurbits
30.	Phosphamidon	–	–	2.0	–	–	–	–
31.	Propargite	–	–	2.0	–	–	–	–
32.	Pyridalyl	–	–	0.2	0.02	0.02	–	–
33.	Quinalphos 20 AF	–	–	0.2	–	–	–	–
34.	Spinosad 2.5 SC	–	–	–	0.02	0.02	–	–
35.	Spinosad 45 SC			0.001	–	–	–	
36.	Spiromesifen			–	–	–	–	–
37.	Thiacloprid			0.02	–	–	–	–
38.	Thiodicarb			0.01	–	–	–	–
39.	Thiamethoxam 25 WG	0.01	0.3	0.01	0.5	–	–	–
40.	Tolfenpyrad			–	0.7	0.01	–	–

17.5.3 ACCEPTABLE DAILY INTAKE

The acceptable daily intake is an estimate of the amount of a food additive, expressed on body weight basis that can be ingested daily over a lifetime without any appreciable health risk (Macholz, 1988). From NOAEL, by dividing the figure normally by a safety factor of 100, the acceptable daily intake (ADI) is calculated. The figure 100 is taken into consideration as a multiple of 10 (10×10), in which the first 10 provides for inter-species variation, whereas the second 10 provides for intra-species variation. Therefore ADI, which is expressed in terms of mg/kg body weight, is an indication of the fact that if a human being consumes that amount of pesticide every day, throughout his lifetime, it will not cause appreciable health risk on the basis of well-known facts at the time of the evaluation of that particular pesticide. ADI is measured usually in milligrams of the substance, per kilogram of body weight of the exposed person, per day (mg/kg/day). The concept of ADI is internationally accepted today as the basis for estimation of safety of food additives and pesticides, for evaluation of contaminants and for legislation in the area of food and drinking water.

17.5.4   NO-OBSERVED-ADVERSE-EFFECT LEVEL

The No-Observed-Adverse-Effect Level (NOAEL) is the highest concentration of an agent, found by study or observation that causes no detectable adverse alteration of morphology, functional capacity, growth, development or lifespan of the target population.

By using different animal models, pesticides, being toxic in nature, are supposed to be thoroughly screened for their safety. For this purpose, studies on acute toxicity, chronic toxicity, allergenicity and so on are undertaken. These data are evaluated and the No-Observed-Adverse-Effect Level (NOAEL) is calculated from the chronic toxicity studies. The acute reference dose is also taken into consideration in the case of toxic pesticides. This NOAEL and acute reference dose are supposed to be taken as the starting information for prescribing the tolerance limits of pesticides in food commodities. The NOAEL is usually referred in terms of milligrams of that particular pesticide per kilogram of body weight.

17.5.5   MAXIMUM PERMISSIBLE INTAKE

The acceptable daily intake is the maximum residue of the toxic chemicals permissible per kilogram body weight per day. Therefore, maximum permissible intake depends on the total body weight. Thus, it is calculated by taking the body weight into consideration. In general, an average body weight of 16 kg of a child is considered instead of adults considering higher degree of vulnerability of the sub-population towards pesticide poisoning.

$$MPI = acceptable\ daily\ intake \times average\ body\ weight$$

17.5.6   THEORETICAL MAXIMUM RESIDUE CONTRIBUTION

In the daily diet of an average person, it is the theoretical maximum amount of a pesticide. It assumes that the diet is composed of all food items for which there are tolerance-level residues of the pesticide. The TMRC is expressed as milligrams of pesticide/kilograms of body weight/day.

17.6 FACTORS AFFECTING PERSISTENCE AND DISSIPATION OF PESTICIDE RESIDUES

After pesticides are applied to the crops, they may interact with the plant surfaces, be exposed to the environmental factors such as wind, sunlight and may be washed off during rainfall. The pesticides may be absorbed by the plant surface (waxy cuticle and root surfaces) and may enter the plant transport system (systemic) or may stay on the surface of the plant (contact). On the surface of the crop, a pesticide can undergo volatilization, photolysis, chemical and microbial degradation and so on. On account of degradation and transfer to another compartment, dissipation can be defined as the loss of pesticide residues from a defined compartment. Volatilization, photolysis and microbial degradation are the main processes by which pesticides get dissipated on fruit or vegetables. The dissipation half-life of commonly used pesticides in major vegetables is mentioned in the Table 17.3.

TABLE 17.3 Dissipation Half-life of Pesticides in/on Vegetables.

Commodity	Pesticide	Half-life (days)	References
Tomato	Flubendiamide	0.72–1.32	Sharma and Parihar (2013)
	Thiacloprid	0.83–1.79	
	Flubendiamide	1.64–1.98	Paramashivam and Banerjee (2013)
	Bifenthrin	1.83–2.32	Chauhan et al. (2012)
	Chlorpyrifos	4.38–4.43	Rani et al. (2013a)
	Mancozeb	3.76 and 4.14	Rani et al. (2013b)
	Metalaxyl	1.29 and 0.41	
	Chlothianidin	7.0–11.9 days	Li et al. (2012)
Brinjal	Thiacloprid	0.47 and 0.50	Sahoo et al. (2013)
	Profenofos	2.15–2.31	Mukherjee et al. (2012)
	Cypermethrin	0.91–1.86	
	Chlorpyrifos	3.27–3.10	
	Cypermethrin	2.19–3.27	
	Quinalphos	2–3	Pathan et al. (2012)
	Flubendiamide	0.62 and 0.54	Takkar et al. (2012)

TABLE 17.3 *(Continue)*

Commodity	Pesticide	Half-life (days)	References
Cabbage	Emamectin Benzoate	1.34–1.72	Wang et al. (2012)
	Fipronil	3.21–3.43	Bhardwaj et al. (2012)
	Trichlorfon	1.80	Li et al. (2011)
	Quinalphos	4.8–5.3	Mohapatra and Deepa (2013)
	Fubendiamide	3.4–3.6	
	Flubendiamide	3.4–3.6	Paramashivam and Banerjee (2013)
	Thiacloprid	12.3–13.1	Dutta et al. (2012)
	Spinosad	1.4 and 1.5	Singh and Battu (2012)
	Metaflumizone	1.7–2.1	Chatterjee and Gupta (2013)
	Emamectin Benzoate	<5	Singh et al. (2013)
Okra	Flubendiamide	4.7–5.1	Das et al. (2012)
	Spiromesifen	1.68 and 1.65	Raj et al. (2012)
	Imidacloprid	1.07 and 2.41	Banerjee et al. (2012)
	Beta-Cyfluthrin	1.98 and 3.30	
	Chlorpyriphos	0.6	Samriti et al. (2012)
Cauliflower	Chlorantraniliprole	1.36	Kar et al. (2013)
	Cypermethrin	1.5–2.1	Gupta et al. (2013)
	Deltamethrin	2.9–3.3	
	Profenofos	2.6–3.0	
	Triazophos	2.2–2.6	
Chilli	Chlorpyriphos	4.43 and 2.01	Jyoti et al. (2012)
	Cypermethrin	2.51 and 2.64	Jyoti et al. (2012)
	Trifloxystrobin	1.81 and 1.58	Sahoo et al. (2012)
	Tebuconazole	1.37 and 1.41	
	Deltamethrin	0.36–1.99	Pandher et al. (2012)
	Acetamiprid	2.24–4.84	Sanyal et al. (2008)
	Chlorfenapyr	2.93–2.96	Ditya et al. (2010)

17.7   STRATEGIES TO REDUCE PRE- AND POST-HARVEST CONTAMINATION

17.7.1   ECO-FRIENDLY PEST MANAGEMENT TECHNIQUES

17.7.1.1   SELECTION OF PEST TOLERANT VARIETIES

Selection of resistant/less susceptible varieties holds well in pest management when the diversity and intensity of pests in any particular location is kept in mind. Plant resistance provides a built-in ability to allow less pest infestation and reduces the load of pesticides in the crop. Being completely safe, host plant resistance fits well with all other components. Unlike cereals, in vegetables, very less number of resistant/less susceptible varieties have been developed (Table 17.4). Pests of sucking nature can be combated to a greater extent with the adoption of resistant varieties.

TABLE 17.4   Tolerant Varieties of Some Vegetable Crops Against Major Insect Pests.

Crop	Pest	Varieties
Tomato	Fruit borer (*H. armigera*)	Arka Vikash, Pusa Gaurav, Pusa Early Dwarf, Punjab Keshri, Punjab Chhuhara, Pant Bahar, Azad, BT 1, T 32, T 27
Brinjal	Shoot and fruit borer (*L. orbonalis*), aphid, jassid, thrips, whitefly	SM 17-4, PBr 129-5 Punjab Barsati, ARV 2-C, Pusa Purple Round, Punjab Neelam, Kalyanpur-2, Punjab Chamkila, Gote-2, PBR-91, GB-1, GB-6
Cabbage	Aphid (*Brevicoryne brassicae*)	All season, Red Drum Head, Sure Head, Express Mail
Cauliflower	Stem borer (*Hellula undalis*)	Early Patna, EMS-3, KW-5, KW-8, Kathmandu Local
Okra	Jassid (*Amrasca biguttula biguttula*)	IC-7194, IC-13999 New Selection, Punjab Padmini
	Shoot and fruit borer (*Earias vittella*)	AE 57, PMS 8, Parkins long green, PKX 9275, Karnual special
Onion	Thrips (*Thripstabaci*)	PBR-2, PBR-6, Arka Niketan, Pusa Ratnar, PBR-4, PBR-5, PBR-6
Round gourd, Pumpkin, Bitter gourd	Fruit fly (*B. cucurbitae*)	Arka Tinda Arka Suryamukhi Hissar-II

17.7.1.2 PLANTING/SOWING TIME

Attack of some insect pests can be reduced in vegetable production by careful consideration of sowing or planting date. For example, an early planting of cucurbits in November might escape the attack of the red pumpkin beetles, whereas flowering beyond October (e.g. in bitter gourd) might cause less infestations of fruit fly. Sowing of okra during the week of June retains less population of borers, thereby incur maximum, healthy yield, whereas July-planted brinjal faces the ravages of shoot and fruit borers. Thus, synchronization of the most susceptible stage of the crop with the inactive period of insect pest reduces the infestation and requirements of chemical intervention for pest management.

17.7.1.3 INTERCROPPING

Intercropping of crops with diverse plant geometry and insect pests breaks the standard mono-cropping and limits the infestation from the pest (Table 17.5). Diverse nature of plant not only obstructs the adults from egg laying but also the release of volatile allelochemicals from a particular crop deters the adult insects from damaging the other. Such effects do not provide appropriate conditions for development of microclimate favouring the multiplication of any particular type of pest. All such planting combinations also enhance the activity of predators and parasites.

TABLE 17.5 Combination of Different Intercropping Effective in Vegetable Pest Management.

Crop combination	Target pest	Source
Cabbage + Carrot	Diamondback moth	Buranday and Raros (1973); Bach and Tabashnik (1990)
Broccoli + Faba bean	Flea beetle	Garcia and Altieri (1992)
Cabbage + French bean	Root fly	Hofsvang (1991)
Cabbage + Tomato	Diamondback moth	Srinivasan and Veeresh (1986)
Okra + Corn	Yellow vein Mosaic	Adhikary et al. (2015)
Bitter gourd + Maize	Fruit fly	Shooker et al. (2006)
Cabbage + Indian Mustard	Diamondback moth	Srinivasan and Moorthy (1992)
Cabbage + Chinese Cabbage	Diamondback moth	Satpathy et al. (2010a & 2009)
Brinjal + Coriander/Fennel	Shoot and Fruit borer	Khorsheduzzaman et al. (1997); Satpathy and Mishra (2011)

17.7.1.4 TRAP CROPS

A trap crop must be distinctively attractive to the target pest than the main crop. It provides protection either by preventing the pest from reaching the main crop or by concentrating them in certain parts of the field where they can be economically destroyed. For example, mustard trap crop along with cabbage has been successfully utilized for the management of diamondback moth, aphid and leaf webber on cabbage. This technology developed was in 1989 (Srinivasan and Moorthy, 1991, 1992), and it recommended two rows of bold-seeded Indian mustard after 25 rows of cabbage. The first row of mustard is sown 15 days prior to the cabbage planting and second row is sown in 25 days after planting. Mustard attracts more than 80% of the cabbage pests. However, in addition to the 2–3 sprays of neem seed kernel extract (NSKE), the mustard foliage is to be sprayed with dichlorvos. Recent studies indicated Chinese cabbage to be the potential trap crop for diamondback moth (Satpathy et al., 2010a). African marigold in tight bud stage functions as a good trap crop to attract the adults of *H. armigera* (Srinivasan, 1994), and it also attracts the adults of leafminer for egg laying on the leaves. When sown in combination with bitter gourd, maize plants applied with bait spray trap kill the fruit fly adults. Planting castor as a trap crop diverts the population of *Spodoptera litura* from cowpea.

17.7.1.5 BIO-PESTICIDES

Bio-pesticides are the formulated products of microbial (bacteria, virus, fungi, nematode etc.) and plant origin (neem etc.) based pesticides. They have foremost advantage that these pesticides degrade or dissipate rapidly in the environment and have satisfactory action on the target pests. Bio-pesticides are often viewed as a replacement for the harmful chemicals and as an effective component of IPM programmes. As of late, bio-pesticides usage has shown a steady increase with a consumption level of more than 6000 MTs during 2011–2012 (Anonymous, 2012). Several microbial and botanical insecticides are included in the schedule of the Insecticides Act, 1968. For use in agriculture, a total of 18 bio-pesticides are currently registered in India, which includes 16 microbial and 2 plant-based bio-pesticides. The neem and microbial-based bio-pesticides are currently used and approved for vegetable pest management in India and are mentioned in Tables 17.6 and 17.7.

TABLE 17.6 Neem-based Insecticides for Insect Pest Control in Vegetable Crops.

Sl. no.	Neem-based insecticides and formulation	Target pests target organism/host	Physiological stage for application	Recommended dose formulation (g/ml) and mode of application
1.	Azadirachtin 0.03% (300 ppm)	Fruit borer, whitefly, leafhopper in okra	Vegetative, flowering and fruiting	2500–5000 ml and spraying
		Fruit and shoot borer, beetles in brinjal	Vegetative, flowering and fruiting	2500–5000 ml and spraying
		Aphids, DBM, cabbage worm, cabbage – looper	Vegetative, head formation	2500–5000 ml and spraying
		Powdery mildew in okra	Vegetative	2–2.5 ml and spraying
2.	Azadirachtin 0.15% (1500 ppm)	Aphids, jassids	Vegetative	1–2 l and spraying
		Aphids, DBM	Vegetative, head formation	2–2.5 l and spraying
		Fruit borer and whitefly	Flowering, fruiting and vegetative	3.25 l and spraying
		Pod borer	Fruiting and vegetative	2 l and spraying
3.	Azadirachtin 0.3% (3000 ppm)	DBM	Head formation	1.67–3.34 l and spraying
4.	Azadirachtin 5% (50,000 ppm)	DBM, aphids, *Spodoptera litura*	Head Formation and Vegetative	200 ml and spraying
		Whitefly, jassids, aphids and shoot and fruit borer	Vegetative, flowering and fruiting	200 ml and spraying
5.	Azadirachtin 1% (10,000 ppm)	Tomato fruit borer and brinjal fruit and shoot borer	Flowering and fruiting	1000–1500 ml and spraying

TABLE 17.7 Microbial-Based Bio-pesticides for Vegetable Crops.

Sl. no.	Biopesticide and formulation	Target pests target organism/ host	Physiological stage of application	Recommended dose formulation (g/ml) and mode of application
1.	*Bacillus thuringiensis var. kurstaki, 3a, 3b, SA-II* WP	Diamond back moth	Head formation	0.5 kg and spraying
2.	*Bacillus thuringiensis var. kurstaki, BMPn 123 (2×) WDG, 3a, 3b*	Brinjal shoot and fruit borer	Vegetative, flowering and fruiting	0.25–0.5 kg and spraying
3.	*Bacillus thuringiensis var. kurstaki, HP* WP	Diamond back moth	Head formation	300–500 g and spraying
4.	*Bacillus thuringiensis var. galleriae Serotype, 3a, 3b,* WP	Diamond back moth	Head formation	0.60–1.0 kg and spraying
		Tomato fruit borer	Flowering and fruiting	1–1.5 kg and spraying
		Okra fruit and shoot borer	Vegetative, flowering and fruiting	1–1.5 kg and spraying
5.	*Bacillus thuringiensis var. kurstaki, strain Z-523, serotype H3a, 3b* WP	Okra fruit and shoot borer	Vegetative , flowering and fruiting	0.4–1.0 kg and spraying
6.	*Bacillus thuringiensis var. kurstaki,* WP	*Spodoptera litura, Spilosoma,* semi-looper, leaf miner in Soybean	Vegetative	0.75–1.0 kg and spraying
		Pod borer in legumes	Flowering and fruiting	0.75–1.0 kg and spraying
7.	*Beauveria bassiana* Strain No. IPL/BB/MI/01 $(1 \times 10^9$ CFU/g min) 1% WP	Fruit borer/spotted bollworm in okra	Vegetative , flowering and fruiting	3.75–5.0 kg and spraying
8.	Nuclear Polyhedrosis Virus of *Spodoptera litura* 0.5% AS	*Spodoptera litura* in tomato	Vegetative and flowering and fruiting	1500 ml and spraying
9.	NPV of *Helicoverpa armigera* Strain No. IBH-17268 $(1 \times 10^9$ POB/ml min) 2.0% AS	*Helicoverpa armigera* in tomato	Flowering and fruiting	250–500 ml and spraying
10.	NPV of *Helicoverpa armigera* 0.43% AS	*Helicoverpa armigera* in tomato	Flowering and fruiting	1500 ml and spraying

TABLE 17.7 *(Continued)*

Sl. no.	Biopesticide and formulation	Target pests target organism/ host	Physiological stage of application	Recommended dose formulation (g/ml) and mode of application
11.	*Pseudomonas fluorescens* 0.5% WP	Damping off in chilli	At sowing	10 g/kg of seeds as seed treatment
		Wilt in tomato		
		Wilt in tomato	Vegetative	2.5 kg and as soil application
12.	*Trichoderma viride* 1% WP	Root rot in cowpea	At sowing	5 g/kg of seed as seed treatment
		Damping off in chilli		
		Stalk rot in cauliflower		
		Root Rot/Wilt/ Damping off in brinjal		
		Seedling wilt in tomato	At sowing	9 g/kg of seed as Seed treatment
		Seedling wilt in tomato	Vegetative	2.5 kg and as soil application at Root zone application
		Root rot in cowpea	Vegetative stage	2.5 kg as soil treatment
		Stalk rot in cauliflower		
		Root rot/Wilt/ Damping off/Collar rot in cabbage		
		Root rot/Wilt/ Damping off	Nursery stage	250 g/50 l of water/400 m². as nursery treatment
		Root rot/Wilt/ Damping off	At transplanting	10 g/l of water as seedling root dip treatment
		Root rot/Collar rot in cabbage	At transplanting	10 g/l water as seedling root dip

17.7.1.6 NATURAL ENEMIES

Vegetables ecosystem is endowed with a large complex of natural enemies (predator and parasitoid) attacking different stages of the insect pests at varying extents (Table 17.8). For pest management in vegetables, till now, except a few, most of these natural enemies are underexploited. Among the egg parasitoids, *Trichogramma* spp. has been utilized to some extent for control of tomato fruit borer (Jalali, 2005). Innundative release of egg parasitoids *Trichogramma brasilensis* @ 250,000 ha^{-1} are also recommended for control of fruit borers on okra and tomato. Six releases of *Trichogramma* at weekly interval @ 50,000/ha with the first release coinciding with 50% flowering in tomato is recommended. *Chrysoperla zastrowi arabica* is an effective predator for control of whitefly, aphid, jassid and eggs of some lepidopteran borers (Khulbe et al., 2005), when the first instar larvae are released @ 50,000 ha^{-1}. On account of their potential in suppressing the pest larvae, the larval parasitoids of diamondback moth, *Cotesia plutellae* and *Diadegma semiclausum*, can be incorporated into biological pest management (Saucke et al., 2000). These natural enemies will prove effective in the protected vegetable production and in areas where insect pests pose a serious problem because of insecticide resistance.

TABLE 17.8 Some Important Natural Enemies of Insect Pests of Vegetable Crops.

Pest	Parasitoid/Predator
Crucifers	
Plutella xylostella	*Cotesia plutellae, Diadegma semiclausum, Brachymeria excarinata*
	Trichogrammatoidea bactrae
Crocidolomia binotalis	*Alanteles crocidolomia, Palexorista solannis, Bracon hebetor*
Hellula undalis	*Bracon* spp.
Tomato	
H. armigera	*Trichogramma chilonis, T. brasiliensis , T. pretiosum, Campoletis chlorideae*
Okra	
Earias spp.	*Trichogramma chilonis, T. ahaeae, T. brasiliensis , Chelonus blackburni*
Spodoptera litura	*Telenomus remus, Peribaea orbata*
Aphis spp.	*Coccinella septempunctata, Chrysoperla zastrowi arabica*

TABLE 17.8 *(Continued)*

Pest	Parasitoid/Predator
Brinjal	
Brinjal shoot and fruit borer	*Eriborus argentiopilosus, Trathala flavo orbitalis*
Gall midge *Asphondylia* sp.	*Eurytoma* sp.
Mealybug (*Phenacoccus solenopsis*)	*Aenasius bambawalei*

17.7.2 JUDICIOUS USE OF CHEMICAL PESTICIDES

17.7.2.1 CONSIDERATION OF ECONOMIC THRESHOLD LEVEL FOR PESTICIDE APPLICATION

Instead of going for routine schedule application of insecticides, need-based spraying on the basis of economic threshold level (ETL) will help to reduce pesticide consumption and environmental abuse. ETL-based application schedule helps maintaining minimum pest residue for survival of natural enemies. ETL for many vegetable insect pests has been worked out (Table 17.9).

TABLE 17.9 Economic Threshold Level (ETL) for Major Insect Pests of Vegetable Crops.

Crop/Pest	Economic threshold level (ETL)
Tomato	
Fruit borer (*H. armigera*)	8 eggs/15 plants or 1 larva/plant or 1 damaged fruit/plant
Whitefly (*B. tabaci*)	3 nymphs/leaf or 4 adults/leaf
Leaf miner (*Liriomyza trifolii*)	26 mines/trifoliates or 6 adults/6 rows
Brinjal	
Fruit and shoot borer (*L. orbonalis*)	0.5–5% shoot and fruit damage
Chillies	
Thrips (*S. dorsalis*)	2 thrips/leaf
Mites (*P. latus*)	1 mite/leaf
Okra	
Leafhopper (*A. biguttula biguttula*)	4.66 hoppers/leaf
Shoot and fruit bore (*Earias vittella*)	5.3% of fruit infestation

TABLE 17.9 *(Continued)*

Crop/Pest	Economic threshold level (ETL)
Cabbage	
Diamondback moth (*P. xylostella*)	2 larvae/plant at 1–4 weeks after transplanting or 5 larvae/plant at 5–10 weeks after transplanting
Cabbage leaf webber (*Crocidolomia binotalis*)	0.3 egg mass/plant
Pea	
Pea aphid (*Acyrthosiphon pisum*)	3–4 aphids/stem tip

17.7.2.2 CHOICE OF CHEMICAL PESTICIDES AND FOLLOWING PRE-HARVEST INTERVAL

Residues of insecticides are a major concern for vegetable crops. Choice of insecticide plays a crucial role to avoid the persistent residue. Pre-harvest interval (PHI) or waiting period of insecticides should be considered in chemical control of insect pests (Table 17.10). As a thumb rule, to minimize residue accumulations, the insecticides with less PHI should be given preference against the longer persistent alternatives. In the plant protection schedule as well, the applications of pesticides are required to be staggered as per their relative PHIs so that the initial residue deposits dissipate to below the MRLs at the stage of harvest. Insecticides of plant origin, insect growth regulators (IGRs) have shown good efficacy against insect pests and can fit well in integrated pest management programmes (IPM).

TABLE 17.10 Waiting Period/Pre-harvest Intervals of Insecticides Recommended for Insect Control in Vegetables.

Sl. no.	Insecticides	Waiting period or pre-harvest interval (days)						
		Tomato	Brinjal	Chilli	Okra	Cabbage	Cauliflower	Cucurbits
1.	Acetamiprid 20 SP	–	–	3	3	7	–	–
2.	Azadirachtin 1%	3	3	–	–	–	–	–
3.	Azadirachtin 0.03%	–	7	–	7	7	–	–
4.	Azadirachtin 5%	5	–	–	5	5	–	–
5.	Buprofezin	–	–	5	5	–	–	–

TABLE 17.10 *(Continued)*

Sl. no.	Insecticides	Tomato	Brinjal	Chilli	Okra	Cabbage	Cauliflower	Cucurbits
		Waiting period or pre-harvest interval (days)						
6.	Carbaryl 5 DP	–	–	–	8	8	–	–
7.	Carbaryl 50 WP	8	5	–	3	8	5	–
8.	Carbosulfan	–	–	8	–	–	–	–
9.	Chlorantranilprole	3	22	3	5	3	–	7 (Bittergourd)
10.	Chlorfenpyre	–	–	5	–	7	–	–
11.	Chlorfluazoron	–	–	–	–	7	–	–
12.	Cyantraniliprole	3	–	3	–	5	–	5 (Gherkins)
13.	Cypermethrin 0.25 DP	–	3	–	–	–	–	–
14.	Cypermethrin 10 EC	–	3	–	3	7	–	–
15.	Cypermethrin 25 EC	–	1	–	3	–	–	–
16.	Deltamethrin 2.8 EC	–	3	5	1	–	–	–
17.	Dicofol	–	15–20	–	15–20	–	–	–
18.	Difenthiuron 50 WP	–	3	3	–	7	–	–
19.	Emamectin benzoate	–	3	3	5	3	–	–
20.	Ethion	–	–	5	–	–	–	–
21.	Etoxazole	–	5		–	–	–	–
22.	Fenazaquin	7	7	10	7	–	–	–
23.	Fenpropathrin	–	10	7	7	–	–	–
24.	Fenpyroximate	–	–	7		–	–	–
25.	Fenvalrate	–	5	–	7	–	7	–
26.	Fipronil	–	–	7	–	7	–	–
27.	Flubendamide 20 WG	5	–		–	7	–	–
28.	Flubendamide 40 SC	–	–	7	–	7	–	–
29.	Flufenoxuron	–	–	–	–	7	–	–
30.	Flumite/Flufenzine	–	5	–	–	–	–	–
31.	Hexythiazox	–	–	3	–	–	–	–
32.	Imidacloprid 70 WG	–	–	–	3	–	–	5 (Cucumber)
33.	Imidacloprid 17.8 SL	3	–	40	3	–	–	–
34.	Indoxacarb 14.5 SC	5	–	5	–	7	–	–

TABLE 17.10 *(Continued)*

Sl. no.	Insecticides	Waiting period or pre-harvest interval (days)						
		Tomato	Brinjal	Chilli	Okra	Cabbage	Cauliflower	Cucurbits
35.	Indoxacarb 15.8 EC	–	–	–	–	5	–	–
36.	λ-Cyhalothrin 4.9 CS	5	5	5	5	–	–	–
37.	λ-Cyhalothrin 5 EC	4	4	5	4	–	–	–
38.	Lufenuron	–	–	5	–	14	5	–
39.	Metaflumizone	–	–		–	3	–	–
40.	Methomyl	5–6	–	5–6	–	–	–	–
41.	Milibectin	–	–	7	–	–	–	–
42.	Novaluron	1–3	–	3	–	5	–	–
43.	Phosphamidon	–	10	–	–	–	–	–
44.	Propergite	–	6	7	–	–	–	–
45.	Pyridalyl	–	–	–	3	3	–	–
46.	Quinalphos 20 AF	7	–	–	7	–	–	–
47.	Spinosad 2.5 SC	–	–	–	–	3	3	–
48.	Spinosad 45 SC	–	–	3	–	–	–	–
49.	Spiromesifen	3	5	7	3	–	–	–
50.	Thiacloprid	–	5	5	–	–	–	–
51.	Thiodicarb	–	6	6	–	7	–	–
52.	Thiamethoxam 25 WG	5	5	–	5	–	–	–
53.	Tolfenpyrad	–	–	–	3	5	–	–
54.	Betacyfluthrin 8.49 + Imidacloprid1 9.81 OD	–	7	–	–	–	–	–
55.	Cypermethrin 3+ Quinalphos 20 EC	–	7	–	–	–	–	–
56.	Deltamethrin 1 + Trizophos 35 EC	–	21	–	–	–	–	–
57.	Indoxacarb 14.5 + Acetamiprid 7.7 SC	–	–	5	–	–	–	–
58.	Novaluron 5.25 + Indoxacarb 4.5 SC	5	–	–	–	–	–	–
59.	Pyriproxyfen 5 + Fenpropathrin 15 EC	–	7	7	7	–	–	–

17.7.2.3   USE OF NEWER GREEN CHEMISTRY MOLECULES

Several new molecules with green chemistries have been developed in the recent years and are released into the market for chemical pest management in vegetables crops against sucking and borer insect pests. These novel insecticides having unique modes of action to target pests are often designated as 'bio-rational' or 'low risk' insecticides (Hara, 2000). There is a paradigm shift in the use of insecticides with the introduction of these new molecules, in which the quantities of newer pesticide molecules and formulations required per hectare are almost 1/4–1/53 times less than the conventional molecules (Kodandaram et al., 2014a). Most of the newer insecticides have several advantages over conventional insecticides. They have high level of selectivity to target pests, excellent efficacy at low dosage, non-persistent in the environment, low mammalian toxicity, less harm to the natural enemies, helpful for delaying insecticide resistance and have no cross-resistance with the conventional insecticides (Kodandaram et al., 2010). All these merits render many of these new insecticides as safer and more suitable to fit well into the integrated pest management (IPM) or insect resistance management (IRM) programmes in vegetable crops. The adoption of these insecticides is likely to increase as farmers and pest control advisors become familiar with their unique characteristics. Some of the new molecules belong to neonicotinoids, oxadiazines, diamides, tetramic/tetronic acid derivatives, phenylpyrazoles, pyridine, avermectins, spinosyns, pyrroles and insect growth regulators (IGRs) (Table 17.11).

TABLE 17.11   Insecticides with New Chemistries for Control of Vegetable Insect Pests.

Sl. no.	Insecticide group	Target site	Mode of action	Active ingredients
1.	Neonicotinoids	Nerve	Agonists of nicotinic acetylcholine receptor (nAChR)	Imidacloprid, acetamiprid, thiamethoxam, thiacloprid
2.	Pyridine-Carboxamide	Nerve	Modulators of chordontal organs	Flonicamid*
3.	Diamide	Nerve and muscle action	Ryanodine receptor modulators	Flubendamide Cyantraniliprole Chlorantraniliprole
4.	Tetronic and tetramic acid derivatives	Lipid synthesis	Inhibitors of acetyl CoA carboxylase	Spiromesifen

TABLE 17.11 *(Continued)*

Sl. no.	Insecticide group	Target site	Mode of action	Active ingredients
5.	Pyrrole insecticides	Energy metabolism	Uncouplers of oxidative phosphorylation via disruption of proton gradient	Chlorfenapyr
6.	Thiourea insecticides	Energy metabolism	Inhibitors of mitochondrial ATP synthase	Diafentiuron
7.	Avermectins	Nerve	Glutamate-gated chloride channel modulators	Emamectin Benzoate Milbemectin
8.	Spinosyns	Nerve	Nicotinic acetylcholine receptor (nAChR) allosteric activators	Spinosad
9.	Phenylpyrazoles	Nerve	GABA gated chloride channels antagonists	Fipronil
10.	Benzoylureas	Growth regulation	Chitin biosynthesis Inhibitors type 0	Flufenoxuron
11.	Chitin synthesis inhibitors	Energy metabolism	Chitin biosynthesis inhibitors type I	Buprofezin, novaluron
12.	METI acarcides	Energy metabolism	Mitochondrial complex I electron transport inhibitors	Fenpyroximate, fenzaquin, tolfenpyrad
13.	Mite growth inhibitor	Growth regulation	Regulates the growth of mite	Hexthiazox, etoxazole, flufenzine
14.	Sulfite ester acaricides	Energy metabolism	Inhibitors of mitochondrial ATP synthase	Propargite

17.7.3 METHODS OF PESTICIDE APPLICATION

To reduce the possibility of spray drift, application of pesticides should be under the right environmental conditions; run-off or leaching that may contaminate other crops. Maintenance and proper calibration of the application equipment should be done to deliver the correct rate uniformly over the field. Calculations should be double-checked to make sure the application rate is the same as recommended. To save labour and money for field applications, farmers often mix various pesticides on their own, but such mixing are not officially permitted as per the label claims, and this might also degrade the active ingredients and result in poor bio-efficacy. Selection of proper pesticide application techniques/methods is vital to the success of any pest control operations. The application of pesticide is not merely the spraying operation; it has to be coupled with a thorough knowledge of the target pest. The method of application depends on nature of pesticide, formulation, pests to be managed, site of application, availability of water and so on. The mode of action of a pesticide, its relative toxicity and other physicochemical properties, helps to decide the handling precautions, agitation requirement and so on. The main purpose of pesticide application technique is to ensure judicious use of active ingredients, to cover the target with maximum efficiency and minimum efforts to keep the pest under control as well as minimum contamination of non-targets. By prudent selection of application by application, a high degree of selectivity and dosage reduction could be achieved. There is tremendous scope in the improvement of application technology through even distribution of lethal dose among many drops. In this connection, a recent advance in electrostatic spraying is worth to mention. Apart from the spraying, by using a very small amount of insecticide without any detrimental effects on natural enemies, the other application techniques such as seed treatment, seedling root dip, spot application, soil drenching and chemigation could be used to control pest more efficiently (Kodandaram, 2014).

17.7.4   ADOPTION OF INTEGRATED PEST MANAGEMENT TECHNOLOGY

IPM is an eco-friendly approach which uses cultural, mechanical and biological tools for keeping pest population below the economic threshold levels. This approach emphasizes more on the use of biocontrol agents and bio-pesticides (Rai et al., 2010, 2013). However, need-based and judicious use of chemical pesticides is permitted and is very critical in pest management and often appears to be the last resort. IPM helps in maximizing crop protection with minimum input costs; minimizing pollution in soil, water and air; reducing occupational health hazards; conserving ecological equilibrium and reducing pesticide residue loads in the food. Good Agricultural Practices (GAPs) amalgamated with IPM takes into account the application of minimum quantities of pesticides necessary to achieve adequate pest control in such a manner that the amount of residues remained in food is at smallest possible. Various government agencies are actively promoting IPM after recognizing the imperative of safe and judicious use of pesticides. To assist farmers and extension workers in the adoption of less-chemical pest control approaches, several IPM package of practices have been developed for important insect pests of vegetables (Table 17.12) under the All India Coordinated Research Programme on Vegetable Crops (AICRP-VC), Indian Institute of Vegetable Research, Varanasi and Indian Institute of Horticultural Research, Bangalore. However, to accommodate new scientific knowledge and experiences, many of these packages of practices need to be reviewed and updated regularly.

TABLE 17.12 Integrated Pest Management (IPM) Modules/Technology for Major Insect Pests of Vegetable Crops.

Crop	Target pest	IPM module or technology
Brinjal	Shoot and fruit borer (*Leucinodes orbonalis*)	Avoid monoculture and follow crop rotation
		Grow the seedlings in raised nursery bed covered with 30-mesh nylon net/muslin cloth to prevent the initial attack of the pest at seedling stage and get healthy seedlings
		Intercropping with coriander/fennel as a single line, double line or border crop
		Seedling root dip in chlorantraniliprole 18.5 SC 0.5 ml/l for 3 h
		Installation of plastic funnel traps @ 100 ha^{-1} baited with sex pheromone of brinjal shoot and fruit borer at 25–30 DAT.
		Weekly removal of infested shoots and fruits.
		Five inoculative releases of *Trichogramma chilonis* 50,000 parasitized eggs/ha at weekly interval coinciding with infestation of the pest.
		Application of NSKE 4% or *Bt*.500 g/ha at the time of flowering.
		Need-based application one or two sprays of rynaxpyr 18.5 SC 0.2–0.4 ml/l or emamectin benzoate 5 SG 0.35 g/l or cartap hydrochloride 500 g/ha or lamda-cyhalothrin 2.5 EC 1.25 ml/l in rotation at vegetative, flowering and fruiting period

TABLE 17.12 *(Continued)*

Crop	Target pest	IPM module or technology
Cucurbits	Fruit fly (*Bactrocera cucurbitae*)	Deep summer ploughing or raking up the soil around the plants to expose the pupae to sunlight and predation by birds
		Collection and destruction of infested fruits
		Installation of used mineral water bottle trap, baited with cue lure (as MAT) saturated wood blocks (ethanol:cuelure:carbaryl in a ratio 6:1:2) @ 25–30 traps/ha prior to flower initiation
		Apply bait spray containing 20 ml/gmalathion 50 EC or carbaryl 50 WP + 20 l water +500 g molasses randomly on 250 plants covering entire field to attract adult flies and control the population
		Use of repellent (NSKE 4%) enhances the trapping and luring in bait spots
Okra	Leafhoppers (*Amrasca biguttula biguttula*)	Seed treatment with imidacloprid 48 FS @ 3–5 ml/kg or imidacloprid 70 WS @ 3–5 g/kg or thiamethoxam 70 WS @ 3 g/kg seed + polymer @ 40 ml/l at the time of sowing which provides protection for 20–25 days after germination
		Soil application of neem cake at 250 kg/ha immediately after germination and repeat it after 30 days
		Installation of yellow sticky traps at random in the field to monitor the insects activity
		Spraying of neem or pongamia soaps at 0.5% or pulverized neem seed powder extract (NSPE) 4% at 10 days interval
		Need-based foliar sprays of any of the insecticides such as imidacloprid 17.8 SL @ 0.35 ml/l or thiamethoxam 25 WG @ 0.35 ml/l or fenpropathrin 30 EC @ 0.75 ml/l or lambda-cyhalothrin5 EC @ 0.6 ml/l at 10–15 days interval at the appearance of the pest
Chilli	Thrips, *Scirtothrips dorsalis* and yellow mite, *Polyphagotarsonemus latus*	Seedling dip with imidacloprid 17.8 SL @ 1 ml/l of water
		Spray of buprofezin 1 ml/l at 25 DAT followed by fipronil 0.2 g/l at 35 DAT, *Verticillium lecanii* 5 g/l at 45 DAT, chlorfenapyr 1 ml/l at 55 DAT, neem oil 1% at 65 DAT and subsequent rotation of the same

TABLE 17.12 *(Continued)*

Crop	Target pest	IPM module or technology
Tomato	Fruit borer (*Helicoverpa armigera*)	Deep summer ploughing to expose the larvae and pupae to sunlight and predation by birds
		Planting of marigold (40 days old) as trap crop with every 16 rows of tomato (25 days old) attracts the pest. Collection and destruction of larvae from marigold flowers
		Install sex pheromone traps @ 5 traps/ha for early pest detection. Change the lure at every 25–30 days
		Two inundative release of *Trichogramma brasiliense* @ 250,000 parasitized eggs/ha during peak flowering stage
		Foliar spray of HNPV @ 250 LE with jaggery (10 g/l), soap powder (5 g/l) and tinopal (1 ml/l) during evening hours
		Spraying of azadirachtin 10,000 ppm @ 3 ml/l or azadirachtin 5% @ 0.5 ml/l at 10 days interval
		Need-based spray of any insecticide such as chlorantranilprole (rynaxpyr) 20 SC @ 0.35 ml/l or cyantraniliprole (cyzapyr) 10 OD @ 1.8 ml/l or indoxacarb 14 SC @ 1 ml/l or novaluron 10 EC @ 1.5 ml/l or methomyl 40 SP @ 1 g/l or lambda cyhalothrin 2.5 SC @ 0.6 ml/l at 10 days interval
Cabbage	DBM, *Plutella xylostella*	Avoid early or late planting of the crop.
		Use Chinese cabbage or paired rows of mustard as trap crop after every 16 rows of cabbage (sowing of first row of mustard 15 days prior to cabbage planting and second row 25 days after cabbage planting) and spray of dichlorvos (0.1%) on the mustard plant
		Application of *B.t.* @ 1 kg/ha
		Need base spray of any insecticide such as chlorantranilprole 20 SC @ 0.1 ml/l or cyantraniliprole 10 OD @ 1.8 ml/l or novaluron 10 EC @ 1.5 ml/l or indoxacarb 15.8 EC @ 0.75 ml/l, flubendamide 40 SC @ 0.5 ml/l and emamectin benzoate 5 SG @ 0.35 gm/l

17.7.5   DECONTAMINATION METHODS

Fruits and vegetables such as other foods pass through culinary and food-processing treatments before they are consumed. The effects of these culinary and food-processing techniques have been investigated by various researchers, and they have been found to reduce the pesticide residue levels except in cases where there is concentration of the product such as in juicing, drying and oil production. During processing treatments, some toxic metabolites may be produced, especially thermal processing. However, the consumers can still be encouraged to employ those processing methods that reduce pesticide residues. Food processing studies often results in transfer factors or food processing factors (PFs) of the pesticide residues in the transition from raw agriculture commodity to the processed product. These processing factors are expressed as the concentration of pesticide after processing divided by the concentration before processing. Some processing factors are available in public literature, whereas others are only available with the pesticide registering bodies. Processing studies have become a part of pesticide registration requirements. Few processing techniques and average processing factor are summarized in Table 17.13. Kaushik et al. (2009) reviewed the effect of processing on dissipation of pesticide residues in fruits and vegetables.

TABLE 17.13   Processing Techniques and Average Processing Factor (PF).

Processing technique	Processing factor
Baking	1.38
Blanching	0.21
Boiling	0.82
Canning	0.71
Frying	0.1
Juicing	0.59
Peeling	0.41
Washing	0.68

Few process techniques and their effect on pesticide dissipation are given below.

17.7.5.1 BAKING

Baking is the technique of prolonged cooking of food by using dry heat normally in an oven. It is primarily used for the preparation of bread, cakes, pastries and pies, tarts and quiches. It is also used for the preparation of baked potatoes, baked apples and baked beans. The loss of pesticide residues may occur through various physicochemical processes during baking, for example evaporation, co-distillation and thermal degradation which may vary with the chemical nature of the individual pesticides. During the process the water contained in the tissue could entrain pesticide molecules (co-distillation), while heat causes evaporation and degradation.

17.7.5.2 BLANCHING

Hot water blanching increases pesticide removal and may hydrolyse substantial fractions of non-persistent compounds. For example, the blanching operation (after peeling) of contaminated potatoes (at level of 1 ppm), resulted in reduction of the residues by 28.3, 22.9, 26.0, 47.3, 46.3 and 45.9% for HCB, lindane, p.p-DDT, dimethoate, pirimiphos-methyl and malathion, respectively. Blanching affects organophosphorous pesticides more than organochlorine, which withstands 100°C. This may be due to the high stability of organochlorines to heat treatment. With the help of blanching, tomatoes fortified with 25 ppm of parathion were processed which resulted in 50% reduction in parathion level.

17.7.5.3 BOILING

Pesticide residues might get degraded during cooking. In particular, pesticide compounds with functionalities such as carbamate, amide, urea, thiocarbonyl and imino group are readily hydrolysed in the presence of trace amounts of acid and/or base during heating or boiling. For example, processing of spinach for 66 min at 252°F reduced the residue loads of diazinon by 58%, azinphos-methyl by 100%, malathion by 96%, methyl parathion by 100% and carbophenothion by 17%.

17.7.5.4 CANNING

This commercial process in its various forms combines elements of washing, peeling, juicing, cooking and concentration. Processing whole tomatoes with incurred vinclozolin residues of 0.73 mg/kg in canned juice, puree and ketchup reduced to 0.18, 0.73 and 0.22 mg/kg, respectively. A canning process that did not include peeling for cherries removed 95% of the tetrachlorvinphos residues (4.3 ppm).

17.7.5.5 FRYING

Processes involving heat can increase volatilization, hydrolysis or other chemical degradation and thus reduces residue levels. Blanching and frying of brinjal for 5 min completely removed the profenofos residues which were initially present at the level of 0.27 ppm. Effect of frying (after peeling) had a more prominent effect on the organophosphorous residues with reduction of residues by 49–53% against the organochlorines with the level of reduction ranging between 30.1 and 35.3%.

17.7.5.6 WASHING

To decrease the residue load and intake of pesticide residues, washing with water and various chemical solutions for domestic and commercial use are necessary. Washing is the most common form of processing which is a preliminary step in both household and commercial preparation. Loosely held residues of several pesticides are removed with reasonable efficiency by varied types of washing processes. Bitter gourds with endosulfan residues of 18.97 and 26.01 ppm on washing for 30 s reflected decontamination to the extent of 59.05 and 42.66%, respectively.

17.7.5.7 PEELING

Peeling of the fruits and vegetables removes the surface residues. Freezing as well as juicing and peeling are necessary to remove the pesticide residues in the skins. A decrease of about 77% of the initial procymidone residue level (0.86 ppm) on tomatoes was observed after peeling (Kaushik et al., 2009).

As indicated, food processing usually results a decrease in pesticide or contaminant levels. However, in some cases, residue levels may increase in the final product due to concentration factors of raw commodities in the process of the final product. This concentration effect can be related with water removal for example in tomatoes used for tomato ketchup. In such cases, a PF of greater than one is observed.

17.7.6   REGULATORY CONTROL

To minimize residue contaminations in vegetables, regulatory control is a necessity. In India, still not many pesticides have label claim for usage in vegetable crops. It should be kept in mind that a pesticide approved for usage in any particular vegetable is not officially allowed for application in another crop without label claim. The dissipation pattern and persistence of a pesticide may be completely different in different crops. It is extremely important to replace the usage of extreme and high toxic chemicals with those having safer mammalian and environmental toxicity profiles to improve the food safety image of Indian vegetables in the international market. For this purpose, new generation chemicals should be brought into the purview of label claim through appropriate bio-efficacy and residue studies. In absence of sufficient choice chemicals, farmers are sometimes compelled to use unapproved chemicals in field and such limitations can only be resolved by including more and more number of choice chemicals through the process of label claim with the CIB and RC. Strict regulatory control needs to be exercised for use of restricted and banned chemicals and those pesticides which do not have any label claims as per the CIB and RC. This can be achieved through the stringent execution of regulatory controls and extension programmes through state governments and other stakeholders with appropriate linkage with the scientific community.

17.8   CONCLUSION

The consumer awareness is growing with respect to the right to have safe food with the advent of information explosion. No one is ready to tolerate the contamination of the food intended for his/her nurturing. On the other hand, our country being in tropical belt, we do not have the luxury of avoiding the use of the pesticides in the pest management programmes. Pesticides are used in the vegetable production to ensure a good harvest to meet the food and nutritional security of the nation. It is the dose and time of application, which differentiates the safe and unsafe use of pesticides. The presence of high pesticide residue levels in some vegetables has been considered to be an issue which warrants sincere attention from all sectors, the growers, the consumers and the various government agencies dealing with advisory, regulatory and law enforcement services. With the Good Agricultural Practices along with integration of all possible effective techniques, including both chemical and non-chemical methods and following pre-harvest intervals of label claim pesticides need to be sincerely followed to minimize the pre- and post-harvest contamination in the quality production of vegetables in a way that protects human health and our environment.

REFERENCES

Adhikary, S. A.; Koundinya, V. V.; Pandit, M. K.; Bhattacharya, B. Evaluation of Efficiency of Baby Corn Based Vegetable Intercropping System. *Int. J. Plant Soil Sci.* **2015,** *5,* 366–374.

Anonymous. Farm Inputs and management. In State of Indian Agriculture 2011–12, Department of Agriculture and Cooperation, Ministry of Agriculture, Government of India, New Delhi, 2012, p 273.

Anonymous. Production and availability of pesticides. In. 36th report on standing committee on chemicals and Fertilizers, Ministry of Chemicals and Fertilisers (Department of Chemicals and Petrochemicals), Government of India, New Delhi, 2013, p 49.

Bach, C. E.; Tabashnik, B. E. Effects of Nonhost Plant Neighbors on Population Densities and Parasitism Rates of the Diamondback Moth (Lepidoptera: Plutellidae). *Environ. Entomol.* **1990,** *19,* 987–994.

Bakore, N.; John, P. J.; Bhatnagar, P. Evaluation of Organochlorine Insecticide Residue Levels in Locally Marketed Vegetables of Jaipur City, Rajasthan. India. *J. Environ. Bio.* **2002,** *23*(3), 247–52.

Bambawale, O. M. Expanding Dimensions of Plant Protection Through Convergence of Chemical, Transgenic and Biological Means. SPS India Foundation Day Lecture, Society of Pesticide Science, India, IARI, New Delhi November 16, 2007.

Banerjee, T.; Banerjee, D.; Roy, S.; Banerjee, H.; Pal, S. A Comparative Study on the Persistence of Imidacloprid and Beta-cyfluthrin in Vegetables. *Bull. Environ. Contam. Toxicol.* **2012,** *89*(1), 193–196.

Bankar, R.; Ray, A. K.; Kumar, A.; Adeppa, K.; Puri, S. Organochlorine Pesticide Residues in Vegetables of Three Major Markets in Uttar Pradesh, India. *Acta Biologica Indica* **2012,** *1,* 77–80.

Bhanti, M.; Taneja, A. Contamination of Vegetables of Different Seasons with Organo-phosphorous Pesticides and Related Health Risk Assessment in Northern India. *Chemosphere* **2007**, *69,* 63–68.

Bhanti, M.; Shukla, G.; Taneja, A. Contamination Levels of Organochlorine Pesticides and Farmers' Knowledge, Perception, Practices in Rural India—a Case Study. *Bull. Environ. Contam. Toxicol.* **2000**, *73,* 787–793.

Bhardwaj, U.; Kumar, R.; Kaur, S.; Sahoo, S. K.; Mandal, K.; Battu, R. S.; Singh, B. Persistence of Fipronil and its Risk Assessment on Cabbage, Brassica oleracea var. Capitata L. *Ecotoxicol. Environ. Saf.* **2012**, *79*(1), 301–308.

Buranday, R. P.; Raros, R. S. Effects of Cabbage-tomato Intercropping on the Incidence and Oviposition of the Diamondback Moth, *Plutella xylostella* (L.). *Philipp. Entomol.* **1973**, *2,* 369–374.

Charan, P. D.; Ali, S. F.; Kachhawa, Y.; Sharma, K. C. Monitoring of Pesticide Residues in Farmgate Vegetables of Central Aravalli Region of Western India. *J. Agri. Environ. Sci.* **2010**, *7,* 255–258.

Chatterjee, N. S.; Gupta, S. Persistence of Metaflumizone on Cabbage (*Brassica oleracea Linne*) and Soil, and its Risk Assessment. *Environ. Monit. Assess.* **2013**, *185,* 6201–6208.

Chauhan, R.; Monga, S.; Kumari, B. Dissipation and Decontamination of Bifenthrin Residues in Tomato (*Lycopersicon esculentum Mill*). *Bull. Environ. Contam. Toxicol.* **2012**, *89,* 181–186.

Das, S. K.; Mukherjee, I.; Das, S. K. Dissipation of Flubendiamide in/on Okra [Abelmoschus esculenta (L.) Moench] Fruits. *Bull. Environ. Contam. Toxicol.* **2012**, *88,* 381–384.

Ditya, P.; Das, S. P.; Sarkar, P. K.; Bhattacharyya, A. Degradation Dynamics of Chlorfenapyr Residue in Chili, Cabbage and Soil. *Bull. Environ. Contam. Toxicol.* **2010**, *84,* 602–605.

Dutta, D.; Niwas, R.; Gopal, M. Comparative Persistence of Thiacloprid in Bt-transgenic Cabbage (*Brassica oleracea cv. capitata*) vis-à-vis Non-transgenic Crop and its Decontamination. *Bull. Environ. Contam. Toxicol.* **2012**, *89,* 1027–1031.

Garcia, M. A.; Altieri, M. A. Explaining Differences in Flea Beetle Phyllotreta Cruciferae Goeze Densities in Simple and Mixed Broccoli Cropping Systems as a Function of Individual Behavior. *Entomol. Exp. Appl.* **1992**, *62,* 201–209.

Gowda, S. R. A.; Somashekar, R. K. Evaluation of Pesticide Residues in Farmgate Samples of Vegetables in Karnataka, India. *Bull. Environ. Contam. Toxicol.* **2012**, *89,* 626–632.

Gupta, S.; Sharma, R. K.; Gajbhiye, V. T.; Gupta, R. K. Persistence of Insecticides in Ready-mix Formulations and Their Efficacy Against *Lipaphis erysimi* (Kalt) in Cauliflower. *Environ. Monit. Assess.* **2013**, *185,* 2107–2114.

Hara, A. H. Finding Alternative Ways to Control Alien Pests—Part 2: New Insecticides Introduced to Fight Old Pests. *Hawaii Landscape* **2000**, *4, 5.*

Hofsvang, T. The Influence of Intercropping and Weeds on the Oviposition of the Brassica Root Flies (Delia Radicum and D. Floralis). *Norw. J. Agri. Sci.* **1991**, *5,* 349–356.

Jalali, S. K. Predators and Parasitoids in Pest Suppression. In Biopesticides: Emerging trends-2005, 11–13th November, 2005 at Palampur, Himachal Pradesh, 2005, pp 27–28.

Jyoti, G. K.; Mandal, B.; Singh, R. S. Balwinder Estimation of Chlorpyriphos and Cypermethrin Residues in Chilli (*Capsicum annum L.*) by Gas–liquid Chromatography. *Environ. Monit. Assess.* **2012**, 185, 5703–5714.

Kapoor, U.; Srivastava, M. K.; Srivastava, A. K.; Patel, D. K.; Garg, V.; Srivastava, L. P. Analysis of Imidacloprid Residues in Fruits, Vegetables, Cereals, Fruit Juices, and Baby Foods and Daily Intake Estimation in and Around Lucknow India. *Environ. Toxicol. Chem.* **2013,** *32,* 723–727.

Kar, A.; Mandal, K.; Singh, B. Environmental Fate of Chlorantraniliprole Residues on Cauliflower using QuEChERS Technique. *Environ. Monit. Assess.* **2013,** *185,* 1255–1263.

Kaushik, G.; Satya, S.; Naik, S. N. Food Processing a Tool to Pesticide Residue Dissipation—A Review. *Food Res. Int.* **2009,** *42,* 26–40.

Khorsheduzzaman, A. K. M.; Mannan, M. A.; Ahmed, A. Brinjal-coriander Intercropping: An Effective IPM Component Against Brinjal Shoot and Fruit Borer, *Leucinodes Orbonalis* Guen (Pyralidae: Lepidoptera). *Bangladesh J. Entomol.* **1997,** *20,* 85–91.

Khulbe, P.; Ravi, P.; Maurya, R. P.; Khan, M. A. Biology of *Chrysoperla carnea* (Stephens) on Different Host Insects. *Annals Plant Prot. Sci.* **2005,** *13*(2), 241–243.

Kodandaram, M. H. Pesticide Application Methods and Dose Calculations. In *Emerging Trends in Plant Protection Inputs and Appliances for Safe and Quality Vegetable Production. Technical Bull No. 56.* Indian Institute of Vegetable Research, Varanasi, India, 2014, pp 264–271.

Kodandaram, M. H.; Rai, A. B.; Halder, J. Novel Insecticides for Management of Insect Pests in Vegetable Crops: A Review. *Veg. Sci.* **2010,** *37,* 109–123.

Kodandaram, M. H.; Saha, S.; Rai, A. B.; Naik, P. S. *Compendium on pesticide use in vegetables;* IIVR Extension Bull No. 50; Indian Institute of Vegetable Research: Varanasi, 2013; p 133.

Kodandaram, M. H.; Rai, A. B.; Halder, J. Novel Insecticide Molecules for Vegetable Insect Pest Management. In *Emerging Trends in Plant Protection Inputs and Appliances for Safe and Quality Vegetable Production;* Technical Bull No. 56; Indian Institute of Vegetable Research: Varanasi, India, 2014a; p 98–110.

Kodandaram, M. H.; Rai, A. B.; Halder, J. Plant Protection Inputs for Safe Vegetable Production: An Overview In: Emerging trends in plant protection inputs and appliances for safe and quality vegetable production. *Technical Bull No. 56,* Indian Institute of Vegetable Research, Varanasi, India, 2014b, pp 1–20.

Kumari, B.; Madan, V. K.; Kumar, R.; Kathpal, T. S. Monitoring of Seasonal Vegetables for Pesticide Residues. *Environ. Monit. Assess.* **2002,** 74, 263–270.

Kumari, B.; Kumar, R.; Madan, V. K.; Singh, R.; Singh, J.; Kathpal, T. S. Magnitude of Pesticidal Contamination in Winter Vegetables from Hisar, Haryana. *Environ. Monit. Assess.* **2003,** *87,* 311–318.

Kumari, B.; Madan, V. K.; Singh, J.; Singh, S.; Kathpal, T. S. Monitoring of Pesticidal Contamination of Farmgate Vegetables from Hisar. *Environ. Monit. Assess.* **2004,** *90,* 65–71.

Li, W.; Ma, Y.; Li, L.; Qin, D.; Wu, Y. The Dissipation Rates of Trichlorfon and its Degradation Product Dichlorvos in Cabbage and Soil. *Chemosphere* **2011,** *82,* 829–833.

Li, L.; Jiang, G.; Liu, C.; Liang, H.; Sun, D.; Li, W. Clothianidin Dissipation in Tomato and Soil, and Distribution in Tomato Peel and Flesh. *Food Control* **2012,** *25,* 265–269.

Macholz, R. Principles for the Safety of Food Additives and Contaminants in Food (Environmental Health Criteria 70). 174 Seiten. World Health Organization: Geneva 1987. Preis: 14,—Sw. fr.; 8, 40 US$." Food/Nahrung 32.6 1988; 638–638.

Mohapatra, S.; Deepa, M. Persistence and Dissipation of Quinalphos in/on Cauliflower and Soil Under the Semi-Arid Climatic Conditions of Karnataka, India. *Bull. Environ. Contam. Toxicol.* **2013,** *90,* 489–493.

Mukherjee, I.; Kumar, A.; Kumar, A. Persistence Behavior of Combination Mix Crop Protection Agents in/on Eggplant Fruits. *Bull. Environ. Contam. Toxicol.* **2012,** *88,* 338–43.

National Horticulture Board (NHB). Indian Horticulture Database. National Horticulture Board, Ministry of Agriculture, Government of India, Gurgaon, India. 278, 2014.

Pandher, S.; Sahoo, S. K.; Battu, R. S.; Singh, B.; Saiyad, M. S.; Patel, A. R.; Shah, P. G.; Reddy, C. N. D.; Reddy, J.; Reddy, K. N.; Rao, Ch. S.; Banerjee, T.; Banerjee, D.; Hudait, R.; Banerjee, H.; Tripathy, V.; Sharma, K. K. Persistence and Dissipation Kinetics of Deltamethrin on Chili in Different Agro-Climatic Zones of India. *Bull. Environ. Contam. Toxicol.* **2012,** *88,* 764–768.

Paramashivam, M.; Banerjee, H. Dissipation of Flubendiamide Residues in/on Cabbage (*Brassica oleracea* L.) *Environ. Monit. Assess.* **2013,** *185,* 1577–1581.

Pathan, A. R.; Parihar, N. S.; Sharma, B. N. Dissipation Study of Quinalphos (25 EC) in/on Brinjal and Soil. *Bull. Environ. Contam. Toxicol.* **2012,** *88,* 894–896.

Rai, A. B.; Swamy, T. M. S.; Satpathy, S.; Pandey, K. K.; Rai, M. In: Emerging trends and strategies for the management of pests and diseases in vegetable crops. Indian Institute of Vegetable Research: Varanasi, 2010; p 216.

Rai, A. B.; Loganathan, M.; Kodandaram, M. H.; Saha, S.; Halder, J. *Modern Approaches in Diagnostics and Management of Pest and Diseases in Vegetable Crops under Protected conditions;* Technical Manual No. 49. Indian Institute of Vegetable Research, Varanasi, India, 2013; p 281.

Rai, A. B.; Halder, J.; Kodandaram, M. H. Emerging Insect Pest Problems in Vegetable Crops and their Management in India: An Appraisal. *Pest Management in Horticultural Ecosystems* **2014a,** *20,* 113–122.

Rai, A. B.; Loganathan, M.; Halder, J.; Venkataravanappa, V.; Naik, P. S. *Eco-friendly Approaches for Sustainable Management of Vegetable Pests;* IIVR Technical Bull No. 53, Indian Institute of Vegetable Research: Varanasi, 2014b; p 104.

Rai, A. B.; Kodandaram, M. H.; Halder, J. Development and evaluation of pest management modules to minimize pre-harvest losses due to leafhopper *Amrasca bigutella bigutella* in okra. In National conference on "Pre-/Post-Harvest Losses and Value Addition in Vegetables. Association of Promotion of Innovations in Vegetables, Indian Institute of Vegetable Research, Varanasi, 2014c, p 43.

Rai, A. B.; Bijen K. Y.; Saha, S. Safe use of Pesticides in Vegetables. In National conference on "Pre-/Post-Harvest Losses and Value Addition in Vegetables. Association of Promotion of Innovations in Vegetables, Indian Institute of Vegetable Research, Varanasi, 2014d, p 131.

Raj, M. F.; Solanki, P. P.; Singh, S.; Vaghela, K. M.; Shah, P. G.; Patel, A. R.; Diwan, K. D. Dissipation of Spiromesifen in/on Okra under Middle Gujarat Conditions. *Pesti. Res. J.* **2012,** *24,* 25–27.

Rani, M.; Saini, S.; Kumari, B. Persistence and Effect of Processing on Chlorpyriphos Residues in Tomato (*Lycopersicon esculantum* Mill.) *Ecotoxicol. Environ. Safety* **2013a,** *95,* 247–252.

Rani, R.; Sharma, V. K.; Rattan, G. S.; Singh, B.; Sharma, N. Dissipation of Residues of Mancozeb and Metalaxyl in Tomato (*Solanum lycopersicum L.*). *Bull. Environ. Contam. Toxicol.* **2013b,** *90,* 248–251.

Sahoo, S. K.; Jyot, G.; Battu, R. S.; Singh, B. Dissipation Kinetics of Trifloxystrobin and Tebuconazole on Chili and Soil. *Bull. Environ. Contam. Toxicol.* **2012,** *88,* 368–371.

Sahoo, S. K.; Mandal, K.; Kaur, R.; Battu, R. S.; Singh, B. Persistence of Thiacloprid Residues on Brinjal (*Solanum melongena L.*). *Environ. Monit. Assess.* **2013,** *185*(9), 7935–7943.

Sanyal, D.; Chakma, D.; Alam, S. Persistence of a Neonicotinoid Insecticide, Acetamiprid on Chili (*Capsicum annum L.*). *Bull. Environ. Contam. Toxicol.* **2008,** *81,* 365–368.

Satpathy, S.; Mishra, D. S. Use of Intercrops and Antifeedants for Management of Eggplant Shoot and Fruit Borer *Leucinodes orbonalis* (Lepidoptera: Pyralidae). *Int. J. Trop. Insect. Sci.* **2011,** *31,* 52–58.

Satpathy, S.; Shivalingaswamy, T. M.; Kumar, A.; Rai, A. B. Evaluation of Crucifers for Trap Crop of Diamondback Moth, *Plutella xylostella. Insect Environ.* **2009,** *15,* 22–24.

Satpathy, S.; Shivalingaswamy, T. M.; Kumar, A.; Rai, A. B.; Mathura, R. Potentiality of Chinese Cabbage (*Brassica rapa* subsp. *Pekinensis*) as Trap Crop for Diamondback Moth (*Plutella xylostella*) Management in Cabbage. *Indian J. Agri. Sci.* **2010a,** *80,* 238–241.

Saucke, H.; Dori, D.; Schmutterer, H. Biological and Integrated Control of *Plutella xylostella* (Lep., Yponomeutidae) and Crocidolomia Pavonana (Lep., Pyralidae) in Brassica Crops in Papua New Guinea. *Biocontrol Sci. Technol.* **2000,** 10, 595–606.

Sharada, R. Interaction of Seed Protectants and Biological Systems with Special Reference to Abatement of Residues. Ph.D. Thesis, University of Mysore, Mysore, Karnataka, India, 1988.

Sharma, B. N.; Parihar, N. S. Dissipation and Persistence of Flubendiamide and Thiacloprid in/on Tomato and Soil. *Bull. Environ. Contam. Toxicol.* **2013,** *90,* 252–255.

Shooker, P.; Khayrattee, F.; Permalloo, S. Use of maize as a trap crops for the control of melon fly, B. Cucurbitae (Diptera:Tephritidae) with GF-120. Bio-control and other control methods [Online]. 2006. Available on: http\\www.fcla.Edu/flaent/fe87p354.pdf.

Singh, S.; Battu, R. S. Dissipation Kinetics of Spinosad in Cabbage (*Brassica oleracea L.var. capitata*). *Toxicol. Environ. Chem.* **2012,** *94,* 319–326.

Singh, G.; Chahil, G. S.; Jyot, G.; Battu, R. S.; Singh, B. Degradation Dynamics of Emamectin Benzoate on Cabbage under Subtropical Conditions of Punjab, India. *Bull. Environ. Contam. Toxicol.* **2013,** *91,* 129–133.

Srinivasan, K. Recent Trends in Insect Pest Management in Vegetable Crops. In *Trends in Agricultural Insect Pests Management;* Dhaiwal, G. S., Arora, R., Eds.; Commonwealth Publishers: New Delhi, India, 1994; pp 345–372.

Srinivasan, K.; Veeresh, G. K. Economic Analysis of a Promising Cultural Practice in the Control of Moth Pests of Cabbage. *Insect Sci. Appl.* **1986,** *7*(4), 559–563.

Srinivasan, K.; Moorthy, P. N. K. Indian Mustard as a Trap Crop for Management of Major Lepidopterous Pests on Cabbage. *Trop. Pest Manage.* **1991,** *37,* 26–32.

Srinivasan, K.; Moorthy, P. N. K. The Development and Adoption of Integrated Pest Management for Major Pests of Cabbage using Indian Mustard as a Trap Crop, In *Diamondback Moth and Other Cruciferous Pests: Proceedings of the Second International Workshop;* Talekar, N. S., Eds.; Asian Vegetable Research and Development Center: Shunhua, Taiwan, 1992; pp 10–14.

Srivastava, A.; Trivedi, K. P.; Srivastava, M. K.; Lohani, M.; Srivastava, L. P. Monitoring of Pesticide Residues in Market Basket Samples of Vegetable from Lucknow City, India: QuEChERS Method. *Environ. Monit. Assess.* **2011,** *176,* 465–472.

Swarnam, T. P.; Velmurugan, A. Pesticide Residues in Vegetable Samples from the Andaman Islands, India. *Environ. Monit. Assess.* **2012,** *185,* 6119–6127.

Takkar, R.; Sahoo, S. K.; Singh, G.; Battu, R. S.; Singh, B. Dissipation Pattern of Flubendiamide in/on Brinjal (*Solanum melongena L.*). *Environ. Monit. Assess.* **2012,** *184,* 5077–5083.

Wang, L.; Zhao, P.; Zhang, F.; Li, Y.; Fengpei, D.; Pan, C. Dissipation and Residue Behavior of Emamectin Benzoate on Apple and Cabbage Field Application. *Ecotoxicol. Environ. Safety* **2012,** *78,* 260–264.

CHAPTER 18

ENTREPRENEURSHIP OPPORTUNITIES IN PROCESSED VEGETABLES IN EMERGING ECONOMIES

P. S. BADAL[1], B. JIRLI[2] AND RAKESH SINGH[1]

[1]Department of Agricultural Economics, BHU, Varanasi 221005, UP, India

[2]Department of Extension Education, BHU, Varanasi 221005, UP, India

[1]Department of Agricultural Economics, Institute of Agricultural Sciences, BHU, Varanasi 221005, UP, India,
E-mail: rsingh66bhu@gmail.com

CONTENTS

18.1   INTRODUCTION

When we consider the position of the farmers in India, we see that they are taken for granted as primary producers and suppliers of raw materials for the industries. Agriculture is also a production system which possesses similar features of industrial production systems but with a major lacuna. What is that lacuna? The primary producer decides the price of the commodity produced/manufactured in case of industrial production system. But in case of agricultural production system, the buyer decides the price of commodity. Hence the farmer, the primary producer, is always in a disadvantageous situation. The primary reasons behind the existing situation include small holdings of farmers (more than 70%), limited individual production, lack of group/cooperative approach, especially in marketing of the agricultural production. Being a biological system, the time of crop maturity is the same throughout the nation, as the simple economic theories prove it, when there is more supply in the market, the prices of commodity comes down. Farmers are the readymade preys of this vicious cycle. Is it that each and every farmer in India is facing the same situation? The answer is obviously 'NO', there are a section of farmers throughout the country who are behaving differently and they are harvesting handsome profits. If so, why majority of the farming community is suffering? This is because they did not change their identity from mere farmer to an entrepreneur.

An individual is willing to be an entrepreneur if he is ready to change his identity. As, if an individual wishes to turn out to be an entrepreneur,

he/she has to transform their personality. As majority of Indian population still derives livelihood from agriculture and allied enterprises,
are they ready to change their identity yet? It is a trillion dollar question to be addressed by the professionals. The efforts are going in this
direction.

As per the data of National Horticultural Board, the area under vegetables is almost stabilized. On the other hand, spices and aromatic plants
have shown marginal increase in area and production. It can be understood from Table 18.1 that India is the largest producer and consumer of
food products. According to FICCI report of 2010, food market in India is
expected to grow nearly 40% in the current year. It is estimated that nearly
21% of GDP is accumulating from expenditure on food products.

TABLE 18.1   Area and Production of Horticulture Crops—all India. (*Source:* Indian
Horticulture Database, 2015–2016.)

Area in '000 Ha/Production in '000 MT					
	2013–2014 (Final)		2014–2015 (Final)		2015–2016 (1st Estimate)	
Crops	Area	Production	Area	Production	Area	Production
Fruits	7216	88,977	6110	86,602	6188	89,018
Vegetables	9396	162,897	9542	169,478	9465	168,506
Aromatic	493	895	659	1000	588	1136
Flowers	255	2297	249	2143	249	2157
Honey	–	76	–	81	–	83
Plantation	3675	16,301	3534	15,575	3532	15,480
Spices	3163	5908	3317	6108	3317	6108

At the global level, food industry is valued at US $ 3.2 trillion and
accounts for over 3/4th of global food sales. Despite the large size of the
industry, only 6% of the processed food is traded around the world as
compared with bulk agricultural commodities in which 16% of produce is
traded. (Source: www.nsdcindia.org/sites/default/files/files/food-process-
ings-2009.pdf) The United States of America is the largest consumer of
processed foods (31%), Europe (39%), Asia Pacific (31%) and rest of
the world consume only 9% of processed foods. It can be concluded that
consumption of processed foods is associated with income of the population and culture of the society.

In the food sector, the concept of entrepreneurship is yet to be understood by Indian stakeholders with these attractive figures on food production and consumption at national level. To address the issue, FICCI through its research unit has identified the following factors which are influencing the food sector:

- Comprehensive national level policy on food-processing sector
- Availability of trained manpower
- Processing plants with cost-effective technologies
- Cost-effective food machinery and packaging technologies
- Constraints in raw material production
- Inadequate infrastructural facilities
- Access to credit
- Market intelligence
- Inconsistency in central and state policies
- Lack of applied research
- Adequate value addition
- Lack of specific plan to attract private sector investment across the value chain
- Food safety laws
- Weights and Measures Act and Packaging Commodity rules
- Taxation

When we analyse the issues associated with promotion of entrepreneurship, similar issues comes in the way. Development of entrepreneurship needs to address the listed issues so that the interested entrepreneurs are attracted towards the industry. There are some efforts in the direction also. According to FICCI report 2010, the key initiatives taken by the government include 'amendment of the Agriculture Produce Marketing Committee Act, rationalization of food laws, implementation of the National Horticulture mission and so forth. The government has also outlined a plan to address the low scale of processing activity in the country by setting up the mega food parks, with integrated facilities for procurement, processing, storage and transport. To promote private sector activity and invite foreign investments in the sector the Government allows 100% FDI in the food processing and cold chain infrastructure. The recent budget has announced several policy measures, especially for the cold chain infrastructure, to encourage private sector activity across the entire value chain'.

When we analyse the food processing industry among the various segments, food grain milling occupies the major share (34%), followed by bread and bakery (20%), dairy products (16%), meat and poultry processing (10%), significantly, next in the order of highest share is aerated water and soft drinks (9%), then comes the fruits and vegetables (4%) and equal amount is being shared by fish processing as well. Last but not the least, alcoholic beverages occupies 3% in food-processing industry. When we compare India with other developing and/or developed nations, the figures of processing agricultural produce are China 40%, Thailand 30%, Brazil 70%, Philippines 78% and Malaysia 80%.

As per estimations, only about 2% of fruits and vegetables are processed in India. On the contrary, the wastage is to the extent of 25%. The various product categories of processing of fruits and vegetables include under-organized sector, juices and drinks constitute 33% followed by pulps/concentrates. Potato chips processing is one promising industry giving constant returns (15%) to the investors. Next in the order are pickles (9%). On the contrary, in under unorganized sector, pickles are dominating with 50% share followed by potato chips 20%. Sauce and ketchup are also dominating in unorganized sector. These are some possibilities wherein we can think of development of entrepreneurship.

According to ASI, APEDA, Vision 2015 of MOFPI, other secondary sources and IMaCS analysis, compound annual growth rate of processed foods shall be varying between 12–16% up to 2022. Fruits and vegetable sector is estimated to show a growth of 13% by 2022. Highest growth is expected in meat and poultry products (16%).

Food processing sector, in general, and fruits and vegetables sector, in particular, is generating huge amount of employment. As per the outcomes of Annual Survey of Industry, NSSO 62nd round—unorganized manufacturing sector in India—employment, assets and borrowings and IMaCS analysis shows that a total of 8.5 million people are employed in food processing sector. Of which, seven million in unorganized sector and 1.5 million in organized sector. In general, the data of Ministry of Micro, Small and Medium Enterprises data also reveals that about 90% of micro, small and medium enterprises are not having any formal registration;

largely, they can be categorized under unorganized sector. Moreover, similar situation prevails in food processing industry.

From the point of view of entrepreneurship, if we analyse the breakup of human recourse requirement across the segments, it is evident from the table that production is having maximum share (55%). It is evident that we need a considerable number of skilled manpower in production aspects of food processing industry. Quality and conducting appropriate tests occupy another major share in food industry wherein qualified and skilled manpower is essential. These figures herald opportunities in food processing industry (Tables 18.2 and 18.3).

TABLE 18.2   Functional Distribution of Human Resource Across the Segments. (*Source: Primary Research and IMaCS Analysis.*)

Functions	Percentage of employee
Procurement	10
Testing and quality	20
Production	55
R&D	1–2
Storage	2–3
Other(sales and other support function)	10

TABLE 18.3   Some of the Successful Enterprises in Fruit and Vegetable Processing.

Sl. no.	Company	Products
1.	Capital Foods	Frozen foods
2.	Dabur India Ltd.	Jams, pickles, fruit beverages
3.	Godrej F&B	Fruit juices, Fresh F&V (Retail)
4.	Green Valley	Frozen fruits and vegetables
5.	Hindustan Unilever Limited	Jams, ketchups, fruit beverages
6.	Mafco	Frozen fruits and vegetables
7.	Mother Dairy (Safal)	Frozen processed F&V
8.	MTR Foods	Pickles, chutneys (dips)
9.	Priya Foods	Pickles, fruit Juices
10.	Temptation Foods	Frozen fruits and vegetables, purees
11.	V.P. Bedekar and Sons Pvt. Ltd	Spices, pickles, fruit and vegetables foods, gravy mixes

These are only some examples, such things can happen in reality at the farmers' level also. If we want to establish an enterprise, we need to pass through various phases. The phases include the following.

18.2 PHASES OF ENTREPRENEURIAL VENTURE

Framework for entrepreneurial venture has been developed in line with Ramachandran and Ray (1998). Entrepreneurial activities originate from individuals; the framework of entrepreneurial venture is described in the following subsections.

18.2.1 PHASE-I—IDENTIFICATION OF ENTERPRISE

Identification of enterprise by an entrepreneur may come from unsatisfied need of people as well as unsatisfied personal need. This is reflected by the alertness of entrepreneur by him/herself, his ability to identify opportunity, family and educational background, professional experience, formal and informal networks and so forth.

18.2.2 PHASE-II—INITIATION/CREATION OF ENTERPRISE

By accumulating and combining physical and other resources, the entrepreneur needs to create an organization for transforming the concept into a marketable product. The ideas of entrepreneurs and environment interact to turn the ideas into reality. The new venture takes away large portion of time and attention of the entrepreneur from the management of environment to the management of organization. While interacting with several components such as bankers, regulatory agencies, experts and advisors, suppliers and so on, an entrepreneur alone has to constantly interact with them to have complete control over the enterprise that he/she created.

18.2.3 PHASE-III—NURTURING OF ENTERPRISE

Organization translates the business concept into marketable product and offers it to the customer. The entrepreneur gets feedback on market response in terms of profitability, sales and so forth. The feedback is measured in terms of efficiency, competitiveness, effectiveness, innovativeness, flexibility and so forth.

18.2.4   PHASE-IV—CONCLUDING/TRANSITION PHASE

Entrepreneurial process does not end with achieving stability and reaches success. At this stage, organizations require different managerial style as venture competition gradually builds up more and more.

18.3   THE ASSOCIATED CHAMBERS OF COMMERCE AND INDUSTRY OF INDIA REPORT

According to KMPG—ASSOCHAM report (2009) on consumer markets—food processing and agribusiness survey the opportunities in vegetable and fruit processing in India and include the following.

18.3.1   PROCESSABLE VARIETIES

Processable varieties of crops India is blessed with multiple agri-climatic zones and has advantages of round-the-year cultivation capability. India still produces varieties of crops, tomatoes and oranges for example that are not commercially viable for producers. The lack of production and identification of items with commercial viability contributes to the significant wastage. We import apples from the United States of America as the apples of Himachal Pradesh lack enough juice for processing. India needs to focus on more processable variety of crops to reduce wastage, increase processing levels and hence, value addition.

18.3.2   CONTRACT FARMING

Eighty per cent of India' 115 million farms are situated on plots of less than 2 ha. A little over 1% of all farms are larger than 10 ha and these constitute 15% of the cultivated land. With organized retail penetration increasing and government's proposed mega food parks obviating the need for an intermediary, Contract Farming is an opportunity for the processors that will help in better handling, price realization and minimizing wastage. PepsiCo and Cadbury's examples of contract farming in India can be a model for others to follow. The potato farming initiated by Pepsi in Jharkhand can be useful in making potato chips that can be supplied to food chains such as McDonald's and others.

18.3.3   INVESTMENT IN INFRASTRUCTURE THROUGH PUBLIC–PRIVATE PARTNERSHIP

To augment the storage and processing capacity, government has announced various policy measures and tax incentives for cold chain and warehousing, Food Parks and Agri Export Zones. Lack of infrastructure is a major roadblock impeding the growth of the processed food segment. With a lot of emphasis on value addition in food processing, it is imperative that investments in infrastructure are a must and should have favourable policy framework and fiscal incentives.

18.3.4   MEGA FOOD PARKS

Lot many State governments have taken the policy directive from the centre and have planned Mega Food Parks with associated fiscal incentives. The proposed food parks aim to bring together all players in the value chain together so as to minimize waste and improve value addition in the industry. This move will also help farmers to realize better prices by removing intermediaries. The processors also will be benefited by a better inventory management and production planning. The proposed Mega Food Parks will be between 10 and 100 ha in size, and 30 locations across India had already been identified. The parks would be set up through private consultants with the government providing grants of up to INR 1500 million each. Each Mega Food Park will have a minimum catchment area of five districts. With a chain developing from the farm gate to the retail shelves with collection and distribution centres and central processing centres in between, in which functions such as sorting, grading and packaging along with irradiation, food incubation cum development will take place, the food processing ministry hopes this initiative to be a commercial success.

18.3.5   INTEGRATED COLD CHAIN

India wastes more fruits and vegetables than it consumes. Due to gaps in the cold chain such as poor infrastructure, insufficient cold storage capacity, unavailability of cold storages in close proximity to farms, poor transportation infrastructure and so forth, about 30% of the fruits and vegetables grown in India (40 million tons amounting to USD 13 billion) get wasted annually. This results in instability in prices, farmers not getting two remunerative prices and rural impoverishment. Government plans to encourage setting up of integrated cold-chain facilities to improve storage and reduce waste by not missing any link in the value chain from the farmer to the consumer/retailer.

18.4   FOOD SAFETY MANAGEMENT SYSTEMS

Food safety is a growing concern across both developed and developing nations. The introduction of Sanitary and Phytosanitary Agreement and stringent safety regulations means the Indian processors/producers should be knowledgeable of the food safety requirements. Food retailers/independent bodies/exporters should take the initiative in educating the backward linkages. Going forward, there will be a need for food safety certifying agencies (for ISO 22000) authorized by the importing countries or standards setting boards.

It is essential to understand the concept of entrepreneurship against this backdrop. When we refer to entrepreneurship, many times we take it synonymously with business. There are distinctive differences between entrepreneurship and business, which include the fact that entrepreneur creates opportunities, whereas businessman makes use of those opportunities. Entrepreneur creates needs, whereas businessman satisfies needs. Businessman is a low or no risk taker, whereas entrepreneur is calculated risk taker. Entrepreneur is always in search of innovations, whereas businessman believes in managing existing establishment. With this brief background, let's understand the issues associated with entrepreneurship.

18.5 MEANING OF ENTREPRENEURSHIP

The term 'entrepreneurship' originates from the French term 'entreprendre' and the German word 'unternehmen'—both mean 'to undertake'. The earliest attempt to define entrepreneurship was by the French writer Bernard F. de Belidor, who defines entrepreneurship 'as buying labour and materials at uncertain prices and selling the resultant output at contracted prices'. However, the concept gathered prominence in economics literature mainly through the writings of Richard Cantillon (1680–1734), who gave the concept of some analytical treatment and assigned the entrepreneur an economic role by emphasizing on 'risk' as a prominent entrepreneurial function (Gopakumar, 1995). Despite a number of developments in the later years, a dynamic theory of entrepreneurship was first developed by Schumpeter (1934). In economic growth and development, innovation and the entrepreneur were considered the central figure by him. Innovations introduced by the entrepreneur are a source of creative disequilibrium. As other firms adopt the innovation, less efficient firms are forced out of business until a new equilibrium is reached. The Post-Schumpeterian developments have proceeded along two different schemes—the Harvard tradition and the Neo-Austrian School; the former is portrayed as an extension of Schumpeterian view (Leibenstein, 1968), whereas the latter is represented as an alternative approach.

Proponents of the Austrian School emphasize the entrepreneur's ability to take advantage of imperfections in information to make innovations. To introduce an innovation and to earn profits, the entrepreneur uses superior information. Unlike Schumpeter, the Austrians see a market before innovation as being in disequilibrium. The introduction of innovation increases the amount of knowledge in the market moving it toward new equilibrium (Kirzner, 1979).

In summary, entrepreneurship is often viewed as a function which involves the exploitation of opportunities that exist within a market. Such exploitation is most commonly associated with the direction and/or combination of productive inputs. Entrepreneurs usually are considered to bear risk while pursuing opportunities and often are associated with creative and innovative actions. In addition, entrepreneurs undertake a managerial role in their activities, but routine management of on-going operations is not considered to be entrepreneurship. An individual may perform an entrepreneurial function in creating an organization, but later, he is relegated to the role of managing it without performing an entrepreneurial role.

In this sense, many small-business owners would not be considered to be entrepreneurs. Finally, individuals within organizations (i.e. non-founders) can be classified as entrepreneurs as they pursue the exploitation of opportunities. This form of entrepreneurship is called intrapreneurship.

A farmer does not become an entrepreneur only by adopting a new agricultural technology in case of agriculture, but he becomes an entrepreneur only when he comes to be an operator of a farm business. A business involves rational decisions on investment after assessing risk, other alternatives and possibilities of profit and loss (De, 1986).

18.6   ELEMENTS OF ENTREPRENEURIAL DISPOSITION

There are certain key elements that define the entrepreneurial disposition, namely personal resourcefulness, achievement orientation, strategic vision, opportunity seeking and innovativeness (Balakrishnan et al., 1998; Knudson et al., 2004).

18.6.1   PERSONAL RESOURCEFULNESS

Personal resourcefulness is the belief in one's own capability for initiating actions directed towards creation and growth of enterprises. Such initiating process requires cognitively mediated self-regulations of internal feelings and emotions, thoughts and actions (Kanungo and Misra, 1992).

18.6.2   STRATEGIC VISION

Milton (1989) suggests that entrepreneurs have a knack for looking at the usual and finding the unusual. This leads to the commonly known entrepreneurial vision. This entrepreneurial vision or the abstract image of the kind of business they wish to create is one that guides their own intentions.

18.6.3   ACHIEVEMENT ORIENTATION

Entrepreneurs have a need for achievement, or a strong ego-drive. Entrepreneurs strive to make a difference in their own lives and in the lives of others. They are determined, persistent and committed to a job until it is finished.

18.6.4   OPPORTUNITY SEEKING

Opportunity seeking involves one's ability to see situations in terms of unmet needs, identifying markets or gaps for which product concepts are to be evolved and the search for creating and maintaining a competitive advantage to derive benefits on a sustained basis.

18.6.5   INNOVATIVENESS

Although a capitalist is one who seeks market return for the capital he invests, an entrepreneur is one who invests skills, assumes responsibilities and controls the activities of the enterprise to seek residual surplus for the 'human industry' and 'coordination' he puts in for the creation and survival of the enterprise. Schumpeter (1949) conceptualized that entrepreneurs are persons who are not necessarily capitalists or those who have command over resources, but they are the ones who create new combinations of factors of production and the market to derive profit.

It is evident that each of the above given disposition can be found in each and every individual with varying intensities. Therefore, it is assumed that a person with a very high intensity of the above predispositions could emerge as an entrepreneur. This gives rise to a process view of entrepreneurship that is entrepreneurship is more a state of becoming than a steady state event. Table 18.4 gives a relationship between entrepreneurial predispositions and enterprise life cycle.

TABLE 18.4   Entrepreneurial Dispositions in Enterprise Life Cycle. (*Source:* Balakrishnan et al. 1998.)

Entrepreneurial dispositions	Enterprise life cycle		
	Start-up	**Stabilization**	**Diversification**
Resourcefulness	High	High	Medium
Achievement-orientation	Medium	High	High
Visioning	Medium	Low	High
Opportunity-seeking	High	Medium	High
Innovativeness	High	Low	Medium

18.6.6 ENTREPRENEURSHIP UNDER CONSTRAINED ENVIRONMENT

Absence of resources with poor households of developing countries is considered an insurmountable obstacle to developing an entrepreneurial culture. However, availability of resources under control has not been found necessarily to be an impediment to growth and success of an entrepreneur (Sexton and Bowman-Upton, 1991). Bryant (1989) argues that entrepreneurs are characteristically people who go beyond the limits of resources over which they have direct control. Bygrave (1994) similarly contends that entrepreneurs find ways to control critical resources without owning them. Indeed, an important quality of entrepreneurs is their ability to be creative with limited resources (Saylor, 1987). According to Saylor (1987), ownership of resources is not a mandatory requirement for entrepreneurs to make use of them. This seems to comply with Kirzner's (1973) argument that ownership of capital is not necessary to provoke its movement or change of application. Thus, it could be concluded that entrepreneurship is a process by which individuals pursue opportunities without regard to resources they control (Kodithuwakku and Rosa, 2002).

18.7 INTERACTION BETWEEN ENTREPRENEURSHIP AND INNOVATION

Knudson et al. (2004) depicted the connection between entrepreneurship and innovation as displayed in Table 18.5. It is hypothesized that two drives propel the behaviour of economic actors (firms and individuals): (i) the drive to commercialize an idea, that is bring it to market realization; and (ii) the drive to innovate. These two drives interact to form four possible entrepreneurial/innovation types: (i) master entrepreneurs who are skilled managers, and risk bearers, but not innovators, (ii) innovative entrepreneurs, skilled entrepreneurs, who are also innovators, (iii) entrepreneurial innovators, skilled innovators who are also entrepreneurs and (iv) master innovators, who are skilled innovators, but not entrepreneurs.

TABLE 18.5 The Interaction Between Entrepreneurship and Innovation.

Type of Entrepreneur/ Innovator	Drive to bring to market	Drive to innovate	Examples
Master entrepreneur	Dominant	Minimal	Individual producer or agri-food processor
Innovative entrepreneur	Primary	Secondary	Innovative farmers
Entrepreneurial innovator	Secondary	Primary	First-generation organic food firms
Master innovator	Minimal	Dominant	Bench engineers and scientists

18.8 MASTER ENTREPRENEURS

A dominant entrepreneurial trait is expressed by master entrepreneurs with little or no desire to innovate. Although decidedly entrepreneurial in their approach (e.g. willingness to take risk, high energy, desire to control their environment and so on), they are not motivated by desire to innovate. Master entrepreneurs see market gaps and fill them with existing business models, products and services. Privately held grain merchandising firms and food manufacturers often are master entrepreneurs. Local grocery store managers fit this description as they often need to be entrepreneurial in their approach to developing local business, but are not innovating as they follow corporate guidelines and make the most out of given resources.

18.9 INNOVATIVE ENTREPRENEURS

Innovative entrepreneurs tend to be entrepreneurial, first and innovative, second. Innovative entrepreneurs constantly seek out new challenges, enjoy taking calculated risks and are driven by a vision of what future could be. They also look out for new ways of doing things unlike master entrepreneurs. Innovative entrepreneurs often grow restless with the pace of change and lack of control over their future. Eventually, they start their own ventures. Unlike master entrepreneurs, they are content to use true and tried business models, products and processes. They are driven to improve these products and processes in an effort to carve out a niche in the market-place. On the basis of their perception of market opportunities, innovative entrepreneurs see market gaps and are willing to take risk. They use a mix of tried and true and new business models, products and processes.

18.9.1   ENTREPRENEURIAL INNOVATORS

Entrepreneurial innovators exhibit primary innovator and secondary entrepreneur traits. Such individuals are never satisfied with the status quo. They are constantly developing improvements to processes. Unlike master innovators who are content with innovation in and of itself, entrepreneurial innovators seek out change, grow tired of bureaucracy and are driven to take risks to see that their innovations reach the marketplace. Entrepreneurial innovators see market gaps and fill them with new business models, products and processes that they are willing to take personally to market place. The first generation firms in the organic food industry started with a conviction about what products to produce. This conviction motivated production innovation that, in turn, triggered the need to market the products so created.

18.9.2   MASTER INNOVATORS

Master innovators express dominant innovator traits with little or no interest to take these innovations to the marketplace. They are content to innovate in a given area, particularly in the one where they specialize or have expertise. Scientists and research and development engineers spend their entire career improving products, processes and formulae with little interest in bringing these innovations to the marketplace.

In development of entrepreneurship in the agribusiness sector, all of the above four combinations have to play a major role. Moreover, it is not essential that a particular individual will be stuck in a one of the above given stylized types. Personal dispositions, life events, changes in technology, events affecting the market and education and training may allow individuals and firms that are flexible enough to switch from one type to another. Barriers to entrepreneurship and innovation must be overcome in order for market realization of an entrepreneur's innovation.

18.10   FACTOR AFFECTING ENTREPRENEURSHIP: A REVIEW OF EMPIRICAL STUDIES

Numerous studies have been done to identify factors that influence entrepreneurial orientation and these factors vary over cultures, gender and environmental contexts. Three key components are important in analysing the growth of SMEs—(1) the characteristics of the entrepreneurs which included age, gender, work experience, education; (2) the characteristics of the small and marginal enterprises (SMEs) comprising of origin of enterprise, length of time in operation, size of enterprise, capital source; and (3) contextual variables which included marketing, technology, information access, entrepreneurial readiness, social network, legality, capital access, government support and business plan.

18.11   CHARACTERISTICS OF THE ENTREPRENEUR

Several previous studies found that demographic characteristics, such as age and gender, and individual background, such as education and former work experience, had an impact on entrepreneurial intention and endeavour (Kolvereid, 1996; Mazzarol et al., 1999).

18.11.1   AGE

Reynolds et al. (2000) found that individuals aged 25–44 years were the most entrepreneurially active. Findings from another study by Sinha (1996) disclosed that successful entrepreneur were relatively younger in age. In their study on Internet café entrepreneurs in Indonesia, Kristiansen et al. (2003) found a significant correlation between age of the entrepreneur and business success. The older (>25 years old) entrepreneurs were more successful than the younger ones.

18.11.2   GENDER

Mazzarol et al. (1999) found that female were generally less likely to be founders of new business than male. Similarly, Kolvereid (1996) found that males had significantly higher entrepreneurial intentions than females.

18.11.3   WORK EXPERIENCE

It is discovered by Kolvereid (1996) that individuals with prior entrepreneurial experience had significantly higher entrepreneurial intentions than those without such experience. Conversely, Mazzarol et al. (1999) found that respondents with previous government employment experience were less likely to be successful founders of small businesses. But in private companies and entrepreneurial intentions, they did not investigate the relationship between previous employment experiences.

18.11.4   EDUCATION

A research by Charney and Libecap (2000) revealed that entrepreneurship education produces self-sufficient enterprising individuals. Furthermore, they found that entrepreneurship education increases the formation of new ventures, the likelihood of self-employment, the likelihood of developing new products and the likelihood of self-employed graduates owning a high-technology business. Moreover, the study revealed that entrepreneurship education of employee increases the sales growth rates of emerging firms and graduates' assets. Similarly, Sinha (1996) who analysed the educational background of the entrepreneur revealed that 72% of the successful entrepreneurs had a minimum of technical qualification, whereas most (67%) of the unsuccessful entrepreneurs did not have any technical background. She summed up that entrepreneurs with business and technical educational background are in a better position to appreciate and analyse hard reality and deal with it intuitively, which seems to play a critical role in entrepreneurial effectiveness.

18.12   CHARACTERISTICS OF THE SMALL AND MARGINAL ENTERPRISES

18.12.1   ORIGIN OF ENTERPRISE

According to Smallbone et al. (1995), in small firms, in which ownership and management were typically combined in one or more individuals, future goals for the business might be determined as much by personal lifestyle and family factors as by commercial considerations. Further, they concluded that one characteristic which did distinguish the best performing firms from other firms in the study was their commitment to growth. In addition, they found another characteristic that did distinguish high growth firms from others and was their propensity to acquire other businesses.

18.12.2   LENGTH OF TIME IN OPERATION

Length time in operation may be associated with learning curve. Old players most probably have learned much from their experiences than have done by new comers. Kristiansen et al. (2003) found that length time in operation was significantly linked to business success.

18.12.3   SIZE OF ENTERPRISE

Size of enterprise reflects how large an enterprise in employment terms is. McMahon (2001) found that enterprise size significantly linked to better business performance. Larger enterprises were found to have a higher level of success.

18.12.4   CAPITAL SOURCE

In a study in Australia, McMahon (2001) discovered that better business growth is associated with greater dependence upon external finance. In a later study, Kristiansen et al. (2003) found that financial flexibility was significantly correlated to business success. Higher level of success was experienced by the SMEs that took advantage of family and third-party investment.

18.12.5   CONTEXTUAL VARIABLES

18.12.5.1   MARKETING

Most of the SMEs operate along traditional lines in marketing. By performing market development, stiffer competition in the market should be responded proactively by SMEs. SMEs faced problems in accessing the market (Mead and Liedholm, 1998; Swierczek and Ha, 2003). Market development is, therefore, crucial for preserving high growth in the business. Smallbone et al. (1995) in their study in the United Kingdom found that the vast majority of the high growth SMEs had identified and responded to new market opportunities. New market opportunities included finding new products or services to offer to the existing customers and obtaining new customers for the existing product or services. In a slightly different term, market stability (i.e. high proportion of regular customers) was found to be significant in determining business success (Kristiansen et al., 2003). Furthermore, market orientation defined as organization culture creates the necessary behaviour for the creation of superior value to customers and was found to be significantly correlated with company performance (Verhees and Meulenberg, 2004). More specifically, they pointed out that market orientation is helpful in selection of an attractive product assortment when the SMEs operate in markets with relatively homogenous product.

18.12.5.2   TECHNOLOGY

Rapid changes in technology should be responded by the SMEs to find alternative ways to sustain their competitive advantage by deploying new process and new growth methods. Technology may play an important role in this respect. In this context, technology has a close relationship with improvement of production process. Previous study has revealed that lack of equipment and out-dated technology are among the hindrances of SME development (Swierczek and Ha, 2003). In their study in the United States, Gundry et al. (2003) disclosed that technological changes in innovations had significant relationship with market growth. A study in Ireland unearthed that technological posture, automation and process innovation were significantly linked to satisfaction on return on investment (ROI) (Gibbons and O'Connor, 2003).

18.12.5.3   INFORMATION ACCESS

For the intention to initiate a new enterprise, availability of business information is similarly important. Singh and Krishna (1994), in their studies of entrepreneurship, pointed out that eagerness in information seeking is one of the major entrepreneurial characteristics. Information seeking refers to the frequency of contact that an individual makes with various sources of information. The result of this activity is most often dependent on information accessibility, either through individual efforts and human capital or as a part of a social capital and networking. Access to new information is indispensable for the initiation, survival and growth of firms (Duh, 2003; Kristiansen, 2002; Mead and Liedholm, 1998; Swierczek and Ha, 2003). The availability of new information is found to be dependent on personal characteristics such as the level of education, infrastructure qualities such as media coverage and telecommunication systems and on social capital such as networks (Kristiansen, 2003b).

18.12.5.4   ENTREPRENEURIAL READINESS

In this study, entrepreneurial readiness refers to self-efficacy. The term self-efficacy, derived from Bandura's (1977) social learning theory, refers to a person's belief in his or her capability to perform a given task. According to Ryan (1970), self-perception plays an important role in the development of intention. Intentions and their underlying attitudes are perception-based, which should mean that they are learned and can be continuously influenced, and not fixed by personality traits formed in early childhood. They will vary across historical and cultural contexts accordingly. Cromie (2000) declared that self-efficacy affects a person's beliefs regarding whether or not certain goals may be attained. The attitude provides the foundation for human motivation (Pajares, 2002) and personal accomplishment: unless people believe that their actions can produce the outcomes they desire, they have little incentive to act or to persevere in the face of adversities (Pajares, 2002). Consequently, people behave according to beliefs about their capabilities rather than on real facts based on their competence and capabilities. In their study among Norwegian and Indonesian students, Kristiansen and Indarti (2004) ascertained a significant correlation between self-efficacy and entrepreneurial intention.

18.12.5.5   SOCIAL NETWORK

Social networks have an impact on the likelihood of successful entrepreneurial endeavour. The study of entrepreneurship has increasingly reflected the general agreement that entrepreneurs and new companies must engage in networks to survive (Huggins, 2000). Networks represent a means for entrepreneurs to reduce risks and transaction costs and also to improve access to business ideas, knowledge and capital (Aldrich and Zimmer, 1986). Between the central actor and other actors, a social network consists of a series of formal and informal ties in a circle of acquaintances and represents channels through which entrepreneurs get access to the necessary resources for business start-up, growth and success

(Kristiansen, 2004). In his study in Tanzania among small-scale garment and carpentry industries, Kristiansen (2003a) found that social network has significant relationship with business adaptability.

18.12.5.6   LEGALITY

Not many studies have been conducted to examine direct relationship between legal aspect readiness and business success. On the account of bribery practices, dealing with legal aspects has forced the SMEs to allocate significant amount of financial resources. Legal aspect is often also used in selection operating decision to ensure future business success (Mazzarol and Choo, 2003).

18.12.5.7   CAPITAL ACCESS

Access to capital is obviously one of the typical obstacles to the start-up of new businesses, not least in developing economies with weak credit and venture capital institutions. Several empirical studies have concluded that as main hindrances to business innovation and success in developing economies, the lack of access to capital and credit schemes and the constraints of financial systems are regarded by potential entrepreneurs (Marsden, 1992; Meier and Pilgrim, 1994; Steel, 1994). Potential sources of capital may be personal savings, extended family networks, community saving and credit systems, or financial institutions and banks. Robinson (1993) found that informal sources of credit, though with high interest rates, constitute very substantial contributions to business start-ups in developing countries, where the capital to labour ratio is normally low, and small amounts of capital may be sufficient for a business start-up. In developed economies with efficient financial infrastructure, access to capital may represent similar restrictions to individuals' perception of entrepreneurial options because of the high entry barrier ensuing from high capital to labour ratios in most industries.

18.12.5.8   GOVERNMENT SUPPORT

Many governments in the world (e.g. Chaston, 1992; Mulhern, 1996; Patrianila, 2003) have been paying a more attention to SME development to strengthen national economy. Indian government through Ministry of Micro, Small and Medium Enterprises have launched many programmes (e.g. giving financial assistance) dedicated to SME development. In short, government support is of necessary condition to foster SMEs development.

18.12.5.9   BUSINESS PLAN

Insufficient awareness of the need for a business plan was identified as one problem at the start-up phase among SMEs (Chaston, 1992). In this context, business plan can also be regarded as development orientation. McMahon (2001) revealed that greater development orientation significantly linked to better business growth. In line with this, Smallbone et al. (1995) pointed out that one characteristic that distinguished the best performing enterprises from other was their commitment to change. Therefore, it is believed that well-planned business activities as manifested in a business plan will yield a better business performance.

18.13   INNOVATIONS AND ENTREPRENEURSHIP THROUGH E-COMMERCE

Entrepreneurship and innovations will be triggered through India's e-commerce, which is expanding at a very faster rate. E-commerce industry has been directly impacting micro, small and medium enterprises. Increasing internet and smart phones penetration are expected to boost expansion of e-commerce sector in tier-II and tier-III cities. It is expected that by 2020, online shoppers of total internet users will be increased up to 36% as against 11% in 2015 and only 0.9% in 2013 (ASSOCHAM, 2016). The number of online shoppers in the country grew by 95% between 2013 and 2015, and the number is expected to reach 140 million by 2018 and 220 million by 2020.

Indian agriculture is dominated by small and marginal producers, who grow vegetables on small scale resulting low marketable surplus. These farmers are characterized by lack of scale economies, poor bargaining power and rudimentary state of storage and logistic infrastructure. To provide better prices for their produce and linking farmers directly to the consumers, the Government of India has launched e-NAM (National

Agricultural Market) portal, which aims to connect 585 wholesale mandis by 2018. This e-commerce ecosystem has led the emergence of new business model.

18.14 NEW ENTREPRENEURIAL OPPORTUNITIES IN EXPORT OF PROCESSED VEGETABLES

After the liberalization of Indian economy, the international trade has got importance. The agricultural sector is not able to achieve 4% of targeted growth but the export of processed fruits and vegetables is growing at the rate of more than 13. It is evident from the Tables 18.6 and 18.7 that there is a significant growth in quantity and value terms in both the product groups namely processed and fresh.

Therefore, there is huge potential in doing business in export of fresh and processed vegetables. To excel in this sector, the entrepreneur should have good knowledge about Sanitary and Phytosanitary (SPS) measures and must comply with it.

TABLE 18.6 Export of Processed Fruits and Vegetables—Quantity in 1000'MT and Value in Rs Billion.

	2012–2013		2013–2014		2014–2015	
	Quantity	Value	Quantity	Value	Quantity	Value
Cucumber and gherkins (prepared and preserved)	238.62	85.66	218.75	95.520	251.18 (14.83)	120.24 (25.88)
Dried and preserved vegetables	68.52	63.80	56.16	74.272	63.70 (13.43)	84.71 (14.1)
Mango pulp	147.82	60.86	174.86	77.295	154.82	84.14
Other processed fruits and vegetables	269.22	173.31	287.38	226.660	316.06	256.99
Pulses	202.75	128.50	345.05	174.637	220.91	120.95
Total	926.93	512.12	1082.20	648.39	1006.68	667.04

Figures in parenthesis indicate per cent change over previous year.

TABLE 18.7 Export of Fresh Fruits and Vegetables—Quantity in 1000'MT and Value in Rs Billion.

	2012–2013		2013–2014		2014–2015	
	Quantity	Value	Quantity	Value	Quantity	Value
Fresh onions	1666.87	196.66	1482.50	316.96	1238.10	230.05
Other fresh vegetables	768.63	151.63	953.73	229.33	835.50	240.22
Walnuts	5.29	19.98	6.73	32.45	2.66	13.64
Fresh mangoes	55.58	26.47	41.28	28.54	43.00	30.25
Fresh grapes	172.74	125.94	192.62	166.65	107.26	108.65
Other fresh fruits	263.97	77.97	240.55	102.16	274.44	124.59
Total	2933.09	598.67	2917.40	876.10	2500.96	747.41

18.15 PRAISING THE GOURDS

It is the largest family of fruit (cucurbitaceae)—even if we generally look down on its members. *Bitter gourd (karela), pointed gourd (parval), tinda, torai* and other gourds and squashes have traditionally been a huge part of summer cuisines in India. Both bitter gourd and pointed gourds are indigenous.

Bitter improve digestion and promote good health according to ayurveda. In many communities, the traditional advent of New Year in mid-April is marked by ritualistic consumption of *neem* leaves, to promote healthy living throughout the year. In a season marked by disease, the bitter gourd was prized in traditional kitchens as much for therapeutic use for the dining. The most popular method of cooking karela reflects the syncretism that defines all Indian food. Stuffed with onion and spices, it obviously takes after the Turkish *dolma* variety of dishes. Deserts made from gourds and squashes have been a Mughal-based tradition. Delicacies such as Agra petha (made from ash gourd) and the vegetable halwa (bottle gourd etc.) of Lucknow and Delhi reflect a Persian–Turkish influence. In Banaras and some eastern UP part, the inventiveness goes a step further. Parval ki mithai is made by slitting the whole fruit and stuffing it with khoya and nuts. The entire fruit is then coated with sugar syrup.

The above examples are modern Indian cooking and have large potential of business opportunities through value addition. Besides creating employment and income to the entrepreneur, this will improve the income of small and marginal farmers who grow vegetables, but during peak seasons, they do not get remunerative prices of their product.

18.16   CONSTRAINTS IN PROCESSING AND MARKETING OF VEGETABLE LEGUMES

18.16.1   LEGUME PROCESSING AND INCOME GENERATION

Processing techniques at the household or cottage level for processing needs are to be upgraded. Equipment utilized in household and local processing needs to be upgraded to allow for increased productivity, reduced drudgery and efficient time management. This will enhance productivity and quality, food security and income generation. Utilization pattern of grain legumes is given in Table 18.8.

TABLE 18.8   Utilization Pattern of Grain Legumes. (*Source:* Ali, 2003.)

Grain legume	Utilization pattern
Chickpea	This is found as immature grains in pod after roasting. De-husked green and tender grains are consumed after boiling or frying or as crushed and cooked snack. It is also used in curry. As mature grains, it is used as whole seed or splits (dhal) or flour. The chickpea flour is used for making the fermented dish *dhokla* and in a number of fried products/snacks. It is most versatile among pulses
Black gram	It is used as whole seed or de-husked splits (cotyledons) for making dhal. Black gram is extensively used in fermented products such as *idli*, *dosa*, hopper and *papad*
Green gram	It is used in curry. Sprouted beans are also consumed. Roasting and frying of whole seeds or splits (cotyledons) are popular ways in India to prepare snack products which are usually spiced. Unspiced products are also available
Pigeon pea	It is primarily used for making dhal (cooked splits) and *sambar* (cooked and spiced splits). Both dishes are soup-like in consistency and appearance and are prepared from de-husked cotyledons that are cooked until tender and seasoned with spices. *Sambar* contains added vegetables (optional) such as eggplant, drum stick, carrots, tomato, potatoes, green peas and so forth
Peas, lentil, kidney bean and so on.	These are generally used as whole seed for making curry. Dhal is prepared from de-husked peas and lentil
Crop residue and by-products	The crop residues such as leaves and grain by-products such as seed coat, germ and broken cotyledons are used as animal feed

18.16.2   *IMPROVED PRODUCT QUALITY AND SAFETY*

Most household- and cottage-processing activities in the tropics are usually carried out.

18.16.2.1   *WITHOUT PRODUCT QUALITY CONTROL*

There is no standardization of product quality. Many local processors also produce under non-hygienic conditions. The products from different process batches have slight differences in product quality. This is on account of the lack of measurement of quality parameters during processing. Most operations are done using subjective judgments for example hand feeling to estimate temperature, ingredients added to the taste of the processor (not according to a standard formula) and so forth. Local processors need to be sensitized and trained on the benefit of ensuring consistent quality products for increased income generation.

18.16.2.2   *NEED FOR IMPROVING NUTRITIONAL QUALITY*

Need for improving nutritional quality of local staple legumes is important in foods and also in addressing protein–energy malnutrition concerns in developing countries. They contain the essential amino acid lysine which is deficient in cereals. On the other hand, cereals contain amino acid methionine which is limiting in legumes. Legumes can be supplemented with cereals to give a balance of amino acid called protein complementation. There is still the need to develop novel food recipes from such composite flours to increase nutrient diversification.

18.16.2.3   *NEED FOR VALUE ADDITION THROUGH PROPER PACKAGING*

Dried legume seeds in the tropics are usually sold in the market places using different measuring weights which can indicate inconsistency in size. Proper packaging, labelling and branding are important considerations that local processors can add to boost sales for supermarket and export that is pre-packaging. It minimizes time wastage, ensures uniform packed weights, eliminates access by rodents in the store, facilitates quick shopping but requires trust on the part of the buyers or consumers.

18.17   INSTITUTIONAL SUPPORT

There are various government bodies which support micro, small, medium as well as large enterprises in agricultural sector. Availability of institutional support is also an opportunity for an entrepreneur. A sample list of institutions is as follows:

- Ministry of Food Processing Industries (MOFPI)
- Investment Commission of India
- Coffee Board of India
- Tea Board of India
- Directorate of Cashew and Cocoa Development (DCCD)
- National Research Development Centre (NRDC)
- National Bank for Agriculture and Rural Development (NABARD)
- Gujarat Agro Industries Corporation (GAIC)
- Coconut Development Board (CDB)
- Spices Board
- National Centre for Trade Information (NCTI)
- Department of Agriculture and Cooperation
- National Oilseeds and Vegetable Oils Development (NOVOD) Board
- Department of Animal Husbandry, Dairying and Fisheries (DAHD)
- All State Agricultural Boards
- National Fisheries Development Board (NFDB)
- The Orissa State Cooperative Milk Producers' Federation Limited (OMFED)
- National Dairy Development Board (NDDB)
- National Agriculture Cooperative Marketing Federation of India (NAFED)
- Commission of Agricultural Costs and Prices
- Andhra Pradesh Dairy Development Cooperative Federation Ltd
- Haryana Dairy Development Cooperative Federation Ltd.
- Karnataka Cooperative Milk Producers Federation Ltd.
- Kerala Cooperative Milk Marketing Federation Ltd.

- Madhya Pradesh State Cooperative Dairy Federation Limited
- Maharashtra Rajya Sahakari Dudh Mahasangh Maryadit
- Punjab State Cooperative Milk Producers' Federation Ltd.
- Rajasthan Cooperative Dairy Federation Ltd., Saras Sankul
- Tamil Nadu Cooperative Milk Producers' Federation Ltd.
- Pradeshik Cooperative Dairy Federation Ltd
- West Bengal Cooperative Milk Producers' Federation Ltd.
- National Horticulture Board (NHB)

In addition to government intuitions, the existing industry associations are also supporting budding entrepreneurs, to quote a few

- All India Food Producers Association (AIFPA)
- Confederation of Indian Food Trade and Industry (CIFTI)
- Indian Dairy Association (IDA)
- Indian Oilseed and Produce Export Promotion Council (IOPEPC)
- Indian Tea Association (ITA)
- Soybean Processors Association of India (SOPA)
- UP Rice Millers Association
- Assam Branch Indian Tea Association (ABITA)
- Darjeeling Tea
- Federation of Biscuit Manufacturers Association (FBMI)
- Indian Industries Association (IIA)
- Laghu Udyog Bharati
- Federation of Indian Small, Micro and Medium Enterprises (FISME)

To establish a strong backward linkage, research institutions play a key role. Entrepreneurs needs research inputs. There are some exclusive institutions which support entrepreneurs which include

- Directorate of Cashew Research (DCR)
- Indian Council of Agricultural Research (ICAR)
- Indian Institute Of Crop Processing Technology (IICPT)
- Protein Foods and Nutrition Development Association of India (PFNDAI)
- National Institute of Fisheries Post Harvest Technology and Training (NIFPHATT)

- Tea Research Association (TRA)
- Central Institute of Fisheries Education (CIFE)
- National Institute of Food Technology and Management (NIFTEM)
- The National Institute of Agricultural Marketing

Additional information on various schemes and policies of Government of India can be obtained from http://smallb.sidbi.in/%20/sectors%20/food-processing%20/food-processing-related-policies-and-schemes.

18.18 CONCLUSION

In development of entrepreneurship, vegetable processing and export sector is having immense opportunities. The basic idea behind is that the primary producer should get his due, which is the demand lasting for many decades. Hence the scenario at national level is analysed highlighting the opportunities and possibilities. Analysis of current scenario reveals many gaps, and fulfilling the gaps satisfies the needs of target community, that is primary producers.

To have the holistic understanding of the concept of entrepreneurship, the chapter compiles the fragmented literature on conceptual development in entrepreneurship development, and research presents some of the most important conceptual models of entrepreneurship. Given the producer demands and policy imperatives that exist to support value-added ventures, particularly in agribusiness development, the concepts presented herein give guidance to researchers and policy makers to foster entrepreneurship and innovation. First, establishment of agribusiness incubation centres can play a key role in strengthening the linkages between entrepreneurship and innovation. They can also assist in the creation of business plans, marketing, feasibility studies and other aspects of product and business development. These centres can also help in reducing the potential for market failure. Market failure occurs when would-be entrepreneurs or innovators let good ideas and intentions die or they do not know how to foster their entrepreneurial or innovation abilities. Another market failure occurs when a poorly conceived idea goes forward and fails, leading to lost jobs, the misallocation of capital and in the worst cases, bankruptcy. University programmes in entrepreneurship can empower the would-be entrepreneur to answer difficult questions and consider factors that the would-be entrepreneur may be overlooking before committing financial resources. To be effective, these centres and programmes need to be geared toward new firms, new products, innovation processes and entrepreneurial education. Traditional services such as situation and outlook analysis, or farm's financial management, will not be sufficient in themselves to meet the new needs.

Second, to enhance the probability of successful entrepreneurship and innovation more broadly, apart from innovation centres, agricultural economists need to develop innovations in curricula and extension activities. Undergraduate and graduate students need exposure and experience with these concepts. Producers and agri-food managers could likely benefit from education about their entrepreneurial/innovation type and how to transition from one type to another. In addition, we may need to create marriages between those who are more strongly entrepreneurs and those who are more strongly innovators. University outreach programmes and centres have a key role to play by establishing structures that foster mentorship and or act as a clearing house in which entrepreneurs and innovators interact and benefit from one another. Such entrepreneurial/innovation teams may be much more effective than individuals working alone. Such teams may help solve the commercialization gap that exists between our laboratory-based science innovations and getting these innovations (and their potential patent revenues) to market. Various institutions supporting and developing entrepreneurship among farmers and others are discussed as well, and a range of schemes of government and some research institutions for the purpose of establishment of strong backward linkages are also listed for the purpose.

REFERENCES

Aldrich, H.; Zimmer, C. Entrepreneurship Through Social Network. In *The Art and Science of Entrepreneurship;* Sexton, D. L., Smilor, R. W., Eds.; Ballinger Publishing: Cambridge 1986; pp 3–25.

Ali, N. *Processing and Utilisation of Legumes 2003;* APO: Tokyo, 2003; pp 116–145.

ASSOCHAM Report. E-commerce in India: A Game Changer for the Economy, 2016.

Balakrishnan, S.; Gopakumar, K.; Kanungo, R. N. Entrepreneurship Development: Concept and Context. In *Entrepreneurship and Innovation;* Kanungo, R. N., Ed.; Sage Publications: New Delhi, 1998; pp 1–38.

Bandura, A. *Social Learning Theory;* Prentice-Hall: Englewood Cliffs, New Jersey, 1977.

Bryant, C. R. Entrepreneurs in the Rural Environment. *J. Rural. Stud.* **1989,** *5*(4), 337–347.

Bygrave, W. D. *Portable MBA in Entrepreneurship;* John Wiley and Sons: New York, 1994.

Charney, A.; Libecap, G. D. Impact of Entrepreneurship Education: Kauffman Center for Entrepreneurial leadership. 2000.

Chaston, I. Supporting New Small Business Start-ups. *J. Eur. Ind. Train.* **1992,** *16*(10), 3–8.

Cromie, S. Assessing Entrepreneurial Inclinations: Some Approaches and Empirical Evidence. *Eur. J Work Organ. Psychol.* **2000,** *9*(1), 7–20.

De, D. Factors Affecting Entrepreneur Characteristics of Farmers. *Indian J. Soc. Work* **1986,** *46*(4), 541–546.

Duh, M. Family Enterprises as an Important Factor of the Economic Development: The Case of Slovenia. *J Enterprising Cult.* **2003,** *11*(2), 111–130.

Federation of Indian Chambers of commerce and Industry (FICCI) Report (2010).

Gibbons, P. T., O'Connor, T. Strategic Posture, Technology Strategy and Performance Among Small Firms. *J Entreprising Cult.* **2003,** *11*(2), 131–146.

Gopakumar, K. Entrepreneurship in Economic Thought: A Thematic Review. *J. Entrepreneurship* 1995, *4*(1), 1–17.

Gundry, L.; Kickul, J.; Welsch, H. P.; Posig, M. Technological Innovation in Women-Owned Firms: Influence of Entrepreneurial Motivation and Strategic Intention. *Int. J. Entrepreneurship Innovation* **2003,** *4*(1), 265–274.

Huggins, R. The Success and Failure of Policy-Implanted Inter-Firm Network Initiatives: Motivations, Processes and Structure. *Entrepreneurship Reg. Dev.* **2000,** *12*(2), 211–236.

Indian Horticulture Database. NHB, Sector-18, Gurgaon, 2015–2016.

Kanungo, R. N.; Misra, S. Managerial Resourcefulness: A Re-conceptualisation of Managerial Skills. *Hum. Relat.* **1992,** *45*(12), 1311–1332.

Kirzner, I. M. *Competition and Entrepreneurship;* University of Chicago Press: Chicago, 1973.

Kirzner, I. *Perception, Opportunity and Profit;* Chicago University Press: Chicago, 1979.

Knudson, W.; Wysocki, A.; Champagne, J.; Christopher Peterson, H. Entrepreneurship and innovation in the Agri-food System. *Am. J. Agric. Econ.* **2004,** *86,* 1330–1336.

Kodithuwakku, S. S.; Rosa, P. The Entrepreneurial Process and Economic Success in a Constrained Environment, *J Bus. Venturing* **2002,** *17*(2), 431–465.

Kolvereid, L. Prediction of Employment Status Choice Intentions. Entrepreneurship Theory and Practice, Fall, 1996, pp 47–57.

Kristiansen, S. Competition and Knowledge in Javanese Rural Business. *Singapore J. Trop. Geogr.* **2002,** *23*(1), 52–70, 13.

Kristiansen, S. Information, Adaptation, and Survival: A Study of Small-Scale Garment and Carpentry Industries in Tanzania. Paper presented at the MU-AUC Conference, Dar es Salaam, Tanzania, 2003a.

Kristiansen, S. Linkages and Rural Non-Farm Employment Creation: Changing Challenges and Policies in Indonesia, Rome, 2003b.

Kristiansen, S. Social Network and Business Success: The Role of Sub-Cultures in an African Context. *Am. J. Econ. Sociol.* **2004,** *63,* 1149–1171.

Kristiansen, S.; Indarti, N. Entrepreneurial Intention Among Indonesian and Norwegian Students. *J. Enterprising Cult.* **2004,** *12*(1), 55.

Kristiansen, S.; Furuholt, B.; Wahid, F. Internet Cafe Entrepreneurs: Pioneers in Information Dissemination in Indonesia. *Int. J. Entrepreneurship Innovation* **2003,** *4*(4), 251–263.

Leibenstein, H. Entrepreneurship and Development. *Am. Econ. Rev.* **1968,** *58*(2), 75.

Marsden, K. African Entrepreneurs—Pioneers of Development. *Small Enterp. Dev.* **1992,** *3*(2), 15–25.

Mazzarol, T.; Choo, S. A Study of the Factors Influencing the Operating Location Decisions of Small Firms. *Prop. Manage.* **2003,** *21*(2), 190–208.

Mazzarol, T.; Volery, T.; Doss, N.; Thein, V. Factors Influencing Small Business Startups. *Int. J. Entrepreneurial Behav. Res.* **1999,** *5*(2), 48–63.

McMahon, R. G. P. Growth and Performance of Manufacturing SMEs: The Influence of Financial Management Characteristics. *Int. Small Bus. J.* **2001,** *19*(3), 10–28.

Mead, D.; Liedholm, C. The Dynamics of Micro and Small Enterprises in Developing Countries. *World Dev.* **1998,** *26*(1), 61–74.

Meier, R.; Pilgrim, M. Policy-Induced Constraints on Small Enterprise Development in Asian Developing Countries. *Small Enterprise Dev.* **1994,** *5*(2), 66–78.

Milton, D. G. The Complete Entrepreneur. *Entrepreneurship Theory Practice* **1989,** *13*, 9–19.

Mulhern, A. Venezuelan Small Businesses and the Economic Crisis: Reflections from Europe. *Int. J. Entrepreneurial Behav. Res.* **1996,** *2*(2), 69–79.

Pajares, F. Overview of Social Cognitive Theory and Self-efficacy from 2002. http://www.emory.edu/EDUCATION/mfp/eff.html.

Patrianila, V. N. Italia Besarkan UKM Dengan. Pengusaha **2003,** *32*, 70–71.

Ramachandran, K.; Ray, S. A Framework and Entrepreneurial Typology for Developing a Comprehensive Theory of Entreprenurship. In *Entrepreneurship and Innovation: Models for Development*; Kanungo, R. N., Ed.; Sage publications: California, United States, 1998; pp 40–63.

Reynolds, P. D.; Hay, M.; Bygrave, W. D.; Camp, S. M.; Autio, E. Global Entrepreneurship Monitor 2000 Executive Report: Babson College, Kauffman Center for Entrepreneurial Leadership, and London Business School. 2000.

Robinson, M. S. Beberapa Strategi Yang Berhasil Untuk Mengembangkan Bank Pedesaan: Pengalaman Dengan Bank Rakyat Indonesia 1970–1990. In *Bunga Rampai Pembiyaan Pertanian Pedesaan;* Sugianto, S. P., Robinson, M. S. Eds.; Institut Bankir Indonesia: Jakarta, 1993; pp 31–224.

Ryan, T. R. *Intentional Behavior: An Approach to Human Motivation;* The Ronald Press Company: New York, 1970.

Saylor, M. Home-Basedenterprise Development and Craft Mareting. *Proceedings of International Rural Entrepreneurship Symposium.* Economic Research Service, USDA, Knoxville, Tenesse, 1987, pp 57–70.

Schumpeter, J. A. *The Theory of Economic Development;* Harvard University Press: Cambridge, Ma, 1934.

Schumpeter J. A. *Economic Theory and Entrepreneurial History—Change and the Entrepreneur, Postulates and Patterns for Entrepreneurial History.* Harvard University Press: Cambridge, MA, 1949.

Sexton, D. L.; Bowman-Upton, N. *Entrepreneurship, Creativity and Growth;* Macmillan: New York, 1991.

Singh, K. A.; Krishna, K. V. S. M. Agricultural Entrepreneurship: The Concept and Evidence. *J. Entrepreneurship* **1994,** *3*(1), 97–111.

Sinha, T. N. Human Factors in Entrepreneurship Effectiveness. *J. Entrepreneurship* **1996,** *5*(1), 23–39.

Smallbone, D.; Leig, R.; North, D. The Characteristics and Strategies of High Growth SMEs. *Int. J. Entrepreneurial Behav. Res.* **1995,** *1*(3), 44.

Steel, W. F. Changing the Institutional and Policy Environment for Small Enterprise Development in Africa. *Small Enterprise Dev.* **1994,** *5*(2), 4–9. 14

Swierczek, F. W.; Ha, T. T. Entrepreneurial Orientation, Uncertainty Avoidance and Firm Performance: An Analysis of Thai and Vietnamese SMEs. *Int. J. Entrepreneurship Innovation* **2003,** *4*(1), 46–58.

Verhees, F. J. H. M.; Meulenberg, M. T. G. Market Orientation, Innovativeness, Product Innovation, and Performance in Small Firms. *Journal of Small Bus. Manage.* **2004,** *42*(2), 134–154.

www.nsdcindia.org/sites/default/files/files/food-processings-2009.pdf.

http://smallb.sidbi.in/%20/sector%20/sector%/food–processing%20/food-processing-related-policies-and-schemes.

INDEX